Multi-Objective Optimization
using Evolutionary Algorithms

Multi-Objective Optimization using Evolutionary Algorithms

Kalyanmoy Deb
Department of Mechanical Engineering,
Indian Institute of Technology, Kanpur, India

JOHN WILEY & SONS, LTD

Chichester • New York • Weinheim • Brisbane • Singapore • Toronto

To Ronny and Rini, two inspirations of my life

Contents

Foreword

Writing a foreword for a book by a former student runs many of the same risks faced by a parent who is talking to others about his or her adult children. First, the parent risks locking the child into something of a time warp, thinking largely about what the child has done in the early part of his or her life and ignoring much of what the child has done since leaving the nest. Second, the parent risks amplifying the minor annoyances of living with a person day to day and ignoring or missing the big picture of virtue and character that the child has developed through the maturation process aided by years of careful parenting. Fortunately, in this instance, the chances of these kinds of type I and II errors are slim, because (1) even as a student Kalyanmoy Deb displayed the combination of virtues that now let us count him among the top researchers in the field of genetic and evolutionary computation, and (2) thinking back to his time as a student, I have a difficult time recalling even a single minor annoyance that might otherwise overshadow his current deeds.

In reviewing this book, *Multi-Objective Optimization using Evolutionary Algorithms*, I find that it is almost a perfect reflection of the Kalyanmoy Deb I knew as my student and that I know now. First, the book chooses to tackle a current, important, and difficult topic multi-objective optimization using genetic and evolutionary algorithms (GEAs)—and the Kalyan I know has always shown good intuition for where to place his efforts, having previously pioneered early studies in niching, competent GAs, coding design, and multi-objective GEAs. Second, the book is readable and pedagogically sound, with a carefully considered teaching sequence for both novice and experienced genetic algorithmists and evolutionaries alike, and this matches well with the Kalyan I know who put teaching, learning, and understanding ahead of any thought of impressing his colleagues with the esoteric or the obscure. Third and finally, the book is scholarly, containing as broad and deep a review of multi-objective GEAs as any contained in the literature to date, and the presentation of that information shows a rich understanding of the interconnections between different threads of this burgeoning literature. Of course, this comes as no surprise to those of us who have worked with Kalyan in the past. He has always been concerned with knowing who did what when, and then showing us how to connect the dots.

For these reasons, and for so many more, I highly recommend this book to readers interested in genetic and evolutionary computation in general, and multi-objective genetic and evolutionary optimization, in particular. The topic of multi-objective

GEAs is one of the hottest topics in the field at present, and practitioners of many stripes are now turning to population-based schemes as an increasingly effective way to determine the necessary tradeoffs between conflicting objectives. Kalyanmoy Deb's book is complete, eminently readable, and accessible to the novice and expert, alike, and the coverage is scholarly, thorough, and appropriate for specialists and practitioners, both. In short, it is my pleasure and duty to urge you to buy this book, read it, use it, and enjoy it. There is little downside risk in predicting that you'll be joining a whole generation of researchers and practitioners who will recognize this volume as the classic that marked a new era in multi-objective genetic and evolutionary algorithms.

David E. Goldberg
University of Illinois at Urbana-Champaign

Preface

Optimization is a procedure of finding and comparing feasible solutions until no better solution can be found. Solutions are termed good or bad in terms of an objective, which is often the cost of fabrication, amount of harmful gases, efficiency of a process, product reliability, or other factors. A significant portion of research and application in the field of optimization considers a single objective, although most real-world problems involve more than one objective. The presence of multiple conflicting objectives (such as simultaneously minimizing the cost of fabrication and maximizing product reliability) is natural in many problems and makes the optimization problem interesting to solve. Since no one solution can be termed as an optimum solution to multiple conflicting objectives, the resulting multi-objective optimization problem resorts to a number of trade-off optimal solutions. Classical optimization methods can at best find one solution in one simulation run, thereby making those methods inconvenient to solve multi-objective optimization problems.

Evolutionary algorithms (EAs), on the other hand, can find multiple optimal solutions in one single simulation run due to their population-approach. Thus, EAs are ideal candidates for solving multi-objective optimization problems. This book provides a comprehensive survey of most multi-objective EA approaches suggested since the evolution of such algorithms. Although a number of approaches were outlined sparingly in the early years of the subject, more pragmatic multi-objective EAs (MOEAs) were first suggested about a decade ago. All such studies exist in terms of research papers in various journals and conference proceedings, which thus force newcomers and practitioners to search different sources in order to obtain an overview of the topic. This fact has been the primary motivation for me to take up this project and to gather together most of the MOEA techniques in one text.

This present book provides an extensive discussion on the principles of multi-objective optimization and on a number of classical approaches. For those readers unfamiliar with multi-objective optimization, Chapters 2 and 3 provide the necessary background. Readers with a classical optimization background can take advantage of Chapter 4 to familiarize themselves with various evolutionary algorithms. Beginning with a detailed description of genetic algorithms, an introduction to three other EAs, namely evolution strategy, evolutionary programming, and genetic programming, is provided. Since the search for multiple solutions is important in multi-objective optimization, a detailed description of EAs, particularly designed to solve multi-modal optimization problems, is also presented. Elite-preservation or emphasizing currently elite solutions is an important operator in an EA. In this book, we classify MOEAs according to whether they preserve elitism or not. Chapter 5 presents a number of non-elitist MOEAs. Each algorithm is described by presenting a step-by-step procedure, showing a hand calculation, discussing advantages and disadvantages of the algorithm, calculating its computational complexity, and finally presenting a computer simulation on a test problem. In order to obtain a comparative evaluation of different algorithms, the same test problem with the same parameter settings is used for most MOEAs presented in the book. Chapter 6 describes a number of elitist MOEAs in an identical manner.

Constraints are inevitable in any real-world optimization problem, including multi-objective optimization problems. Chapter 7 presents a number of techniques specializing in handling constrained optimization problems. Such approaches include simple modifications to the MOEAs discussed in Chapters 5 and 6 to give more specialized new MOEAs.

Whenever new techniques are suggested, there is room for improvement and further research. Chapter 8 discusses a number of salient issues regarding MOEAs. This chapter amply emphasizes the importance of each issue in developing and applying MOEAs in a better manner by presenting the current state-of-the-art research and by proposing further research directions.

Finally, in Chapter 9, the usefulness of MOEAs in real-world applications is demonstrated by presenting a number of applications in engineering design. This chapter also discusses plausible hybrid techniques for combining MOEAs with a local search technique for developing an even better and a pragmatic multi-objective optimization tool.

This book would not have been completed without the dedication of a number of my students, namely Sameer Agrawal, Amrit Pratap, Tushar Goel and Thirunavukkarasu Meyarivan. They have helped me in writing computer codes for investigating the performance of the different algorithms presented in this book and in discussing with me for long hours various issues regarding multi-objective optimization. In this part of the world, where the subject of evolutionary algorithms is still a comparative fad, they were my colleagues and inspirations. I also appreciate the help of Dhiraj Joshi, Ashish Anand, Shamik Chaudhury, Pawan Nain, Akshay Mohan, Saket Awasthi and Pawan Zope. In any case, I must not forget to thank Nidamarthi Srinivas who took up the challenge to code the first viable MOEA based on the non-domination concept. This ground-breaking study on non-dominated sorting GA (NSGA) inspired many MOEA researchers and certainly most of our MOEA research activities at the Kanpur Genetic Algorithms Laboratory (KanGAL), housed at the Indian Institute of Technology Kanpur, India.

The first idea for writing this book originated during my visit to the University of Dortmund during the period 1998–1999 through the Alexander von Humboldt (AvH) Fellowship scheme. The resourceful research environment at the University of Dortmund and the ever-supportive sentiments of AvH organization were helpful in formulating a plan for the contents of this book. Discussions with Eckart Zitzler, Lothar Thiele, Jürgen Branke, Frank Kursawe, Günter Rudolph and Ian Parmee on various issues on multi-objective optimization are acknowledged. Various suggestions given by Marco Laumanns and Eckart Zitzler in improving an earlier draft of this book are highly appreciated. I am privileged to get continuous support and encouragement from two stalwarts in the field of evolutionary computation, namely David E. Goldberg and Hans-Paul Schwefel. The help obtained from Victoria Coverstone-Carroll, Bill Hartmann, Hisao Ishibuchi and Eric Michelssen was also very useful. I also thank David B. Fogel for pointing me towards some of the early multi-objective EA studies.

Besides our own algorithms for multi-objective optimization, this book also presents a number of algorithms suggested by other researchers. Any difference between what is presented here and the original version of these algorithms is purely unintentional. Wherever in doubt, the original source can be referred. However, I would be happy to receive any such comments, which would be helpful to me in preparing the future editions of this book.

The completion of this book came at the expense of my long hours of absence from home. I am indebted to Debjani, Debayan, Dhriti, and Mr and Mrs S. K. Sarkar for their understanding and patience.

Kalyanmoy Deb
Indian Institute Technology Kanpur
deb@iitk.ac.in

In the second printing of the book, exercise problems to Chapters 1 to 8 are added, in order for the book to be used as a course text. Some minor changes in the text are also made, mainly to improve readability.

February 2002   *Kalyanmoy Deb*

1

Prologue

Optimization refers to finding one or more feasible solutions which correspond to extreme values of one or more objectives. The need for finding such optimal solutions in a problem comes mostly from the extreme purpose of either designing a solution for minimum possible cost of fabrication, or for maximum possible reliability, or others. Because of such extreme properties of optimal solutions, optimization methods are of great importance in practice, particularly in engineering design, scientific experiments and business decision-making.

When an optimization problem modeling a physical system involves only one objective function, the task of finding the optimal solution is called *single-objective optimization*. Since the time of the Second World War, most efforts in the field have been made to understand, develop, and apply single-objective optimization methods. Now-a-days, however, there exist single-objective optimization algorithms that work by using gradient-based and heuristic-based search techniques. Besides deterministic search principles involved in an algorithm, there also exist stochastic search principles, which allow optimization algorithms to find globally optimal solutions more reliably. In order to widen the applicability of an optimization algorithm in various different problem domains, natural and physical principles are mimicked to develop *robust* optimization algorithms. Evolutionary algorithms and simulated annealing are two examples of such algorithms.

When an optimization problem involves more than one objective function, the task of finding one or more optimum solutions is known as *multi-objective optimization*. In the parlance of management, such search and optimization problems are known as multiple criterion decision-making (MCDM). Since multi-objective optimization involves multiple objectives, it is intuitive to realize that single-objective optimization is a degenerate case of multi-objective optimization. However, there is another reason why more attention is now being focused on multi-objective optimization – a matter we will discuss in the next paragraph.

Most real-world search and optimization problems naturally involve multiple objectives. The extremist principle mentioned above cannot be applied to only one objective, when the rest of the objectives are also important. Different solutions may produce trade-offs (conflicting scenarios) among different objectives. A solution that is extreme (in a better sense) with respect to one objective requires a compromise in other objectives. This prohibits one to choose a solution which is optimal with respect to only one objective. Let us consider the decision-making involved in buying an automobile car. Cars are available at prices ranging from a few thousand to few hundred thousand dollars. Let us take two extreme hypothetical cars, i.e. one costing about ten thousand dollars (solution 1) and another costing about a hundred thousand dollars (solution 2), as shown in Figure 1. If the cost is the only objective of this decision-making process, the optimal choice is solution 1. If this were the only objective to all buyers, we would have seen only one type of car (solution 1) on the road and no car manufacturer would have produced any expensive cars. Fortunately, this decision-making process is not a single-objective one. Barring some exceptions, it is expected that an inexpensive car is likely to be less comfortable. The figure indicates that the cheapest car has a hypothetical comfort level of 40%. To rich buyers for whom comfort is the only objective of this decision-making, the choice is solution 2 (with a hypothetical maximum comfort level of 90%, as shown in the figure). Between these two extreme solutions, there exist many other solutions, where a trade-off between cost and comfort exists. A number of such solutions (solutions A, B, and C) with differing costs and comfort levels are shown in the figure. Thus, between any two such solutions, one is better in terms of one objective, but this betterment comes only from a sacrifice on the other objective.

1.1   Single and Multi-Objective Optimization

Right from standards nine or ten in a secondary school, students are taught how to find the minimum or maximum of a single-variable function. Beginning with the first and second-order derivative-based techniques of finding an optimum (a generic term

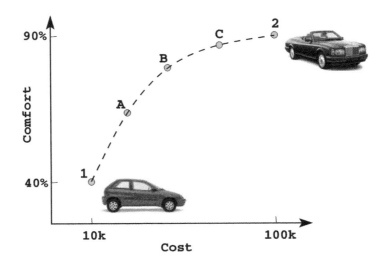

Figure 1   Hypothetical trade-off solutions are illustrated for a car-buying decision-making problem.

meaning either a minimum, maximum or saddle point), students in higher-level classes are taught how to find an optimum in a multi-variable function. In undergraduate level courses focusing on optimization-related studies, they are taught how to find the true optimum in the presence of constraints. Constrained optimization is important in practice, since most real-world optimization problems involve constraints restricting some properties of the system to lie within pre-specified limits. An advanced-level course on optimization which is usually taught to undergraduate and graduate students focuses on theoretical aspects of optimality, convergence proofs and special-purpose optimization algorithms for nonlinear problems, such as integer programming, dynamic programming, geometric programming, stochastic programming, and various others. Not enough emphasis is usually given to multi-objective optimization. There is, however, a reason for this.

As in the case of single-objective optimization, multi-objective optimization has also been studied extensively. There exist many algorithms and application case studies involving multiple objectives. However, there is one matter common to most such studies. The majority of these methods avoid the complexities involved in a true multi-objective optimization problem and transform multiple objectives into a single objective function by using some user-defined parameters. Thus, most studies in classical multi-objective optimization do not treat multi-objective optimization any differently than single-objective optimization. In fact, multi-objective optimization is considered as an application of single-objective optimization for handling multiple objectives. The studies seem to concentrate on various means of converting multiple objectives into a single objective. Many studies involve comparing different schemes of such conversions, provide reasons in favor of one conversion over another, and suggest better means of conversion. This is contrary to our intuitive realization that single-objective optimization is a degenerate case of multi-objective optimization and multi-objective optimization is not a simple extension of single-objective optimization.

It is true that theories and algorithms for single-objective optimization are applicable to the optimization of the transformed single objective function. However, there is a fundamental difference between single and multi-objective optimization which is ignored when using the transformation method. We will discuss this important matter in the following subsection.

1.1.1 Fundamental Differences

Without the loss of generality, let us discuss the fundamental difference between single- and multi-objective optimization with a two-objective optimization problem. For two conflicting objectives, each objective corresponds to a different optimal solution. In the above-mentioned decision-making problem of buying a car, solutions 1 and 2 are these optimal solutions. If a buyer is willing to sacrifice cost to some extent from solution 1, the buyer can probably find another car with a better comfort level than this solution. Ideally, the extent of sacrifice in cost is related to the gain in comfort. Thus, we can visualize a set of optimal solutions (such as solutions 1, 2, A, B and C) where a gain in one objective calls for a sacrifice in the other objective.

Now comes the big question. With all of these trade-off solutions in mind, can one say which solution is the best with respect to both objectives? The irony is that none of these trade-off solutions is the best with respect to both objectives. The reason lies in the fact that no solution from this set makes both objectives (cost and comfort) look better than any other solution from the set. Thus, in problems with more than one conflicting objective, there is no single optimum solution. There exist a number of solutions which are all optimal. Without any further information, no solution from the set of optimal solutions can be said to be better than any other. Since a number of solutions are optimal, in a multi-objective optimization problem many such (trade-off) optimal solutions are important. This is the fundamental difference between a single-objective and a multi-objective optimization task. Except in multi-modal single-objective optimization, the important solution in a single-objective optimization is the lone optimum solution, whereas in multi-objective optimization, a number of optimal solutions arising because of trade-offs between conflicting objectives are important.

1.2   Two Approaches to Multi-Objective Optimization

Although the fundamental difference between these two optimizations lies in the cardinality in the optimal set, from a practical standpoint a user needs only one solution, no matter whether the associated optimization problem is single-objective or multi-objective. In the case of multi-objective optimization, the user is now in a dilemma. Which of these optimal solutions must one choose? Let us try to answer this question for the case of the car-buying problem. Knowing the number of solutions that exist in the market with different trade-offs between cost and comfort, which car does one buy? This is not an easy question to answer. It involves many other considerations, such as the total finance available to buy the car, distance to be driven each day, number of passengers riding in the car, fuel consumption and cost, depreciation value, road conditions where the car is to be mostly driven, physical health of the passengers, social status, and many other factors. Often, such higher-level information is non-technical, qualitative and experience-driven. However, if a set of many trade-off solutions are already worked out or available, one can evaluate the pros and cons of each of these solutions based on all such non-technical and qualitative, yet still important, considerations and compare them to make a choice. Thus, in a multi-objective optimization, ideally the effort must be made in finding the set of trade-off optimal solutions by considering all objectives to be important. After a set of such trade-off solutions are found, a user can then use higher-level qualitative considerations to make a choice. In view of these discussions, we therefore suggest the following principle for an *ideal multi-objective optimization procedure*:

Step 1 Find multiple trade-off optimal solutions with a wide range of values for objectives.

Step 2 Choose one of the obtained solutions using higher-level information.

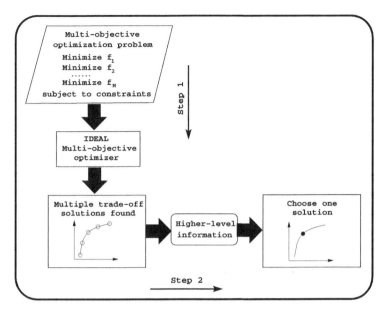

Figure 2   Schematic of an ideal multi-objective optimization procedure.

Figure 2 shows schematically the principles in an ideal multi-objective optimization procedure. In Step 1 (vertically downwards), multiple trade-off solutions are found. Thereafter, in Step 2 (horizontally, towards the right), higher-level information is used to choose one of the trade-off solutions. With this procedure in mind, it is easy to realize that single-objective optimization is a degenerate case of multi-objective optimization, as argued earlier. In the case of single-objective optimization with only one global optimal solution, Step 1 will find only one solution, thereby not requiring us to proceed to Step 2. In the case of single-objective optimization with multiple global optima, both steps are necessary to first find all or many of the global optima and then to choose one from them by using the higher-level information about the problem.

If thought of carefully, each trade-off solution corresponds to a specific order of importance of the objectives. It is clear from Figure 1 that solution A assigns more importance to cost than to comfort. On the other hand, solution C assigns more importance to comfort than to cost. Thus, if such a relative preference factor among the objectives is known for a specific problem, there is no need to follow the above principle for solving a multi-objective optimization problem. A simple method would be to form a composite objective function as the weighted sum of the objectives, where a weight for an objective is proportional to the preference factor assigned to that particular objective. This method of scalarizing an objective vector into a single composite objective function converts the multi-objective optimization problem into a single-objective optimization problem. When such a composite objective function is optimized, in most cases it is possible to obtain one particular trade-off solution. This procedure of handling multi-objective optimization problems is much simpler, yet still being more subjective than the above ideal procedure. We call this procedure a *preference-based* multi-objective optimization. A schematic of this procedure is shown in Figure 3. Based on the higher-level information, a preference vector **w** is first chosen. Thereafter, the preference vector is used to construct the composite function, which is then optimized to find a single trade-off optimal solution by a single-objective optimization algorithm. Although not often practiced, the procedure can be used to find multiple trade-off solutions by using a different preference vector and repeating the above procedure.

It is important to realize that the trade-off solution obtained by using the preference-based strategy is largely sensitive to the relative preference vector used in forming the composite function. A change in this preference vector will result in a (hopefully) different trade-off solution. We shall show in the next chapter that any arbitrary preference vector need not result in a trade-off optimal solution to all problems. Besides this difficulty, it is intuitive to realize that finding a relative preference vector itself is highly subjective and not straightforward. This requires an analysis of the non-technical, qualitative and experience-driven information to find a quantitative relative preference vector. Without any knowledge of the likely trade-off solutions, this is an even more difficult task. Classical multi-objective optimization methods which convert multiple objectives into a single objective by using a relative preference vector of objectives work according to this preference-based strategy. Unless a reliable and accurate preference vector is available, the optimal solution obtained by such methods is highly subjective to the particular user.

The ideal multi-objective optimization procedure suggested earlier is less subjective. In Step 1, a user does not need any relative preference vector information. The task there is to find as many different trade-off solutions as possible. Once a well-distributed set of trade-off solutions is found, Step 2 then requires certain problem information in

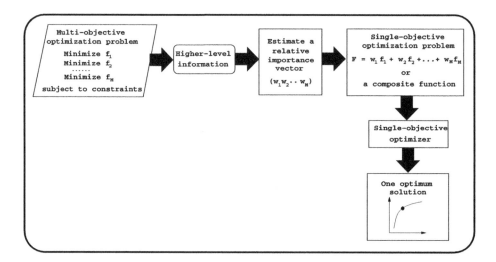

Figure 3   Schematic of a preference-based multi-objective optimization procedure.

order to choose one solution. It is important to mention that in Step 2, the problem information is used to evaluate and compare each of the obtained trade-off solutions. In the ideal approach, the problem information is not used to search for a *new* solution; instead, it is used to choose one solution from a set of already obtained trade-off solutions. Thus, there is a fundamental difference in using the problem information in both approaches. In the preference-based approach, a relative preference vector needs to be supplied without any knowledge of the possible consequences. However, in the proposed ideal approach, the problem information is used to choose one solution from the obtained set of trade-off solutions. We argue that the ideal approach in this matter is more methodical, more practical, and less subjective. At the same time, we highlight the fact that if a reliable relative preference vector is available to a problem, there is no reason to find other trade-off solutions. In such a case, a preference-based approach would be adequate.

Since Step 2 requires various subjective and problem-dependent considerations, we will not discuss this step further in this present book. Most of this book is devoted to finding multiple trade-off solutions for multi-objective optimization problems. However, we will also discuss a number of techniques to find a preferred *distribution* of trade-off solutions, particularly for cases where information about the relative importance of the objectives is available.

1.3   Why Evolutionary?

It is mentioned above that the classical way to solve multi-objective optimization problems is to follow the preference-based approach, where a relative preference vector is used to scalarize multiple objectives. Since classical search and optimization methods use a point-by-point approach, where one solution in each iteration is modified to a different (hopefully better) solution, the outcome of using a classical optimization method is a single optimized solution. Thinking along this working principle of available optimization methods, it would not be surprising to apprehend that the development of preference-based approaches was motivated by the fact that available optimization methods could find only a single optimized solution in a single simulation run. Since only a single optimized solution could be found, it was, therefore, necessary to convert the task of finding multiple trade-off solutions in a multi-objective optimization to one of finding a single solution of a transformed single-objective optimization problem.

However, the field of search and optimization has changed over the last few years by the introduction of a number of non-classical, unorthodox and stochastic search and optimization algorithms. Of these, the *evolutionary algorithm* (EA) mimics nature's evolutionary principles to drive its search towards an optimal solution. One of the most striking differences to classical search and optimization algorithms is that EAs use a population of solutions in each iteration, instead of a single solution. Since a population of solutions are processed in each iteration, the outcome of an EA is also a population of solutions. If an optimization problem has a single optimum, all EA population members can be expected to converge to that optimum solution. However, if an optimization problem has multiple optimal solutions, an EA can be used to capture multiple optimal solutions in its final population.

This ability of an EA to find multiple optimal solutions in one single simulation run makes EAs unique in solving multi-objective optimization problems. Since Step 1 of the ideal strategy for multi-objective optimization requires multiple trade-off solutions to be found, an EA's population-approach can be suitably utilized to find a number of solutions in a single simulation run.

1.4   Rise of Multi-Objective Evolutionary Algorithms

Starting with multi-objective studies from the early days of evolutionary algorithms, this book presents various techniques of finding multiple trade-off solutions using evolutionary algorithms. Early applications to multi-objective optimization problems were mainly preference-based approaches, although the need for finding multiple trade-off solutions was clearly stated. The first real application of EAs in finding multiple trade-off solutions in one single simulation run was suggested and worked out in 1984 by David Schaffer in his doctoral dissertation (Schaffer, 1984). His vector-evaluated genetic algorithm (VEGA) made a simple modification to a single-objective GA and demonstrated that GAs can be used to capture multiple trade-off solutions for a few iterations of a VEGA. However, if continued for a large number of iterations, a VEGA population tends to converge to individual optimal solutions. After this study, EA researchers did not pay much attention to multi-objective optimization for almost half a decade. In 1989, David E. Goldberg, in his seminal book (Goldberg, 1989), suggested a 10-line sketch of a plausible multi-objective evolutionary algorithm (MOEA) using the concept of domination. Taking the clue from his book, a number of researchers across the globe have since developed different implementations of MOEAs. Of these, Fonseca and Fleming's multi-objective GA (Fonseca and Fleming, 1995), Srinivas and Deb's non-dominated sorting GA (NSGA) (Srinivas and Deb, 1994), and Horn, Nafploitis and Goldberg's niched Pareto-GA (NPGA) (Horn et al., 1994) were immediately tested for different real-world problems to demonstrate that domination-based MOEAs can be reliably used to find and maintain multiple trade-off solutions. More or less at the same time, a number of other researchers suggested different ways to use an EA to solve multi-objective optimization problems. Of these, Kursawe's diploidy approach (Kursawe, 1990), Hajela and Lin's weight-based approach (Hajela and Lin, 1992), and Osyczka and Kundu's distance-based GA (Osyczka and Kundu, 1995) are just a few examples. With the rise of interest in multi-objective optimization, EA journals began to bring out special issues (Deb and Horn, 2000), EA conferences started holding tutorials and special sessions on evolutionary multi-objective optimization, and an independent international conference on evolutionary multi-criterion optimization (Zitzler et al., 2001) is recently arranged.

In order to assess the volume of studies currently being undertaken in this field, we have made an attempt to count the number of studies year-wise from our collection of research papers on this topic. Such a year-wise count is presented in Figure 4. The latter shows the increasing trend in the number of studies in the field. The slight reduction in the number of studies in 1999 shown in this figure is most likely to result from the non-availability of some material to the author until after the date of publication of this book. Elsewhere (Veldhuizen and Lamont, 1998), studies up until 1998 are classified according to various other criteria. Interested readers can refer to that work for more statistical information.

Despite all of these studies, to date, there exists no book or any other monograph where different MOEA techniques are gathered in one convenient source. Both researchers and newcomers to the field have to collect various research papers and dissertations to find out the various techniques which have already been attempted. This book fills that gap and will hopefully serve to provide a comprehensive survey of multi-objective evolutionary algorithms up until the year 2000.

1.5 Organization of the Book

For those readers not so familiar with the concept of multi-objective optimization, Chapter 2 provides some useful definitions, discusses the principle of multi-objective optimization, and briefly mentions optimality conditions. Chapter 3 describes a number of classical optimization methods which mostly work according to the preference-based approach. Starting with the weighted approach, discussions on the ϵ-constraint method, Tchebycheff methods, value function methods and goal programming methods are highlighted. In the light of these techniques, a review of many classical techniques is also presented.

Since this book is about multi-objective evolutionary algorithms, it is, therefore, necessary to discuss the single-objective evolutionary algorithms, particularly for those readers who are not familiar with evolutionary computation. Chapter 4 begins with outlining the difficulties associated with classical methods and highlights the

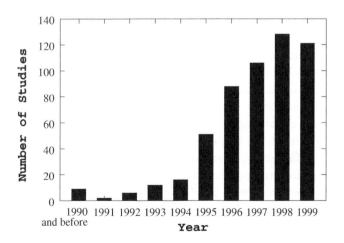

Figure 4 Year-wise growth of number of studies on MOEAs.

need for a population-based optimization method for multi-objective optimization. Thereafter, four different evolutionary algorithms, i.e. genetic algorithms (GAs), evolution strategy (ES), evolutionary programming (EP) and genetic programming (GP), are discussed. Since most existing MOEAs are GA-based, we have provided rather detailed descriptions of the working principles of both a binary-coded and a real-parameter GA. Constraints are evident in real-world problem solving. We have also highlighted a number of constraint handling techniques which are used in a GA. In the context of multi-objective optimization, we have discussed the need for finding a well-distributed set of trade-off solutions. Since finding and maintaining a well-distributed set of optimal solutions is equivalent to solving a multi-modal problem for finding and maintaining multiple optimal solutions simultaneously, we also elaborate various EA techniques used in solving multi-modal problems.

Discussions on the fundamentals of multi-objective optimization in Chapter 2, the classical methods used to solve multi-objective problems in Chapter 3, and the working principles of various evolutionary optimization techniques in Chapter 4 provide a platform for presenting different MOEAs. Since the number of existing MOEAs are many, we have classified them in two categories, namely non-elitist and elitist MOEAs. In the context of single-objective EAs, the need of an elite-preserving operator has been amply demonstrated both theoretically and experimentally in the literature. Since elite-preservation is found to be important in multi-objective optimization, we group the algorithms based on whether they use elitism or not. First, non-elitist MOEAs are described in Chapter 5. Starting from early suggestions, most of the commonly used MOEAs are discussed. In order to demonstrate the working principles of these algorithms, they are clearly stated in a step-by-step format. The computational complexities of most algorithms are also calculated. Based on the descriptions of these algorithms, the advantages and disadvantages of each are also outlined. Finally, simulation results are shown on a simple test problem to show the performance of different MOEAs. Since an identical test function is used for most algorithms, a qualitative comparison of the algorithms can be obtained from such simulation results.

Chapter 6 presents a number of elitist MOEAs. Highlighting the need of elitism in multi-objective optimization, different elitist MOEAs are then described in turn. Again, the algorithms are described in a step-by-step manner, the computational complexities are worked out, and the advantages and disadvantages are highlighted. Simulation results on the same test problem are also shown, not only to have a comparative evaluation of different elitist MOEAs, but also to comprehend the differences between elitist and non-elitist MOEAs.

Constraints are inevitable in practical multi-objective optimization problems. In Chapter 7, we will discuss two constraint-handling methods which have been recently suggested. In addition, an approach based on a modified definition of domination is suggested and implemented with an elitist MOEA. Three approaches are applied on a test problem to show the efficacy of each approach.

Chapter 8 discusses a number of salient issues related to multi-objective optimization. Some of these issues are common to both evolutionary and classical optimization methods. Various means of representing trade-off solutions in problems having more than two objective problems are discussed. The scaling of two-objective optimization algorithms in larger-dimensional problems is also discussed. Thereafter, various issues related only to multi-objective evolutionary algorithms have been particularly highlighted in order to provide an outline of future research directions. Of these, the performance measures of an MOEA, the design of difficult multi-objective test problems, the comparison of MOEAs on difficult test problems, the maintenance of diversity in objective versus decision spaces, convergence issues and implementations of controlled elitism are of immediate interest to research in the field. From a practical standpoint, techniques for finding a preferred set of trade-off solutions instead of the complete Pareto-optimal set, and the use of multi-objective optimization techniques to solve single-objective constrained optimization problems and to solve unbiased goal programming problems are some suggested directions for future research.

No description of an algorithm is complete unless its performance is tested and applied to practical problems. In Chapter 9, we will present a number of case studies where elitist and non-elitist MOEAs are applied to a number of engineering problems and a space trajectory design problem. Although these application case studies are not the only applications which exist in the literature, they amply demonstrate the purpose with which we have begun this chapter. All application case studies show how MOEAs can find a number of trade-off solutions in various problems in one single simulation run. Chapter 9 also proposes a hybrid MOEA approach along with a local search technique for finding better-converged and better-distributed trade-off solutions. Thus, in addition to providing a number of MOEA techniques to find multiple trade-off solutions as required in Step 1 of the ideal approach of multi-objective optimization, this book has also addressed Step 2 of this approach by outlining a number of rational techniques for choosing a compromised solution. Reasonable methods are suggested to reduce the cardinality of the set of trade-off solutions so as to allow the decision-making task easier for the user.

Exercise Problems

1. Clearly state the advantages and disadvantages of the ideal and preference-based multi-objective optimization procedures.

2. Consider the following two-objective optimization problem:

$$\text{Minimize} \quad f_1(x) = x_1^2 + x_2^2,$$
$$\text{Minimize} \quad f_2(x) = (x_1 - 1)^2 + x_2^2,$$
$$-2 \le x_1 \le 2,$$
$$-2 \le x_2 \le 2.$$

By using the preference-based procedure, calculate the optimum solutions for each of the following three weight vectors: (i) $w^{(1)} = (1,0)^T$, (ii) $w^{(2)} = (0.5, 0.5)^T$, (iii) $w^{(3)} = (0,1)^T$.

3. In problem 2, write the optimum solution vector x as a function of the chosen weight vector w (where $w_1 + w_2 = 1$).

4. State the differences in the working principles of the following optimization problems:

   (a) single-objective optimization problems,
   (b) multi-modal optimization problems,
   (c) multi-objective optimization problems.

5. Using an ideal multi-objective optimization algorithm, explain how the following single-objective optimization problem can be solved to find the sole minimum?

$$\text{Minimize} \quad f(x_1, x_2) = 10 - x_1 + x_1 x_2 + x_2^2.$$

6. Using an ideal multi-objective optimization algorithm, explain how the following single-objective optimization problem can be solved to find multiple minima?

$$\text{Minimize} \quad f(x_1, x_2) = (x_1^2 + x_2 - 11)^2 + (x_1 + x_2^2 - 7)^2.$$

   If the user prefers a solution having non-negative values of variables, which is the preferred solution?

7. Using an ideal multi-objective optimization algorithm, explain how the following multi-objective optimization problem can be solved to find multiple Pareto-optimal solutions:

$$\text{Minimize} \quad f_1(x_1, x_2) = (x_1 - 2)^2 + (x_2 - 1)^2,$$
$$\text{Minimize} \quad f_2(x_1, x_2) = 9x_1 - (x_2 - 1)^2.$$

   If the user prefers a solution with the smallest absolute value of f_2, which is the preferred Pareto-optimal solution?

2

Multi-Objective Optimization

As the name suggests, a multi-objective optimization problem (MOOP) deals with more than one objective function. In most practical decision-making problems, multiple objectives or multiple criteria are evident. Because of a lack of suitable solution methodologies, an MOOP has been mostly cast and solved as a single-objective optimization problem in the past. However, there exist a number of fundamental differences between the working principles of single and multi-objective optimization algorithms. In a single-objective optimization problem, the task is to find one solution (except in some specific multi-modal optimization problems, where multiple optimal solutions are sought) which optimizes the sole objective function. Extending the idea to multi-objective optimization, it may be wrongly assumed that the task in a multi-objective optimization is to find an optimal solution corresponding to each objective function. In this chapter, we will discuss the principles of multi-objective optimization and present optimality conditions for any solution to be optimal in the presence of multiple objectives.

2.1 Multi-Objective Optimization Problem

A multi-objective optimization problem has a number of objective functions which are to be minimized or maximized. As in the single-objective optimization problem, here too the problem usually has a number of constraints which any feasible solution (including the optimal solution) must satisfy. In the following, we state the multi-objective optimization problem (MOOP) in its general form:

$$\left.\begin{array}{ll} \text{Minimize/Maximize} & f_m(\mathbf{x}), \qquad m = 1, 2, \ldots, M; \\ \text{subject to} & g_j(\mathbf{x}) \geq 0, \qquad j = 1, 2, \ldots, J; \\ & h_k(\mathbf{x}) = 0, \qquad k = 1, 2, \ldots, K; \\ & x_i^{(L)} \leq x_i \leq x_i^{(U)}, \quad i = 1, 2, \ldots, n. \end{array}\right\} \qquad (2.1)$$

A solution $\mathbf{x}$ is a vector of n decision variables: $\mathbf{x} = (x_1, x_2, \ldots, x_n)^{\mathsf{T}}$. The last set of constraints are called variable bounds, restricting each decision variable x_i to take a value within a lower $x_i^{(L)}$ and an upper $x_i^{(U)}$ bound. These bounds constitute a *decision variable space* $\mathcal{D}$, or simply the decision space. Throughout this book, we will use the terms *point* and *solution* interchangeably to mean a solution vector $\mathbf{x}$.

Associated with the problem are J inequality and K equality constraints. The terms $g_j(\mathbf{x})$ and $h_k(\mathbf{x})$ are called constraint functions. The inequality constraints are treated as 'greater-than-equal-to' types, although a 'less-than-equal-to' type inequality constraint is also taken care of in the above formulation. In the latter case, the constraint must be converted into a 'greater-than-equal-to' type constraint by multiplying the constraint function by -1 (Deb, 1995). A solution $\mathbf{x}$ that does not satisfy *all* of the $(J + K)$ constraints and *all* of the 2N variable bounds stated above is called an *infeasible solution*. On the other hand, if any solution $\mathbf{x}$ satisfies all constraints and variable bounds, it is known as a *feasible solution*. Therefore, we realize that in the presence of constraints, the entire decision variable space $\mathcal{D}$ need not be feasible. The set of all feasible solutions is called the *feasible region*, or $\mathcal{S}$. In this present book, sometimes we will refer to the feasible region as simply the search space.

There are M objective functions $\mathbf{f}(\mathbf{x}) = (f_1(\mathbf{x}), f_2(\mathbf{x}), \ldots, f_M(\mathbf{x}))^\top$ considered in the above formulation. Each objective function can be either minimized or maximized. The duality principle (Deb, 1995; Rao, 1984; Reklaitis et al., 1983), in the context of optimization, suggests that we can convert a maximization problem into a minimization one by multiplying the objective function by -1. The duality principle has made the task of handling mixed type of objectives much easier. Many optimization algorithms are developed to solve only one type of optimization problems, such as e.g. minimization problems. When an objective is required to be maximized by using such an algorithm, the duality principle can be used to transform the original objective for maximization into an objective for minimization.

Although there is a subtle difference in the way that a criterion function and an objective function is defined (Chankong et al., 1985), in a broad sense we treat them here as identical. One of the striking differences between single-objective and multi-objective optimization is that in multi-objective optimization the objective functions constitute a multi-dimensional space, in addition to the usual decision variable space. This additional space is called the *objective space*, $\mathcal{Z}$. For each solution $\mathbf{x}$ in the decision variable space, there exists a point in the objective space, denoted by $\mathbf{f}(\mathbf{x}) = \mathbf{z} = (z_1, z_2, \ldots, z_M)^\top$. The mapping takes place between an n-dimensional solution vector and an M-dimensional objective vector. Figure 5 illustrates these two spaces and a mapping between them.

Multi-objective optimization is sometimes referred to as *vector optimization*, because a vector of objectives, instead of a single objective, is optimized.

2.1.1   Linear and Nonlinear MOOP

If all objective functions and constraint functions are linear, the resulting MOOP is called a multi-objective linear program (MOLP). Like the linear programming problems, MOLPs also have many theoretical properties. However, if any of the objective or constraint functions are nonlinear, the resulting problem is called a nonlinear multi-objective problem. Unfortunately, for nonlinear problems the solution

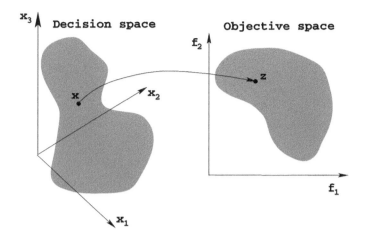

Figure 5   Representation of the decision variable space and the corresponding objective space.

techniques often do not have convergence proofs. Since most real-world multi-objective optimization problems are nonlinear in nature, we do not assume any particular structure of the objective and constraint functions here.

2.1.2 Convex and Nonconvex MOOP

Before we discuss a convex multi-objective optimization problem, let us first define a convex function.

Definition 2.1. *A function* $f : \mathbb{R}^n \to \mathbb{R}$ *is a convex function if for any two pair of solutions* $x^{(1)}, x^{(2)} \in \mathbb{R}^n$, *the following condition is true:*

$$f\left(\lambda x^{(1)} + (1-\lambda)x^{(2)}\right) \leq \lambda f(x^{(1)}) + (1-\lambda)f(x^{(2)}), \tag{2.2}$$

for all $0 \leq \lambda \leq 1$.

The above definition gives rise to the following properties of a convex function:

1. The linear approximation of $f(x)$ at any point in the interval $[x^{(1)}, x^{(2)}]$ always *underestimates* the actual function value.
2. The Hessian matrix of $f(x)$ is positive definite for all x.
3. For a convex function, a local minimum is always a global minimum.[1]

Figure 6 illustrates a convex function. A function satisfying the inequality shown in

---

[1] In the context of single-objective minimization problems, a solution having the smallest function value in its neighborhood is called a local minimum solution, while a solution having the smallest function value in the feasible search space is called a global minimum solution.

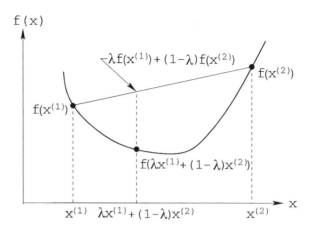

Figure 6   A convex function is illustrated. A line joining function values at two points $x^{(1)}$ and $x^{(2)}$ always estimates a large value of the true convex function.

equation (2.2) with a '>' sign instead of a '$\leq$' sign is called a nonconvex function. To test if a function is convex within an interval, the Hessian matrix $\nabla^2 f$ is calculated and checked for its positive-definiteness at all points in the interval. One of the ways to check the positive-definiteness of a matrix is to compute the eigenvalues of the matrix and check to see if all eigenvalues are positive. To test if a function f is nonconvex in an interval, the Hessian matrix $-\nabla^2 f$ is checked for its positive-definiteness. If it is positive-definite, the function f is nonconvex.

It is interesting to realize that if a function $g(x)$ is nonconvex, the set of solutions satisfying $g(x) \geq 0$ represents a convex set. Thus, a feasible search space formed with nonconvex constraint functions will enclose a convex region. Now, we are ready to define a convex MOOP.

Definition 2.2. *A multi-objective optimization problem is convex if all objective functions are convex and the feasible region is convex (or all inequality constraints are nonconvex and equality constraints are linear).*

According to this definition, an MOLP is a convex problem. The convexity of an MOOP is an important matter, which we shall see in subsequent chapters. There exist many algorithms which can handle convex MOOPs well, but face difficulty in solving nonconvex MOOPs. Since an MOOP has two spaces, the convexity in each space (objective and decision variable space) is important to a multi-objective optimization algorithm. Moreover, although the search space can be nonconvex, the Pareto-optimal front may be convex.

2.2   Principles of Multi-Objective Optimization

We illustrate the principles of multi-objective optimization through an airline routing problem. We all are familiar with the intermediate stopovers that most airlines force us to take, particularly when flying long distance. Airlines try different strategies to compromise on the number of intermediate stopovers and earn a large business mileage by introducing 'direct' flights. Let us take a look at a typical, albeit hypothetical, airline routing for some cities in the United States of America, as shown in Figure 7. If we look carefully, it is evident that there are two main 'hubs' (Los Angeles and New York) for this airline. If these two hubs are one's cities of origin and destination, the traveler is lucky. This is because there are likely to be densely packed schedules of flights between these two cities. However, if one has to travel between some other cities, let us say between Denver and Houston, there is no direct flight. The passenger has to travel to one of these hubs first and then take more flights from there to reach the destination. In the Denver–Houston case, one has to fly to Los Angeles, fly on to New York and then make the final lap to Houston.

To an airline, such modular networks of routes is easiest to maintain and coordinate. Better service facilities and ground staff need only be maintained at the hubs, instead of at all airports. Although one then travels longer distance than the actual geographical distance between the cities of origin and destination, this helps an airline to reduce the cost of its operation. Such a solution is ideal from the airline's point of view, but not so convenient from the point of view of a passenger's comfort. However, the situation is not that biased against the passenger's point of view either. By reducing the cost of operation, the airline is probably providing a cheaper ticket. However, if comfort or convenience is the only consideration to a passenger, the latter would like to have a network of routes which would be entirely different to that shown in Figure 7. A hypothetical routing is shown in Figure 8. In such a network, any two airports would be connected by a route, thereby allowing a direct flight between all airports. Since the operation cost for such a scenario will be exorbitantly high, the cost of flying with such a network would also be high.

Thus, we see a trade-off between two objectives in the above problem – cost versus convenience. A less-costly flight is likely to have more intermediate stopovers causing

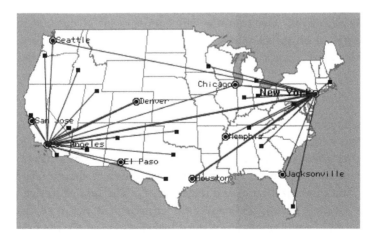

Figure 7 A typical network of airline routes showing hub-like connections.

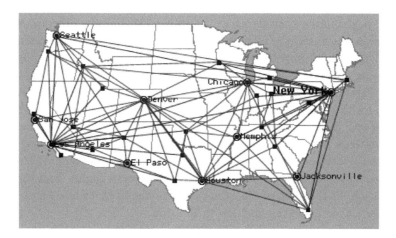

Figure 8   A hypothetical (but convenient) airline routing.

more inconvenience to a passenger, while a high-comfort flight is likely to have direct
routes, thus causing an expensive ticket. The important matter is that between any
two arbitrary cities in the first map (which resembles the routing of most airlines)
there does not exist a flight which is less costly as well as being largely convenient. If
there was, that would have been the only solution to this problem and nobody would
have complained about paying more or wasting time by using a 'hopping' flight. The
above two solutions of a hub-like network of routes and a totally connected network
of routes are two extreme solutions to this two-objective optimization problem. There
exist many other compromised solutions which have lesser hub-like network of routes
and more expensive flights than the solution shown in Figure 7. Innovative airlines
are constantly on the lookout for such compromises and in the process making the
network of routes a bit less hub-like, so giving the passengers a bit more convenience.

The 'bottomline' of the above discussion is that when multiple conflicting objectives
are important, there cannot be a single optimum solution which simultaneously
optimizes all objectives. The resulting outcome is a set of optimal solutions with
a varying degree of objective values. In the following subsection, we will make this
qualitative idea more quantitative by discussing a simple engineering design problem.

2.2.1   Illustrating Pareto-Optimal Solutions

We take a more concrete engineering design problem here to illustrate the concept of
Pareto-optimal solutions. Let us consider a cantilever design problem (Figure 9) with
two decision variables, i.e. diameter (d) and length (l). The beam has to carry an end
load P. Let us also consider two conflicting objectives of design, i.e. minimization
of weight f_1 and minimization of end deflection f_2. The first objective will resort to
an optimum solution having the smaller dimensions of d and l, so that the overall
weight of the beam is minimum. Since the dimensions are small, the beam will not be
adequately rigid and the end deflection of the beam will be large. On the other hand, if

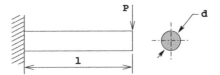

Figure 9   A schematic of a cantilever beam.

the beam is minimized for end deflection, the dimensions of the beam are expected to
be large, thereby making the weight of the beam large. For our discussion, we consider
two constraints: the developed maximum stress σ_{max} is less than the allowable strength
S_y and the end deflection δ is smaller than a specified limit δ_{max}. With all of the
above considerations, the following two-objective optimization problem is formulated
as follows:

$$\left.\begin{array}{rl} \text{Minimize} & f_1(d,l) = \rho\frac{\pi d^2}{4}l, \\ \text{Minimize} & f_2(d,l) = \delta = \dfrac{64Pl^3}{3E\pi d^4}, \\ \text{subject to} & \sigma_{max} \leq S_y, \\ & \delta \leq \delta_{max}, \end{array}\right\} \qquad (2.3)$$

where the maximum stress is calculated as follows:

$$\sigma_{max} = \frac{32Pl}{\pi d^3}. \qquad (2.4)$$

The following parameter values are used:

$$\begin{array}{lll} \rho = 7800 \text{ kg/m}^3, & P = 1 \text{ kN}, & E = 207 \text{ GPa}, \\ S_y = 300 \text{ MPa}, & \delta_{max} = 5 \text{ mm}. \end{array}$$

The left plot in Figure 10 marks the feasible decision variable space in the overall
search space enclosed by $10 \leq d \leq 50$ mm and $200 \leq l \leq 1000$ mm. It is clear that
not all solutions in the rectangular decision space are feasible. Every feasible solution
in this space can be mapped to a solution in the feasible objective space shown in
the right plot. The correspondence of a point in the left figure with that in the right
figure is also shown.

  This figure shows many solutions trading-off differently between the two objectives.
Any two solutions can be picked from the feasible objective space and compared. For
some pairs of solutions, it can be observed that one solution is better than the other
in both objectives. For certain other pairs, it can be observed that one solution is
better than the other in one objective, but is worse in the second objective. In order
to establish which solution(s) are optimal with respect to both objectives, let us hand-
pick a few solutions from the search space. Figure 11 is drawn with many such solutions
and five of these solutions (marked A to E) are presented in Table 1. Of these solutions,
the minimum weight solution (A) has a diameter of 18.94 mm, while the minimum
deflection solution (D) has a diameter of 50 mm. It is clear that solution A has a

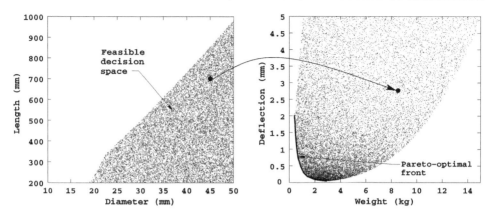

Figure 10   The feasible decision variable space (left) and the feasible objective space (right).

smaller weight, but has a larger end-deflection than solution D. Hence, none of these two solutions can be said to be better than the other with respect to both objectives. When this happens between two solutions, they are called *non-dominated* solutions. If both objectives are equally important, one cannot say, for sure, which of these two solutions is better with respect to both objectives. Two other similar solutions (B and C) are also shown in the figure and in the table. Of these four solutions (A to D), any pair of solutions can be compared with respect to both objectives. Superiority of one over the other cannot be established with both objectives in mind. There exist many such solutions (all solutions, marked using circles in the figure, are obtained by using NSGA-II – a multi-objective EA – to be described later) in the search space. For clarity, these solutions are joined with a curve in the figure. All solutions lying on this curve are special in the context of multi-objective optimization and are called *Pareto-optimal* solutions. The curve formed by joining these solutions is known as a Pareto-optimal front. The same Pareto-optimal front is also marked on the right plot of Figure 10 by a continuous curve. It is interesting to observe that this front lies in the bottom-left corner of the search space for problems where all objectives are to be minimized.

Table 1   Five solutions for the cantilever design problem.

| | d | l | Weight | Deflection |
Solution	(mm)	(mm)	(kg)	(mm)
A	18.94	200.00	0.44	2.04
B	21.24	200.00	0.58	1.18
C	34.19	200.00	1.43	0.19
D	50.00	200.00	3.06	0.04
E	33.02	362.49	2.42	1.31

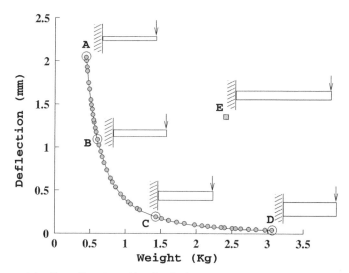

Figure 11 Four Pareto-optimal solutions and one non-optimal solution.

It is important to note that the feasible objective space not only contains Pareto-optimal solutions, but also solutions that are not optimal. We will give a formal definition of Pareto-optimal solutions later. Here, we simply mention that the entire feasible search space can be divided into two sets of solutions – a Pareto-optimal and a non-Pareto-optimal set. Consider solution E in Figure 11 and also in Table 1. By comparing this with solution C, we observe that the latter is better than solution E in both objectives. Since solution E has a larger weight and a larger end-deflection than solution C, the latter solution is clearly the better of the two. Thus, solution E is a sub-optimal solution and is of no interest to the user. When this happens in the comparison of two solutions, solution C is said to *dominate* solution E or that solution E is *dominated* by solution C. There exist many such solutions in the search space, which can be dominated by at least one solution from the Pareto-optimal set. In other words, there exists at least one solution in the Pareto-optimal set, which will be better than any non-Pareto-optimal solution. It is clear from the above discussion that in multi-objective optimization the task is to find the Pareto-optimal solutions.

Instead of considering the entire search space for finding the Pareto- and non-Pareto-optimal sets, such a division based on domination can also be made for a finite set of solutions P chosen from the search space. Using a pair-wise comparison as above, one can divide the set P into two non-overlapping sets P_1 and P_2, such that P_1 contains all solutions that do not dominate each other and at least one solution in P_1 dominates any solution in P_2. The set P_1 is called the *non-dominated* set, while the set P_2 is called the *dominated* set. In Section 2.4.6, we shall discuss different computationally efficient methods to identify the non-dominated set from a finite set of solutions.

There is an interesting observation about dominated and non-dominated sets, which is worth mentioning here. Let us compare solutions D and E. Solution D is better in the second objective but is worse in the first objective compared to solution E.

Thus, in the absence of solutions A, B, C, and any other non-dominated solution, we would be tempted to put solution E in the same group with solution D. However, the presence of solution C establishes the fact that solutions C and D are non-dominated with respect to each other, while solution E is a dominated solution. Thus, the non-dominated set must be collectively compared with any solution x for establishing whether the latter solution belongs to the non-dominated set or not. Specifically, the following two conditions must be true for a non-dominated set P_1:

1. Any two solutions of P_1 must be non-dominated with respect to each other.
2. Any solution not belonging to P_1 is dominated by at least one member of P_1.

2.2.2   Objectives in Multi-Objective Optimization

It is clear from the above discussion that, in principle, the search space in the context of multiple objectives can be divided into two non-overlapping regions, namely one which is optimal and one which is non-optimal. Although a two-objective problem is illustrated above, this is also true in problems with more than two objectives. In the case of conflicting objectives, usually the set of optimal solutions contains more than one solution. Figure 11 shows a number of such Pareto-optimal solutions denoted by circles. In the presence of multiple Pareto-optimal solutions, it is difficult to prefer one solution over the other without any further information about the problem. If higher-level information is satisfactorily available, this can be used to make a *biased* search (see Section 8.6 later). However, in the absence of any such information, all Pareto-optimal solutions are equally important. Hence, in the light of the ideal approach, it is important to find as many Pareto-optimal solutions as possible in a problem. Thus, it can be conjectured that there are two goals in a multi-objective optimization:

1. To find a set of solutions as close as possible to the Pareto-optimal front.
2. To find a set of solutions as diverse as possible.

The first goal is mandatory in any optimization task. Converging to a set of solutions which are not close to the true optimal set of solutions is not desirable. It is only when solutions converge close to the true optimal solutions that one can be assured of their near-optimality properties. This goal of multi-objective optimization is common to the similar optimality goal in a single-objective optimization.

On the other hand, the second goal is entirely specific to multi-objective optimization. In addition to being converged close to the Pareto-optimal front, they must also be sparsely spaced in the Pareto-optimal region. Only with a diverse set of solutions, can we be assured of having a good set of trade-off solutions among objectives. Since MOEAs deal with two spaces – decision variable space and objective space – 'diversity' among solutions can be defined in both of these spaces. For example, two solutions can be said to be diverse in the decision variable space if their Euclidean distance in the decision variable space is large. Similarly, two solutions are diverse in the objective space, if their Euclidean distance in the objective space is large. Although in most problems diversity in one space usually means diversity in the other space, this may not be so in all problems. In such complex and nonlinear problems, it is then the task to find a set of solutions having a good diversity in the desired space (Deb, 1999c).

2.2.3 Non-Conflicting Objectives

Before we leave this section, it is worth pointing out that there exist multiple Pareto-optimal solutions in a problem only if the objectives are conflicting to each other. If the objectives are not conflicting to each other, the cardinality of the Pareto-optimal set is one. This means that the minimum solution corresponding to any objective function is the same. For example, in the context of the cantilever design problem, if one is interested in minimizing the end-deflection δ and minimizing the maximum developed stress in the beam, σ_{max}, the feasible objective space is different. Figure 12 shows that the Pareto-optimal front reduces to a single solution (solution A marked on the figure). A little thought will reveal that the minimum end-deflection happens for the most rigid beam with the largest possible diameter. Since this beam also corresponds to the smallest developed stress, this solution also corresponds to the minimum-stress solution. In certain problems, it may not be obvious that the objectives are not conflicting to each other. In such combinations of objectives, the resulting Pareto-optimal set will contain only one optimal solution.

2.3   Difference with Single-Objective Optimization

Besides having multiple objectives, there are a number of fundamental differences between single-objective and multi-objective optimization, as follows:

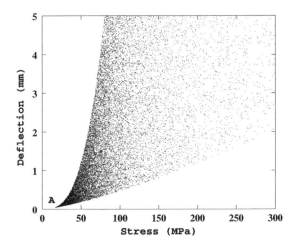

Figure 12 End-deflection and developed maximum stress are two non-conflicting objectives leading to one optimal solution (A).

- two goals instead of one;
- dealing with two search spaces;
- no artificial fix-ups.

We will discuss these in the following subsections.

2.3.1   Two Goals Instead of One

In a single-objective optimization, there is one goal – the search for an optimum solution. Although the search space may have a number of local optimal solutions, the goal is always to find the global optimum solution. However, there is an exception. In the case of multi-modal optimization (see Section 4.6 later), the goal is to find a number of local and global optimal solutions, instead of finding one optimum solution. However, most single-objective optimization algorithms aim at finding one optimum solution, even when there exist a number of optimal solutions. In a single-objective optimization algorithm, as long as a new solution has a better objective function value than an old solution, the new solution can be accepted.

However, in multi-objective optimization, there are clearly two goals. Progressing towards the Pareto-optimal front is certainly an important goal. However, maintaining a diverse set of solutions in the non-dominated front is also essential. An algorithm that finds a closely packed set of solutions on the Pareto-optimal front satisfies the first goal of convergence to the Pareto-optimal front, but does not satisfy maintenance of a diverse set of solutions. Since all objectives are important in a multi-objective optimization, a diverse set of obtained solutions close to the Pareto-optimal front provides a variety of optimal solutions, trading objectives differently. A multi-objective optimization algorithm that cannot find a diverse set of solutions in a problem is as good as a single-objective optimization algorithm.

Since both goals are important, an efficient multi-objective optimization algorithm must work on satisfying both of them. It is important to realize that both of these tasks are somewhat *orthogonal* to each other. The achievement of one goal does not necessarily achieve the other goal. Explicit or implicit mechanisms to emphasize convergence near the Pareto-optimal front and the maintenance of a diverse set of solutions must be introduced in an algorithm. Because of these dual tasks, multi-objective optimization is more difficult than single-objective optimization.

2.3.2   Dealing with Two Search Spaces

Another difficulty is that a multi-objective optimization involves two search spaces, instead of one. In a single-objective optimization, there is only one search space – the decision variable space. An algorithm works in this space by accepting and rejecting solutions based on their objective function values. Here, in addition to the decision variable space, there also exists the objective or criterion space. Although these two spaces are related by an unique mapping between them, often the mapping is nonlinear and the properties of the two search spaces are not similar. For example, a proximity of two solutions in one space does not mean a proximity in the other space. Thus, while achieving the second task of maintaining diversity in the obtained set of solutions, it is important to decide the space in which the diversity must be achieved.

In any optimization algorithm, the search is performed in the decision variable space. However, the proceedings of an algorithm in the decision variable space can be traced in the objective space. In some algorithms, the resulting proceedings in the objective space are used to steer the search in the decision variable space. When this happens, the proceedings in both spaces must be coordinated in such a way that the creation of new solutions in the decision variable space is complimentary to the diversity needed in the objective space. This, by no means, is an easy task and more importantly is dependent on the mapping between the decision variables and objective function values.

2.3.3 No Artificial Fix-Ups

It is needless to say that most real-world optimization problems are naturally posed as a multi-objective optimization problem. However, because of the lack of a suitable means of handling multi-objective problems as a true multi-objective optimization problem in the past, designers had to innovate different fix-ups. We will discuss a number of such methods in the next chapter. Of these, the weighted sum approach and the ϵ-constraint method are the most popularly used. In the weighted sum approach, multiple objectives are weighted and summed together to create a composite objective function. Optimization of this composite objective results in the optimization of individual objective functions. Unfortunately, the outcome of such an optimization strategy depends on the chosen weights. The second approach chooses one of the objective functions and treats the rest of the objectives as constraints by limiting each of them within certain pre-defined limits. This fix-up also converts a multi-objective optimization problem into a single-objective optimization problem. Unfortunately here too, the outcome of the single-objective constrained optimization results in a solution which depends on the chosen constraint limits.

Multi-objective optimization for finding multiple Pareto-optimal solutions eliminates all such fix-ups and can, in principle, find a set of optimal solutions corresponding to different weight and ϵ-vectors. Although only one solution is needed for implementation, a knowledge of such multiple optimal solutions may help a designer to compare and choose a compromised optimal solution. It is true that a multi-objective optimization is, in general, more complex than a single-objective optimization, but the avoidance of multiple simulation runs, no artificial fix-ups, availability of efficient population-based optimization algorithms, and above all, the concept of dominance helps to overcome some of the difficulties and give a user the practical means to handle multiple objectives, a matter which was not possible to achieve in the past.

2.4 Dominance and Pareto-Optimality

Most multi-objective optimization algorithms use the concept of dominance in their search. Here, we define the concept of dominance and related terms and present a number of techniques for identifying dominated solutions in a finite population of solutions.

2.4.1   Special Solutions

We first define some special solutions which are often used in multi-objective optimization algorithms.

Ideal Objective Vector

For each of the M conflicting objectives, there exists one different optimal solution. An objective vector constructed with these individual optimal objective values constitutes the ideal objective vector.

Definition 2.3. *The m-th component of the ideal objective vector $\mathbf{z}^*$ is the constrained minimum solution of the following problem:*

$$\left.\begin{array}{ll} Minimize & f_m(\mathbf{x}) \\ subject\ to & \mathbf{x} \in \mathcal{S}. \end{array}\right\} \tag{2.5}$$

Thus, if the minimum solution for the m-th objective function is the decision vector $\mathbf{x}^{*(m)}$ with function value f_m^*, the ideal vector is as follows:

$$\mathbf{z}^* = \mathbf{f}^* = (f_1^*, f_2^*, \cdots, f_M^*)^\mathsf{T}.$$

In general, the ideal objective vector corresponds to a non-existent solution. This is because the minimum solution of equation (2.5) for each objective function need not be the same solution. The only way an ideal objective vector corresponds to a feasible solution is when the minimal solutions to all objective functions are identical. In this case, the objectives are not conflicting to each other and the minimum solution to any objective function would be the only optimal solution to the MOOP. Figure 13 shows the ideal objective vector ($\mathbf{z}^*$) in the objective space of a hypothetical two-objective minimization problem.

It is interesting to ponder the question: 'If the ideal objective vector is non-existent, what is its use?' In most algorithms which are seeking to find Pareto-optimal solutions, the ideal objective vector is used as a reference solution (we are using the word 'solution' corresponding to the ideal objective vector loosely here, realizing that an ideal vector represents a non-existent solution). It is also clear from Figure 13 that solutions closer to the ideal objective vector are better. Moreover, many algorithms require the knowledge of the lower bound on each objective function to normalize objective values in a common range, a matter which we shall discuss later in Chapter 5.

Utopian Objective Vector

The ideal objective vector denotes an array of the lower bound of all objective functions. This means that for every objective function there exists at least one

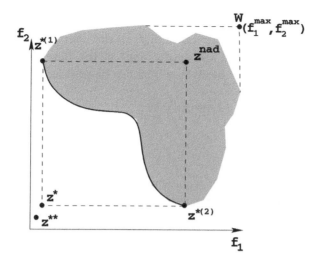

Figure 13 The ideal, utopian, and nadir objective vectors.

solution in the feasible search space sharing an identical value with the corresponding element in the ideal solution. Some algorithms may require a solution which has an objective value strictly better than (and not equal to) that of any solution in the search space. For this purpose, the utopian objective vector is defined as follows.

Definition 2.4. *A utopian objective vector* z^{**} *has each of its components marginally smaller than that of the ideal objective vector, or* $z_i^{**} = z_i^* - \epsilon_i$ *with* $\epsilon_i > 0$ *for all* $i = 1, 2, \ldots, M$.

Figure 13 shows a utopian objective vector. Like the ideal objective vector, the utopian objective vector also represents a non-existent solution.

Nadir Objective Vector

Unlike the ideal objective vector which represents the lower bound of each objective in the entire feasible search space, the nadir objective vector, z^{nad}, represents the upper bound of each objective in the entire Pareto-optimal set, and not in the entire search space. A nadir objective vector must not be confused with a vector of objectives (marked as 'W' in Figure 13) found by using the worst feasible function values, f_i^{max}, in the entire search space.

Although the ideal objective vector is easy to compute (except in complex multi-modal objective problems), the nadir objective vector is difficult to compute in practice. However, for well-behaved problems (including linear MOOPs), the nadir objective vector can be derived from the ideal objective vector by using the *payoff table* method described in Miettinen (1999). For two objectives (Figure 13), if $z^{*(1)} = \left(f_1(x^{*(1)}), f_2(x^{*(1)})\right)^T$ and $z^{*(2)} = \left(f_1(x^{*(2)}), f_2(x^{*(2)})\right)^T$ are coordinates of the minimum solutions of f_1 and f_2, respectively, in the objective space, then the nadir objective vector can be estimated as $z^{\mathrm{nad}} = \left(f_1(x^{*(2)}), f_2(x^{*(1)})\right)^T$.

The nadir objective vector may represent an existent or a non-existent solution, depending on the convexity and continuity of the Pareto-optimal set. In order to normalize each objective in the entire range of the Pareto-optimal region, the knowledge of nadir and ideal objective vectors can be used as follows:

$$f_i^{\text{norm}} = \frac{f_i - z_i^*}{z_i^{\text{nad}} - z_i^*}. \tag{2.6}$$

2.4.2 Concept of Domination

Most multi-objective optimization algorithms use the concept of domination. In these algorithms, two solutions are compared on the basis of whether one dominates the other solution or not. We will describe the concept of domination in the following paragraph.

We assume that there are M objective functions. In order to cover both minimization and maximization of objective functions, we use the operator $\lhd$ between two solutions i and j as $i \lhd j$ to denote that solution i is better than solution j on a particular objective. Similarly, $i \rhd j$ for a particular objective implies that solution i is worse than solution j on this objective. For example, if an objective function is to be minimized, the operator $\lhd$ would mean the '<' operator, whereas if the objective function is to be maximized, the operator $\lhd$ would mean the '>' operator. The following definition covers mixed problems with minimization of some objective functions and maximization of the rest of them.

Definition 2.5. *A solution* $x^{(1)}$ *is said to dominate the other solution* $x^{(2)}$, *if both conditions 1 and 2 are true:*

1. *The solution* $x^{(1)}$ *is no worse than* $x^{(2)}$ *in all objectives, or* $f_j(x^{(1)}) \not\rhd f_j(x^{(2)})$ *for all* $j = 1, 2, \ldots, M$.
2. *The solution* $x^{(1)}$ *is strictly better than* $x^{(2)}$ *in at least one objective, or* $f_{\bar{j}}(x^{(1)}) \lhd f_{\bar{j}}(x^{(2)})$ *for at least one* $\bar{j} \in \{1, 2, \ldots, M\}$.

If any of the above condition is violated, the solution $x^{(1)}$ does not dominate the solution $x^{(2)}$. If $x^{(1)}$ dominates the solution $x^{(2)}$ (or mathematically $x^{(1)} \preceq x^{(2)}$), it is also customary to write any of the following:

- $x^{(2)}$ is dominated by $x^{(1)}$;
- $x^{(1)}$ is non-dominated by $x^{(2)}$, or;
- $x^{(1)}$ is non-inferior to $x^{(2)}$.

Let us consider a two-objective optimization problem with five different solutions shown in the objective space, as illustrated in Figure 14. Let us also assume that the objective function 1 needs to be maximized while the objective function 2 needs to be minimized. Five solutions with different objective function values are shown in this figure. Since both objective functions are of importance to us, it is usually difficult to find one solution which is best with respect to both objectives. However, we can use

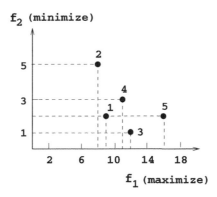

Figure 14   A population of five solutions.

the above definition of domination to decide which solution is better among any two given solutions in terms of both objectives. For example, if solutions 1 and 2 are to be compared, we observe that solution 1 is better than solution 2 in objective function 1 and solution 1 is also better than solution 2 in objective function 2. Thus, both of the above conditions for domination are satisfied and we may write that solution 1 dominates solution 2. We take another instance of comparing solutions 1 and 5. Here, solution 5 is better than solution 1 in the first objective and solution 5 is no worse (in fact, they are equal) than solution 1 in the second objective. Thus, both the above conditions for domination are also satisfied and we may write that solution 5 dominates solution 1.

It is intuitive that if a solution $x^{(1)}$ dominates another solution $x^{(2)}$, the solution $x^{(1)}$ is better than $x^{(2)}$ in the parlance of multi-objective optimization. Since the concept of domination allows a way to compare solutions with multiple objectives, most multi-objective optimization methods use this domination concept to search for non-dominated solutions.

2.4.3   Properties of Dominance Relation

The definition 2.5, described on page 28, defines the dominance relation between any two solutions. There are three possibilities that can be the outcome of the dominance check between two solutions 1 and 2. i.e. (i) solution 1 dominates solution 2, (ii) solution 1 gets dominated by solution 2, or (iii) solutions 1 and 2 do not dominate each other. Let us now discuss the different binary relation properties (Cormen et al., 1990) of the dominance operator.

Reflexive: The dominance relation is *not reflexive*, since any solution p does not dominate itself according to Definition 2.5. The second condition of dominance relation in Definition 2.5 does not allow this property to be satisfied.

Symmetric: The dominance relation is also *not symmetric*, because $p \preceq q$ does not imply q $\preceq$ p. In fact, the opposite is true. That is, if p dominates q, then q does not dominate p. Thus, the dominance relation is *asymmetric*.

Antisymmetric: Since the dominance relation is not symmetric, it cannot be antisymmetric as well.

Transitive: The dominance relation is *transitive*. This is because if p $\preceq$ q and q $\preceq$ r, then p $\preceq$ r.

There is another interesting property that the dominance relation possesses. If solution p does not dominate solution q, this does not imply that q dominates p.

In order for a binary relation to qualify as an ordering relation, it must be at least transitive (Chankong and Haimes, 1983). Thus, the dominance relation qualifies as an ordering relation. Since the dominance relation is not reflexive, it is a *strict partial order*. In general, if a relation is reflexive, antisymmetric, and transitive, it is loosely called a *partial order* and a set on which a partial order is defined is called a *partially ordered set*. However, it is important to note that the dominance relation is not reflexive and is not antisymmetric. Thus, the dominance relation is not a partial order relation in its general sense. The dominance relation is only a strict partial order relation.

2.4.4 Pareto-Optimality

Continuing with the comparisons in the previous subsection, let us compare solutions 3 and 5 in Figure 14, because this comparison reveals an interesting aspect. We observe that solution 5 is better than solution 3 in the first objective, while solution 5 is worse than solution 3 in the second objective. Thus, the first condition is not satisfied for both of these solutions. This simply suggests that we cannot conclude that solution 5 dominates solution 3, nor can we say that solution 3 dominates solution 5. When this happens, it is customary to say that solutions 3 and 5 are non-dominated with respect to each other. When both objectives are important, it cannot be said which of the two solutions 3 and 5 is better.

For a given finite set of solutions, we can perform all possible pair-wise comparisons and find which solution dominates which and which solutions are non-dominated with respect to each other. At the end, we expect to have a set of solutions, any two of which do not dominate each other. This set also has another property. For any solution outside of this set, we can always find a solution in this set which will dominate the former. Thus, this particular set has a property of dominating all other solutions which do not belong to this set. In simple term, this means that the solutions of this set are better compared to the rest of solutions. This set is given a special name. It is called the *non-dominated set* for the given set of solutions. In the example problem, solutions 3 and 5 constitutes the non-dominated set of the given set of five solutions. Thus, we define a set of non-dominated solutions as follows.

Definition 2.6 (Non-dominated set). *Among a set of solutions* P, *the non-dominated set of solutions* P' *are those that are not dominated by any member of the set* P.

When the set P is the entire search space, or $P = S$, the resulting non-dominated set P' is called the *Pareto-optimal set*. Figure 15 marks the Pareto-optimal set with continuous curves for four different scenarios with two objectives. Each objective can be minimized or maximized. In the top-left figure, the task is to minimize both objectives f_1 and f_2. The solid curve marks the Pareto-optimal solution set. If f_1 is to be minimized and f_2 is to be maximized for a problem having the same search space, the resulting Pareto-optimal set is different and is shown in the top-right figure. Here, the Pareto-optimal set is a union of two disconnected Pareto-optimal regions. Similarly, the Pareto-optimal sets for two other cases ((maximizing f_1, minimizing f_2) and (maximizing f_1, maximizing f_2)) are also shown in the bottom-left and bottom-right figures, respectively. In any case, the Pareto-optimal set always consists of solutions from a particular edge of the feasible search region.

It is important to note that an MOEA can be easily used to handle all of the above cases by simply using the domination definition. However, to avoid any confusion, most applications use the duality principle (discussed earlier) to convert a maximization problem into a minimization problem and treat every problem as a combination of minimizing all objectives.

Like global and local optimal solutions in the case of single-objective optimization, there could be global and local Pareto-optimal sets in multi-objective optimization.

Definition 2.7 (Globally Pareto-optimal set). *The non-dominated set of the entire feasible search space* S *is the globally Pareto-optimal set.*

On many occasions, the globally Pareto-optimal set is simply referred to as the Pareto-optimal set. Since solutions of this set are not dominated by any feasible member of the search space, they are optimal solutions of the MOOP. We define a locally Pareto-optimal set as follows (Deb, 1999c; Miettinen, 1999).

Definition 2.8. *If for every member* x *in a set* P *there exists no solution* y *(in the neighborhood of* x *such that* $\|y - x\|_\infty \le \epsilon$, *where* ϵ *is a small positive number) dominating any member of the set* P, *then solutions belonging to the set* P *constitute a locally Pareto-optimal set.*

Figure 16 shows two locally Pareto-optimal sets (marked with continuous curves). When any solution (say 'B') in this set is perturbed locally in the decision variable space, no solution can be found dominating any member of the set. It is interesting to note that for continuous search space problems, the locally Pareto-optimal solutions need not be continuous in the decision variable space and the above definition will still hold good. Later in Section 8.3, we shall present instances of MOOPs having many locally Pareto-optimal sets. Zitzler (1999) added a neighborhood constraint on the objective space in the above definition to make it more generic.

By the above definition, it is also true that a globally Pareto-optimal set is also a locally Pareto-optimal set.

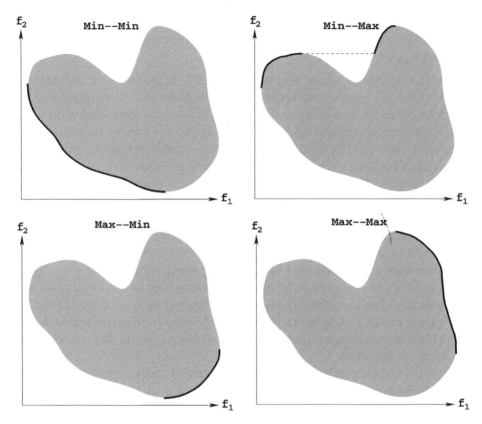

Figure 15   Pareto-optimal solutions are marked with continuous curves for four combinations of two types of objectives.

2.4.5   Strong Dominance and Weak Pareto-Optimality

The dominance relationship between the two solutions defined in Definition 2.5 is sometimes referred to as a *weak* dominance relation. This definition can be modified and a strong dominance relation can be defined as follows.

Definition 2.9. *A solution* $x^{(1)}$ *strongly dominates a solution* $x^{(2)}$ *(or* $x^{(1)} \prec x^{(2)}$ *), if solution* $x^{(1)}$ *is strictly better than solution* $x^{(2)}$ *in all* M *objectives.*

Referring to Figure 14, we now observe that solution 5 does not strongly dominate solution 1, although we have seen earlier that solution 5 weakly dominates solution 1. However, solution 3 strongly dominates solution 1, since solution 3 is better than solution 1 in both objectives. Thus, if a solution $x^{(1)}$ strongly dominates a solution $x^{(2)}$, the solution $x^{(1)}$ also weakly dominates solution $x^{(2)}$, but not vice versa. The strong dominance operator has the same properties as that described in Section 2.4.3 for the weak dominance operator.

The above definition of strong dominance can be used to define a *weakly nondominated set*.

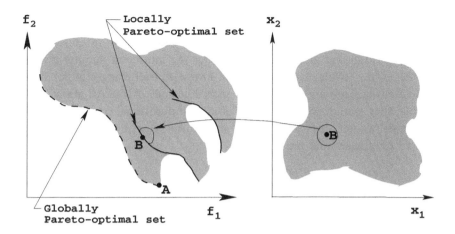

Figure 16 Locally and globally Pareto-optimal solutions.

Definition 2.10 (Weakly non-dominated set). *Among a set of solutions* P, *the weakly non-dominated set of solutions* P' *are those that are not strongly dominated by any other member of the set* P.

The above definition suggests that a weakly non-dominated set found from a set of P solutions contains all members of the non-dominated set obtained by using Definition 2.6 from the same set P. In other words, for a given population of solutions, the cardinality of the weakly non-dominated set is greater than or equal to the cardinality of the non-dominated set obtained by using the usual Definition 2.6. The definition of a globally or locally weakly Pareto-optimal set can also be defined similarly by using the definition of the weakly non-dominated set.

From the above discussion it is clear that a Pareto-optimal set is always a non-dominated set. But there may exist non-dominated sets containing some Pareto-optimal solutions and some non-Pareto-optimal solutions. Thus, it is important to realize that the non-dominated solutions found by an optimization algorithm need not represent the true Pareto-optimal set. However, it can be hoped that population-based search algorithms which adequately emphasize the non-dominated set of a given population to steer its search are likely to approach the true Pareto-optimal region. The important question is: 'How would we find the non-dominated set in a given population of solutions?' Since a non-dominated set may be required to be identified in each iteration of a multi-objective optimization algorithm, we are interested in a computationally efficient procedure of identifying the non-dominated set from a population of solutions. We will discuss a few such procedures in the next subsection.

2.4.6 *Procedures for Finding a Non-Dominated Set*

Finding the non-dominated set of solutions from a given set of solutions is similar in principle to finding the minimum of a set of real numbers. In the latter case, when two numbers are compared to identify the smaller number, a '<' relation operation is used. In the case of finding the non-dominated set, the dominance relation $\preceq$ (or $\prec$, as the case may be) can be used to identify the better of two given solutions. Like the existence of different algorithms for finding the minimum number from a finite set (Cormen et al., 1990), many approaches have also been suggested for finding the non-dominated set from a given population of solutions. Although many approaches are possible, they would usually have different computational complexities. To make matters simple, we will discuss here three procedures, starting from a naive and slow approach to an efficient and fast approach.

Approach 1: Naive and Slow

In this approach, each solution i is compared with every other solution in the population to check if it is dominated by any solution in the population. If the solution i is found to be dominated by any solution, this means that there exists at least one solution in the population which is better than i in all objectives. Hence the solution i cannot belong to the non-dominated set. We mark a flag against the solution i to denote that it does not belong to the non-dominated set. However, if no solution is found to dominate solution i, it is a member of the non-dominated set. This is how any other solution in the population can be checked to see if it belongs to the non-dominated set. The following approach describes a step-by-step procedure for finding the non-dominated set in a given set P of size N.

Identifying the Non-Dominated Set: Approach 1

Step 1 Set solution counter $i = 1$ and create an empty non-dominated set P'.

Step 2 For a solution $j \in P$ (but $j \neq i$), check if solution j dominates solution i. If yes, go to Step 4.

Step 3 If more solutions are left in P, increment j by one and go to Step 2; otherwise, set $P' = P' \cup \{i\}$.

Step 4 Increment i by one. If $i \leq N$, go to Step 2; otherwise stop and declare P' as the non-dominated set.

We illustrate the working principle of the above procedure on the same set of five $(N = 5)$ solutions, as shown in Figure 14. Ideally, the exact objective vector for each solution will be used in executing the above procedure, but here we use the figure to compare different solutions. We follow the above step-by-step procedure in the following.

Step 1 We set $i = 1$ and $P' = \emptyset$.

Step 2 We compare solution 1 with all other solutions for domination, starting from solution 2. We observe that solution 2 does not dominate solution 1.

Step 3 However, solution 3 dominates solution 1. Thus, we move to Step 4.

Step 4 Solution 1 does not belong to the non-dominated set and we increment i to 2 and move to Step 2 to check the fate of solution 2.

Step 2 We observe that solution 1 dominates solution 2. We therefore move to Step 4.

Step 4 Thus, solution 2 does not belong to the non-dominated set. Next, we check solution 3.

Steps 2 and 3 Starting from solution 1, we observe that neither solution 1 nor 2 dominates solution 3. In fact, solutions 4 and 5 also do not dominate solution 3. Thus, we include solution 3 in the non-dominated set, $P' = \{3\}$.

Step 4 We now check solution 4.

Step 2 Solution 5 dominates solution 4. Thus, it cannot be a member of P'.

Step 4 Now we check the final solution (solution 5).

Step 2 We observe that none of the solutions (1 to 4) dominates solution 5.

Step 3 So, solution 5 also belongs to the non-dominated set. Thus, we update $P' = \{3, 5\}$.

Step 4 We have now considered all five solutions and found the non-dominated set $P' = \{3, 5\}$.

It is clear that Step 2 requires $O(N)$ comparisons for domination and each comparison for domination needs M function value comparisons. Thus, the total complexity (we define complexity here as the total number of function value comparisons) of Step 2 is $O(MN)$. Step 4 recursively calls Step 2, thereby requiring at *most* $O(MN^2)$ computations of the above procedure. It is important to realize that this is the worst case complexity. Since in Step 2, not all $(N-1)$ solutions may have to be checked for domination with every solution i, the average case complexity may be smaller.

In order to investigate how the average case complexity of this implementation grows with the population size and the number of objectives, we perform a systematic experimental study. For three different M values and a number of population sizes (N), we create 100 random populations. For each population, we use the above approach and record the total number of function value comparisons needed to find the non-dominated set. The average of this comparisons is plotted in Figure 17. Similar plots for two other approaches are also shown. These two approaches will be described in the next subsections. We have used $M = 10$ in this plot, and population sizes in the range 100 to 1000 are used. An average of 100 simulation runs are shown here, and the figure is plotted using a log–log graph. Since the results obtained fall more or less on a straight line, this suggests a polynomial increase in the complexity with the population size. When we compute the slope of this plot for this log-log graph, it is found to be 1.943, thereby suggesting a near-quadratic complexity.

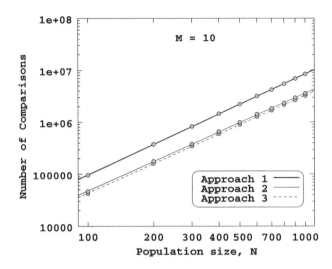

Figure 17 Average number of comparisons versus population size N shown for ten objectives ($M = 10$).

Although the worst case complexity varies linearly with the number of objective functions (M), the average case scenario may be different. This is because as the number of objective functions increases, the Pareto-optimal front becomes a higher-dimensional surface. For example, with $M = 2$, the Pareto-optimal front is at most a two-dimensional curve. For three objectives ($M = 3$), the Pareto-optimal region is at most a three-dimensional surface, and so on. Thus, for the same population size an increase in M may result in a larger proportion of solutions in the non-dominated set. When most solutions lie in the non-dominated front, the algorithms will have different computational complexity compared to the worst case calculations. However, we do not worry much about how the complexity depends on M; instead, we are more interested in knowing how the complexity depends on N. This is because for a given multi-objective optimization problem, M remains fixed. It then becomes important to know what population size is appropriate for obtaining a particular accuracy in the obtained solutions and what is the resulting run-time complexity. We will discuss more about how the computational complexity depends on M later in Section 8.8.

Approach 2: Continuously Updated

This approach is similar in principle to Approach 1, except that better bookkeeping is used to make the algorithm faster. In this approach, every solution from the population is checked with a partially filled population for domination. To start with, the first solution from the population is kept in an empty set P'. Thereafter, each solution i (the second solution onwards) is compared with all members of the set P', one by one. If the solution i dominates any member of P', then that solution is removed from

P'. In this way non-members of the non-dominated solutions get deleted from P'. Otherwise, if solution i is dominated by any member of P', the solution i is ignored. If solution i is not dominated by any member of P', it is entered in P'. This is how the set P' grows with non-dominated solutions. When all solutions of the population are checked, the remaining members of P' constitute the non-dominated set.

Identifying the Non-Dominated Set: Approach 2

> **Step 1** Initialize $P' = \{1\}$. Set solution counter $i = 2$.
>
> **Step 2** Set $j = 1$.
>
> **Step 3** Compare solution i with j from P' for domination.
>
> **Step 4** If i dominates j, delete the j-th member from P' or update $P' = P' \backslash \{P'^{(j)}\}$. If $j < |P'|$, increment j by one and then go to Step 3. Otherwise, go to Step 5. Alternatively, if the j-th member of P' dominates i, increment i by one and then go to Step 2.
>
> **Step 5** Insert i in P' or update $P' = P' \cup \{i\}$. If $i < N$, increment i by one and go to Step 2. Otherwise, stop and declare P' as the non-dominated set.

We illustrate the above procedure of identifying the non-dominated set by using the previous example (in Figure 14) of five population members.

Step 1 $P' = \{1\}$ and we set $i = 2$.

Step 2 We set the solution counter of P' as $j = 1$ (which refers to solution 1).

Step 3 We now compare solution 2 ($i = 2$) with the lone member of P' (solution 1) for domination. We observe that solution 1 dominates solution 2 . Since the j-th member of P' dominates solution i, we increment i to 3 and go to Step 2. This means that solution 2 does not belong to the non-dominated set.

Step 2 The set P' still has solution 1 ($j = 1$) only.

Step 3 Now we compare solution 3 with solution 1. We observe that solution 3 ($i = 3$) dominates solution 1. Thus, we delete the j-th (or the first) member from P' and update $P' = \emptyset$. Thus, $|P'| = 0$. This depicts that a non-member of the non-dominated set gets deleted from P'. We now move to Step 5.

Step 5 We insert $i = 3$ in P' or update $P' = \{3\}$. Since $i < 5$ here, we increment i to 4 and move to Step 2.

Step 2 We set $j = 1$, which refers to the lone element (solution 3) of P'.

Step 3 By comparing solution 4 with solution 3, we observe that solution 3 dominates solution 4. Thus, we increment i to 5 and move to Step 2.

Step 2 We still have solution 3 in P'.

Step 3 Now, we compare solution 5 with solution 3. We observe that neither of them dominates the other. Thus, we move to Step 5.

Step 5 We insert solution 5 in P' and update P' = {3,5}. Since i = 5, we stop and declare P' = {3,5} as the non-dominated set.

Here, we observe that the second element of the population is compared with only one solution P', the third solution with at most two solutions of P', and so on. This requires a maximum of $1+2+\cdots+(N-1)$ or $N(N-1)/2$ domination checks. Although this computation is also $O(MN^2)$, the actual number of computations is about half of that required in Approach 1. It is interesting to note that the size of P' may not always increase (dominated solutions will get deleted from P') and not every solution in the population may be required to be checked with all solutions in the current P' set (the solution may get dominated by a solution of P'). Thus, the actual computational complexity may be smaller than the above estimate.

Figure 17 shows that Approach 2 has an average number of comparisons much smaller than that of Approach 1. Since the slope of the fitted lines of both approaches are more or less the same, both approaches have a similar order of computational complexity. However, there is a remarkable difference in the actual number of computations. Next, we will discuss an approach which is computationally faster than the above two approaches.

Approach 3: Kung et al.'s Efficient Method

This approach first sorts the population according to the descending order of importance to the first objective function value. Thereafter, the population is recursively halved as top (T) and bottom (B) subpopulations. Knowing that the top-half of the population is better in terms of the first objective function, the bottom-half is then checked for domination with the top-half. The solutions of B that are not dominated by any member of T are combined with members of T to form a merged population M. The merging and the domination check starts with the innermost case (when there is only one member left in either T or B in recursive divisions of the population) and the proceeds in a bottom-up fashion. This approach is not as easy to visualize as the previous two approaches, but, as discussed in Kung et al. (1975), it is the most computationally efficient method.

Identifying the Non-Dominated Set: Approach 3

> **Step 1** Sort the population according to the descending order of importance in the first objective function and rename the population as P of size N.
>
> **Step 2, Front(P)** If |P| = 1, return P as the output of **Front**(P). Otherwise, T = **Front**($P^{(1)}$–$P^{(|P|/2)}$) and B = **Front**($P^{(|P|/2+1)}$–$P^{(|P|)}$). If the i-th non-dominated solution of B is not dominated by *any* non-dominated solution of T, create a merged set M = T∪{i}. Return M as the output of **Front**(P).

What finally returns from Step 2 is the non-dominated set.

It is important to mention that Step 2 uses a self-recursion. We illustrate the working of the above procedure on the same example problem (Figure 14) as used in the previous approaches with five population members.

Step 1 Since the first objective is to be maximized in this problem, we sort the population according to the decreasing value of the first objective function. We obtain the following sequence:

$$P = \{5, 3, 4, 1, 2\}$$

Step 2 When the entire set P enters Front( ), the size of P is not one. Thus, we set $T = \text{Front}(\{5, 3\})$ and $B = \text{Front}(\{4, 1, 2\})$. Recursively, the $\text{Front}(\{5, 3\})$ would divide the set $\{5, 3\}$ into two halves and make $\text{Front}(\{5\})$ and $\text{Front}(\{3\})$ as the next inner T and B sets. Since the sizes of these two inner sets are one each, these two solutions are sent as output to the above Front( ) calls. Moving up a step to $\text{Front}(\{5, 3\})$ with $T = \{5\}$ and $B = \{3\}$, the domination check and merging operation are then executed. We observe that solution 3 is not dominated by solution 5. Therefore, the merged set contains both solutions, or $M = \{5, 3\}$. Thus, the outcome of the $T = \text{Front}(\{5, 3\})$ is the set $\{5, 3\}$.

Similarly, the $\text{Front}(\{4, 1, 2\})$ operation makes two calls, i.e. $\text{Front}(\{4\})$ and $\text{Front}(\{1, 2\})$. The output from the former is $\{4\}$. The latter makes two more calls, i.e. $\text{Front}(\{1\})$ and $\text{Front}(\{2\})$, with outputs as $\{1\}$ and $\{2\}$, respectively. Now, to complete the $\text{Front}(\{1, 2\})$ operation, we need to check whether solution 2 is indeed dominated by solution 1. As we already know that solution 2 is dominated by solution 1, we have $M = \{1\}$ as the output of $\text{Front}(\{1, 2\})$. Now, to complete the $\text{Front}(\{4, 1, 2\})$ operation, we need to check the domination of solutions 4 (T) and 1 (B). Since solution 1 is not dominated by solution 4, the merged set is $M = \{4, 1\}$, which is also the output of the $\text{Front}(\{4, 1, 2\})$ call.

Now, for the overall call to Front(P), the sets are $\{5, 3\}$ and $\{4, 1\}$, respectively. Checking the domination of the non-dominated solution (solution 1 in the second objective) in B with the non-dominated solution (solution 3 in the second objective) of T, we observe that both solution 1 gets dominated by solution 3 in the second objective. Thus, the output of the Front(P) call is the set $\{5, 3\}$, and this is the non-dominated set.

After some lengthy computations, Kung et al. (1975) found that the complexity of this approach is $O\left(N(\log N)^{M-2}\right)$ for $M \geq 4$ and $O(N \log N)$ for $M = 2$ and 3. Interested readers may refer to the original study for details of this complexity calculation. Figure 17 shows that Approach 3 performed the best among all three approaches.

With $M = 10$ objectives, we observe a similar order of the three approaches in terms of their computational complexities (see Table 2). The corresponding rates of increase in complexity are 1.9543, 1.8834, and 1.8826 for Approaches 1, 2 and 3, respectively. For $M = 20$ objectives, the complexity measures are similar. Although

Table 2   Computational complexity of finding the non-dominated set for different values of M.

M	Approach 1	Approach 2	Approach 3
4	$14.714N^{1.5447}$	$25.354N^{1.1248}$	$14.120N^{1.1228}$
10	$12.614N^{1.9543}$	$8.145N^{1.8834}$	$7.431N^{1.8826}$
20	$19.525N^{2.0034}$	$9.772N^{2.0032}$	$9.413N^{2.0013}$

the complexity of Approach 3 is shown in a polynomial order, this is primarily done to investigate the effect of population size alone. It is important to note that the above time complexities reflect the average time complexity of an approach and that the orders of computational complexity presented in each approach are calculated for the worst case scenarios. Nevertheless, the experimental results show that with large values of M, the complexity approaches $O(N^2)$ in the range $100 \leq N \leq 1000$.

It is clear that the non-dominated solutions found by any of the above approaches are the *best* solutions in a population. Thus, these solutions are also called the solutions of the best non-dominated set.

2.4.7   Non-Dominated Sorting of a Population

We shall see later in Chapter 5 that most evolutionary multi-objective optimization algorithms require us to find only the best non-dominated front in a population. These algorithms classify the population into two sets – the non-dominated set and the remaining dominated set. However, there exist some algorithms which require the entire population to be classified into various non-domination levels. In such algorithms, the population needs to be sorted according to an ascending level of non-domination. The best non-dominated solutions are called non-dominated solutions of level 1. In order to find solutions for the next level of non-domination, there is a simple procedure which is usually followed. Once the best non-dominated set is identified, they are temporarily disregarded from the population. The non-dominated solutions of the remaining population are then found and are called non-dominated solutions of level 2. In order to find the non-dominated solutions of level 3, all non-dominated solutions of levels 1 and 2 are disregarded and new non-dominated solutions are found. This procedure is continued until all population members are classified into a non-dominated level. It is important to reiterate that non-dominated solutions of level 1 are better than non-dominated solutions of level 2, and so on.

Non-Dominated Sorting

Step 1 Set all non-dominated sets P_j, $(j = 1, 2, \ldots)$ as empty sets. Set non-domination level counter $j = 1$.

Step 2 Use any one of the Approaches 1 to 3 to find the non-dominated set P' of population P.

Step 3 Update $P_j = P'$ and $P = P \backslash P'$.

Step 4 If $P \neq \emptyset$, increment j by one and go to Step 2. Otherwise, stop and declare all non-dominated sets P_i, for $i = 1, 2, \ldots, j$.

We illustrate the working of the above procedure on the same set of five solutions (Figure 14) by using Approach 1 for identifying the non-dominated set.

Step 1 We first set $j = 1$ to identify the non-dominated solutions of the first level.

Step 2 As shown earlier, the first non-dominated set is $P' = \{3, 5\}$.

Step 3 Update $P = P' = \{3, 5\}$ and modify P by deleting solutions 3 and 5 from it, or $P = \{1, 2, 4\}$.

Step 4 Since P is a non-empty set, we move to Step 2 in search of solutions of the second non-domination level ($j = 2$).

Step 2 We now move to Approach 1, outlined above on page 34.

> **Step 1** Set $i = 1$ and $P' = \emptyset$.
>
> **Step 2** Solution 2 does not dominate solution 1.
>
> **Step 3** We now check solution 4 with solution 1.
>
> **Step 2** Solution 4 does not dominate solution 1 either.
>
> **Step 3** We set $P' = \{1\}$.
>
> **Step 4** We increment i to 2 (check for solution 2) and move to Step 2.
>
> **Step 2** Solution 1 dominates solution 2.
>
> **Step 3** Thus, we move to Step 4.
>
> **Step 4** We now check solution 4 for its inclusion in the second non-dominated set.
>
> **Step 2** Solution 1 does not dominate solution 4.
>
> **Step 3** Check with solution 2.
>
> **Step 2** Solution 2 does not dominate solution 4 either.
>
> **Step 3** We include solution 4 in the set or $P' = \{1, 4\}$.
>
> **Step 4** All three solutions are checked for non-domination. Thus, we declare the non-dominated set as $P' = \{1, 4\}$.

Step 3 Update $P_2 = P' = \{1, 4\}$ and modify P by deleting P' from it, or $P = \{2\}$.

Step 4 We increment j to 3 and move to Step 2 in search of solutions in the third non-dominated set.

Since there is only one solution left in P, this must belong to the third non-dominated set. Thus, there are three non-dominated sets in the population with the following classification in a decreasing order of non-domination (or an increasing order of domination):

$$((3,5), (1,4), (2))$$

The corresponding sets are marked in Figure 18 with hypothetical curves. The other two approaches can also be used and an identical outcome will be found.

The complexity of this complete classification procedure is the sum of the individual complexities involved in identifying each non-dominated set. It is important to note that after the first non-dominated set is identified, the number of solutions left in P is smaller than the original population size. Thus, subsequent identifications will require smaller computational complexity. In an average scenario, the number of non-domination levels is usually much smaller than the population size. We shall show later in Section 8.8 that as M increases the number of different non-domination levels in a random population reduces drastically. Thus, for all practical purposes, the computational complexity of the overall non-dominated sorting procedure is governed by the procedure of identifying the first non-dominated set.

In a similar experimental study, we find the overall computational complexity needed in sorting several populations according to different non-domination levels. The rate of increase in the complexity is slightly more compared to just finding the best non-dominated set. Interestingly, even in the complete non-dominated sorting problem, Approach 3 turns out to be better than the other two approaches. Table 3 shows the estimated complexity measures (from simulation results) for different population sizes (N) and number of objectives (M). It is clear from these computations that Approach 3 requires the least amount of computational effort to sort a population into different non-domination levels.

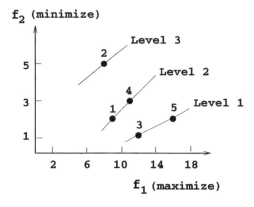

Figure 18   Solutions classified into various non-domination classes.

Table 3 Computational complexities for a complete non-dominated sorting procedure.

M	Approach 1	Approach 2	Approach 3
4	$3.876N^{1.9722}$	$3.274N^{1.7463}$	$3.145N^{1.7135}$
10	$10.942N^{1.9722}$	$6.786N^{1.9194}$	$6.369N^{1.9133}$
20	$19.526N^{2.0034}$	$9.768N^{2.0033}$	$9.408N^{2.0014}$

An $O(MN^2)$ Non-Dominated Sorting Procedure

Using the above procedure, each front can be identified with at most $O(MN^2)$ computations. In certain scenarios, this procedure may demand more than $O(MN^2)$ computational effort for the overall non-dominated sorting of a population. Here, we suggest a completely different procedure which uses a better bookkeeping strategy requiring $O(MN^2)$ overall computational complexity.

First, for each solution we calculate two entities: (i) *domination count* n_i, the number of solutions which dominate the solution i, and (ii) S_i, a set of solutions which the solution i dominates. This requires $O(MN^2)$ comparisons. At the end of this procedure, all solutions in the first non-dominated front will have their domination count as zero. Now, for each of these solutions (each solution i with $n_i = 0$), we visit each member (j) of its set S_i and reduce its domination count by one. In doing so, if for any member j the domination count becomes zero, we put it in a separate list P'. After such modifications on S_i are performed for each i with $n_i = 0$, all solutions of P' would belong to the second non-dominated front. The above procedure can be continued with each member of P' and the third non-dominated front can be identified. This process continues until all solutions are classified.

An $O(MN^2)$ Non-Dominated Sorting Algorithm

> **Step 1** For each $i \in P$, $n_i = 0$ and initialize $S_i = \emptyset$. For all $j \neq i$ and $j \in P$, perform Step 2 and then proceed to Step 3.
>
> **Step 2** If $i \preceq j$, update $S_p = S_p \cup \{j\}$. Otherwise, if $j \preceq i$, set $n_i = n_i + 1$.
>
> **Step 3** If $n_i = 0$, keep i in the first non-dominated front P_1 (we called this set P' in the above paragraph). Set a front counter $k = 1$.
>
> **Step 4** While $P_k \neq \emptyset$, perform the following steps.
>
> **Step 5** Initialize $Q = \emptyset$ for storing next non-dominated solutions. For each $i \in P_k$ and for each $j \in S_i$,
>
> > **Step 5a** Update $n_j = n_j - 1$.
> > **Step 5b** If $n_j = 0$, keep j in Q, or perform $Q = Q \cup \{j\}$.
>
> **Step 6** Set $k = k + 1$ and $P_k = Q$. Go to Step 4.

Steps 1 to 3 find the solutions in the first non-dominated front and require $O(MN^2)$ computational complexity. Steps 4 to 6 repeatedly find higher fronts and require at most $O(N^2)$ comparisons, as argued below.

For each solution i in the second or higher-level of non-domination, the domination count n_i can be at most $N-1$. Thus, each solution i will be visited at most $N-1$ times before its domination count becomes zero. At this point, the solution is assigned a particular non-domination level and will never be visited again. Since there are at most $N-1$ such solutions, the complexity of identifying second and more fronts is $O(N^2)$. Thus, the overall complexity of the procedure is $O(MN^2)$. It is important to note that although the time complexity has reduced to $O(MN^2)$, the storage requirement has increased to $O(N^2)$.

2.5   Optimality Conditions

In this section, we will outline the theoretical Pareto-optimality conditions for a constrained MOOP. Here, we state the optimality conditions for the MOOP given in equation (2.1). We assume that all objectives and constraint functions are continuously differentiable. As in single-objective optimization, there exist first- and second-order optimality conditions for multi-objective optimization. Here, we state the first-order necessary condition and a specific sufficient condition. Interested readers may refer to more advanced classical books on multi-objective optimization (Chankong and Haimes, 1983; Ehrgott, 2000; Miettinen, 1999) for a more comprehensive treatment of different optimality conditions.

The following condition is known as the necessary condition for Pareto-optimality.

Theorem 2.5.1. *(Fritz–John necessary condition). A necessary condition for* $\mathbf{x}^*$ *to be Pareto-optimal is that there exist vectors* $\lambda \geq 0$ *and* $\mathbf{u} \geq 0$ *(where* $\lambda \in \mathbb{R}^M$, $\mathbf{u} \in \mathbb{R}^J$ *and* $\lambda, \mathbf{u} \neq 0$) *such that the following conditions are true:*

1. $\sum_{m=1}^{M} \lambda_m \nabla f_m(\mathbf{x}^*) - \sum_{j=1}^{J} u_j \nabla g_j(\mathbf{x}^*) = 0$, *and*
2. $u_j g_j(\mathbf{x}^*) = 0$ *for all* $j = 1, 2, \ldots, J$.

For a proof, readers may refer to Cunha and Polak (1967). Miettinen (1999) argues that the above theorem is also valid as the necessary condition for a solution to be weakly Pareto-optimal. Those readers familiar with the Kuhn–Tucker necessary conditions for single-objective optimization will immediately recognize the similarity between the above conditions and that of the single-objective optimization. The difference is in the inclusion of a λ-vector with the gradient vector of the objectives.

For an unconstrained MOOP, the above theorem requires the following condition:

$$\sum_{m=1}^{M} \lambda_m \nabla f_m(\mathbf{x}^*) = 0$$

to be necessary for a solution to be Pareto-optimal. Writing the above vector equation in matrix form, we have the following necessary condition for an M-objective and n-variable unconstrained MOOP:

$$
\begin{bmatrix}
\dfrac{\partial f_1}{\partial x_1} & \dfrac{\partial f_2}{\partial x_1} & \cdots & \dfrac{\partial f_M}{\partial x_1} \\
\dfrac{\partial f_1}{\partial x_2} & \dfrac{\partial f_2}{\partial x_2} & \cdots & \dfrac{\partial f_M}{\partial x_2} \\
\cdots & \cdots & \cdots & \cdots \\
\dfrac{\partial f_1}{\partial x_n} & \dfrac{\partial f_2}{\partial x_n} & \cdots & \dfrac{\partial f_M}{\partial x_n}
\end{bmatrix}
\begin{bmatrix}
\lambda_1 \\ \lambda_2 \\ \vdots \\ \lambda_M
\end{bmatrix}
=
\begin{bmatrix}
0 \\ 0 \\ \vdots \\ 0
\end{bmatrix}.
\tag{2.7}
$$

For nonlinear objective functions, the partial derivatives are expected to be nonlinear. For a given λ-vector, the nonexistence of a corresponding Pareto-optimal solution can be checked by using the above equation. If the above set of equations are not satisfied, a Pareto-optimal solution corresponding to the given λ-vector does not exist. However, the existence of a Pareto-optimal solution is not guaranteed by the above necessary condition. That is, any solution which satisfies equation (2.7) is not necessarily a Pareto-optimal solution.

For problems with $n = M$ (identical number of decision variables and objectives), the Pareto-optimal solutions must satisfy the following:

$$
\begin{vmatrix}
\dfrac{\partial f_1}{\partial x_1} & \dfrac{\partial f_2}{\partial x_1} & \cdots & \dfrac{\partial f_M}{\partial x_1} \\
\dfrac{\partial f_1}{\partial x_2} & \dfrac{\partial f_2}{\partial x_2} & \cdots & \dfrac{\partial f_M}{\partial x_2} \\
\cdots & \cdots & \cdots & \cdots \\
\dfrac{\partial f_1}{\partial x_n} & \dfrac{\partial f_2}{\partial x_n} & \cdots & \dfrac{\partial f_M}{\partial x_n}
\end{vmatrix}
= 0.
\tag{2.8}
$$

The determinant of the partial derivative matrix must be zero for Pareto-optimal solutions. For a two-variable, two-objective MOOP, the above condition reduces to the following:

$$
\frac{\partial f_1}{\partial x_1}\frac{\partial f_2}{\partial x_2} = \frac{\partial f_1}{\partial x_2}\frac{\partial f_2}{\partial x_1}.
\tag{2.9}
$$

The following theorem offers sufficient conditions for a solution to be Pareto-optimal for convex functions.

Theorem 2.5.2. *(Karush–Kuhn–Tucker sufficient condition for Pareto-optimality). Let the objective functions be convex and the constraint functions of the problem shown in equation (2.1) be nonconvex. Let the objective and constraint functions be continuously differentiable at a feasible solution* x^**. A sufficient condition for* x^* *to be Pareto-optimal is that there exist vectors* $\lambda > 0$ *and* $u \geq 0$ *(where* $\lambda \in \mathbb{R}^M$ *and* $u \in \mathbb{R}^J$ *) such that the following equations are true:*

1. $\sum_{m=1}^{M}\lambda_m \nabla f_i(x^*) - \sum_{j=1}^{J}u_j \nabla g_j(x^*) = 0$, *and*
2. $u_j g_j(x^*) = 0$ *for all* $j = 1, 2, \ldots, J$.

For a proof, see Miettinen (1999). If the objective functions and constraints are not convex, the above theorem does not hold. However, for pseudo-convex and non-differentiable problems, different necessary and sufficient conditions do exist (Bhatia and Aggarwal, 1992).

2.6 Summary

Most real-world search and optimization problems are naturally posed as multi-objective optimization problems. However, due to the complexities involved in solving multi-objective optimization problems and due to the lack of suitable and efficient solution techniques, they have been transformed and solved as single-objective optimization problems. In this chapter, we have argued that such a transformation is often subjective to the user. Moreover, because of the presence of conflicting multiple objectives, a multi-objective optimization problem results in a number of optimal solutions, known as Pareto-optimal solutions. Faced with multiple objectives, ideally a user is interested in finding multiple Pareto-optimal solutions. Thus, there are two tasks of an ideal multi-objective optimization algorithm, namely (i) to find multiple Pareto-optimal solutions and (ii) to seek for Pareto-optimal solutions with a good diversity in objective and/or decision variable values.

Most multi-objective optimization methods use the concept of domination in their search. Thus, we have presented definitions for domination, a non-dominated set and a Pareto-optimal set of solutions. Thereafter, we have suggested three approaches for identifying the non-dominated set from a population of solutions. Since some multi-objective evolutionary algorithms requires a complete non-dominated sorting of a set of N solutions, we have discussed two approaches, of which one guarantees performing the task in $O(MN^2)$ computational effort. Finally, the optimality conditions for solutions to become Pareto-optimal are presented. In the next chapter, we will present a number of classical approaches for solving multi-objective optimization problems.

Exercise Problems

1. Determine if the following functions are convex?
   (a) $f(x) = x(1 - x)$,
   (b) $f(x_1, x_2) = x_1^2 + x_2^2 - 10$,
   (c) $f(x_1, x_2) = x_2^4 - 4x_1$ where $x_1, x_2 \geq 0$.

2. For what values of x_1 and x_2 is the following function convex?
$$f(x_1, x_2) = x_1 - x_1^3 x_2 - x_2^2.$$

3. Determine if the feasible region bounded by the following constraints constitute a convex region? Give reasons.
$$g_1(x) \equiv 4x_1 - x_1^2 + 4x_2 - x_2^2 \geq 4,$$
$$g_2(x) \equiv 4x_1 - x_2^2 \geq 0.$$

4. Consider the following two objective functions:
$$\text{Minimize} \quad x_1^4 + x_2^4,$$
$$\text{Maximize} \quad \exp(x_1^2) + \exp(x_2^2).$$

   If both objectives are to be optimized, does the resulting problem give rise to multiple Pareto-optimal solutions? Give reasons.

5. For the following three-objective optimization problem, identify the ideal and nadir objective vectors:

$$\text{Minimize} \quad f_1(x) = (1 + g(x_3))\cos(\theta_1\pi/2)\cos(\theta_2\pi/2),$$
$$\text{Minimize} \quad f_2(x) = (1 + g(x_3))\cos(\theta_1\pi/2)\sin(\theta_2\pi/2),$$
$$\text{Minimize} \quad f_3(x) = (1 + g(x_3))\sin(\theta_1\pi/2),$$
$$\text{where} \quad \theta_i = \frac{\pi}{4(1+g(x_3))}(1 + 2g(x_3)x_i), \quad \text{for } i = 1,2,$$
$$g(x_3) = (x_3 - 0.5)^2,$$
$$0 \le x_i \le 1, \quad \text{for } i = 1,2,3.$$

Is the nadir objective vector a feasible solution?

6. Determine if the first solution dominates the second solution?

(a) (min, min): $\mathbf{f}^{(1)} = (1.2, 3.5)^\mathsf{T}$, $\mathbf{f}^{(2)} = (1.5, 3.0)^\mathsf{T}$.
(b) (min, max, min): $\mathbf{f}^{(1)} = (10.5, 1.5, -10.0)^\mathsf{T}$, $\mathbf{f}^{(2)} = (5.0, 0.5, -12)^\mathsf{T}$.
(c) (max. max, min): $\mathbf{f}^{(1)} = (5,5,3)^\mathsf{T}$, $\mathbf{f}^{(2)} = (2,5,4)^\mathsf{T}$.

7. Consider the following three objective functions:

$$\text{Minimize} \quad f_1(x) = x_1^2 + 2\sin\pi x_2,$$
$$\text{Maximize} \quad f_2(x) = x_1 x_2 - 10,$$
$$\text{Minimize} \quad f_3(x) = 5x_1^3 - x_1 x_2^2.$$

and the following points:

(i) $(1,0)^\mathsf{T}$,   (ii) $(0,1)^\mathsf{T}$,   (iii) $(1,1)^\mathsf{T}$,   (iv) $(2,1)^\mathsf{T}$.

Which solutions are non-dominated solutions?

8. If *strict* domination is used, which solutions are non-dominated solutions in the previous problem?

9. For the following problem

$$\text{Minimize} \quad f_1(x) = x_1,$$
$$\text{Minimize} \quad f_2(x) = x_2,$$
$$\text{subject to} \quad g(x) \equiv (x_1 - 2)^2 + (x_2 - 2)^2 \le 4,$$

(a) What proportion of the feasible region is dominated by the solution $(1,1)^\mathsf{T}$?
(b) What proportion of the feasible region dominates the solution $(1,1)^\mathsf{T}$?
(c) If the solution $(1,1)^\mathsf{T}$ is present in a set of non-dominated solutions, what is the minimum proportion of the undiscovered Pareto-optimal region?

10. For the following problem

$$\text{Minimize} \quad f_1(x) = x_1,$$
$$\text{Minimize} \quad f_2(x) = x_2,$$
$$\text{subject to} \quad g_1(x) \equiv x_1^2 + x_2^2 \le 1,$$
$$g_2(x) \equiv (x_1 - 1)^2 + (x_2 - 1)^2 \le 1,$$

identify the Pareto-optimal set when

(a) both f_1 and f_2 are minimized,
(b) f_1 is minimized and f_2 is maximized,
(c) f_1 is maximized and f_2 is minimized,
(d) both f_1 and f_2 are maximized.

11. Use the continuously updated method to identify the non-dominated set from the following points (all objectives are minimized):

Soln. Id.	f_1	f_2	f_3
1	2	3	1
2	5	1	10
3	3	4	10
4	2	2	2
5	3	3	2
6	4	4	5

12. Use the $O(MN^2)$ sorting procedure to sort the above points according to increasing levels of non-domination.

13. By using the Fritz-John necessary condition, identify the candidate Pareto-optimal solutions of the following problems:

(a) Minimize $f_1(x_1, x_2) = x_1^2 + x_2^2,$
     Minimize $f_2(x_1, x_2) = x_2^2 - 4x_1.$
(b) Minimize $f_1(x_1, x_2) = x_2 \sin x_1,$
     Minimize $f_2(x_1, x_2) = x_1 + x_2.$
(c) Minimize $f_1(x_1, x_2) = x_1^4 - 4x_1 x_2,$
     Minimize $f_2(x_1, x_2) = x_1 + 2x_2^2.$

3

Classical Methods

In this chapter, we will describe a few commonly used classical methods for handling multi-objective optimization problems. We refer to these methods as *classical* methods, mainly to distinguish them from evolutionary methods, which we will discuss in the remaining chapters of this book. Classical multi-objective optimization methods have been around for at least the past four decades. During this period, many algorithms have also been suggested. Many researchers have attempted to classify algorithms according to various considerations. Cohon (1985) classified them into the following two types:

- Generating methods;
- Preference-based methods.

In the generating methods, a few non-dominated solutions are generated for the decision-maker, who then chooses one solution from the obtained non-dominated solutions. No a priori knowledge of relative importance of each objective is used. On the other hand, in the preference-based methods, some known preference for each objective is used in the optimization process. Hwang and Masud (1979) and later Miettinen (1999) fine-tuned the above classification and suggested the following four classes:

- No-preference methods;
- Posteriori methods;
- A priori methods;
- Interactive methods.

The no-preference methods do not assume any information about the importance of objectives, but a heuristic is used to find a single optimal solution. It is important to note that although no preference information is used, these methods do not make any attempt to find multiple Pareto-optimal solutions. Posteriori methods use preference information of each objective and iteratively generate a set of Pareto-optimal solutions. Although this method or the Cohon's generating methods are similar in principle to the ideal approach proposed in Chapter 1, the classical method of generating Pareto-optimal solutions require some knowledge on algorithmic parameters which will ensure finding a Pareto-optimal solution. On the other hand, a priori methods use more information about the preferences of objectives and usually find one preferred Pareto-optimal solution. Interactive methods use the preference information progressively during the optimization process.

Although there exist other classifications, the above classification based on the extent of preference information is an appropriate one. We will outline here a number of classical methods in the order of increasing use of preference information.

3.1   Weighted Sum Method

The weighted sum method, as the name suggests, scalarizes a set of objectives into a single objective by pre-multiplying each objective with a user-supplied weight. This method is the simplest approach and is probably the most widely used classical approach. Faced with multiple objectives, this method is the most convenient one that comes to mind. For example, if we are faced with the two objectives of minimizing the cost of a product and minimizing the amount of wasted material in the process of fabricating the product, one naturally thinks of minimizing a weighted sum of these two objectives.

Although the idea is simple, it introduces a not-so-simple question. What values of the weights must one use? Of course, there is no unique answer to this question. The answer depends on the importance of each objective in the context of the problem and a scaling factor, which we will address a little later. The weight of an objective is usually chosen in proportion to the objective's relative importance in the problem. For example, in the above-mentioned two-objective minimization problem, the cost of the product may be more important than the amount of wasted material. Thus, the user can set a higher weight for the material cost than for the amount of wasted material. Although there exist ways to quantify the weights from this qualitative information (Parmee et al., 2000), the weighted sum approach requires a precise value of the weight for each objective.

However, setting up an appropriate weight vector also depends on the scaling of each objective function. It is likely that different objectives take different orders of magnitude. In the above example again, the cost of the product may vary between 100 to 1000 dollars, whereas the amount of wasted material may vary between 0.01 and 0.1 m^3. When such objectives are weighted to form a composite objective function, it would be better to scale them appropriately so that each has more or less the same order of magnitude. For example, we may multiply the product cost by $1(10^{-3})$ and the amount of wasted material by $1(10^1)$ to make them equally important. This process is called *normalization* of objectives.

After the objectives are normalized, a composite objective function $F(x)$ can be formed by summing the weighted normalized objectives and the MOOP given in equation (2.1) is then converted to a single-objective optimization problem as follows:

$$\left.\begin{array}{lll} \text{Minimize} & F(\mathbf{x}) = \sum_{m=1}^{M} w_m f_m(\mathbf{x}), & \\ \text{subject to} & g_j(\mathbf{x}) \geq 0, & j = 1, 2, \ldots, J; \\ & h_k(\mathbf{x}) = 0, & k = 1, 2, \ldots, K; \\ & x_i^{(L)} \leq x_i \leq x_i^{(U)}, & i = 1, 2, \ldots, n. \end{array}\right\} \qquad (3.1)$$

Here, w_m ($\in [0, 1]$) is the weight of the m-th objective function. Since the minimum of the above problem does not change if all weights are multiplied by a constant, it is the usual practice to choose weights such that their sum is one, or $\sum_{m=1}^{M} w_m = 1$.

Mathematically oriented readers may find a number of interesting theorems regarding the relationship between the optimal solution of the above problem to the true Pareto-optimal solutions in classical texts (Chankong and Haimes, 1983; Ehrgott, 2000; Miettinen, 1999). For the above problem, two interesting theorems are reproduced below from Miettinen (1999).

Theorem 3.1.1. *The solution to the problem represented by equation (3.1) is Pareto-optimal if the weight is positive for all objectives.*

The proof of this theorem is achieved by first assuming that the optimal solution obtained with all positive weights is not a Pareto-optimal solution and then showing a contradiction of this assumption. This theorem is true for any MOOP. However, it conveys only a part of the story. With a positive weight vector, the obtained optimum solution to the problem in equation (3.1) is a Pareto-optimal solution. However, it does not imply that any Pareto-optimal solution can be obtained by using a positive weight vector. The next theorem confirms this for convex problems.

Theorem 3.1.2. *If x* is a Pareto-optimal solution of a convex multi-objective optimization problem, then there exists a non-zero positive weight vector* **w** *such that* x* *is a solution to the problem given by equation (3.1).*

For the proof, refer to Miettinen (1999). This theorem suggests that for a convex MOOP, any Pareto-optimal solution can be found by using the weighted sum method.

Let us now illustrate how the weighted sum approach can find Pareto-optimal solutions of the original problem. For simplicity, we consider the two-objective problem shown in Figure 19. The feasible objective space and the corresponding Pareto-optimal solution set are shown. With two objectives, there are two weights w_1 and w_2, but only one is independent. Knowing any one, the other can be calculated by simple subtraction.

Knowing the weights, we can also calculate the composite function F. Its contour surfaces can then be visualized in the objective space, as shown by lines 'a', 'b', 'c', and 'd' in Figure 19. Since F is a linear combination of both objectives f_1 and f_2, we would expect a straight line as the contour line of F on the objective space. This is because any solution on the contour line will have the same F value. If considered carefully, this contour line is not an arbitrary line; its slope is related to the choice of the weight vector. In fact, for two objectives, its slope is $-w_1/w_2$. The location of the line depends on the value of F on any point on the line. The effect of lowering the contour line from 'a' to 'b' is in effect jumping from solutions of higher F values to a lower one.

Since the problem represented by equation (3.1) requires minimization of F, the task is to find the contour line with the minimum F value. This happens with the contour line which is tangential to the search space and also lies in the bottom-left corner

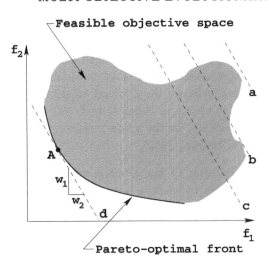

Figure 19 Illustration of the weighted sum approach on a convex Pareto-optimal front.

of this space. In the figure, this line is marked as 'd'. The tangent point 'A' is the minimum solution of F, and is consequently the Pareto-optimal solution corresponding to the weight vector.

If a different weight vector is used, the slope of the contour line would be different and the above procedure, in general, would result in a different optimal solution. Using Theorem 3.1.2, we can argue that for a convex MOOP, multiple Pareto-optimal solutions can be found by solving the corresponding problem (equation (3.1)) with multiple positive weight vectors, one at a time, each time finding a Pareto-optimal solution. The above theorem can be extended to nonconvex MOOPs, particularly when the Pareto-optimal front is convex.

3.1.1   Hand Calculations

To illustrate, we choose the following two-objective optimization problem:

$$\begin{aligned} \text{Minimize} \quad & f_1(\mathbf{x}) = x_1, \\ \text{Minimize} \quad & f_2(\mathbf{x}) = 1 + x_2^2 - x_1 - a\sin(b\pi x_1), \\ \text{subject to} \quad & 0 \le x_1 \le 1, \quad -2 \le x_2 \le 2. \end{aligned} \right\} \tag{3.2}$$

The problem has two parameters a and b, which control the convexity of the search space. Figure 20 shows the objective space for $a = 0.2$ and $b = 1$. For illustration, the complete objective space is not shown; instead, the complete Pareto-optimal region and its neighboring space are presented.

We form the composite function F with a weight vector $(w_1, w_2)^\mathsf{T}$:

$$F(\mathbf{x}) = w_1 x_1 + w_2 \left[1 + x_2^2 - x_1 - a\sin(b\pi x_1) \right]. \tag{3.3}$$

We shall substitute the values of a and b later. Using the first-order optimality

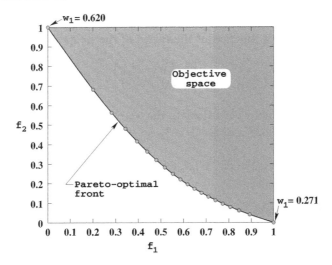

Figure 20   Illustration of the objective space and the Pareto-optimal set. Optimal solutions of the composite function F for equi-spaced values of $w_1 \in [0.271, 0.620]$ are shown by using circles.

condition (Deb, 1995), we obtain:

$$\frac{\partial F}{\partial x_1} = w_1 + w_2\left[-1 - ab\pi \cos(b\pi x_1)\right], \tag{3.4}$$

$$\frac{\partial F}{\partial x_2} = 2w_2 x_2. \tag{3.5}$$

Equating the right sides of the above two equations to zero, we obtain the stationary solutions as follows:

$$x_1^* = \frac{1}{b\pi} \cos^{-1}\left[\frac{1}{ab\pi}\left(\frac{w_1}{w_2} - 1\right)\right], \tag{3.6}$$

$$x_2^* = 0. \tag{3.7}$$

The next task is to satisfy the second-order optimality conditions. Calculating the Hessian matrix and then finding the conditions for positive semi-definiteness of the matrix, we obtain the second-order condition for the minimum of F as follows:

$$\sin(b\pi x_1^*) \geq 0. \tag{3.8}$$

This condition suggests that x_1^* must lie in discrete regions of constant width $1/b$, or $2i/b \leq x_1^* \leq (2i+1)/b$, for all $i = 0, 1, 2, \ldots$, while satisfying the condition that the upper bound of x_1^* is one (in order to satisfy the variable upper bound). Thus, the implications of all of the above calculations is that for the chosen values of w_1 and w_2, the optimal solution is given by equations (3.6) and (3.7), provided that equation (3.8) is satisfied.

The above calculations give us the exact minimum of the weighted function F for any a and b. Now, we set $a = 0.2$ and $b = 1$ to find the optimal solutions of F. Since,

$b = 1$, the optimal solution is valid for $i = 0$ only. By using equation (3.6), we obtain the optimal solution as a function of the ration of the weights as follows:

$$x_1^* = \frac{1}{\pi} \cos^{-1} \left[\frac{5}{\pi} \left(\frac{w_1}{w_2} - 1 \right) \right], \quad x_2^* = 0. \qquad (3.9)$$

The above equation allows us to calculate the weight vector at two extreme solutions. At $x_1^* = 0$, $w_1/w_2 = 1.628$ and at $x_1^* = 1$, $w_1/w_2 = 0.372$. Using $w_1 + w_2 = 1$, we obtain

$$x_1^* = 0: \quad w_1 = 0.620, \quad w_2 = 0.380.$$
$$x_1^* = 1: \quad w_1 = 0.271, \quad w_2 = 0.729.$$

Thus, choosing any weight vector within the above bounds ($0.271 \le w_1 \le 0.620$ and $w_2 = 1 - w_1$), we can find the corresponding optimal solution using equation (3.9). According to Theorem 3.1.2, these solutions are Pareto-optimal solutions to the original problem.

It is interesting to note that if a single-objective optimization algorithm is applied a number of times with a uniformly distributed set of weight vectors, the resulting Pareto-optimal solutions may not be uniformly spaced either in the decision variable space or in the objective space. This can be understood by noting the nonlinear relationship of x_1 and w_1/w_2 in equation (3.9). Figure 20 above shows the Pareto-optimal solutions corresponding to 20 uniformly distributed slopes, with w_1 values in the range $[0.271, 0.620]$. Other values of w_1 (that is, $w_1 \in [0, 0.271]$ and $[0.620, 1]$) will result in one of the two extreme solutions: $x_1^* = 0$ or $x_1^* = 1$.

3.1.2   Advantages

This is probably the simplest way to solve an MOOP. The concept is intuitive and easy to use. For problems having a convex Pareto-optimal front, this method guarantees finding solutions on the entire Pareto-optimal set.

3.1.3   Disadvantages

However, there are a number of difficulties with this approach. In handling mixed optimization problems, such as those with some objectives of the maximization type and some of the minimization type, all objectives have to be converted into one type. Although different conversion procedures can be adopted, the duality principle is convenient and does not introduce any additional complexity.

In most nonlinear MOOPs, a uniformly distributed set of weight vectors need not find a uniformly distributed set of Pareto-optimal solutions. Since this mapping is not usually known, it becomes difficult to set the weight vectors to obtain a Pareto-optimal solution in a desired region in the objective space.

Moreover, different weight vectors need not necessarily lead to different Pareto-optimal solutions. In the above example problem, many weight vectors (with $w_1 \in [0, 0.271]$) lead to the same solution ($x_1^* = 1$, $x_2^* = 0$). This information about which

be obtained as optimal solutions for these weight values. Thus, the region BC (with $f_1^* \in (1/6, 5/6)$) will be undiscovered by using the weighted sum approach.

3.2 ϵ-Constraint Method

In order to alleviate the difficulties faced by the weighted sum approach in solving problems having nonconvex objective spaces, the ϵ-constraint method is used. Haimes et al. (1971) suggested reformulating the MOOP by just keeping one of the objectives and restricting the rest of the objectives within user-specified values. The modified problem is as follows:

$$\left.\begin{array}{ll} \text{Minimize} & f_\mu(\mathbf{x}), \\ \text{subject to} & f_m(\mathbf{x}) \leq \epsilon_m, \qquad m = 1, 2, \ldots, M \text{ and } m \neq \mu; \\ & g_j(\mathbf{x}) \geq 0, \qquad j = 1, 2, \ldots, ; \\ & h_k(\mathbf{x}) = 0, \qquad k = 1, 2, \ldots, K; \\ & x_i^{(L)} \leq x_i \leq x_i^{(U)}, \quad i = 1, 2, \ldots, n. \end{array}\right\} \qquad (3.10)$$

In the above formulation, the parameter ϵ_m represents an upper bound of the value of f_m and need not necessarily mean a small value close to zero. Let us first illustrate the working of the ϵ-constraint method. Let us say that we retain f_2 as an objective and treat f_1 as a constraint: $f_1(\mathbf{x}) \leq \epsilon_1$. Figure 23 shows four scenarios with different ϵ_1 values.

Let us consider the third scenario with $\epsilon_1 = \epsilon_1^c$ first. The resulting problem with this constraint divides the original feasible objective space into two portions, $f_1 \leq \epsilon_1^c$ and $f_1 > \epsilon_1^c$. The left portion becomes the feasible solution of the resulting problem stated in equation (3.10). Now, the task of the resulting problem is to find the solution which minimizes this feasible region. From Figure 23, it is clear that the minimum

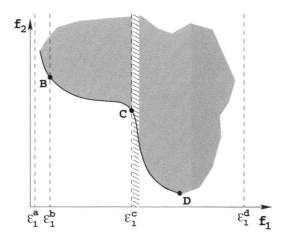

Figure 23   The ϵ-constraint method.

solution is 'C'. In this way, intermediate Pareto-optimal solutions can be obtained in the case of nonconvex objective space problems by using the ϵ-constraint method.

For convex or nonconvex objective spaces, the following proof says all about the utility of the ϵ-constraint method.

Theorem 3.2.1. *The unique solution of the ϵ-constraint problem stated in equation (3.10) is Pareto-optimal for any given upper bound vector* $\epsilon = (\epsilon_1, \ldots, \epsilon_{\mu-1}, \epsilon_{\mu+1}, \ldots, \epsilon_M)^T.$

The proof comes from the assumption that the unique solution to the ϵ-constraint problem is not Pareto-optimal and then showing that this assumption violates the definition of Pareto-optimality (Miettinen, 1999).

3.2.1 Hand Calculations

Let us now illustrate the working of the ϵ-constraint method on the problem stated in equation (3.2) with $a = 0.1$ and $b = 3$ (Figure 22). We have discussed earlier that this problem has a nonconvex objective space. Keeping f_2 as the only objective and limiting f_1 to ϵ_1, we have the following single-objective constrained optimization problem:

$$\left.\begin{array}{rl} \text{Minimize} & f_2(x_1, x_2) = 1 + x_2^2 - x_1 - 0.1\sin(3\pi x_1), \\ \text{subject to} & x_1 \leq \epsilon_1, \\ & 0 \leq x_1 \leq 1, \quad -2 \leq x_2 \leq 2. \end{array}\right\} \qquad (3.11)$$

Allowing the variable bounds and considering only one constraint, $g_1 \equiv \epsilon_1 - x_1 \geq 0$, we write the Kuhn–Tucker optimality conditions (Deb, 1995) by using a Lagrange multiplier u_1 for the constraint g_1:

$$\begin{array}{rcl} \nabla f_2 - u_1 \nabla g_1 & = & 0, \\ g_1 & \geq & 0, \\ u_1 g_1 & = & 0, \\ u_1 & \geq & 0. \end{array}$$

We have the following equations:

$$\begin{array}{rcl} -1 - 0.3\pi\cos(3\pi x_1) + u_1 & = & 0, \\ 2x_2 & = & 0, \\ \epsilon_1 - x_1 & \geq & 0, \\ u_1(\epsilon_1 - x_1) & = & 0, \\ u_1 & \geq & 0. \end{array}$$

The first equation suggests that $u_1 > 0$. Thus, the fourth equation suggests that $x_1 = \epsilon_1$. The second equation suggests $x_2 = 0$. Thus, the optimum solution is $x_1^* = \epsilon_1$ and $x_2^* = 0$, which lies on the Pareto-optimal front. This clearly shows that by changing ϵ_1 values, different Pareto-optimal solutions can be found.

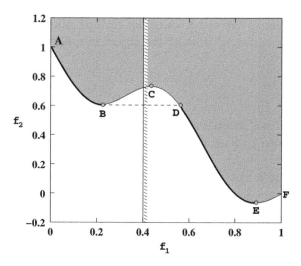

Figure 24 A two-objective problem with disjointed Pareto-optimal sets.

Let us change the problem by using $a = 0.2$ and $b = 3$ (Figure 24), before leaving this subsection. Once again, we treat the first objective as a constraint by limiting its value within $[0, \epsilon_1]$. The Kuhn–Tucker conditions are as follows:

$$-1 - 0.6\pi \cos(3\pi x_1) + u_1 = 0,$$
$$2x_2 = 0,$$
$$\epsilon_1 - x_1 \geq 0,$$
$$u_1(\epsilon_1 - x_1) = 0,$$
$$u_1 \geq 0.$$

Here too, $x_2^* = 0$. However, u_1 need not be greater than zero. When $u_1 = 0$, the optimal value of x_1 satisfies $\cos(3\pi x_1^*) = -1/0.6\pi$, or $x_1^* = 0.226$ and 0.893 in the specified range of x_1. In the complete range of $\epsilon_1 \in [0, 1]$, we find the following optimum values of x_1^* (with $x_2^* = 0$):

$$x_1^* = \begin{cases} \epsilon_1, & \text{if } 0 \leq \epsilon_1 \leq 0.226 \text{ (AB);} \\ 0.226 & \text{if } 0.226 < \epsilon_1 < 0.441 \text{ (BC);} \\ 0.226 & \text{if } f_2(\epsilon_1, 0) > f_2(0.226, 0) \text{ and } 0.441 \leq \epsilon_1 \leq 0.893 \text{ (CD);} \\ \epsilon_1 & \text{if } f_2(\epsilon_1, 0) \leq f_2(0.226, 0) \text{ and } 0.441 \leq \epsilon_1 \leq 0.893 \text{ (DE);} \\ 0.893, & \text{if } 0.893 < \epsilon_1 < 1 \text{ (EF).} \end{cases}$$

When ϵ_1 is chosen in the range CE, there are two cases: $u_1 = 0$ or $u_1 > 0$. The minimum of the two cases is to be accepted. The switch-over takes place at $x_1 = 0.562$. Figure 24 shows that if $\epsilon_1 = 0.4$ (or any value in the range BCD, or $\epsilon_1 \in [0.226, 0.562]$) is used, the optimal solution is solution B: $x_1^* = 0.226$ and $x_2^* = 0$. However, when ϵ_1 is chosen at any point on the true Pareto-optimal region (AB or DE), that point is found as the optimum. Similarly, when ϵ_1 is chosen in the range EF, the solution E is the optimal solution. This is how the ϵ-constraint method can identify the true Pareto-optimal region independent of whether the objective space is convex, nonconvex, or discrete.

3.2.2   Advantages

Different Pareto-optimal solutions can be found by using different ϵ_m values. The same method can also be used for problems having convex or nonconvex objective spaces alike.

   In terms of the information needed from the user, this algorithm is similar to the weighted sum approach. In the latter approach, a weight vector representing the relative importance of each objective is needed. In this approach, a vector of ϵ values representing, in some sense, the location of the Pareto-optimal solution is needed. However, the advantage of this method is that it can be used for any arbitrary problem with either convex or nonconvex objective space.

3.2.3   Disadvantages

The solution to the problem stated in equation (3.10) largely depends on the chosen ϵ vector. It must be chosen so that it lies within the minimum or maximum values of the individual objective function. Let us refer to Figure 23 again. Instead of choosing ϵ_1^c, if ϵ_1^a is chosen, there exists no feasible solution to the stated problem. Thus, no solution would be found. On the other hand, if ϵ_1^d is used, the entire search space is feasible. The resulting problem has the minimum at 'D'. Moreover, as the number of objectives increases, there exist more elements in the ϵ vector, thereby requiring more information from the user.

3.3   Weighted Metric Methods

Instead of using a weighted sum of the objectives, other means of combining multiple objectives into a single objective can also be used. For this purpose, weighted metrics such as l_p and l_∞ distance metrics are often used. For non-negative weights, the weighted l_p distance measure of any solution x from the ideal solution z^* can be minimized as follows:

$$
\left.
\begin{array}{lll}
\text{Minimize} & l_p(\mathbf{x}) = \left(\sum_{m=1}^{M} w_m |f_m(\mathbf{x}) - z_m^*|^p\right)^{1/p}, & \\
\text{subject to} & g_j(\mathbf{x}) \geq 0, & j = 1, 2, \ldots, J; \\
& h_k(\mathbf{x}) = 0, & k = 1, 2, \ldots, K; \\
& x_i^{(L)} \leq x_i \leq x_i^{(U)}, & i = 1, 2, \ldots, n.
\end{array}
\right\} \quad (3.12)
$$

The parameter p can take any value between 1 and ∞. When $p = 1$ is used, the resulting problem is equivalent to the weighted sum approach. When $p = 2$ is used, a weighted Euclidean distance of any point in the objective space from the ideal point is minimized.

When a large p is used, the above problem reduces to a problem of minimizing the largest deviation $|f_m(\mathbf{x}) - z_m^*|$. This problem has a special name – the *weighted Tchebycheff* problem:

$$
\begin{array}{lll}
\text{Minimize} & l_\infty(\mathbf{x}) = \max_{m=1}^M w_m |f_m(\mathbf{x}) - z_m^*|, & \\
\text{subject to} & g_j(\mathbf{x}) \geq 0, & j = 1, 2, \ldots, J; \\
& h_k(\mathbf{x}) = 0, & k = 1, 2, \ldots, K; \\
& x_i^{(L)} \leq x_i \leq x_i^{(U)}, & i = 1, 2, \ldots, n.
\end{array} \right\} \quad (3.13)
$$

However, the resulting optimal solution obtained by the chosen l_p depends on the parameter p. We illustrate the working principle of this method in Figures 25, 26, and 27 for p = 1, 2 and ∞, respectively. In all of these figures, optimum solutions for two different weight vectors are shown. It is clear that with p = 1 or 2, not all Pareto-optimal solutions can be obtained. In these cases, the figures show that no solution in the region BC can be found by using p = 1 or 2. However, when the weighted Tchebycheff metric is used (Figure 27), any Pareto-optimal solution can be found. There exists a special theorem for the weighted Tchebycheff metric formulation (Miettinen, 1999):

Theorem 3.3.1. *Let* $\mathbf{x}^*$ *be a Pareto-optimal solution. There then exists a positive weighting vector such that* $\mathbf{x}^*$ *is a solution of the weighted Tchebycheff problem shown in equation (3.13), where the reference point is the utopian objective vector* $\mathbf{z}^{**}$.

However, it is also important to note that as p increases the problem becomes non-differentiable and many gradient-based methods cannot be used to find the minimum solution of the resulting single-objective optimization problem.

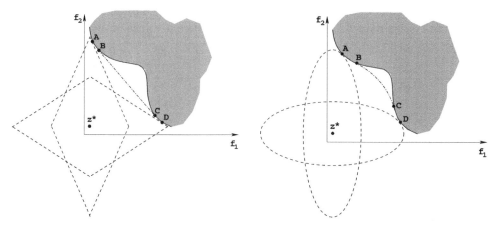

Figure 25 The weighted metric method with p = 1.

Figure 26 The weighted metric method with p = 2.

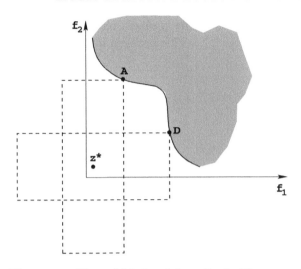

Figure 27 The weighted metric method with $p = \infty$.

3.3.1 Hand Calculations

Let us consider the problem shown in equation (3.2) with $a = 0.1$ and $b = 3$ (Figure 22). We have seen earlier that this problem has a nonconvex objective space. We use the weighted metric method with $p = 2$ and assume $z_1^* = z_2^* = 0$. Thus, the resulting l_2^2 is as follows:[1]

$$l_2^2 = w_1 x_1^2 + w_2 \left[1 + x_2^2 - x_1 - 0.1\sin(3\pi x_1)\right]^2.$$

The minimum of this function corresponds to $x_2^* = 0$ and x_1^*, satisfying the following equation:

$$w_1 x_1^* - w_2 \left[1 + x_2^{*2} - x_1^* - 0.1\sin(3\pi x_1^*)\right]\left[1 + 0.3\pi\cos(3\pi x_1^*)\right] = 0.$$

There exists a number of roots of this equation. However, considering the one which minimizes l_2 yields the following value of x_1^* for different values of w_1:

w_1	1.00	0.95	0.90	0.85	0.80	0.75	0.70
x_1^*	0.000	0.077	0.122	0.151	0.172	0.190	0.204
w_1	0.65	0.60	0.55	0.50	0.45	0.40	0.35
x_1^*	0.216	0.227	0.237	0.246	0.688	0.709	0.727
w_1	0.30	0.25	0.20	0.15	0.10	0.05	0.00
x_1^*	0.745	0.762	0.779	0.798	0.819	0.849	1.000

Figure 28 plots these optimal solutions. It is clear that this method with $p = 2$ cannot identify certain regions (BC, marked on the figure) in the Pareto-optimal front. It is interesting to note that this algorithm can find more diverse solutions when compared to the weighted sum approach, or the l_1-metric method (Figure 22).

---

[1] Minimization of l_2 and l_2^2 will produce the same optimum.

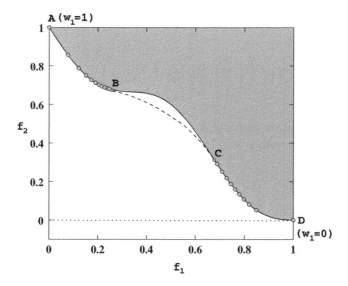

Figure 28 Optimal solutions corresponding to 21 equi-spaced w_1 values between zero and one.

3.3.2 Advantages

The weighted Tchebycheff metric guarantees finding each and every Pareto-optimal solution when z^* is a utopian objective vector (Miettinen, 1999). Although in the above discussions only l_p metrics are suggested, other distance metrics are also used. Shortly, we will suggest two improvements to this approach, which may be useful in solving problems having a nonconvex objective space.

3.3.3 Disadvantages

Since different objectives may take values of different orders of magnitude, it is advisable to normalize the objective functions. This requires a knowledge of the minimum and maximum function values of each objective.

Moreover, this method also requires the ideal solution z^*. Therefore, all M objectives need to be independently optimized before optimizing the l_p metric.

3.3.4 Rotated Weighted Metric Method

Instead of directly using the l_p metric as it is stated in equation (3.12), the l_p metric can be applied with an arbitrary rotation from the ideal point. Let us say that the rotated objective reference axes $\tilde{\mathbf{f}}$ is related to the original objective axes $\mathbf{f}$ by the following relationship:

$$\tilde{\mathbf{f}} = R\mathbf{f}, \tag{3.14}$$

where R is the rotation matrix of size $M \times M$. The modified l_p metric then becomes:

$$\tilde{l}_p = \left(\sum_{m=1}^{M} w_m |\tilde{f}_m(\mathbf{x}) - z_m^*|^p \right)^{1/p}. \tag{3.15}$$

By using different rotation matrices, the above function can be minimized. Figure 29 illustrates the scenario with $p = 2$. In this case, the metric is equivalent to:

$$\left[(f(\mathbf{x}) - \mathbf{z}^*)^\top C (f(\mathbf{x}) - \mathbf{z}^*) \right]^{1/2},$$

where $C = R^\top \mathrm{Diag}(w_1, \cdots, w_M) R$. The rotation matrix R will transform the objective axes into another set of axes dictated by the rotation matrix. In this way, iso-l_2 solutions become aligned in a rotated ellipsoid, as shown in Figure 29. With this strategy and by changing each member of the C matrix, any Pareto-optimal solution can be obtained.

Let us now apply this method to the problem used earlier in the hand calculation. We introduce here the parameter α, i.e. the angle of rotation of the f_1 axis. With respect to this rotation angle, we have the following rotation matrix:

$$C = \begin{bmatrix} \cos \alpha & \sin \alpha \\ -\sin \alpha & \cos \alpha \end{bmatrix}^\top \begin{bmatrix} w_1 & 0 \\ 0 & w_2 \end{bmatrix} \begin{bmatrix} \cos \alpha & \sin \alpha \\ -\sin \alpha & \cos \alpha \end{bmatrix}.$$

Using $w_1 = 0.01$ and $w_2 = 0.99$, and varying α between zero and 90 degrees, we obtain the following minimum value of l_2 (with $x_2^* = 0$):

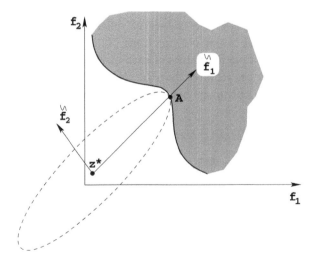

Figure 29 The proposed correlated weighted metric method with $p = 2$.

α	0	5	10	15	20	25	30
x_1^*	0.898	0.822	0.775	0.738	0.706	0.676	0.646
α	35	40	45	50	55	60	65
x_1^*	0.616	0.583	0.546	0.502	0.447	0.379	0.308
α	70	75	80	85	90		
x_1^*	0.247	0.193	0.141	0.085	0.018		

Figure 30 shows these solutions. Any solution in the Pareto-optimal front can be obtained by changing the angle and the weight vector. For more than two objectives, there are $\binom{M}{2}$ angles which constitute the rotation matrix. These can be varied to find different Pareto-optimal solutions.

However, one difficulty with this approach is that there are many parameters, which must be fixed. Moreover, in certain problems, the minimum solution corresponding to the modified l_2 metric may result in a dominated solution, thereby not guaranteeing that we will always find a Pareto-optimal solution.

3.3.5 Dynamically Changing the Ideal Solution

Another remedy to the problem of not being able to find some Pareto-optimal solutions with the original l_p metric (for small p values) would be to update the point z^*, for every time that a Pareto-optimal solution is found. In this way, the l_p distance of the ideal solution comes closer to the Pareto-optimal front and previously undiscovered Pareto-optimal solutions can now be found. With each Pareto-optimal solution found so far, all combinations of objective values may be formed to construct new candidate ideal solutions. Thereafter, a candidate solution which does not dominate any of the other solutions (or, in principle, the candidate solution which is closest to the Pareto-

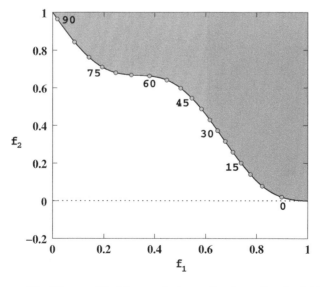

Figure 30 The modified l_2 can find any Pareto-optimal solution.

optimal front) may be chosen as the new z^*.

Figure 31 illustrates this method. After the Pareto-optimal solutions A and B are found, the ideal point may be updated to the new point O. It is then easier to find some Pareto-optimal solutions (such as solution C) which were impossible to find by using the original z^*.

3.4   Benson's Method

This procedure is similar to the weighted metric approach, except that the reference solution is taken as a feasible non-Pareto-optimal solution. A solution z^0 is randomly chosen from the feasible region. Thereafter, the non-negative difference $(z_m^0 - f_m(x))$ of each objective is calculated and their sum is maximized:

$$
\left.
\begin{aligned}
\text{Maximize} \quad & \textstyle\sum_{m=1}^{M} \max\left(0, (z_m^0 - f_m(\mathbf{x}))\right), \\
\text{subject to} \quad & f_m(\mathbf{x}) \le z_m^0, & m = 1, 2 \ldots, M; \\
& g_j(\mathbf{x}) \ge 0, & j = 1, 2, \ldots, J; \\
& h_k(\mathbf{x}) \ge 0, & k = 1, 2, \ldots, K; \\
& x_i^{(L)} \le x_i \le x_i^{(U)}, & i = 1, 2, \ldots, n.
\end{aligned}
\right\}
\tag{3.16}
$$

Figure 32 shows the chosen solution z^0 in the feasible region. For any solution x, the above objective function has a value equal to half of the perimeter of a hypercube having z^0 and $f(x)$ as the diagonal points. The only requirement is that the solution x must weakly dominate the solution at z^0. The maximization of the above objective is similar to finding a hypercube with the maximum perimeter. Since the Pareto-optimal region lies at the extreme of a feasible search space, the optimum solution of the above optimization problem is a member of the Pareto-optimal front. For a number of properties of this approach, refer to the original study (Benson, 1978) or to Ehrgott (2000). Since the algorithm is straightforward, we do not perform any hand

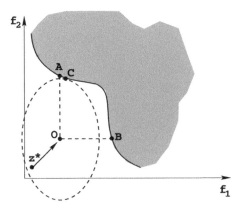

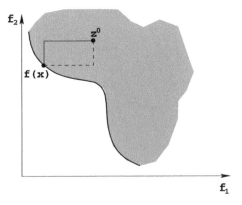

Figure 31   Moving z^* closer to the Pareto-optimal front may allow more Pareto-optimal solutions to be found.

Figure   32   Illustration   of   Benson's method. Half of the perimeter of the box is the objective function.

calculation. Interested readers may take this as an exercise for using the test problem of the previous section and to complete an iteration of the above method for a proper understanding.

3.4.1 Advantages

To avoid scaling problems, individual differences can be normalized before the summation. To obtain different Pareto-optimal solutions, the differences can be weighted before summation. Thereafter, by changing the weight vector, different Pareto-optimal solutions can be obtained. In such a scenario, the use of the nadir point, z^{nad}, as the chosen point may be found suitable. If z^0 is chosen appropriately, this method can be used to find solutions in the nonconvex Pareto-optimal region.

3.4.2 Disadvantages

The optimization problem formulated above has an additional number of constraints needed to restrict the search in the region dominating the chosen solution z^0. Moreover, the objective function is non-differentiable, thereby causing difficulties for gradient-based methods to solve the above problem. Although a modified formulation is suggested in Ehrgott (2000) for differentiable objective functions, the resulting optimization problem has equality constraints which are usually difficult to handle.

3.5 Value Function Method

In the value function (or utility function) method, the user provides a mathematical value function $U : \mathbb{R}^M \rightarrow \mathbb{R}$ relating all M objectives. The value function must be valid over the entire feasible search space. The task is then to maximize the value function as follows:

$$\left.\begin{array}{lll} \text{Maximize} & U(\mathbf{f}(\mathbf{x})) & \\ \text{subject to} & g_j(\mathbf{x}) \geq 0, & j = 1, 2, \ldots, J; \\ & h_k(\mathbf{x}) = 0, & k = 1, 2, \ldots, K; \\ & x_i^{(L)} \leq x_i \leq x_i^{(U)}, & i = 1, 2, \ldots, n. \end{array}\right\} \quad (3.17)$$

Here, $\mathbf{f}(\mathbf{x}) = (f_1(\mathbf{x}), f_2(\mathbf{x}), \ldots, f_M(\mathbf{x}))^\mathsf{T}$. As seen from the above problem, the value function provides interactions among different objectives. Among the two solutions i and j, if $U(\mathbf{f}(i)) > U(\mathbf{f}(j))$ solution i is then preferred to solution j. Rosenthal (1985) states that the value function must be *strongly decreasing* before it can be used in multi-objective optimization. This means that the preference of a solution must increase if one of the objective function values is decreased while keeping other objective function values the same. This led Miettinen (1999) to prove the following theorem:

Theorem 3.5.1. *Let the value function* $U : \mathbb{R}^M \rightarrow \mathbb{R}$ *be strongly decreasing. Let* U *attain its maximum at* $\mathbf{f}^*$. *Then,* $\mathbf{f}^*$ *is Pareto-optimal.*

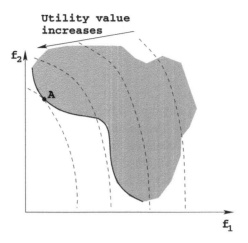

Figure 33 Contours of the value function.

In Figure 33, we show the contours of a nonlinear value function. For the nature of the contours shown in this figure, the solution A, where a contour of the value function is tangential to the Pareto-optimal front, is the preferred solution. It is important to note that this method can find only one solution at a time. By changing the parameters involved in the value function, different Pareto-optimal solutions can be found.

In some sense, the weighted sum approach or the weighted metric approach can be written as special cases of the value function approach. In such approaches, special forms of the value function are used. However, the formulation of the problem described in equation (3.17) allows a generic functional relationship among different objectives. The value function is usually chosen to be a nonlinear function of objectives. A good deal of theory of value functions (or utility functions) can be found in Keeney and Raiffa (1976).

3.5.1 Advantages

This idea is simple and ideal, if adequate value function information is available. The value function methods are mainly used in practice to multi-attribute decision analysis problems with a discrete set of feasible solutions (Keeney and Raiffa, 1976), although the principle can also be used in continuous search spaces.

3.5.2 Disadvantages

As evident from the above discussions, the obtained solution entirely depends on the chosen value function. It also requires users to come up with a value function which is globally applicable over the entire search space. Thus, there is a danger of using an over-simplified value function.

3.6   Goal Programming Methods

Goal programming was first introduced in an application of a single-objective linear programming problem by Charnes et al. (1955). However, goal programming gained popularity after the work of Ignizio (1976, 1978), Lee (1972), and various others. Romero (1991) has presented a comprehensive overview of the technique and listed a plethora of engineering applications where the goal programming technique has been used (Clayton et al., 1982; Sayyouth, 1981).

The main idea in goal programming is to find solutions which attain a predefined target for one or more objective functions. If there exists no solution which achieves pre-specified targets in all objective functions (the user is being optimistic), the task is to find solutions which minimize deviations from the targets. On the other hand, if a solution with the desired target exists, the task of goal programming is to identify that particular solution. In some sense, this task is similar to that in satisficing decision-making and the obtained solution is a satisficing solution, which can be different from an optimal solution. First, we illustrate the working of a goal programming method in a single-objective problem.

Let us consider a design objective $f(\mathbf{x})$, which is a function of a solution vector $\mathbf{x}$. Without loss of generality, we consider a objective function which is to be minimized (such as the fabrication cost of an engineering component). In goal programming, a target value t is chosen for every design objective by the user and the task is then to find a solution which has an objective value equal to t, subject to the condition that the resulting solution is feasible. We formulate the optimization problem, as follows:

$$\left. \begin{array}{ll} \text{goal} & (f(\mathbf{x}) = t), \\ & \mathbf{x} \in \mathcal{S}, \end{array} \right\} \tag{3.18}$$

where $\mathcal{S}$ is the feasible search region. If the target t is smaller than the optimal objective value, $f(\mathbf{x}^*)$, naturally there exists no feasible solution which will attain the above goal exactly. The objective of goal programming is then to find that solution which will minimize the deviation d between the achievement of the goal and the aspiration target, t. The solution for this problem is still $\mathbf{x}^*$, and the overestimate is $d = f(\mathbf{x}^*) - t$. Similarly, if the target cost t is larger than the maximum feasible cost, f_{max}, the solution of the goal programming problem is x, which makes $f(\mathbf{x}) = f_{max}$. However, if the target cost t is within $[f(\mathbf{x}^*), f_{max}]$, the solution to the goal programming problem is that feasible solution x which makes the objective value exactly equal to t. Although this solution may not be the optimal solution of the constrained $f(\mathbf{x})$, this solution is the outcome of the above goal program. In the above example, we have considered a single-objective problem. However, goal programming is mostly applied in multi-objective optimization problems. In fact, goal programming brings interesting scenarios when multiple criteria are considered.

In the above example, an 'equal-to' type goal is discussed. However, there can be four different types of goal criteria, as shown below (Steuer, 1986):

1. Less-than-equal-to $(f(\mathbf{x}) \leq t)$.

2. Greater-than-equal-to $(f(x) \geq t)$.
3. Equal-to $(f(x) = t)$.
4. Within a range $(f(x) \in [t^l, t^u])$.

In order to achieve the above goals, two non-negative deviation variables (n and p) are usually introduced. For the less-than-equal-to type goal, the positive deviation p is subtracted from the objective function, so that $f(x) - p \leq t$. Here, the deviation p quantifies the amount by which the objective value has surpassed the target t. The objective of goal programming is to minimize the deviation p so as to find the solution for which the deviation is minimum. If $f(x) > t$, the deviation p should take a non-zero positive value; otherwise it must be zero.

For the greater-than-equal-to type goal, a negative deviation n is added to the objective function, so that $f(x) + n \geq t$. The deviation n quantifies the amount by which the objective function has not satisfied the target t. Here, the objective of goal programming is to minimize the deviation n. For $f(x) < t$, the deviation n should take a non-zero positive value, otherwise it must be zero. For the equal-to type goal, the objective function needs to have the target value t, and thus both positive and negative deviations are used, so that $f(x) - p + n = t$. Here, the objective of goal programming is to minimize the summation $(p + n)$, so that the obtained solution is minimally away from the target in either direction. If $f(x) > t$, the deviation p should take a non-zero positive value and if $f(x) < t$, the deviation n should take a non-zero positive value. For $f(x) = t$, both deviations p and n must be zero. The fourth type of goal is handled by using two constraints: $f(x) - p \leq t^l$ and $f(x) + n \geq t^u$. The objective here is to minimize the summation $(p + n)$. All of the above constraints can be replaced by a generic equality constraint:

$$f(x) - p + n = t. \tag{3.19}$$

Thus, to solve a goal programming problem, each goal is converted into at least one equality constraint, and the objective is to minimize all deviations p and n. Goal programming methods differ in the ways that the deviations are minimized. Here, we briefly discuss three popular methods. In all of these methods, we assume that there are M objectives $f_j(x)$, each having one of the above four types of goal.

3.6.1 Weighted Goal Programming

A composite objective function with deviations from each of M objectives is used, as described below:

$$\left.\begin{array}{ll} \text{Minimize} & \sum_{j=1}^{M}(\alpha_j p_j + \beta_j n_j), \\ \text{subject to} & f_j(x) - p_j + n_j = t_j, \quad j = 1, 2, \ldots, M, \\ & x \in \mathcal{S}, \\ & n_j, p_j \geq 0, \qquad\qquad j = 1, 2, \ldots, M. \end{array}\right\} \tag{3.20}$$

Here, the parameters α_j and β_j are weighting factors for positive and negative deviations of the j-th objective. For less-than-equal-to type goals, the parameter β_j

is zero. Similarly, for greater-than-equal-to type goals, the parameter α_j is zero. For range-type goals, there exists a pair of constraints for each objective function. Usually, the weight factors α_j and β_j are fixed by the decision-maker, which makes the method subjective to the user. We illustrate this matter through a simple example problem, as follows:

$$
\left.
\begin{aligned}
\text{goal} \quad & (f_1 = 10x_1 \leq 2), \\
\text{goal} \quad & \left(f_2 = \frac{10 + (x_2 - 5)^2}{10x_1} \leq 2\right), \\
\text{subject to} \quad & S \equiv (0.1 \leq x_1 \leq 1, \quad 0 \leq x_2 \leq 10).
\end{aligned}
\right\}
\tag{3.21}
$$

The *decision space*, which is the feasible solution space $(x \in S)$ is shown in Figure 34 (shaded region). The goal lines $(f_1 \leq 2$ and $f_2 \leq 2)$ are shown in Figure 35. It is clear that there exists no feasible solution which achieves both goals. Thus, the resulting solution to this goal programming problem will violate either or both of the above goals, but in a minimum sense. In solving this problem by using the weighted goal programming approach, the following nonlinear programming (NLP) problem is constructed:

$$
\left.
\begin{aligned}
\text{Minimize} \quad & \alpha_1 p_1 + \alpha_2 p_2, \\
\text{subject to} \quad & 10x_1 - p_1 \leq 2, \\
& \frac{10 + (x_2 - 5)^2}{10x_1} - p_2 \leq 2, \\
& 0.1 \leq x_1 \leq 1, \quad 0 \leq x_2 \leq 10, \quad p_1 \geq 0, \quad p_2 \geq 0.
\end{aligned}
\right\}
\tag{3.22}
$$

Note that in the above NLP problem, the deviations n_1 and n_2 in the constraints are eliminated by using a '$\leq$' relation. Figures 34 and 35 make the concept of goal programming clear. Since no solution in the target space lies in the objective space, the objective of goal programming is to find that solution in the objective space which minimizes the deviation from the target space in both objectives. Now comes the

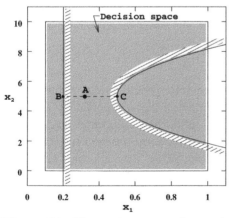

Figure 34 The goal programming problem shown in solution space (Deb, 2001). Reproduced by permission of Operational Research Society Ltd.

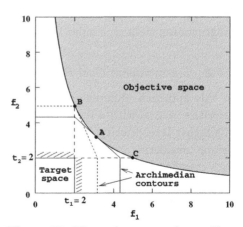

Figure 35 The goal programming problem shown in objective space (Deb, 2001). Reproduced by permission of Operational Research Society Ltd.

dependence of the resulting solution on the weight factors α_1 and α_2. By choosing a value of these weight factors, one, in fact, constructs an artificial penalty function (sometimes known as a utility function) away from the objective space. Thus, the objective $\alpha_1 p_1 + \alpha_2 p_2$ produces contours (known as Archimedian contours), as shown in Figure 35. The minimization of the problem given by equation (3.22) with a set of α_1 and α_2 values finds a solution in which an Archimedian contour makes a tangent to the feasible objective space. If an equal importance to both objectives (that is, $\alpha_1 = \alpha_2 = 0.5$) is given, the minimum contour (marked by the continuous line) is shown in Figure 35 and the resulting solution (marked as 'A') is as follows:

$$x_1 = 0.3162, \quad x_2 = 5.0, \quad p_1 = p_2 = 1.162.$$

At this solution, the objective function values are $f_1 = 3.162$ and $f_2 = 3.162$, thereby violating goals $f_1 \leq 2$ and $f_2 \leq 2$. An interesting scenario emerges when different weight factors are chosen. For example, if $\alpha_1 = 1$ and $\alpha_2 = 0$ are chosen, the resulting contour is shown by a dashed line and the corresponding solution (marked as 'B') is as follows:

$$x_1 = 0.2, \quad x_2 = 5.0, \quad p_1 = 0, \quad p_2 = 3.0.$$

On the other hand, if $\alpha_1 = 0$ and $\alpha_2 = 1$ are chosen, the resulting solution (marked as 'C') is as follows:

$$x_1 = 0.5, \quad x_2 = 5.0, \quad p_1 = 3.0, \quad p_2 = 0.$$

Solutions A, B and C are shown on Figure 34. This figure shows that there exist many more such solutions which lie in the interval $0.2 \leq x_1 \leq 0.5$ and $x_2 = 5.0$, each one of which is the solution to the above goal programming problem for a different set of weight factors α_1 and α_2. Thus, we observe that the solution to the goal programming problem largely depends on the chosen weight factors. Moreover, as outlined elsewhere (Romero, 1991), there exist a number of other difficulties with the weighted goal programming method, similar to those found with the weighted sum approach.

3.6.2   Lexicographic Goal Programming

In this approach, different goals are categorized into several levels of preemptive priorities. A goal with a lower-level priority is infinitely more important than a goal of a higher-level priority. Thus, it is important to fulfill the goals of first-level priority before considering goals of a second level of priority. Some researchers argue that such consideration of goals is most practical (Ignizio, 1976), although there exist some critics of this approach (Zeleny, 1982).

   This approach formulates and solves a number of sequential goal programming problems. First, only goals and corresponding constraints of the first-level priority are considered in the formulation of the goal programming problem and the latter is solved. If there exist multiple solutions to the resulting problem, another goal programming problem is formulated with goals having the second-level priority. In this case, the objective is only to minimize any deviations in the goals of second-level priority. However, the goals of first-level priority are used as hard constraints so that the obtained solution does not violate the goals of first-level priority. This process continues with goals of other higher-level priorities in sequence. The process is terminated as soon as one of the goal programming problems results in a single solution. When this happens, all subsequent goals of higher-level priorities are meaningless and are known as redundant goals (Romero, 1991). Usually, a single Pareto-optimal solution is found by this method.

We illustrate the working principle of the lexicographic goal programming approach in Figure 36. If objective f_1 is more important than objective f_2, we minimize the problem with f_1 first, ignoring f_2. We shall find multiple solutions in AB and CD in the first-level goal programming. Since there exists more than one solution to this problem, we proceed to the second-level optimization, where f_2 is minimized. The search is restricted among solutions found in the first-level goal programming. The solution to the second-level goal programming is D, which is the minimum solution of f_2 among all solutions in AB and CD. Thus, solution D is the solution of the overall lexicographic goal programming. It is interesting to note that if f_2 was considered more important than f_1, then solution E would have been the only optimum solution of the first-level optimization and the procedure would have been stopped there. The solution E would then have been declared as the solution to the overall lexicographic goal programming problem.

3.6.3 Min–Max Goal Programming

This approach is similar to the weighted goal programming approach, but instead of minimizing the weighted sum of the deviations from the targets, the maximum

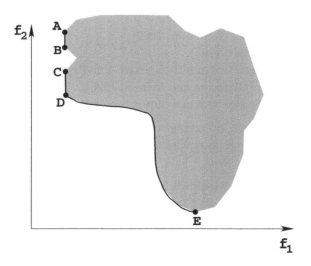

Figure 36   The lexicographic goal programming approach.

deviation in any goal from the target is minimized. The resulting nonlinear programming problem becomes as follows:

$$
\left.
\begin{aligned}
\text{Minimize} \quad & d \\
\text{subject to} \quad & \alpha_j p_j + \beta_j n_j \leq d, && j = 1, 2, \ldots, M, \\
& f_j(\mathbf{x}) - p_j + n_j = t_j, && j = 1, 2, \ldots, M, \\
& \mathbf{x} \in \mathcal{S}, \\
& n_j, p_j \geq 0, && j = 1, 2, \ldots, M.
\end{aligned}
\right\} \tag{3.23}
$$

Here, the parameter d becomes the maximum deviation in any goal. This method requires the choice of weight factors α_j and β_j, thereby making the approach subjective to the user. In some sense, this approach is similar to the weighted Tchebycheff approach, except that the ideal solution z^* is replaced with the target solution.

3.7 Interactive Methods

There exist a number of interactive methods, where minimum knowledge is needed a priori. For example, there is no need to know a value function relating to the objectives before even starting to solve the problem. As and when some Pareto-optimal solutions are found, their location and interactions are analyzed. The main aspect in these approaches is that from time to time during the optimization process the decision-maker is involved in providing some information about the direction of search, weight vector, reference points, and other factors. Since the decision-maker is involved in the optimization process, these techniques are becoming popular in practice; however, for the same reasons these approaches lose their simplicity. Some of the most popular methods include the following:

1. Interactive surrogate worth trade-off (ISWT) method (Chankong and Haimes, 1983).
2. Step method (Benayoun et al., 1971).
3. Reference point method (Wierzbicki, 1980).
4. Guess method (Buchanan, 1997).
5. Nondifferentiable interactive multi-objective bundle-based optimization system (NIMBUS) approach (Miettinen and Mäkelä, 1995).
6. Light beam search (Jaszkiewicz and Slowinsky, 1994).

For a detailed discussion on the above and other interactive approaches, interested readers may refer to the original studies or to Miettinen (1999).

3.8 Review of Classical Methods

From the above descriptions of some of the most popular classical multi-objective optimization algorithms, we observe a number of difficulties, particularly if the user is interested in finding multiple Pareto-optimal solutions:

1. Only one Pareto-optimal solution can be expected to be found in one simulation run of a classical algorithm.

2. Not all Pareto-optimal solutions can be found by some algorithms in nonconvex MOOPs.

3. All algorithms require some problem knowledge, such as suitable weights or ϵ or target values.

All of the classical algorithms described here suggest a way to convert a multi-objective optimization problem into a single-objective optimization problem. The weighted sum approach suggests minimizing a weighted sum of multiple objectives, the ϵ-constraint method suggests optimizing one objective function and use all other objectives as constraints, weighted metric methods suggests minimizing an l_p metric constructed from all objectives, the value function method suggests maximizing an overall value function (or utility function) relating all objectives, and goal programming methods suggest minimizing a weighted sum of deviations of objectives from user-specified targets. These conversion methods result in a single-objective optimization problem, which must be solved by using a single-objective optimization algorithm. In most cases, the optimal solution to the single-objective optimization problem is expected to be a Pareto-optimal solution. Such a solution is specific to the parameters used in the conversion method. In order to find a different Pareto-optimal solution, the parameters must be changed and the resulting new single-objective optimization problem has to be solved again. Thus, in order to find N different Pareto-optimal solutions, at least N different single-objective optimization problems need to be formed and solved.

Even after forming N single-objective optimization problems and solving them, some algorithms do not guarantee finding solutions in the entire Pareto-optimal region. We have argued earlier that if the MOOP is nonconvex, then the weighted sum approach, the weighted metric approach (with small p), or the value function approach have limitations in finding solutions in the nonconvex region of the Pareto-optimal set. Stating this otherwise, this means that no matter what weight vector is used, solutions in the nonconvex Pareto-optimal region cannot be found by these methods.

All methods discussed here require some knowledge about the problem. In the case of the weighted sum approach, we have shown earlier that a uniformly spaced set of weight vectors may not produce uniformly spaced Pareto-optimal solutions. Moreover, different weight vectors may produce an identical Pareto-optimal solution. The extent of non-uniformity in the obtained Pareto-optimal solutions and the range of useful weight vectors largely depend on the problem. Similarly, in the ϵ-constraint method the choice of the ϵ vector is an important one. The resulting Pareto-optimal solution largely depends on the chosen ϵ vector. Once again, a uniformly spaced ϵ vector[2] may not produce a uniformly spaced set of Pareto-optimal solutions. Similar arguments can be made with the weighted metric method. The value function approach depends on the utility function U. The parameters involved in these functions can be changed to

---

[2] A uniformly spaced ϵ vector can be created by systematically varying each member of the ϵ vector uniformly and by keeping the others constant.

find different Pareto-optimal solutions. Here too, the resulting Pareto-optimal solution will depend on the chosen parameters. In the goal programming method, the target values and the chosen weights largely affect which Pareto-optimal solution will result.

Despite these shortcomings, classical methods have a number of advantages, because of which they are used in solving real-world multi-objective optimization problems. The proofs of convergence (which are not discussed adequately in this present text, but can be found in texts on classical methods) to the Pareto-optimal set are their main strength. The weighted sum approach and the weighted metric approach guarantee that every Pareto-optimal solution corresponds to an optimal solution to the resulting single-objective optimization problem for a convex MOOP. The weighted Tchebycheff method and the ϵ-constraint method guarantee that every Pareto-optimal solution corresponds to an optimal solution to the resulting single-objective optimization problem for any convex or nonconvex MOOP. With these properties, it then depends on the efficiency of the chosen single-objective optimization algorithm to find the true Pareto-optimal solutions. Moreover, the methods are simple and easy to implement on a computer.

In order to highlight the difference between the classical preference-based approach and the ideal approach of multi-objective optimization, we show simulation results of an elitist non-dominated sorting genetic algorithm (NSGA-II), discussed later in Section 6.2, on two problems used in illustrating the weighted sum approach earlier in this chapter. The first problem ($a = 0.2$ and $b = 1$ in equation (3.2)) results in a convex Pareto-optimal front and the second problem ($a = 0.1$ and $b = 3$ in the same equation) results in a nonconvex Pareto-optimal region. NSGA-II (with a population size of 20 and a maximum of 25 generations) uses 500 function evaluations in each case. Figures 37 and 38 show all 20 obtained non-dominated solutions for the first and second problem, respectively. These figures demonstrate how multiple solutions, very close to or on the Pareto-optimal front, can be simultaneously obtained using a single

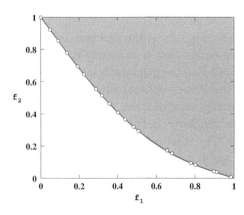

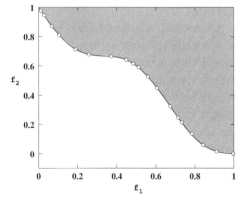

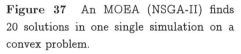

Figure 37 An MOEA (NSGA-II) finds 20 solutions in one single simulation on a convex problem.

Figure 38 The same MOEA (NSGA-II) finds 20 solutions in one single simulation on a nonconvex problem.

simulation run of an MOEA and without using any additional problem parameters. Importantly, the performance of the MOEA used in this study does not seem to matter to the convexity of the Pareto-optimal front. Moreover, a reasonably good distribution of solutions is obtained in both problems. Although we have not presented any MOEA technique yet, the above simulation results, in comparison to those obtained using classical methods discussed in this chapter, provide a clear distinction between the ideal and preference-based approaches.

3.9 Summary

The principle of multi-objective optimization was described in the previous chapter. It was highlighted there that multi-objective optimization is different from single-objective optimization. In the case of conflicting objectives, the resulting multi-objective optimization problem (MOOP) gives rise to a number of optimal solutions, known as Pareto-optimal solutions or non-inferior solutions. Since none of these solutions can be said to be any better than the others, the first task in an ideal multi-objective optimization is to find as many such Pareto-optimal solutions as possible.

In this present chapter, we have reviewed a number of classical multi-objective optimization algorithms. Most algorithms convert the MOOP problem into a single-objective optimization problem by using some user-defined procedures. Of these, the weighted approach converts multiple objectives into a single objective by using a weighted sum of objectives. The weight vector is user-defined. Since this method is incapable of finding trade-off optimal solutions in problems with a nonconvex Pareto-optimal region, the ϵ-constraint approach converts all but one of the objective functions into constraints. Again, user-defined limits are used to constrain the objectives. Another way to solve nonconvex problems is to minimize a Tchebycheff metric constructed by using multiple objectives. This method also requires a weight vector for emphasizing the objectives differently. For some of these algorithms, there exist theorems proving that the optimal solution of the converted single-objective optimization problem is one of the Pareto-optimal solutions. These proofs make classical multi-objective optimization algorithms interesting. On the other hand, in their practical use each of these algorithms may have to be used many times, hopefully each time finding a different Pareto-optimal solution. Moreover, each of these classical methods involve a number of user-defined parameters, which are difficult to set in an arbitrary problem.

In the next chapter, we will present an overview of evolutionary algorithms, and in subsequent chapters explain how such algorithms can be a useful alternative to finding multiple Pareto-optimal solutions simultaneously in a multi-objective optimization problem.

Exercise Problems

1. Consider the two-objective optimization problem:

$$\text{Minimize} \quad f_1(\mathbf{x}) = 2x_1x_2,$$
$$\text{Minimize} \quad f_2(\mathbf{x}) = x_1^2 + x_2^2.$$

   (a) Using the weight vector $\mathbf{w} = (w_1, 1 - w_1)^\mathsf{T}$, find the Pareto-optimal solutions in terms of w_1.
   (b) Does a uniform set of $\mathbf{w}$ vectors produce a uniformly distributed set of Pareto-optimal solutions? Explain.
   (c) What is the relationship between f_1 and f_2 for the Pareto-optimal solutions?
   (d) What weight vectors will produce the following feasible objective values:
      (i) $\mathbf{f} = (-2,3)^\mathsf{T}$ (ii) $\mathbf{f} = (-10,11)^\mathsf{T}$?

2. Consider the following problem:

$$\text{Minimize} \quad f_1(\mathbf{x}) = x^3 + y^2,$$
$$\text{Minimize} \quad f_2(\mathbf{x}) = y^2 - 4x.$$

   (a) Using the weight vector $\mathbf{w} = (w_1, 1 - w_1)^\mathsf{T}$, find the Pareto-optimal solutions in terms of w_1.
   (b) What is the relationship between f_1 and f_2 for the Pareto-optimal solutions?
   (c) What is the Pareto-optimal solution corresponds to $w_1 = 0.5$?
   (d) Show that the weighted-sum approach will not find half of the Pareto-optimal front.

3. In the above problem, can the weighted-sum approach find the objective vector $\mathbf{f} = (-1,4)^\mathsf{T}$? Explain. The above problem is attempted to solve using the ϵ-constraint method and the following reformulation is made:

$$\text{Minimize} \quad y^2 - 4x,$$
$$\text{subject to} \quad x^3 + y^2 \leq \epsilon_1.$$

   (a) Find the optimal solution to the above problem in terms of ϵ_1 using Kuhn-Tucker optimality conditions.
   (b) What value of ϵ_1 will correspond to the objective vector $\mathbf{f} = (-1,4)^\mathsf{T}$?

4. Consider the following problem:

$$\text{Minimize} \quad f_1(\mathbf{x}) = x^3 + y^2,$$
$$\text{Minimize} \quad f_2(\mathbf{x}) = 5(y^2 - x).$$

   Using the weighted l_2 distance metric, find the Pareto-optimal solutions corresponding to the following weight vectors:
   (a) $(w_1, w_2)^\mathsf{T} = (1,0)^\mathsf{T}$,

(b) $(w_1, w_2)^T = (0.5, 0.5)^T$,

(c) $(w_1, w_2)^T = (0, 1)^T$.

Draw a sketch of the objective space and discuss if all Pareto-optimal solutions can be found by the weighted l_2 distance metric method.

5. For the following two-objective problem

$$\begin{aligned} \text{Minimize} \quad & f_1(x) = x^2 + y^2, \\ \text{Minimize} \quad & f_2(x) = 5 + y^2 - x, \\ \text{subject to} \quad & -5 \leq x, y \leq 5, \end{aligned}$$

the utility function $U = 50 - f_1 - f_2$ is used. Find the Pareto-optimal solution of the resulting problem.

6. Consider the following problem:

$$\begin{aligned} \text{Minimize} \quad & f_1(x) = x^2 + y^2, \\ \text{Minimize} \quad & f_2(x) = 25 + y^2 - x^2, \\ \text{subject to} \quad & -5 \leq x, y \leq 5. \end{aligned}$$

Find the Pareto-optimal solution for the utility function $U = 100 - f_2 + f_1^2$.

7. Find the solution to the following goal programming problem:

$$\begin{aligned} \text{goal} \quad & (f_1(x) = x^2 + y^2 \leq 2), \\ \text{goal} \quad & (f_2(x) = y^2 - x \leq -2), \end{aligned}$$

in terms of the weight factors $(\alpha_1, 1 - \alpha_1)^T$. What are the solutions for $\alpha_1 = 1$, 0.5, and 0?

8. Find the solution to the following goal programming problem to have an equal importance to each goal:

$$\begin{aligned} \text{goal} \quad & (f_1(x) = x^3 - y^2 \leq 100), \\ \text{goal} \quad & (f_2(x) = y^2 - x \leq 0). \end{aligned}$$

4

Evolutionary Algorithms

Evolutionary algorithms (EAs) mimic natural evolutionary principles to constitute search and optimization procedures. EAs are different from classical search and optimization procedures in a variety of ways. In this chapter, we will present a number of popular evolutionary algorithms by discussing their differences with classical methods.

Although this book is about multi-objective optimization, most multi-objective evolutionary algorithms modify single-objective evolutionary algorithms in special ways. Thus, an understanding of single-objective evolutionary algorithms will be essential in understanding the working principles of multi-objective evolutionary algorithms. Among different evolutionary algorithms, we will discuss in detail binary-coded and real-parameter genetic algorithms. Thereafter, we will present various procedures of evolution strategy. Finally, brief descriptions of evolutionary programming and genetic programming are presented. However, there exist other evolutionary and nature-inspired algorithms, such as ant-colony optimization, simulated evolution, DNA computing, and cultural algorithms, descriptions of which can all be found in the EA literature.

As discussed in the previous chapter, one of the two tasks in the ideal multi-objective optimization is to find widely spread solutions in the obtained non-dominated front. Finding and maintaining multiple solutions in one single simulation run is a unique feature of evolutionary optimization techniques. At the end of this chapter, we will also describe a number of methods to find multiple optimal solutions in multi-modal function optimization using evolutionary techniques. Some of these methods are directly used in the multi-objective evolutionary methods described in the subsequent chapters.

4.1 Difficulties with Classical Optimization Algorithms

In any growing field of research and application, it becomes difficult and dangerous to call any study 'classical'. In this book, we have loosely called all search and optimization algorithms that use a single solution update in every iteration and that mainly use a deterministic transition rule as classical methods. Many such optimization algorithms can be found in standard textbooks (Arora, 1989; Deb, 1995;

Fox, 1971; Haug and Arora, 1989; Himmelblau, 1972; Reklaitis et al., 1983). Interested readers can refer to these texts for a good understanding of some of those algorithms.

Most classical point-by-point algorithms use a deterministic procedure for approaching the optimum solution. Such algorithms start from a random guess solution. Thereafter, based on a pre-specified transition rule, the algorithm suggests a search direction, which is often arrived at by considering local information. A uni-directional search is then performed along the search direction to find the best solution. This best solution becomes the new solution and the above procedure is continued for a number of times. Figure 39 illustrates this procedure. Algorithms vary mostly in the way the search directions are defined at each intermediate solution.

Classical optimization methods can be classified into two distinct groups: direct methods and gradient-based methods (Deb, 1995). In direct search methods, only the objective function ($f(\mathbf{x})$) and the constraint values ($g_j(\mathbf{x})$, $h_k(\mathbf{x})$) are used to guide the search strategy, whereas gradient-based methods use the first- and/or second-order derivatives of the objective function and/or constraints to guide the search process. Since derivative information is not used, the direct search methods are usually slow, requiring many function evaluations for convergence. For the same reason, they can also be applied to many problems without a major change in the algorithm. On the other hand, gradient-based methods quickly converge near an optimal solution, but are not efficient in non-differentiable or discontinuous problems. In addition, there are some common difficulties with most classical direct and gradient-based techniques, as follows:

- The convergence to an optimal solution depends on the chosen initial solution.
- Most algorithms tend to get *stuck* to a suboptimal solution.
- An algorithm efficient in solving one optimization problem may not be efficient in solving a different optimization problem.
- Algorithms are not efficient in handling problems having a discrete search space.

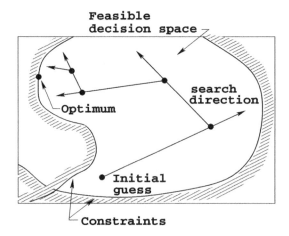

Figure 39   Most classical methods use a point-by-point approach.

- Algorithms cannot be efficiently used on a parallel machine.

Since nonlinearities and complex interactions among problem variables often exist in real-world optimization problems, the search space usually contains more than one optimal solution, of which most are undesired locally optimal solutions having inferior objective function values. While solving these problems, when classical methods get attracted to any of these locally optimal solutions, there is no escape.

Every classical optimization algorithm is designed to solve a specific type of problem. For example, the geometric programming method (Reklaitis et al., 1983) is designed to solve only *posynomial-type* objective function and constraints. Geometric programming is efficient in solving such problems but cannot be applied easily to solve other types of functions. The conjugate direction or conjugate gradient methods have convergence proofs for solving quadratic objective functions having one optimal solution, but they are not expected to work well in problems having multiple optimal solutions. The Frank–Wolfe's successive linear programming method (Reklaitis et al., 1983) works efficiently on linear function and constraints, but for solving nonlinear problems its performance largely depends on the chosen initial conditions. Thus, one algorithm may be best suited for one problem and may not even be applicable to a different problem. This requires users to know a number of optimization algorithms in order to solve different optimization problems.

In most practical optimization problems, some decision variables are restricted to take discrete values only. This requirement often arises to meet the market conditions. For example, if the diameter of a mechanical component is a decision variable and the component is likely to be procured 'off-the-shelf', the optimization algorithm cannot use any arbitrary diameter. A usual practice to tackle such problems is to assume that all variables are continuous during the optimization process. Thereafter, an available size closer to the obtained solution is recommended. However, there are major difficulties with this approach. First, since infeasible values of a decision variable are allowed in the optimization process, the optimization algorithm spends enormous time in computing infeasible solutions (in some cases, it may not even be possible to evaluate an infeasible solution). This makes the search effort inefficient. Secondly, as post-optimization calculations, the nearest lower and upper available sizes need to be checked for each infeasible discrete variable. For n such discrete variables, a total of 2^n additional solutions need to be evaluated. Thirdly, two options checked for each variable may not guarantee formation of the optimal combination with respect to other variables. All of these difficulties can be eliminated if only feasible values of the variables are allowed during the optimization process.

Many real-world optimization problems require the use of simulation software involving the finite element method, the computational fluid mechanics approach, nonlinear equation solving, or other computationally extensive methods to compute the objective function and constraints. Because of the affordability and availability of parallel computing systems, it has now become convenient to use them in solving complex real-world optimization problems. Since most classical methods use the point-by-point approach, where one solution gets updated to a new solution in one iteration, the advantages of parallel systems cannot be fully exploited.

The above discussion suggests that classical methods without major fix-ups may face difficulties in solving practical optimization problems. In this chapter, we will describe a few evolutionary algorithms which may alleviate some of the above difficulties and which are increasingly replacing classical methods in practical problem solving.

4.2   Genetic Algorithms

Over the last decade, genetic algorithms (GAs) have been extensively used as search and optimization tools in various problem domains, including the sciences, commerce and engineering. The primary reasons for their success are their broad applicability, ease of use and global perspective (Goldberg, 1989).

The concept of a genetic algorithm was first conceived by John Holland of the University of Michigan, Ann Arbor. Thereafter, he and his students have contributed much to the development of this field. Most of the initial research work can be found in various International Conference Proceedings. However, there now exist several textbooks on GAs (Goldberg, 1989; Gen and Cheng, 1997; Holland, 1975; Michalewicz, 1992; Mitchell, 1996; Vose, 1999). A more comprehensive description of GAs, along with other evolutionary algorithms, can be found in the recently compiled 'Handbook on Evolutionary Computation' (Bäck et al., 1997). Three journals ('Evolutionary Computation Journal' published by MIT Press, 'Transactions on Evolutionary Computation' published by IEEE and 'Genetic Programming and Evolvable Machines' published by Kluwer Academic Publishers) are now dedicated to promote research in this field. In addition, most GA applications can also be found in various domain-specific journals.

In this section, we will first describe the working principle of a binary-coded GA. Thereafter, we will describe a real-parameter GA, which is ideally suited to handle problems with a continuous search space. Because of the population-approach of GAs, constraints can be handled in a much better way than the way in which they are handled in classical search and optimization algorithms. We will also describe a number of approaches for handling constraints in GAs.

4.2.1   Binary Genetic Algorithms

As the name suggests, genetic algorithms (GAs) borrow their working principle from natural genetics. In this section, we will describe the principles of a GA's operation. To illustrate the working of GAs better, we will also show a hand-simulation of one iteration of GAs on a two-variable problem. A theoretical description of GA parameter interactions and other salient issues are then presented.

Working Principles

GAs are search and optimization procedures that are motivated by the principles of natural genetics and natural selection. Some fundamental ideas of genetics are borrowed and used artificially to construct search algorithms that are robust and require minimal problem information.

The working principle of GAs is very different from that of most classical optimization techniques. We describe the working of a GA by illustrating a simple can design problem. A cylindrical can is considered to have only two parameters – the diameter d and height h. Let us consider that the can needs to have a volume of at least 300 ml and the objective of the design is to minimize the cost of the can material. With this constraint and the objective, we first write the corresponding nonlinear programming problem (NLP problem) (Deb, 1999b):

$$\left.\begin{array}{ll} \text{Minimize} & f(d, h) = c \left(\frac{\pi d^2}{2} + \pi dh \right), \\ \text{subject to} & g_1(d, h) \equiv \frac{\pi d^2 h}{4} \geq 300, \\ \text{Variable bounds} & d_{min} \leq d \leq d_{max}, \\ & h_{min} \leq h \leq h_{max}. \end{array}\right\} \qquad (4.1)$$

The parameter c is the cost of the can material per square cm, and the decision variables d and h are allowed to vary in $[d_{min}, d_{max}]$ and $[h_{min}, h_{max}]$ cm, respectively.

Representing a Solution

In order to use GAs to find the optimal decision variables d and h, which satisfy the constraint g_1 and minimizes f, we first need to represent them in binary strings. Let us assume that we shall use five bits to code each of the two decision variables, thereby making the overall string length equal to 10. The following string represents a can of diameter 8 cm and height 10 cm:

$$\underbrace{01000}_{d} \ \underbrace{01010}_{h}$$

This string and corresponding decision variables are shown in Figure 40. In this representation, the lower and upper bounds of both decision variables are considered to be zero and 31, respectively. With five bits to represent a decision variable, there are exactly 2^5 or 32 different solutions possible. Choosing the lower and upper bounds as described above allows the GA to consider only integer values in the range [0, 31]. However, GAs are not restricted to use only integer values; in fact, GAs can be assigned to use any other integer or non-integer values just by changing the string

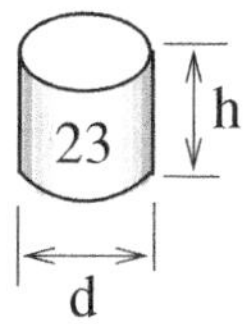

$(d, h) = (8, 10)$ cm

$(Chromosome) = 0\ 1\ 0\ 0\ 0\ \ 0\ 1\ 0\ 1\ 0$

Figure 40   A typical can and its chromosomal representation are shown. The cost of the can is marked as 23 units. This figure is taken from Sādhanā (Deb, 1999b). Reproduced with permission from Indian Academy of Sciences, Bangalore.

length and lower and upper bounds:

$$x_i = x_i^{min} + \frac{x_i^{max} - x_i^{min}}{2^{\ell_i} - 1} DV(s_i), \tag{4.2}$$

where ℓ_i is the string length used to code the i-th variable and $DV(s_i)$ is the decoded value of the string s_i (where the complete string is $s = \cup_{i=1}^{n} s_i$). In the above example, $\ell_i = 5$, $x_i^{min} = 0$ and $x_i^{max} = 31$ for both variables, such that $x_i = DV(s_i)$. The above mapping function allows the following features to exist in the decision variables:

1. Any arbitrary (albeit finite) precision can be achieved in the decision variables by using a long enough string.
2. Different decision variables can have different precisions by simply using different substring lengths.
3. Decision variables are allowed to take positive and negative values.
4. Variable bounds are taken care of by the mapping function given in equation (4.2).

Coding the decision variables in a binary string is primarily used to achieve a pseudo-chromosomal representation of a solution. For example, the 10-bit string illustrated above can be explained to exhibit a biological representation of a can having a diameter of 8 cm and a height of 10 cm. Natural chromosomes are made of many genes, each of which can take one of many different allele values (such as, the gene responsible for the eye color in a person's chromosomes may be expressed as black, whereas it could have been blue or some other color). When we see the person, we see the person's phenotypic representation, but each feature of the person is precisely written in his/her chromosomes – the genotypic representation of the person. In the can design problem, the can itself is the phenotypic representation of an artificial chromosome of 10 genes. To see how these 10 genes control the phenotype (the shape) of the can, let us investigate the leftmost bit (gene) of the diameter (d) parameter. A value of 0 at this bit (the most significant bit) allows the can to have diameter values in the range [0, 15] cm, whereas the other value 1 allows the can to have diameter values in the range [16, 31] cm. Clearly, this bit (or gene) is responsible for dictating the 'slimness' of the can. If the allele value 0 is expressed, the can is slim, while if the value 1 is expressed, the can is 'fat'. Each other bit position or a combination of two or more bit positions can also be explained to support the can's phenotypic appearance, but some of these explanations are interesting and important, and some are not.

After choosing a string representation scheme and creating a population of strings at random, we are ready to apply genetic operations to such strings to hopefully find better populations of solutions. However, before we discuss the genetic operators used in GAs, we shall describe an intermediate step of assigning a 'goodness' measure to each solution represented by a string.

Assigning Fitness to a Solution

It is important to reiterate that binary GAs work with strings representing the decision variables, instead of decision variables themselves. Once a string (or a solution) is

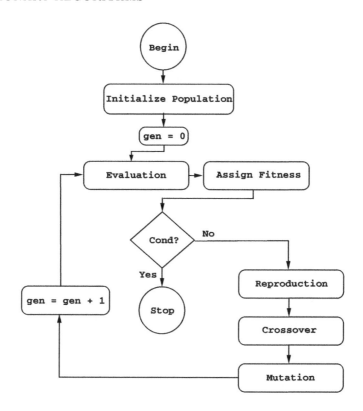

Figure 41   A flowchart of the working principle of a GA.

created by genetic operators, it is necessary to evaluate the solution, particularly in the context of the underlying objective and constraint functions. In the absence of constraints, the fitness of a string is assigned a value which is a function of the solution's objective function value. In most cases, however, the fitness is made equal to the objective function value. For example, the fitness of the above can represented by the 10-bit string s is:

$$F(s) \quad = \quad 0.065 \left[\pi \, (8)^2/2 + \pi \, (8) \, (10) \right],$$
$$= \quad 23,$$

assuming c $=$ 0.065. Since the objective of the optimization here is to minimize the objective function, it is to be noted that a solution with a smaller fitness value compared to another solution is better.

We are now in a position to describe the genetic operators which constitute the main part of a GA cycle. Figure 41 shows a flowchart of the working of a GA. Unlike classical search and optimization methods, a GA begins its search with a random set of solutions, instead of just one solution. Once a random population of solutions (in the above example, a random set of binary strings) is created, each is evaluated in the context of the underlying NLP problem (as discussed above) and a fitness is

assigned to each solution. The evaluation of a solution means calculating the objective
function value and constraint violations. Thereafter, a metric must be defined by
using the objective function value and constraint violations to assign a relative merit
to the solution (called the *fitness*). A termination condition is then checked. If the
termination criterion is not satisfied, the population of the solutions is modified by
three main operators and a new (and hopefully better) population is created. The
generation counter is incremented to indicate that one generation (or, one iteration,
in the parlance of classical search methods) of the GA is completed. The flowchart
shows that the working of a GA is simple and straightforward. We will now discuss
the genetic operators, in the light of the can design problem.

Figure 42 shows the phenotypes of a random population of six cans. The fitness (for
feasible solutions, this is the cost term, while for infeasible solutions it is the cost plus
the penalty in proportion to constraint violation) of each can is marked on the latter.
It is interesting to note that two solutions do not have an internal volume of 300 ml
and thus are penalized by adding an extra artificial *cost*, a matter which is discussed
in Section 4.2.3 later. Currently, it will suffice to note that the extra penalized cost is
large enough to cause all infeasible solutions to have a worse fitness value than that
of any feasible solution. We are now ready to discuss three genetic operators.

Reproduction or Selection Operator

The primary objective of the reproduction operator is to make duplicates of good
solutions and eliminate bad solutions in a population, while keeping the population
size constant. This is achieved by performing the following tasks:

1. Identify good (usually above-average) solutions in a population.
2. Make multiple copies of good solutions.
3. Eliminate bad solutions from the population so that multiple copies of good
   solutions can be placed in the population.

There exists a number of ways to achieve the above tasks. Some common methods
are *tournament selection, proportionate selection* and *ranking selection* (Goldberg

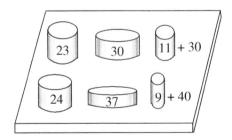

Figure 42   A random population of six cans. This figure is taken from Sādhanā (Deb,
1999b). Reproduced with permission from Indian Academy of Sciences, Bangalore.

and Deb, 1991).

In the tournament selection, tournaments are played between two solutions and the better solution is chosen and placed in the mating pool. Two other solutions are picked again and another slot in the mating pool is filled with the better solution. If carried out systematically, each solution can be made to participate in exactly two tournaments. The best solution in a population will win both times, thereby making two copies of it in the new population. Using a similar argument, the worst solution will lose in both tournaments and will be eliminated from the population. In this way, any solution in a population will have zero, one or two copies in the new population. It has been shown elsewhere (Goldberg and Deb, 1991) that the tournament selection has better or equivalent convergence and computational time complexity properties when compared to any other reproduction operator that exists in the literature.

Figure 43 shows six different tournaments played between old population members (each gets exactly two turns). When cans with a cost of 23 units and 30 units are chosen at random for the first tournament, the can costing 23 units wins and a copy of it is placed in the mating pool. The next two cans are chosen for the second tournament and a copy of the better can is then placed in the mating pool. This is how the mating pool (Figure 44) is formed. It is interesting to note how better solutions (having less costs) have made themselves to have multiple copies in the mating pool and worse solutions have been discarded. This is precisely the purpose of a reproduction or

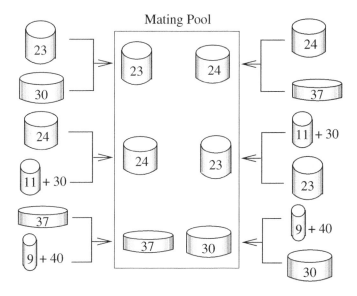

Figure 43 Tournaments are played between the six population members of Figure 42. The population enclosed by the dashed box forms the mating pool. This figure is taken from Sādhanā (Deb, 1999b). Reproduced with permission from Indian Academy of Sciences, Bangalore.

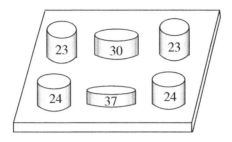

Figure 44   The population after reproduction operation. This figure is taken from Sādhanā
(Deb, 1999b). Reproduced with permission from Indian Academy of Sciences, Bangalore.

a selection operator. An interesting aspect of the tournament selection operator is
that just by changing the comparison operator, the minimization and maximization
problems can be handled easily.

There exist a number of other selection operators. Before we discuss other genetic
operators, let us describe a number of other popular selection operators. In the
proportionate selection method, solutions are assigned copies, the number of which
is proportional to their fitness values. If the average fitness of all population members
is f_{avg}, a solution with a fitness f_i gets an expected f_i/f_{avg} number of copies.
The implementation of this selection operator can be thought of as a roulette-wheel
mechanism, where the wheel is divided into N (population size) divisions, where
the size of each is marked in proportion to the fitness of each population member.
Thereafter, the wheel is spun N times, each time choosing the solution indicated
by the pointer, as shown in Figure 45. This figure shows a roulette wheel for five
individuals having different fitness values. Since the third individual has a higher
fitness value than any other, it is expected that the roulette wheel selection (RWS)
will choose the third solution more often than any other solution. This RWS scheme
can also be easily simulated on a computer. Using the fitness value F_i of all strings, the
probability of selecting the i-th string is $p_i = F_i / \sum_{j=1}^{N} F_j$. Thereafter, the cumulative
probability $(P_i = \sum_{j=1}^{i} p_j)$ of each string can be calculated by adding the individual

Solution, i	F_i	p_i	P_i	$p_i N$
1	25.0	0.25	0.25	1.25
2	5.0	0.05	0.30	0.25
3	40.0	0.40	0.70	2.00
4	10.0	0.10	0.80	0.50
5	20.0	0.20	1.00	1.00

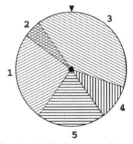

Figure 45   A roulette wheel is marked for five individuals according to their fitness values.
The third individual has a higher probability of selection than any other.

probabilities from the top of the list. Thus, the bottom-most string in the population has a cumulative probability (P_N) equal to 1. The roulette-wheel concept can be simulated by realizing that the i-th string in the population represents the cumulative probability values in the range $[P_{i-1}, P_i]$. The first string represents the cumulative values from zero to P_1. In order to choose N strings, N random numbers between zero to one are created. Thus, a string that represents the chosen random number in the cumulative probability range (calculated from the fitness values) for the string is copied to the mating pool. Figure 46 shows the cumulative probability line (ranging from zero to one) and the corresponding ranges of each of the five solutions for the example problem outlined in Figure 45. A random number r_i shown by the pointer marks solution 3. Hence, solution 3 is selected as a member of the mating pool. In this way, the string with a higher fitness value represents a larger range of cumulative probability values and therefore has a higher probability of being copied into the mating pool. Since the computation of the average fitness requires the fitness of all population members, this selection operator is slow compared to the tournament selection method (Goldberg and Deb, 1991). Moreover, the roulette-wheel selection operator inherently maximizes the fitness function.

The above implementation of the proportionate selection is *noisy* in the sense of introducing a large variance in its realizations. The variance may be reduced by using a somewhat deterministic version of the above roulette-wheel selection operator. In the *stochastic remainder roulette-wheel* selection (SRWS) operator, the probabilities p_i are multiplied by the population size and the expected number of copies is calculated for all solutions. Thereafter, each solution is first assigned a number of copies equal to the integer part of the expected number. Thereafter, the usual roulette-wheel selection (RWS) operator is applied with the fractional part of the expected number of all solutions to assign further copies. Since a part of the assignment is deterministic, this operator is less noisy. For the example problem shown in Figure 45, the expected number of copies ($p_i N$) is also computed. Under the SRWS scheme, we first assign one copy each to solutions 1 and 5, two copies to solution 3, and no copies to solutions 2 and 4. The remaining one slot in the mating pool is filled by using a RWS operation on the entire population with the fractional values as fitness: (0.25, 0.25, 0.00, 0.50, 0.00).

Whether the basic roulette-wheel selection or the stochastic remainder roulette-wheel selection operator is used, $O(N)$ random numbers have to be created to complete the selection process. This computation can be reduced if a different version

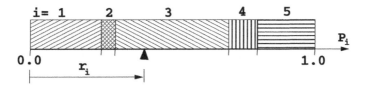

Figure 46   An implementation of the roulette-wheel selection operator.

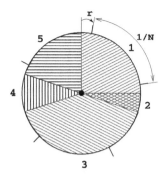

Figure 47   The stochastic universal sampling selection procedure.

– *stochastic universal sampling* (SUS) – is used (Baker, 1985). In this version, only one random number r is chosen for the whole selection process. Since N different solutions have to be chosen, a set of N equi-spaced numbers is created:

$$R = \{r, r + 1/N, r + 2/N, \ldots, r + (N - 1)/N\} \mod 1.$$

Thereafter, a solution corresponding to each member of R is chosen from the cumulative probability values as before. Figure 47 shows the selection procedure for the same five solutions using the SUS method. Assuming that the periphery of the roulette wheel is of one unit, the first solution is chosen according to the location of the random number on the periphery. Other solutions are chosen by traversing identical distances (1/N) along the periphery of the roulette wheel. In the above case, solutions 1 and 3 get two copies each and solution 5 gets one copy after selection operation. Solutions 2 and 4 do not get any copies in the mating pool.

Besides having a large computational complexity, proportionate selection methods have a *scaling* problem. The outcome of this selection operator is dependent on the true value of the fitness, instead of the relative fitness values of the population members. For example, if in a population, one solution has a large fitness value compared to the rest of the population members, the probability of choosing this *super-solution* would be close to one, thereby dominating the mating pool with its copies. On the other hand, if in a population, all solutions have more or less the same fitness value, every solution has a similar probability of selection (p_i). This will lead to a single assignment of each solution in the mating pool. This phenomenon is equivalent to not performing the selection operation at all. In order to circumvent both of these difficulties, raw fitness values are mapped within a predefined range of scaled fitness values (Goldberg, 1989). However, the tournament selection does not have this scaling problem.

The scaling difficulty can also be avoided by using a *ranking* selection method. First, the solutions are sorted according to their fitness, from the worst (rank 1) to the best (rank N). Each member in the sorted list is assigned a fitness equal to the rank of the solution in the list. Thereafter, the proportionate selection operator is applied with the ranked fitness values, and N solutions are chosen for the mating pool.

Crossover Operator

A crossover operator is applied next to the strings of the mating pool. A little thought will indicate that the reproduction operator cannot create any new solutions in the population. It only makes more copies of good solutions at the expense of not-so-good solutions. The creation of new solutions is performed by crossover and mutation operators. Like the reproduction operator, there exists a number of crossover operators in the GA literature (Spears, 1998), but in almost all crossover operators, two strings are picked from the mating pool at random and some portion of the strings are exchanged between the strings to create two new strings. In a single-point crossover operator, this is performed by randomly choosing a crossing site along the string and by exchanging all bits on the right side of the crossing site.

Let us illustrate the crossover operator by picking two solutions (called parent solutions) from the new population created after the reproduction operator. The cans and their genotype (strings) are shown in Figure 48. The third site along the string length is chosen at random and the contents of the right side of this cross site are exchanged between the two strings. The process creates two new strings (called offspring). Their phenotypes (the cans) are also shown in this figure. Since a single cross site is chosen here, this crossover operator is called the *single-point* crossover operator.

It is important to note that the above crossover operator created a solution (having a cost of 22 units) which is better in cost than both of the parent solutions. One may wonder that if a different cross site were chosen or two other strings were chosen for crossover, whether we would have found a better offspring every time. It is true that every crossover between any two solutions from the new population is not likely to find offspring better than both parent solutions, but it will be clear in a while that the chance of creating better solutions is far better than random. This is true because the parent strings being crossed are not any two arbitrary random strings. These strings have survived tournaments played with other solutions during the earlier reproduction phase. Thus, they are expected to have some good bit combinations in their string representations. Since, a single-point crossover on a pair of parent strings

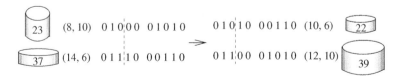

Figure 48 The single-point crossover operator. This figure is taken from Sādhanā (Deb, 1999b). Reproduced with permission from Indian Academy of Sciences, Bangalore.

can only create ℓ different string pairs (instead of all $2^{\ell-1}$ possible string-pairs) with bit combinations from either strings, the created offspring are also *likely* to be good strings. Moreover, every crossover may not create better solutions, but we will not worry too much about this here. If bad solutions are created, they will get eliminated in the next reproduction operation and hence will have a short life. However, let us think about the other possibility, that is, a good solution is created by the crossover operator. Since the offspring is good, it is likely to get more copies in the next reproduction operation and is also likely to get more chances to achieve a crossover with other good solutions in subsequent generations. Thus, more and more solutions in the population are likely to have similar chromosomes. This is exactly how biologists and evolutionists have explained how complex life forms have formed from simple ones (Dawkins, 1976, 1986; Eldredge, 1989).

In order to preserve some good strings selected during the reproduction operator, not all strings in the population are used in a crossover. If a crossover probability of p_c is used, then $100p_c\%$ strings in the population are used in the crossover operation and $100(1-p_c)\%$ of the population are simply copied to the new population.

The above concept of exchanging partial information between two strings can also be achieved with more than one cross sites. In a two-point crossover operator, two different cross sites are chosen at random. This will divide the string into three substrings. The crossover operation is completed by exchanging the middle substring between the strings, as shown in the following example:

	Parents											Offspring								
0	1	0	0	0	0	1	0	1	0		0	1	1	1	0	0	1	0	1	0
0	1	1	1	0	0	0	1	1	0	$\rightarrow$	0	1	0	0	0	0	0	1	1	0

Extending this idea, one can also implement a n-point crossover, where n cross sites (odd or even) can be chosen. If an odd number of cross sites is chosen, the end of each string can be considered as an additional cross site, thereby making the total number of cross sites an even number. The crossover can then proceed by exchanging alternate substrings. The extreme of this process is the *uniform* crossover, where one offspring is constructed by choosing every bit with a probability p (usually $p = 0.5$ is used) from either parent, as shown in the following example:

	Parents											Offspring								
0	1	0	0	0	0	1	0	1	0		0	1	0	1	0	0	0	0	1	0
0	1	1	1	0	0	0	1	1	0	$\rightarrow$	0	1	1	0	0	0	1	1	1	0

Usually, $p = 0.5$ is used. In the example problem, the second, fourth, fifth, seventh and ninth bits are exchanged between the parents. In terms of the extent of exploration (or search) power of a crossover operator, a single-point crossover preserves the structure of the parent strings to the maximum extent in the offspring. The extent of string-preservation reduces with the increase of cross sites in the crossover operator and is a minimum in the case of a uniform crossover operator. Spears (1998) has analyzed a number of these crossover operators and has found several interesting properties for many of them.

Mutation Operator

The crossover operator is mainly responsible for the search aspect of genetic algorithms, even though the mutation operator is also used for this purpose. The bit-wise mutation operator changes a 1 to a 0, and vice versa, with a mutation probability of p_m. The need for mutation is to keep diversity in the population. Figure 49 shows how a string obtained after the use of reproduction and crossover operators has been mutated to another string, thus representing a slightly different can. Once again, the solution obtained in the illustration is better than the original solution. Although it may not happen in all the instances of a mutation operation, mutating a string with a small probability is not a random operation since the process has a bias for creating only a few solutions in the search space.

The bit-wise mutation procedure described above requires the creation of a random number for every bit. In order to reduce the computational complexity, Goldberg (1989) suggested a *mutation clock* operator, where after a bit is mutated, the location of the next mutated bit is determined by an exponential distribution. The mean of the distribution is assumed to be $\mu = 1/p_m$. The procedure is simple. First, create a random number $r \in [0, 1]$. Then, estimate the next bit to be mutated by skipping $\eta = -p_m \ln(1 - r)$ bits from the current bit. In this way, on an average, $O(1/p_m)$ times less random numbers are to be generated. This mutation clock operator has been used in Deb and Agrawal (1999b).

The three operators – selection, crossover, and mutation operators – are simple and straightforward. The reproduction operator selects good strings, while the crossover operator recombines together good substrings from two good strings to hopefully form a better substring. The mutation operator alters a string locally to hopefully create a better string. Since none of these operations are performed deterministically, these claims are not guaranteed, nor explicitly tested, during a GA generation. However, it is expected that if bad strings are created they will be eliminated by the reproduction operator in subsequent generations and if good strings are created, they will be emphasized. Later, we shall see some intuitive reasoning as to why a GA with these simple operators constitutes a potential search and optimization algorithm.

Fundamental Differences

As seen from the above description of a GA's working principles, such algorithms are very different from most of the traditional optimization algorithms. The fundamental differences are described in the following paragraphs.

$$\boxed{22} \quad (10, 6) \quad 0\ 1\ 0\ 1\ 0\ \ 0\ 0\ 1\ 1\ 0 \ \longrightarrow\ 0\ 1\ 0\ 0\ 0\ \ 0\ 0\ 1\ 1\ 0 \quad (8, 6) \quad \boxed{16}$$

Figure 49 The bit-wise mutation operator. The fourth bit is mutated to create a new string. This figure is taken from Sādhanā (Deb, 1999b). Reproduced with permission from Indian Academy of Sciences, Bangalore.

Binary GAs work with a coding of decision variables, instead of the variables themselves. They work with a discrete search space, even though the function may be continuous. On the other hand, since function values at various discrete solutions are required, a discrete or a discontinuous function may be handled by using GAs. This allows GAs to be applied to a wide variety of problem domains. Another advantage is that GA operators exploit the similarities in string-structures to make an effective search. We shall discuss more about this matter a little later. One of the drawbacks of using a coding is that a suitable coding must be chosen for proper working of a GA. Although it is difficult to know beforehand what coding is suitable for a problem, a number of experimental studies (Bäck, 1996; Radcliffe, 1991) suggest that a coding which *respects* the underlying *building block* processing must be used.

The more striking difference between GAs and most classical optimization methods is that GAs work with a population of solutions instead of a single solution. Because there is more than one string being processed simultaneously and used to update every string in the population, it is likely that the expected GA solution may be a global solution. Even though some classical algorithms are population-based, such as Box's algorithm (Box, 1965), they do not use the obtained information efficiently. Moreover, since a population is what is updated at every generation, a set of solutions (in the case of multi-modal optimization, multi-objective optimization and others) can be obtained simultaneously.

In the above discussion of GA operators and their working principles, no word has been mentioned about the gradient or any other auxiliary problem information. In fact, GAs do not require any auxiliary information except the objective function values, although problem information can be used to speed up the GA's search process. The direct search methods used in traditional optimization also do not require gradient information explicitly, although in some of these methods search directions are found by using the objective function values, which are similar in concept to the gradient of the function. Moreover, some classical direct search methods work under the assumption that the function to be optimized is unimodal. GAs do not impose any such restrictions.

The other difference is that GAs use probabilistic rules to guide their search. On the face of it, this may look unnecessary, but careful thinking may provide some interesting properties for this type of search. The basic problem with most of the classical methods is that they use fixed transition rules to move from one solution to another solution. Since the destiny of these methods is pre-determined, they can only be applied to a special class of problems, in particular those for which the transition rules lead most search points towards the correct optimum. Thus, these methods are not robust and cannot be applied to a wide variety of problems. GAs, on the other hand, use probabilistic transition rules and an initial random population. These two features allow GAs to recover from early mistakes, if any, and enable them to handle a wide class of problems.

Another difference to most classical methods is that GAs can be easily and conveniently used in parallel systems. By using the tournament selection operator, where two strings are picked at random and the better string is copied in the mating pool, only two processors are involved at a time. Since any crossover operator requires interaction between only two strings, and since mutation requires alteration in only one string at a time, GAs are suitable for parallel implementations. There is another advantage. Since in real-world design optimization problems, most computational time is spent in evaluating solutions, with multiple processors all solutions in a population can be evaluated in a distributed manner. This will reduce the overall computational time substantially.

Every good optimization method needs to balance the extent of exploration of information obtained up until the current generation through recombination and mutation operators with the extent of exploitation through the selection operator. If the solutions obtained are exploited too much, premature convergence is expected. On the other hand, if too much stress is given on a search, the information obtained thus far has not been used properly. Therefore, the solution time may be enormous and the search exhibits a similar behavior to that of a random search. Most classical methods have fixed transition rules and hence have fixed degrees of exploration and exploitation. Since these issues can be controlled in a GA by varying the parameters involved in the genetic operators, GAs provide an ideal platform for performing a flexible search.

Understanding How GAs Work

The working principle described above is simple, with GA operators involving string copying and substring exchange, plus the occasional alteration of bits. Indeed, it is surprising that with such simple operators and mechanisms, a potential search is possible. We will try to give an intuitive answer to such doubts and also remind the reader that a number of studies have attempted to find a rigorous mathematical convergence proof for GAs (Rudolph, 1994; Vose, 1999; Whitley, 1992). Even though the operations are simple, GAs are highly nonlinear, massively multi-faceted, stochastic and complex. There exist studies using Markov chain analysis which involve deriving transition probabilities from one state to another and manipulating them to find the convergence time and solution. Since the number of possible states for a reasonable string length and population size become unmanageable even with the high-speed computers available today, other analytical techniques (statistical mechanics approaches and diffusion models) have also been used to analyze the convergence properties of GAs.

In order to investigate why GAs work, let us apply the GA for only one-cycle to a numerical maximization problem:

$$\left. \begin{array}{l} \text{Maximize} \quad \sin(x), \\ \text{Variable bound} \quad 0 \leq x \leq \pi. \end{array} \right\} \tag{4.3}$$

We will use five-bit strings to represent the variable x in the range $[0, \pi]$, so that the string (00000) represents the $x = 0$ solution and the string (11111) represents the $x = \pi$ solution. The other 30 strings are mapped in the range $[0, \pi]$ uniformly. Let us also assume that we use a population of size four, the proportionate selection, the single-point crossover operator with $p_c = 1$, and no mutation (or, $p_m = 0$). To start the GA simulation, we create a random initial population, evaluate each string, and then use three GA operators, as shown in Table 4. The first string has a decoded value equal to 9 and this string corresponds to a solution $x = 0.912$, which has a function value equal to $\sin(0.912) = 0.791$. Similarly, the other three stings are also evaluated. Since the proportionate reproduction scheme assigns a number of copies according to a string's fitness, the expected number of copies for each string is calculated in column 5. When the proportionate selection operator is actually implemented, the number of copies allocated to the strings is shown in column 6. Column 7 shows the mating pool. It is noteworthy that the third string in the initial population has a fitness which is very small compared to the average fitness of the population and is eliminated by the selection operator. On the other hand, the second string, being a good string, made two copies in the mating pool. The crossover sites are chosen at random and the four new strings created after crossover is shown in column 3 of the bottom table. Since no mutation is used, none of the bits are altered. Thus, column 3 of the bottom table represents the population at the end of one cycle of a GA. Thereafter, each of these stings is then decoded, mapped and evaluated. This completes one generation of a GA simulation. The average fitness of the new population is found to be 0.710, i.e. an improvement from that in the initial population. It is interesting to note that even though all operators used random numbers, a GA with all three operators produces a directed search, which usually results in an increase in the average quality of solutions from one generation to the next.

The string copying and substring exchange are all interesting and seem to improve the average performance of a population, but let us now ask the question: 'What has been processed in one cycle of a GA?' If we investigate carefully, we observe that among the strings of the two populations there are some similarities in the string positions among the strings. By the application of three GA operators, the number of strings with similarities at certain string positions has been increased from the initial population to the new population. These similarities are called *schema* in the GA literature. More specifically, a schema represents a set of strings with certain similarities at certain string positions. To represent a schema for binary codings, a triplet (1, 0 and ∗) is used; a ∗ represents both 1 or 0. It is interesting to note that a string is also a schema representing only one string – the string itself.

Two definitions are associated with a schema. The *order* of a schema H is defined as the number of defined positions in the schema and is represented as $o(H)$. A schema with full order $o(H) = \ell$ represents a string. The *defining length* of a schema H is defined as the distance between the outermost defined positions. For example, the schema $H = (\ast\ 1\ 0\ \ast\ \ast\ 0\ \ast\ \ast\ \ast)$ has an order $o(H) = 3$ (there are three defined positions: 1 at the second gene, 0 at the third gene, and 0 at the sixth gene) and a defining length $\delta(H) = 6 - 2 = 4$.

A schema $H_1 = (1\ 0\ \ast\ \ast\ \ast)$ represents eight strings with a 1 in the first position and a 0 in the second position. From Table 4, we observe that there is only one string

Table 4 One generation of a GA hand-simulation on the function $\sin(x)$.

	Initial population					
String	DV^a	x	$f(x)$	f_i/f_{avg}	AC^b	Mating pool
01001	9	0.912	0.791	1.39	1	01001
10100	20	2.027	0.898	1.58	2	10100
00001	1	0.101	0.101	0.18	0	10100
11010	26	2.635	0.485	0.85	1	11010
	Average, f_{avg}	0.569				

Mating Pool	CS^c	New population			
		String	DV^a	x	$f(x)$
01001	3	01000	8	0.811	0.725
10100	3	10101	21	2.128	0.849
10100	2	10010	18	1.824	0.968
11010	2	11100	28	2.838	0.299
			Average, f_{avg}		0.710

[a] DV, decoded value of the string.
[b] AC, actual count of strings in the population.
[c] CS, cross site.

representing this schema H_1 in the initial population and that there are two strings representing this schema in the new population. On the other hand, even though there was one representative string of the schema $H_2 = (0\ 0\ *\ *\ *)$ in the initial population, there is not one in the new population. There are a number of other schemata that we may investigate and conclude whether the number of strings they represent is increased from the initial population to the new population or not.

The so-called schema theorem provides an estimate of the growth of a schema H under the action of one cycle of the above tripartite GA. Holland (1975) and later Goldberg (1989) calculated the growth of the schema under a selection operator and then calculated the survival probability of the schema under crossover and mutation operators, but did not calculate the probability of constructing a schema from recombination and mutation operations in a generic sense. For a single-point crossover operator with a probability p_c, a mutation operator with a probability p_m, and the proportionate selection operator, Goldberg (1989) calculated the following lower bound on the schema growth in one iteration of a GA:

$$m(H, t + 1) \geq m(H, t)\frac{f(H)}{f_{avg}}\left[1 - p_c\frac{\delta(H)}{\ell - 1} - p_m o(H)\right], \qquad (4.4)$$

where $m(H, t)$ is the number of copies of the schema H in the population at generation t, $f(H)$ is the fitness of the schema (defined as the average fitness of all strings representing the schema in the population), and f_{avg} is the average fitness of the population. The above inequality leads to the schema theorem (Holland, 1975), as follows.

Theorem 4.2.1. *Short, low-order, and above-average schemata receive exponentially increasing number of trials in subsequent generations.*

A schema represents a number of similar strings. Thus, a schema can be thought of as representing a certain region in the search space. For the above function, the schema $H_1 = (1\ 0\ *\ *\ *)$ represents strings with x values varying from 1.621 to 2.330 with function values varying from 0.999 to 0.725. On the other hand, the schema $H_2 = (0\ 0\ *\ *\ *)$ represents strings with x values varying from 0.0 to 0.709 with function values varying from 0.0 to 0.651. Since our objective is to maximize the function, we would like to have more copies of strings representing schema H_1 than H_2. This is what we have accomplished in Table 4 without having to count all of these competing schema and without the knowledge of the complete search space, but by manipulating only a few instances of the search space. Let us use the inequality shown in equation (4.4) to estimate the growth of H_1 and H_2. We observe that there is only one string (the second string) representing this schema, or $m(H_1, 0) = 1$. Since all strings are used in the crossover operation and no mutation is used, $p_c = 1.0$ and $p_m = 0$. For the schema H_1, the fitness $f(H_1) = 0.898$, the order $o(H_1) = 2$, and the defining length $\delta(H_1) = 1$. In addition, the average fitness of the population is $f_{avg} = 0.569$. Thus, we obtain from equation (4.4):

$$m(H_1, 1) \geq (1) \cdot \frac{0.898}{0.569} \left[1 - (1.0) \frac{1}{5-1} - (0.0)(2) \right],$$
$$= 1.184.$$

The above calculation suggests that the number of strings representing the schema H_1 must increase. We have two representations (the second and third strings) of this schema in the next generation. For the schema H_2, the estimated number of copies using equation (4.4) is $m(H_2, 1) \geq 0.133$. Table 4 shows that no representative string of this schema exists in the new population.

The schema H_1 for the above example has only two defined positions (the first two bits) and both defined bits are tightly spaced (very close to each other) and contain the possible near-optimal solution (the string $(1\ 0\ 0\ 0\ 0)$ is the optimal string in this problem). The schemata that are short, low-order, and above-average are known as the *building blocks*. While GA operators are applied on a population of strings, a number of such building blocks in various parts along the string get emphasized, such as H_1 (which has the first two bits in common with the true optimal string) in the above example. Note that although H_2 is short and low-order, it is not an above-average schema. Thus, H_2 is not a building block. This is how GAs can emphasize different short, low-order and above-average schemata in the population. Once adequate number of such building blocks are present in a GA population, they get combined together due to the action of the GA operators to form bigger and better building blocks. This process finally leads a GA to find the optimal solution. This hypothesis is known as the *Building Block Hypothesis* (Goldberg, 1989).

Significance of Population Size

In a genetic algorithm, a decision about whether to have a '1' or a '0' has to be made at every bit position. The mechanism of using selection, crossover and mutation is one of the many ways of arriving at a decision. Holland viewed this decision-making in each bit independently and compared the process to a two-armed bandit game-playing problem. With one arm being fixed to either a 1 or a 0, the task is to find the arm which makes the maximum payoff (or fitness). With respect to the above definition of schema, every one-bit information is an order-one schema. Since the rest $(\ell - 1)$ bits are not considered, the fitness of one instantiation of any order-one schema is dependent on the exact value of bits at other locations.

In order to demonstrate the significance of an adequate population size in a GA, let us make use of a simple bimodal single-variable objective function for maximization:

$$f(x) = c_1 N(x, a_1, b_1) + c_2 N(x, a_2, b_2), \tag{4.5}$$

where the function $N(x, a_i, b_i)$ is a Gaussian function with a mean at a_i and a standard deviation of b_i. The search space is spanned over $[0, 1]$. By varying these six parameters, one can obtain functions of differing complexity. Let us choose the location of two maxima: $a_1 = 0.25$, and $a_2 = 0.80$. First, we use the following setting:

$$b_1 = 0.05, \quad b_2 = 0.05, \quad c_1 = 0.5, \quad c_2 = 1.0.$$

This makes the second maximum (at $x = 0.80$) the global maximum solution. Figure 50 shows the function. Let us consider two competing schemata ($H_0 \equiv 0 * \ldots *$) and ($H_1 \equiv 1 * \ldots *$) and analyze their growth under genetic operators. Figure 50 also shows that the left region ($x \in [0, 0.5]$) is represented by H_0 and the right region ($x \in [0.5, 1.0]$) is represented by the schema H_1. The average fitness of these two schemata are also shown with dashed lines. It is clear that the schema H_1 has a better fitness than H_0. Since the schema order and defining length are identical for both these schemata and H_1 is above-average, it becomes a building block. Thus, it is expected that a GA will make the right decision of choosing more solutions of H_1 than H_0. Since a GA can decide the correct direction of search by processing order-one schemata alone, a population of size two is essentially enough. However, note that the average fitness values shown by dashed lines in Figure 50 for any of the schemata do not adequately represent the fitness of all the strings which are represented by the schema. There is a large variation in the fitness values among the strings representing a schema. Because of this variability in fitness values, an appropriate number of copies in each region must be present to adequately represent the fitness variations. If we assume that λ (> 1) number of solutions are needed to statistically represent a region, an overall population of size 2λ is enough to solve the above problem by using a binary-coded GA.

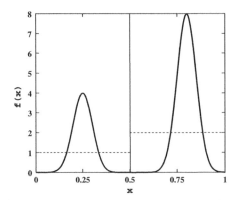

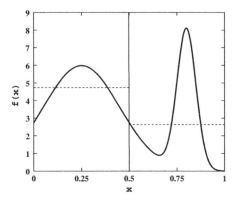

Figure 50   The simple bimodal function and average fitness values of two order-one schemata.

Figure 51   The modified bimodal function and average fitness values of two order-one schemata.

Now let us change the problem by choosing another parameter setting:

$$b_1 = 0.20, \quad b_2 = 0.05, \quad c_1 = 3, \quad c_2 = 1.$$

Figure 51 shows the function having two maxima. Now, the local maximum solution occupies a larger basin of attraction than the global maximum solution. When the fitnesses of H_0 and H_1 are computed (Table 5), it is observed that H_0 has a better fitness than H_1. Thus, if a population size of 2λ is used to solve the modified problem, a GA will be misled and emphasize more solutions of H_0 in a random initial population. Although this does not mean that a GA will not be able to recover from such early mistakes, a reliable application of the GA would be to make sure that it is started in the correct direction from the very first generation. In order to investigate the higher-order schema competitions, we have computed the fitness of four order-two competing schemata in Table 5, where the order of H_{00}, H_{01}, H_{10} and H_{11} is vertically downwards. These values are also shown pictorially in Figure 52. Here also,

Table 5   Schema fitness values for the bimodal problem.

Schema fitness					
Order-one		Order-two		Order-three	
(0.000–0.500)	4.732	(0.000–0.250)	4.732	(0.000–0.125)	3.848
(0.500–1.000)	2.633	(0.250–0.500)	4.732	(0.125–0.250)	5.616
		(0.500–0.750)	1.828	(0.250–0.375)	5.616
		(0.750–1.000)	3.439	(0.375–0.500)	3.848
				(0.500–0.625)	1.808
				(0.625–0.750)	1.848
				(0.750–0.875)	6.324
				(0.875–1.000)	0.553

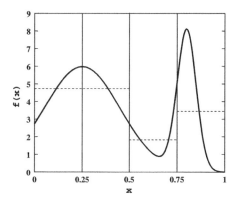

Figure 52 Average fitness values of different order-two schemata shown for the modified bimodal function.

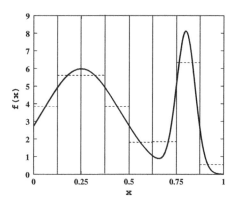

Figure 53 Average fitness values of different order-three schemata shown for the modified bimodal function.

we observe that the schema (representing the region $0.750 \leq x \leq 0.875$) containing the global maximum solution does not win the competition. When we compute the order-three schema competitions (see Table 5), we observe that the schema containing the global maximum wins the competition. Figure 53 clearly shows that the schema containing the global maximum has the maximum fitness value. However, the overall function is favored for the local maximum. In order to recognize the importance of the global maximum solution, a schema with a narrower region must be emphasized. In fact, any higher-order (> 3) schema competition will favor the global maximum solution. Thus, if a GA has to find the global maximum in a reliable way, the GA has to start comparing order-three schemata with the variability in string fitness in mind. Any competition lower than order three will favor the local maximum solution. One of the ways to ensure that a GA will start processing from order-three schemata is to have a population large enough to house strings representing all 2^3 competing schemata. Let us recall that the former bimodal problem (see Figure 50) required us to start processing from order-one schemata (having 2^1 or two competing schemata) in order to proceed in the correct direction. Since here, eight schemata are competing, a population of size 8λ would be adequate to represent a sufficient number of strings from each of the eight competing schemata. Thus, the modified bimodal problem needs four times more population size in order to reliably find the global maximum solution compared to the original bimodal problem.

Thus, we observe that the population size is related to the complexity of the problem. The parameter λ depends on the variability in fitness values of strings representing a schema. For example, when keeping the location of two maxima, if the function is not as smooth (Figure 54), the required sample size λ needed to detect a *signal* from the *noise* would be higher. Based on these considerations, Goldberg et al. (1992) calculated a population sizing estimate for a binary-coded GA. Following this study, Harik et al. (1999) tightened the sizing expression by considering the schema competition from a gambler's ruin model.

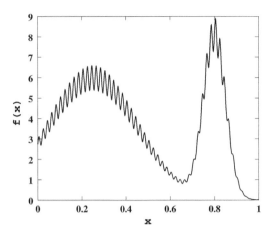

Figure 54   A function with a large variability in function values demands a large population size to find the global optimum.

GA Operator Interactions

Besides choosing an appropriate population size, another important matter is the balance between exploitation caused by the selection operator and the exploration introduced by the chosen recombination and mutation operators. If the selection operator uses too much selection pressure, meaning that it emphasizes the population-best solution too much by assigning many copies of it, the population loses its diversity very quickly. In order to bring back the diversity, the exploration power of the recombination and mutation operators must be large, meaning that these operators must be able to create solutions which are fairly different from the parent solutions. Otherwise, the population can become the victim of excessive selection pressure and eventually converge to a sub-optimal solution. On the other hand, if the selection pressure is very low, meaning that not much emphasis is given to the population-best solutions, the GA's search procedure behaves like a random search process. Although a qualitative argument of the balance between these two issues can be made and understood, a quantitative relationship between them is difficult to achieve.

Goldberg et al. (1993b) balanced the extent of exploitation and exploration issues by calculating the characteristic times of a selection and a crossover operator. In their earlier work (Goldberg and Deb, 1991), the *take-over times* t_s of a number of selection operators were calculated. The take-over time was defined as the number of generations required for the population-best solution to occupy all but one of the population slots by repetitive application of the selection operator alone. This characteristic time of a selection operator provides information about the speed with which the best solution in a population is emphasized. It was observed that binary tournament selection and linear ranking selection (with an assignment of two copies to the best solution) have the same take-over times. Proportionate selection is much slower than tournament selection.

For the uniform crossover operator, investigators have calculated the *mixing time* t_c, which refers to the number of generations required before repetitive application of the crossover operator alone can find a desired solution. This time gives information about how long a GA would have to wait before an adequate mixing of population members can produce the desired solutions.

By comparing the order of magnitudes of these two characteristic times, investigators argued that a GA will work successfully if the following relationship holds:

$$p_c \geq A \ln s, \tag{4.6}$$

where A is a constant which relates to the string length and population size and s is the selection pressure. On the one-max test problem, where the objective is to maximize the number of 1s in a string, Figure 55 shows the theoretical *control map* for successful GAs. A similar control map is obtained using GA simulation results (Figure 56). What is interesting to note from these figures is that a GA with any arbitrary parameter setting is not expected to work well even on a simple problem. A GA with a selection pressure s and a crossover probability p_c falling inside the control map finds the desired optimum. The above problem is a bit-wise linear problem, where a decision can be made in each bit, independent of the decisions taken at other bits. Even for this problem, a tournament selection with a large tournament size (such as $s = 20$ or so) and the uniform crossover operator with a small crossover probability (such as $p_c = 0.2$ or so) and no mutation, is not expected to find the true optimum solution. Although these parameter settings are probably extreme ones to choose in any application, they provide an understanding of the importance of interactions among GA operators and their parameter settings in performing a successful GA run.

For a selection pressure lower than a critical value, GAs can drift to any arbitrary

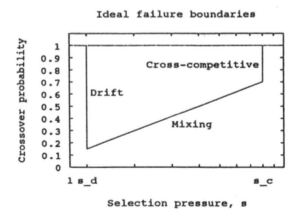

Figure 55   A theoretical control map for the working of a GA on the one-max problem (Goldberg et al., 1993b). Reprinted with permission from the Journal of the Society of Instruments and Control Engineers (SICE).

Simulation results

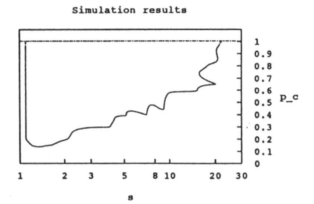

Figure 56   Simulation results show the working region for a GA on the one-max problem (Goldberg et al., 1993b). Reprinted with permission from the Journal of the Society of Instruments and Control Engineers (SICE).

solution and for a very large selection pressure, GAs cause important building blocks to compete among themselves, thereby not finding the desired optimum. These two cases become the two extreme bounds on the selection pressure in the above control map.

GA Operator Interactions under a Fixed Number of Trials

It is clear from the description of the working principles of a GA that such operators are tunable with the parameters associated with each of them. Since there exist flexibilities for changing the importance of one operator over another or flexibilities in using a different representation scheme to suit a problem, GAs are widely applicable to various types of problems. As we have seen in the previous section, along with this flexibility a burden is placed on the part of the user to choose an appropriate combination of GA operators and their parameters. While facing a real-world problem, the overall time to solve the problem is often fixed. This means that whatever algorithm is used to solve the problem at hand, we are only allowed to use a fixed number of trials (or function evaluations). For the sake of our discussion here, let us say that we are allowed to use a total of S trials to solve a problem. Since in a generational GA we create N new offspring at every generation, this means that we are allowed to run a GA having a population size N for a maximum of $t_{max} = S/N$ generations. If N is small, a large number of generations are allowed. On the other hand, if N is large, a small number of generations will be allowed. Then, an important point to ponder is what population size is appropriate to choose when the overall number of trials is fixed. A related question to ask is: 'Given the adequate population size, what combination of GA operators is appropriate?' It is important to realize that the answer to these questions will be different if the maximum number of allowed trials does not have a bound. Several researchers have attempted to find such GA operators and their parameter interactions for fixed number of trials, simply because such considerations are pragmatic in solving real-world problems.

Deb and Agrawal (1999b) conducted a series of experiments with different GA operators and parameter settings and applied these to different problems of varying difficulties. The outcome of this study is important and is outlined in the following:

1. For simpler problems (such as well-behaved unimodal or linear problems), a GA with a selection operator and a crossover or a mutation operator, or a combination of crossover and mutation operators, can all work satisfactorily. However, there is a distinct difference in the required population size in each case. For a selecto-mutation GA, a small population size (such as 3–6) provides the optimum performance. Since a selecto-mutation GA is similar to a local search approach, instead of more population members it requires a larger number of iterations to navigate its search towards the optimum. On the other hand, for a selecto-recombinative GA (with no mutation operator), the population size requirement is rather high. A GA with a crossover requires an adequate population size (described in Section 4.2.1) to steer the search in the right direction. Since the search relies on combining salient building blocks, the building block discovery and their emphasis requires an adequate number of population members. However, once the salient building blocks are found, not many generations are needed to combine them together. Thus, although a larger population size is in order, the number of generations required may be comparatively smaller. This is in contradiction to the common belief that since GAs use a population of solutions in each iteration, they are computationally more expensive than a local search algorithm.

2. For difficult problems (where difficulty can come from multi-modality, dimensionality of the search space, ruggedness of the fitness function, etc.), selecto-mutation GAs do not work successfully in finding the correct optimum solution. However, selecto-recombinative GAs can find the correct optimum with an adequate population size. Oates et al. (1999a, 1999b) observed similar results on a number of other problems.

Elite-Preserving Operator

In order to preserve and use previously found best solutions in subsequent generations, an elite-preserving operator is often recommended. In addition to an overall increase in performance, there is another advantage of using such elitism. In an elitist GA, the statistics of the population-best solutions cannot degrade with generations. There exists a number of ways to introduce elitism. We will discuss here a couple of popular approaches.

In a simple implementation, the best $\epsilon\%$ of the population from the current population is directly copied to the next generation. The rest $((100-\epsilon)\%)$ of the new population is created by the usual genetic operations applied on the entire current population (including the chosen $\epsilon\%$ elite members). In this way, the best solutions of the current population not only get passed from one generation to another, but they also participate with other members of the population in creating other population members.

The above implementation is a generational one. However, elitism is more commonly introduced in the context of steady-state GAs, where after every offspring is created, it is used to modify the entire population. The elitism is introduced by comparing two offspring with two parent solutions and then keeping the two better solutions. A somewhat generational version of this concept is implemented by first creating an offspring population by using usual genetic operations and then choosing the best N solutions from a combined population (of size 2N) of parent and offspring populations (Eshelman, 1991).

Theoretical Studies

Besides 'back-of-the-envelope' calculations and empirical studies, there exists a number of rigorous mathematical studies where a complete processing of a finite population under GA operators is modeled by using Markov chains (Vose, 1999; Vose and Wright, 1996) and statistical mechanics (Prügel-Bennett and Shapiro, 1994; Rogers and Prügel-Bennett, 1998). On different classes of functions, the recent dynamical systems analysis of GAs by Vose and Rowe (2000) as a random heuristic search reveals the complex dynamical behavior of a binary-coded GA with many metastable attractors. The interesting outcome is the effect of population size on the degree of metastability. For larger population sizes, the duration near a metastable state is more. In the statistical mechanics approach, instead of considering microscopic details of the evolving system, several macroscopic variables describing the system are modeled. Prügel-Bennett et al. (1994) studied the evolution of the first four cumulants of the fitness distribution as macroscopic variables. These are the mean, standard deviation, skewness and kurtosis of the fitness distribution. Analyzing different GA implementations with the help of these cumulants provides interesting insights into the complex interactions among GA operators (Rogers and Prügel-Bennett, 1998).

Rudolph (1994) has shown that a simple GA with an elite-preserving operator and non-zero mutation probability converges to the global optimum solution. However, the absence of elitism does not always allow a simple GA to have this convergence property.

Why Use Binary Alphabets?

Binary alphabets must naturally be used in problems having Boolean decision variables. In practice, many problems involve Boolean decision variables, particularly where the presence or absence of an entity is a decision variable. However, in handling a generic discrete search space problem, one can think of using higher-cardinality alphabets as well. Although it is not entirely intuitive why binary alphabets are better, there exist a couple of reasons for choosing binary alphabets. There also exists at least one counter-argument to such a proposal.

Holland (1975) and Goldberg (1989) argued that although a GA with a population size N processes N strings in every generation, in the schema space as many as $O(N^3)$ schemata get processed. This leverage provides a GA with its *implicit parallelism* property. Based on a count of the total number of schema being processed, Goldberg (1991) showed that a binary coding offers the maximum number of schemata per bit of information. Since the schemata count is the maximum for binary alphabets, the leverage achievable from it is also a maximum.

Reeves (1993b) estimated the minimum initial population size needed so that every possible solution in the search space can be reached from this population by crossover alone. The initial population must contain a string such that at every gene position every possible alphabet is present. Since mutation operation is ignored and crossover can only exchange information between strings, this condition will ensure that repetitive operation of crossover to this initial population will eventually create any possible solution in the search space. The conclusion of this study is that the higher the cardinality of the alphabet, then the larger is the minimum population size requirement to achieve the above condition. This means that with larger cardinality alphabets, more storage and computational effort will be necessary.

Antonisse (1989) criticized the schema counting procedure of Holland and Goldberg. He argued that for more than two alphabets, the use of one 'don't care' symbol to represent all combinations of alphabets is flawed. He suggested that all $(2^\nu - \nu - 1)$ different 'don't care' symbols must be used, where ν is the cardinality of the chosen alphabet. This makes $(2^\nu - 1)$ different possibilities in each gene position, thereby making the overall number of schemata as $(2^\nu - 1)^\ell$, and not $(\nu + 1)^\ell$. This modified number of total schemata per alphabet turns out to be monotonically increasing with the cardinality ν; in addition, the greater the cardinality of the alphabet, then the larger is the effect of Antonisse's schema processing. Since a GA does not work on the schema space, it is difficult to evaluate the merits of both of these arguments in an absolute sense. However, it is clear that whether it is Holland's schemata or Antonisse's augmented schemata that get processed in a GA with a higher cardinality alphabet, the implicit parallelism argument holds in both cases.

Other GAs

In addition to the simple binary GA described above, there exist a number of advanced GAs, specifically designed to have advanced features. In order to ensure the convergence properties and to accelerate the rate of convergence, elitist GAs are commonly used. Of these, Eshelman's (Eshelman, 1991) CHC[1] is an elitist GA which selects the best N solutions from a combination of 2N population members formed by

---

[1] The interpretation of C, H and C is derived from three features of the CHC algorithm: cross generation selection, heterogeneous recombination and cataclysmic mutation (Eshelman, 2001).

using the parent and the offspring populations. Moreover, the CHC algorithm uses a restricted mating (with a uniform like crossover operator) so that dissimilar solutions (in terms of bit differences between them) will mate with each other. Because of a high selection pressure introduced by the elite-preserving operation, this GA needs to use a highly disruptive recombination operator.

Genetic Implementor (GENITOR) (Whitley, 1989) is a steady-state elitist GA, where each offspring is introduced into the parent population one at a time. Parents are chosen according to a ranking selection operator and the created offspring replaces the worst individual in the parent population. On a generational framework, Goldberg and Deb (1991) have shown that the GENITOR algorithm has an inherently large selection pressure. Once again, this GA requires a highly disruptive recombination operator to maintain a balance between exploration and exploitation issues.

Messy GAs (Goldberg et al., 1989, 1990) are variable-length GAs where the identification of all salient building blocks and the subsequent mixing of them is temporally separated into two phases. In the primordial phase, important building blocks are identified by comparing different plausible building blocks with one another. In the juxtapositional phase, the necessary building blocks are combined together by the use of a cut-and-splice operator to form optimal solutions in difficult problems. We will describe messy GAs in more detail in Section 6.7. The quest for finding the necessary linkage needed in a problem on the fly, a matter which took shape with the messy GA study, has now been paid adequate attention. The continuing research on gene expression GAs (Kargupta, 1997), linkage learning GAs (Harik and Goldberg, 1996), the Bayesian optimization algorithm (Pelican et al., 1999), the factorial distribution algorithm (Mühlenbein and Mahnig, 1999) and various others represent worthy efforts applied in this direction.

Genetic algorithms have also been extensively used in scheduling and other combinatorial optimization problems. The representation used in such problems is different from the usual binary string representation. This also requires different recombination operators to be devised for creating valid offspring from valid parent solutions. A number of such GA implementations and applications can be found in various books (Goldberg, 1989; Gen and Cheng, 1997; Michalewicz, 1992; Mitchell, 1996), as well as certain journals and conference proceedings.

4.2.2   Real-Parameter Genetic Algorithms

When binary-coded GAs need to be used to handle problems having a continuous search space, a number of difficulties arise. One difficulty is the Hamming cliffs associated with certain strings (such as strings 01111 and 10000) from which a transition to a neighboring solution (in real space) requires the alteration of many bits. Hamming cliffs present in a binary coding cause artificial hindrance to a gradual search in the continuous search space. The other difficulty is the inability to achieve any arbitrary precision in the optimal solution. In binary-coded GAs, the string length must be chosen a priori to enable GAs to achieve a certain precision in the solution.

The more the required precision, then the larger is the string length. For large strings, the population size requirement is also large (Goldberg et al., 1992), thereby increasing the computational complexity of the algorithm. Since a fixed coding scheme is used to code the decision variables, variable bounds must be such that they bracket the optimum variable values. Since in many problems this information is not usually known a priori, this may cause some difficulty in using binary-coded GAs in such problems. Furthermore, a careful thinking of the schema processing in binary strings reveals that not all Holland's schemata are equally important in most problems having a continuous search space. To a continuous search space, the meaningful schemata are those that represent the contiguous regions of the search space. The schema ($\ast\ \ast\ \ast\ \ast$ 1), for example, represents every other point in the discretized search space. Although this schema may be useful in certain periodic or oscillatory functions, the schema (1 $\ast\ \ast\ \ast\ \ast$) signifies a more meaningful schema representing the right-half of the search space in most problems. Thus, the crossover operator used in the binary coding needs to be redesigned in order to increase the propagation of more meaningful schemata pertaining to a continuous search space.

There exists a number of real-parameter GA implementations, where crossover and mutation operators are applied directly to real parameter values. Since real parameters are used directly (without any string coding), solving real-parameter optimization problems is a step easier when compared to the binary-coded GAs. Unlike in the binary-coded GAs, decision variables can be directly used to compute the fitness values. Since the selection operator works with the fitness value, any selection operator used with binary-coded GAs can also be used in real-parameter GAs.

However, the difficulty arises with the search operators. In the binary-coded GAs, decision variables are coded in finite-length strings and exchanging portions of two parent strings is easier to implement and visualize. Simply flipping a bit to perform mutation is also convenient and resembles a natural mutation event. In real-parameter GAs, the main challenge is how to use a pair of real-parameter decision variable vectors to create a new pair of offspring vectors or how to perturb a decision variable vector to a mutated vector in a meaningful manner. As in such cases the term 'crossover' is not that meaningful, they can be best described as *blending* operators (DeJong, 1999). However, most blending operators in real-parameter GAs are known as crossover operators, and we will continue to use the same term here. First, we describe a number of real-parameter crossover operators and then present a few real-parameter mutation operators. Herrara et al. (1998) have provided a good overview of many real-parameter crossover and mutation operators.

Linear Crossover

One of the earliest implementations was reported by Wright (1991), where a linear crossover operator created the three solutions, $0.5(x_i^{(1,t)} + x_i^{(2,t)})$, $(1.5x_i^{(1,t)} - 0.5x_i^{(2,t)})$ and $(-0.5x_i^{(1,t)} + 1.5x_i^{(2,t)})$ from two parent solutions $x_i^{(1,t)}$ and $x_i^{(2,t)}$ at generation t, with the best two solutions being chosen as offspring. Figure 57 shows the three

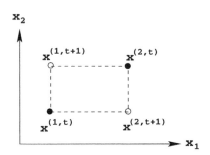

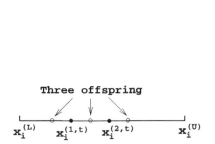

Figure 57 Wright's linear crossover opera-tor. Parents are marked by filled circles.

Figure 58 Naive crossover on two decision variables. Parents are marked by filled circles.

offspring created from the two parent solutions.

A Naive Crossover

This crossover operator is similar to the crossover operators used in binary-coded GAs. Cross sites are only allowed to be chosen at the variable boundaries. For example, a single-point crossover at the third site will produce the following offspring:

$$
\begin{array}{llllllll}
\text{Parent 1:} & (x_1^{(1,t)} & x_2^{(1,t)} & x_3^{(1,t)} & x_4^{(1,t)} & \ldots & x_n^{(1,t)}) \\
\text{Parent 2:} & (x_1^{(2,t)} & x_2^{(2,t)} & x_3^{(2,t)} & x_4^{(2,t)} & \ldots & x_n^{(2,t)}) \\[1ex]
\text{Offspring 1:} & (x_1^{(1,t)} & x_2^{(1,t)} & x_3^{(1,t)} & x_4^{(2,t)} & \ldots & x_n^{(2,t)}) \\
\text{Offspring 2:} & (x_1^{(2,t)} & x_2^{(2,t)} & x_3^{(2,t)} & x_4^{(1,t)} & \ldots & x_n^{(1,t)})
\end{array}
$$

For two decision variables and for the case when the cross site falls in the variable boundary, the offspring can be either parents themselves or the other two diagonal solutions, as shown in Figure 58. Like in the single-point crossover operator, two-point, n-point, or uniform crossover operators can also be used in a similar manner. This crossover operator does not have an adequate search power and thus the search within a decision variable has to mainly rely on the mutation operator.

Blend Crossover and Its Variants

Goldberg introduced the concept of virtual alphabets in the context of real-parameter GAs (Goldberg, 1991), a matter which is discussed on page 124. Eshelman and Schaffer (1993) have introduced the notion of *interval* schemata which is similar in principle to the virtual alphabets. They also suggested a blend crossover (BLX-α) operator for real-parameter GAs. For two parent solutions $x_i^{(1,t)}$ and $x_i^{(2,t)}$ (assuming $x_i^{(1,t)} < x_i^{(2,t)}$), the BLX-$\alpha$ randomly picks a solution in the range $[x_i^{(1,t)} - \alpha(x_i^{(2,t)} - x_i^{(1,t)}),$ $x_i^{(2,t)} + \alpha(x_i^{(2,t)} - x_i^{(1,t)})]$. This crossover operator is illustrated in Figure 59.

Thus, if u_i is a random number between 0 and 1, the following is an offspring:

$$x_i^{(1,t+1)} = (1 - \gamma_i)x_i^{(1,t)} + \gamma_i x_i^{(2,t)}, \tag{4.7}$$

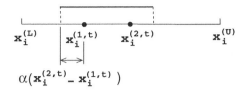

Figure 59 The BLX-α operator. Parents are marked by filled circles.

where $\gamma_i = (1+2\alpha)u_i - \alpha$. If α is zero, this crossover creates a random solution in the range $(x_i^{(1,t)}, x_i^{(2,t)})$. In a number of test problems, the investigators have reported that BLX-0.5 (with $\alpha = 0.5$) performs better than BLX operators with any other α value. However, it is important to note that the factor γ_i is uniformly distributed for a fixed value of α. However, BLX-α has an interesting property: the location of the offspring depends on the difference in parent solutions. This will be clear if we rewrite equation (4.7) as follows:

$$\left(x_i^{(1,t+1)} - x_i^{(1,t)}\right) = \gamma_i \left(x_i^{(2,t)} - x_i^{(1,t)}\right). \tag{4.8}$$

If the difference between the parent solutions is small, the difference between the offspring and parent solutions is also small. This property of a search operator allows us to constitute an adaptive search. If the diversity in the parent population is large, an offspring population with a large diversity is expected, and vice versa. Thus, such an operator will allow the searching of the entire space early on (when a random population over the entire search space is initialized) and also allow us to maintain a focussed search when the population tends to converge in some region in the search space.

It is important to note that there exists a number of other crossover operators which work by using the same principle. The arithmetic crossover (Michalewicz and Janikow, 1991) uses equation (4.7) with a fixed value of γ for all decision variables. However, γ is chosen by carefully calculating its maximum allowed value in all decision variables so that the resulting values do not exceed the lower or upper limits. The extended crossover operator (Voigt et al., 1995) is also similar to BLX-α.

Simulated Binary Crossover

In 1995, this author and his students developed the simulated binary crossover (SBX) operator, which works with two parent solutions and creates two offspring (Deb and Agrawal, 1995; Deb and Kumar, 1995). As the name suggests, the SBX operator simulates the working principle of the single-point crossover operator on binary strings. In these studies, we showed that this crossover operator *respects* the *interval* schemata processing, in the sense that common interval schemata between the parents are preserved in the offspring. The procedure of computing the offspring $x_i^{(1,t+1)}$ and $x_i^{(2,t+1)}$ from the parent solutions $x_i^{(1,t)}$ and $x_i^{(2,t)}$ is described as follows. A spread factor β_i is defined as the ratio of the absolute difference in offspring values to that of the parents:

$$\beta_i = \left| \frac{x_i^{(2,t+1)} - x_i^{(1,t+1)}}{x_i^{(2,t)} - x_i^{(1,t)}} \right|. \tag{4.9}$$

First, a random number u_i between 0 and 1 is created. Thereafter, from a specified probability distribution function, the ordinate β_{q_i} is found so that the area under the probability curve from 0 to β_{q_i} is equal to the chosen random number u_i. The probability distribution used to create an offspring is derived to have a similar *search power* to that in a single-point crossover in binary-coded GAs and is given as follows (Deb and Agrawal, 1995):

$$\mathcal{P}(\beta_i) = \begin{cases} 0.5(\eta_c + 1)\beta_i^{\eta_c}, & \text{if } \beta_i \le 1; \\ 0.5(\eta_c + 1)\dfrac{1}{\beta_i^{\eta_c+2}}, & \text{otherwise.} \end{cases} \tag{4.10}$$

Figure 60 shows the above probability distribution with $\eta_c = 2$ and 5 for creating offspring from two parent solutions ($x_i^{(1,t)} = 2.0$ and $x_i^{(2,t)} = 5.0$) in the real space. In the above expressions, the distribution index η_c is any non-negative real number. A large value of η_c gives a higher probability for creating 'near-parent' solutions and a small value of η_c allows distant solutions to be selected as offspring. Using equation (4.10), we calculate β_{q_i} by equating the area under the probability curve equal to u_i, as follows:

$$\beta_{q_i} = \begin{cases} (2u_i)^{\frac{1}{\eta_c+1}}, & \text{if } u_i \le 0.5; \\ \left(\dfrac{1}{2(1 - u_i)}\right)^{\frac{1}{\eta_c+1}}, & \text{otherwise.} \end{cases} \tag{4.11}$$

After obtaining β_{q_i} from the above probability distribution, the offspring are calculated as follows:

$$x_i^{(1,t+1)} = 0.5\left[(1 + \beta_{q_i})x_i^{(1,t)} + (1 - \beta_{q_i})x_i^{(2,t)}\right], \tag{4.12}$$

$$x_i^{(2,t+1)} = 0.5\left[(1 - \beta_{q_i})x_i^{(1,t)} + (1 + \beta_{q_i})x_i^{(2,t)}\right]. \tag{4.13}$$

Thus, the following step-by-step procedure is followed to create two offspring ($x_i^{(1,t+1)}$ and $x_i^{(2,t+1)}$) from two parent solutions ($x_i^{(1,t)}$ and $x_i^{(2,t)}$):

Step 1: Choose a random number $u_i \in [0, 1)$.

Step 2: Calculate β_{q_i} using equation (4.11).

Step 3: Compute the offspring by using equations (4.12) and (4.13).

Note that two offspring are symmetric about the parent solutions. This is deliberately enforced to avoid a bias towards any particular parent solution in a single crossover operation. Another interesting aspect of this crossover operator is that for

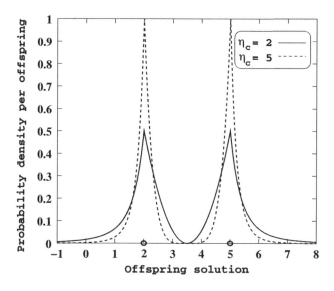

Figure 60   The probability density function for creating offspring under an SBX-η_c operator. Parents are marked with an 'o'. Reproduced from Deb and Beyer (2001) (© 2001 by the Massachusetts Institute of Technology).

a fixed η_c the offspring have a spread which is proportional to that of the parent solutions:

$$\left(x_i^{(2,t+1)} - x_i^{(1,t+1)}\right) = \beta_{q_i}\left(x_i^{(2,t)} - x_i^{(1,t)}\right). \tag{4.14}$$

This has an important implication. Let us consider two scenarios: (i) two parents are far away from each other, and (ii) two parents are closer to each other. For illustration, both of these cases (with parent solutions $x_i^{(1,t)} = 2.0$ and $x_i^{(2,t)} = 5.0$ in the first case and with parent solutions $x_i^{(1,t)} = 2.0$ and $x_i^{(2,t)} = 2.5$ in the second case) and the corresponding probability distributions with $\eta_c = 2$ are shown in Figures 61 and 62, respectively. For an identical random number u_i, the parameter β_{q_i} take the same value in both cases. From equation (4.14) it is clear that in the first case the offspring are likely to be more widely spread than in the second case. Figures 61 and 62 also show the corresponding offspring (marked with a box) for $u_i = 0.9$ or $\beta_{q_i} = 5^{1/3}$. These figures clearly show that if the parent values are far from each other (the first case), it is possible for solutions away from the parents to be created. Compare the offspring $x_i^{(1,t+1)} = 0.935$ with parent $x_i^{(1,t)} = 2.0$. However, if the parent values are close by (the second case), distant offspring are not likely. Compare the offspring $x_i^{(1,t+1)} = 1.822$ with parent $x_i^{(1,t)} = 2.0$ created by using the same random number as in the first case. In initial populations, where the solutions are randomly placed (like the first case), this allows almost any value to be created as an offspring. However, when the solutions tend to converge due to the action of genetic operators (like the second case), distant solutions are not allowed, thereby focusing the search to a narrow region.

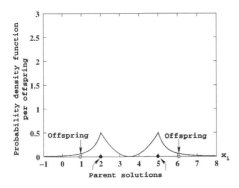 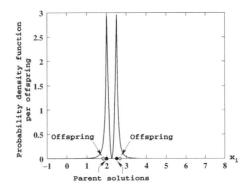

Figure 61　Probability distribution of offspring with distant parents.

Figure 62　Probability distribution of offspring with closely spaced parents.

It is interesting to note that both equations (4.12) and (4.13) can be written in the form of equation (4.7) with the following relationship: $\gamma_i = 0.5(1 \mp \beta_{q_i})$. However, it is important that, unlike in the BLX-α operator, the equivalent γ_i term in the SBX operator is not uniformly distributed (refer to equation (4.11)). The SBX operator biases solutions near each parent more favorably than solutions away from the parents. Essentially, the SBX operator has two properties:

1. The difference between the offspring is in proportion to the parent solutions.
2. Near-parent solutions are monotonically more likely to be chosen as offspring than solutions distant from parents.

The SBX operator is also extended for bounded variables (Deb, 2000). Using the SBX operator, a mixed-integer programming algorithm with GAs (GeneAS) was introduced and applied to a number of engineering design problems (Deb, 1997a).

Fuzzy Recombination Operator

Voigt et al. (1995) suggested a fuzzy recombination (FR) operator, which is similar to the above SBX operator. In fact, the FR operator can be thought of as a special case of the SBX operator with $\eta_c = 1$. Like other recombination operators, this operator also has a tunable parameter d. First, the difference between two parent solutions is computed. Thereafter, a triangular probability distribution is used around each parent. This distribution has its apex located at the parent solution and the base is proportional to the difference in parents or $d(x_i^{(2,t)} - x_i^{(1,t)})$, as shown in Figure 63. Like the SBX operator, this operator also gives importance to the creation of solutions closer to the parents than away from them. The distribution can be changed with the parameter d. As d is large, solutions away from the parents can get created. However, if d is small, only near-parent solutions are allowed. Like the BLX and SBX operators, this operator also creates distant solutions in proportion to the spread of the parent solutions.

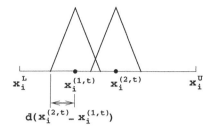

Figure 63   The FR operator. Parents are marked by filled circles.

Unimodal Normally Distributed Crossover

Ono and Kobayashi (1997) suggested a unimodal normally distributed crossover (UNDX) operator, where three or more parent solutions are used to create two or more offspring. Offspring are created from an ellipsoidal probability distribution with one axis formed along the line joining two of the three parent solutions and the extent of the orthogonal direction is decided by the perpendicular distance of the third parent from the axis (Figure 64). Unlike the BLX operator, this operator assigns more probability for creating solutions near the center of the first two parents than near the parents themselves. Recently, a multi-parental UNDX operator with more than three parents has also been suggested (Kita et al., 1998). Like the BLX operator, this operator also assigns offspring proportional to the spread of the parent solutions.

Simplex Crossover

In the simplex crossover (Tsutsui et al., 1999), more than two parents participate in the crossover operation. As many as $(n+1)$ parents (where n is the number of decision variables) are used as a simplex (representing a region enclosed by the points). The centroid of these parents are first calculated. Thereafter, the simplex is enlarged by extending the apex of the simplex at points away from the centroid. Each apex is

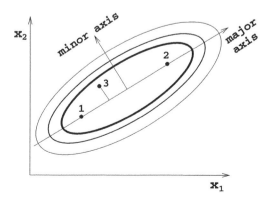

Figure 64   The UNDX operator.

placed on a line joining the centroid and each parent. Thereafter, with a uniform probability distribution, a number of solutions (H = 200 is suggested) are created from this extended simplex. Figure 65 illustrates this procedure. Investigators have suggested replacing only two of the parents by two solutions from a set of H new solutions and the two parent solutions. The first offspring is the best solution of H randomly created solutions and the other is chosen by using a rank-based roulette-wheel selection from the set H, excluding the already chosen first offspring. Since, at least n solutions are needed to *span* an n-dimensional search space, the use of $(n+1)$ solutions in the simplex seems like a good idea.

On a number of different correlated problems, investigators have shown that their GA with the simplex crossover can find increased precision in the obtained solution much faster than GAs with the UNDX operator. Moreover, they argue that this crossover operator is invariant in terms of linear coordinate transformation. Since the choice of two offspring from H new solutions is different from the usual implementation of creating two offspring from two parent solutions, a straightforward comparison of this crossover operator with the UNDX operator or any other real-parameter crossover operator (such as SBX or BLX) is difficult.

Fuzzy Connectives Based Crossover

This operator (Herrara et al., 1995) is also applied variable by variable. For each variable x_i, three regions are marked: region I $(x_i^{(L)}, x_i^{(1,t)})$, region II $(x_i^{(1,t)}, x_i^{(2,t)})$ and region III $(x_i^{(2,t)}, x_i^{(U)})$ (here, we assume that $x_i^{(1,t)} < x_i^{(2,t)}$). The fourth region is an overlapping region $(y_i^{(1)}, y_i^{(2)})$ with all of the above three regions. Here, $y_i^{(1)} \leq x_i^{(1,t)}$ and $y_i^{(2)} \geq x_i^{(2,t)}$. All of these four regions are shown in Figure 66. Once these regions are identified, one solution is chosen from each of them by using user-defined fuzzy

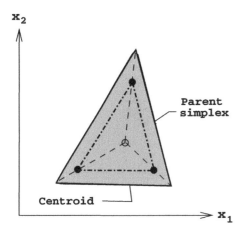

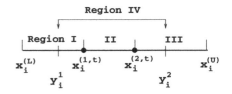

Figure 65   The SPX operator. The parents are marked by filled circles.

Figure 66   Four regions for the FCB crossover operator. The parents are marked by filled circles.

connective functions, which are defined over two normalized parents:

$$s = \frac{x_i^{(1,t)} - x_i^{(L)}}{x_i^{(2,t)} - x_i^{(1,t)}}, \quad s' = \frac{x_i^{(2,t)} - x_i^{(L)}}{x_i^{(2,t)} - x_i^{(1,t)}}.$$

Four different connectives (T, G, P, and $\hat{C}$) are used in different regions:

$$\text{Region I: } x_i^{(1,t+1)} = x_i^{(L)} + (x_i^{(U)} - x_i^{(L)})T(s,s'), \tag{4.15}$$

$$\text{Region II: } x_i^{(2,t+1)} = x_i^{(L)} + (x_i^{(U)} - x_i^{(L)})G(s,s'), \tag{4.16}$$

$$\text{Region III: } x_i^{(3,t+1)} = x_i^{(L)} + (x_i^{(U)} - x_i^{(L)})P(s,s'), \tag{4.17}$$

$$\text{Region IV: } x_i^{(4,t+1)} = x_i^{(L)} + (x_i^{(U)} - x_i^{(L)})\hat{C}(s,s'). \tag{4.18}$$

Investigators have suggested different forms for these connectives. By applying this approach to a number of test problems, they suggested that the following connectives give the best performance:

$$T(s,s') = \min(s,s'), \quad G(s,s') = \max(s,s'),$$
$$P(s,s') = (1-\lambda)s + \lambda s' \quad \hat{C}(s,s') = T^{1-\lambda}G^{\lambda}.$$

A list of other fuzzy connectives can be found from the original study (Herrera et al., 1998). Of the four solutions, two of these directly replace the two parent solutions. The other two solutions replace two random solutions taken from the rest of the population members waiting to undergo a crossover. It is clear that the performance of this crossover depends on the chosen fuzzy connective functions. Compared to other crossover operators, it is interesting to ponder whether such complicated arithmetic operators are needed to make the crossover operator efficient.

Unfair Average Crossover

In the unfair average crossover (Nomura and Miyoshi, 1996), two parents are used and two offspring are created. However, unlike the SBX or BLX operators, this crossover operator does not preserve the mean of the parent solutions, but instead will create offspring towards one of the parent solutions. Unless this favored parent solution is the better of the two parent solutions, introduction of such a bias is hard to justify. Specifically, the following equations can be used to create offspring with $\alpha \in [0, 0.5]$:

$$x_i^{(1,t+1)} = \begin{cases} (1+\alpha)x_i^{(1,t)} - \alpha x_i^{(2,t)}, & \text{for } i = 1, \ldots, j, \\ -\alpha x_i^{(1,t)} + (1+\alpha)x_i^{(2,t)}, & \text{for } i = j+1, \ldots, n. \end{cases} \tag{4.19}$$

$$x_i^{(2,t+1)} = \begin{cases} (1-\alpha)x_i^{(1,t)} + \alpha x_i^{(2,t)}, & \text{for } i = 1, \ldots, j, \\ \alpha x_i^{(1,t)} + (1-\alpha)x_i^{(2,t)}, & \text{for } i = j+1, \ldots, n. \end{cases} \tag{4.20}$$

The parameter j is a randomly chosen integer between 1 and n, indicating the cross site. For two variables ($n = 2$) and with a cross site at $j = 2$, the bias in creating offspring is clearly shown in Figure 67. The shaded region marks the possible location of the offspring.

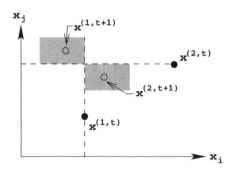

Figure 67 The unfair crossover operator illustrated on a two-dimensional search space. The parents are marked by filled circles.

Similarity in Different Crossovers

The plethora of different real-parameter crossover operators discussed above clearly shows the importance and the need for real-parameter GAs. However, a simple thought also comes to mind: 'Are all these crossover operators different from each other?' There exist a number of theoretical studies involved with trying to understand the properties of different crossover operators (Nomura, 1997; Qi and Palmieri, 1993). A recent study (Beyer and Deb, 2000) has found similarity in all of these operators.

The issue of 'this crossover operator' versus 'that crossover operator' is really context-dependent. The crossover cannot be viewed as an independent operator. The issue of exploitation and exploration makes a crossover operator dependent on the chosen selection operator for a successful GA run. Viewed in this way, Beyer and Deb argued that in most situations the selection operator reduces the population diversity, because selection eliminates some population members by making duplicates of a few others. Figure 68 shows a population, one instance of the effect of a selection operator on this population, and one instance of the effect of a recombination operator on the mating pool. The extent of reduction in diversity due to the selection operator can be related to the exploitation power of the selection operator. Thereafter, when a crossover operator is acted on the population found after the selection operator, it should enhance the population diversity in general. Such a balance between the reduction and enhancement of population diversity will allow a GA to have an adequate search property. It is important to highlight that a balance between exploitation and exploration is also achieved from the characteristic time of each operator for binary GAs (refer to page 104). Based on these arguments, two postulations have been recently suggested (Beyer and Deb, 2000). Under a crossover operator:

1. the population mean should not change;
2. the population diversity should increase, in general.

Although not intuitive, the first postulation is desired. Since most crossover operators

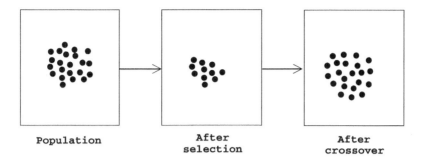

Population   After   After
                   selection   crossover

Figure 68   When selection reduces the diversity, the recombination operator must increase the diversity.

do not use any fitness information, there is no reason for a crossover operator to steer the search in any particular direction. The best that a crossover operator can do is to keep the mean of the offspring population the same as that of the parent population and alter the population variance. The second postulation also makes sense. Since a selection operator reduces the population variance in general, if a crossover operator also reduces the population variance, the combined action may lead to a premature convergence. To counterbalance the reduction in population variance due to the selection operator, the crossover operator should increase the population variance in general. Beyer and Deb (2000) gave arguments for an increase in population diversity for several types of objective functions. In the light of the above two postulates, it will suffice to characterize a crossover operator with two factors: the shift in the population mean and the growth of population variance with generations. The study calculated the population mean and variance under three real-parameter crossover operators in terms of their characteristic parameters. By equating the variance growth of the SBX, BLX and FR operators, the investigators obtained four sets of characteristic parameter values (shown in Table 6) producing identical variance growth in a flat fitness landscape. This means that on a flat fitness landscape, a GA with SBX-2, BLX-0.662, and FR-1.095 will have similar performances. Although in other fitness

Table 6   Parameter values for an identical population variance growth of three crossover operators on a flat fitness landscape.

SBX	BLX	FR
η_c	α	d
2	0.662	1.095
3	0.500	0.707
5	0.419	0.433
10	0.381	0.226

functions, these parameters values may not produce exactly the same performance, the above argument suggests that there exist characteristic parameter values which will make these crossover operators equivalent. In other words, all of the mean-preserving crossover operators can be expected to show similar behavior for certain characteristic parameter settings of each operator. However, the generic FCB crossover operator and the unfair average crossover operator discussed above do not preserve the population mean. There may be a difference in performance of GAs with these crossover operators with GAs with mean-preserving crossover operators. It is mentioned earlier that the motivation for using a crossover operator which does not preserve the mean of the parent population in the created offspring population is weak.

Since real-parameter crossover operators directly manipulate two or more real numbers to create one or more real numbers as offspring, one may wonder whether there is a special need for using another mutation operator. The confusion arises because both operators seem to be doing the same task, i.e. perturb every solution in the parent population to create a new population. The distinction among operators lies in the extent of perturbation allowed in each operator. Although debatable, this author believes the distinction between a crossover and a mutation operator lies mainly in the number of parent solutions used in the perturbation process. If only one parent solution is used, it should be called a mutation operation. With only one parent, a range of perturbations must be predefined. However, if more than one parent solution is used in the perturbation process, the range of perturbation may be adaptive and can be determined from the diversity of the parents on the fly. If the diversity in participating parents is large, the crossover can create offspring which are also diverse with respect to each other. Thus, with a crossover operator, it is possible to achieve adaptively large or small perturbations without any predefined setting of the range of perturbation. As in the binary-coded GA a mutation is meant to have a local perturbation, in real-parameter GAs a local perturbation in a predefined manner can also be useful in maintaining diversity in a population. In the following, we will mention some of the most commonly used real-parameter mutation operators.

Random Mutation

The simplest mutation scheme would be to create a solution randomly from the entire search space (Michalewicz, 1992):

$$y_i^{(1,t+1)} = r_i(x_i^{(U)} - x_i^{(L)}), \tag{4.21}$$

where r_i is a random number in $[0, 1]$. Figure 69 shows this probability distribution with a continuous line. This operator is independent of the parent solution and is equivalent to a random initialization. Instead of creating a solution from the entire search space, a solution in the vicinity of the parent solution with a uniform probability distribution (shown with a dashed line in Figure 69) can also be chosen:

$$y_i^{(1,t+1)} = x_i^{(1,t)} + (r_i - 0.5)\Delta_i, \tag{4.22}$$

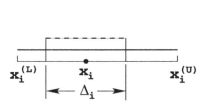

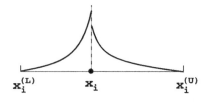

Figure 70 The non-uniform mutation
operator.

Figure 69 The random mutation operator.

where Δ_i is the user-defined maximum perturbation allowed in the i-th decision variable. Care should be taken to check if the above calculation takes $y_i^{(1,t+1)}$ outside of the specified lower and upper limits. Instead of using a uniform probability distribution, non-uniform probability distributions have also been suggested. In the following subsections, we will present a few such mutation operators.

Non-Uniform Mutation

Here, the probability of creating a solution closer to the parent is more than the probability of creating one away from it (Figure 70). However, as the generations (t) proceed, this probability of creating solutions closer to the parent gets higher and higher (Michalewicz, 1992):

$$y_i^{(1,t+1)} = x_i^{(1,t+1)} + \tau(x_i^{(U)} - x_i^{(L)})) \left(1 - r_i^{(1-t/t_{max})^b}\right). \qquad (4.23)$$

Here, τ takes a boolean value, -1 or 1, each with a probability of 0.5. The parameter t_{max} is the maximum number of allowed generations, while b is a user-defined parameter. In this way, from early on the above mutation scheme acts like a uniform distribution, while in later generations it acts like Dirac's function, thus allowing a focussed search.

Normally Distributed Mutation

A simple and popular method is to use a zero-mean Gaussian probability distribution:

$$y_i^{(1,t+1)} = x_i^{(1,t+1)} + N(0, \sigma_i). \qquad (4.24)$$

Here, the parameter σ_i is a fixed, user-defined parameter. This parameter is important and must be correctly set in a problem. Such a parameter can also be adaptively changed in every generation by some predefined rule. This operator is identical to the mutation operator used in evolution strategy (ES). However, the latter also has a self-adaptive mutation operator, where the mutation strength is evolved with the decision variables (Fogel et al., 1995; Hansen and Ostermeier, 1996; Herdy, 1992; Sarvanan et al., 1995; Schwefel, 1987b). For handling bounded decision variables, special care must be taken to make sure that no solution is created outside of the specified lower and upper limits.

Polynomial Mutation

Like in the SBX operator, the probability distribution can also be a polynomial function, instead of a normal distribution (Deb and Goyal, 1996):

$$y_i^{(1,t+1)} = x_i^{(1,t+1)} + (x_i^{(U)} - x_i^{(L)})\bar{\delta}_i, \tag{4.25}$$

where the parameter $\bar{\delta}_i$ is calculated from the polynomial probability distribution $\mathcal{P}(\delta) = 0.5(\eta_m + 1)(1 - |\delta|)^{\eta_m}$:

$$\bar{\delta}_i = \begin{cases} (2r_i)^{1/(\eta_m+1)} - 1, & \text{if } r_i < 0.5, \\ 1 - [2(1 - r_i)]^{1/(\eta_m+1)}, & \text{if } r_i \geq 0.5. \end{cases} \tag{4.26}$$

This distribution is similar to the non-uniform mutation operator, although a fixed value of the parameter η_m is suggested here. It is observed that a value of η_m produces a perturbation of the order $O(1/\eta_m)$ in the normalized decision variable. For handling bounded decision variables (Deb and Agrawal, 1999a), the mutation operator is modified for two regions, i.e. $[x_i^{(L)}, x_i]$ and $[x_i, x_i^{(U)}]$, very similar to the non-uniform mutation described above. The difference is that in this mutation operator the shape of the probability distribution is directly controlled by an external parameter η_m and the distribution is not dynamically changed with generations.

Theory of Real-Parameter Genetic Algorithms

Although real-parameter GAs deal with real parameter values and bring the GA technique a step closer to classical optimization algorithms, there does not exist many theoretical studies on real-parameter GAs. In the following paragraphs, we will discuss a couple of salient studies.

Goldberg (1991) argued that in a real-parameter GA the action of a selection operator causes the above-average solutions to survive in the population. In each decision variable, these sub-regions of above-average solutions constitute *virtual characters* of a virtual alphabet. For example, Figure 71 shows that the first and second decision variables have virtual alphabets with a cardinality of three and two, respectively. Thus, after a few generations, a real-parameter GA treats a continuous search space problem as a discrete search space problem and the search is then similar to a binary-coded GA with low cardinal alphabets. In such an event, the search gets restricted to certain islands in the search space. Although mutation can allow some diffusion, the mutation strength must be large enough to create solutions outside of the width of the virtual characters. Based on this virtual alphabet theory, Goldberg defined a number of *blocking* functions, where the global optimum lies in a region not marked by the virtual characters. Since a real-parameter GA will soon converge to intersections of virtual characters of different decision variables, the global optimum will be difficult to find.

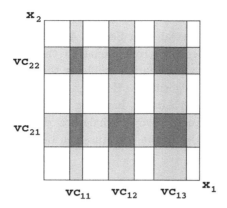

Figure 71 Virtual characters are shown in a two-variable space. The dark-shaded regions are intersections of virtual hyperplanes.

Qi and Palmieri (1993) made an extensive study on different combinations of real-parameter GA operators by using a discrete-time stochastic process. The use of a large population assumption allowed the investigators to find the probability density functions $\pi_t(\cdot)$, characterizing the distribution of the population at any generation t. Thereafter, by letting the population size N go to infinity, the limiting behavior of the GA operator interactions was found. The salient features of their study are as follows:

1. For a GA with selection and mutation operators, the selection operator tends to reduce the variance of $\pi_t(\cdot)$, while mutation spreads the distribution. However, the combined effect is a shift towards the optimum in a unimodal problem. With a large number of generations, the distribution $\pi_t(\cdot)$ approaches a Dirac function, $\delta(\mathbf{x} - \mathbf{x}^*)^3$.
2. The second observation is that a GA with a selection operator alone monotonically increases the density of solutions in the regions with above-average fitness.
3. The third observation is that a repetitive application of a crossover operator alone decorrelates the decision variables, leading to:

$$\lim_{t \to \infty} \pi_t(x_1, x_2, \ldots, x_n) = \Pi_{i=1}^n \pi_0(x_i), \tag{4.27}$$

where $\pi_0(x_i)$ is the marginal density function of x_i.
4. The fourth observation is that a combined effect of selection and crossover allows hyperplanes of higher average fitness value to be sampled more often. The consequence of this is that intersection of these hyperplanes allows better solutions to be sampled more often. This can be considered as the real-coded version of the building-block hypothesis. A hyperplane representing a larger region is equivalent to a low-order schema in binary-coded GAs. Intersection of hyperplanes (low-order schemata) signifies smaller regions (higher-order schemata) in the context of real-parameter GAs.

4.2.3   Constraint-Handling in Genetic Algorithms

In the example problem illustrated in Section 4.2.1, we have handled the volume constraint by adding a penalty proportional to the constraint violation to the objective function value. In minimization problems, this is a common way of handling constraints in evolutionary algorithms. However, in this section we will describe a number of other efficient methods for handling constraints. As outlined elsewhere (Michalewicz and Schoenauer, 1996; Michalewicz et al., 2000), most constraint handling methods which exist in the literature can be classified into five categories, as follows:

1. Methods based on preserving feasibility of solutions.
2. Methods based on penalty functions.
3. Methods biasing feasible over infeasible solutions.
4. Methods based on decoders.
5. Hybrid methods.

We will describe these methods briefly in the following subsections.

Methods Based on Preserving Feasibility of Solutions

In this approach, two feasible solutions, after recombination and mutation operations, will create two feasible offspring. This can be achieved in a number of different ways.

First, one decision variable can be eliminated by using an equality constraint either implicitly or explicitly. In this way, the equality constraint is always satisfied. For example, for the following equality constraint:

$$h(\mathbf{x}) \equiv 2x_1 - x_2^2 x_3 = 0,$$

the variable x_1 can be eliminated by the following equation: $x_1 = 0.5x_2^2 x_3$ in the objective function and in all other constraints. In this way, an EA has to use one decision variable less. This also allows the above equality constraint to be satisfied automatically in all solutions used in the optimization process. Similarly, for K linear simultaneous equalities, K decision variables can be eliminated (only $(n - K)$ decision variables need to be used in an EA), thereby automatically satisfying all K linear equality constraints. In the implicit method of handling an equality constraint $h(\mathbf{x})$, a variable x_k can be calculated from the constraint by finding the root of:

$$h\left(x_k, (\mathbf{x} \backslash x_k)^{(i)}\right) = 0, \tag{4.28}$$

where $(\mathbf{x} \backslash x_k)^{(i)}$ is the $(n-1)$ real-parameter vector (without x_k) corresponding to the i-th solution. Here, an EA uses all n decision variables, but in finding the root of the above equation, the value of x_k is not taken from the individual vector; instead, the estimated root of the above equation is used as a value of x_k. There are two ways to handle this new value of x_k. In the Lamarckian method, the new value is substituted in the i-th individual vector, while in the Baldwinian method, the new value is not substituted. However, both methods use the new value to compute objective function and other constraints. Although the Lamarckian method of repairing an infeasible solution seems desirable (Deb and Goel, 2001b), the Baldwinian method is found to be more efficient in some problems (Liu et al., 2000).

Since in most interesting optimization problems, the optimal solution lies on the intersection of constraint boundaries, operators that search only in the boundary area between feasible and infeasible region are found to be efficient in certain problems (Michalewicz and Schoenauer, 1996).

Methods Based on Penalty Functions

In most applications, an exterior penalty term (Deb, 1995; Reklaitis et al., 1983), which penalizes infeasible solutions, is used. Based on the constraint violation $g_j(x)$ or $h_k(x)$, a bracket-operator penalty term is added to the objective function and a penalized function is formed:

$$F(x) = f(x) + \sum_{j=1}^{J} R_j \langle g_j(x) \rangle + \sum_{k=1}^{K} r_k |h_k(x)|, \qquad (4.29)$$

where R_j and r_k are user-defined penalty parameters. The bracket-operator $\langle\ \rangle$ denotes the absolute value of the operand, if the operand is negative. Otherwise, if the operand is non-negative, it returns a value of zero. Since different constraints may take different orders of magnitude, it is essential to normalize all constraints before using the above equation. A constraint $\underline{g}_j(x) \geq b_j$ can be normalized by using the following transformation:

$$g_j(x) \equiv \underline{g}_j(x)/b_j - 1 \geq 0.$$

Equality constraints can also be normalized similarly. Normalizing constraints in the above manner has an additional advantage. Since all normalized constraint violations take more or less the same order of magnitude, they all can be simply added as the overall constraint violation and thus only one penalty parameter R will be needed to make the overall constraint violation of the same order as the objective function:

$$F(x) = f(x) + R \left[\sum_{j=1}^{J} \langle g_j(x) \rangle + \sum_{k=1}^{K} |h_k(x)| \right]. \qquad (4.30)$$

When an EA uses a fixed value of R in the entire run, the method is called the *static penalty method*. There are two difficulties associated with this static penalty function approach:

1. The optimal solution of $F(x)$ depends on penalty parameters R. Users usually have to try different values of R to find which value would steer the search towards the feasible region. This requires extensive experimentation to find any reasonable solution. This problem is so severe that some researchers have used different values of R depending on the level of constraint violation (Homaifar et al., 1994), while

some have used a sophisticated temperature-based evolution of penalty parameters through generations (Michalewicz and Attia, 1994) involving a few parameters describing the rate of evolution. We will discuss these *dynamically* changing penalty methods a little later.

2. The inclusion of the penalty term *distorts* the objective function (Deb, 1995). For small values of R, the distortion is small, but the optimum of F(x) may not be near the true constrained optimum. On the other hand, if a large R is used, the optimum of F(x) is closer to the true constrained optimum, but the distortion may be so severe that F(x) may have artificial locally optimal solutions. This primarily happens due to interactions among multiple constraints. EAs are not free from the distortion effect caused due to the addition of the penalty term in the objective function. However, EAs are comparatively less sensitive to distorted function landscapes due to the stochasticity in their operators.

In order to investigate the effect of the penalty parameter R on the performance of EAs, we consider a well-studied welded beam design problem (Reklaitis et al., 1983). Figure 72 shows the configuration of the welded-beam problem. The resulting optimization problem has four design variables, $x = (h, \ell, t, b)$, and five inequality constraints, involving different stress, buckling and logical limitations:

$$
\begin{aligned}
\text{Minimize} \quad & f_w(x) = 1.104\,71 h^2 \ell + 0.048\,11 tb(14.0 + \ell), \\
\text{Subject to} \quad & g_1(x) \equiv 13\,600 - \tau(x) \geq 0, \\
& g_2(x) \equiv 30\,000 - \sigma(x) \geq 0, \\
& g_3(x) \equiv b - h \geq 0, \\
& g_4(x) \equiv P_c(x) - 6000 \geq 0, \\
& g_5(x) \equiv 0.25 - \delta(x) \geq 0, \\
& 0.125 \leq h \leq 10, \\
& 0.1 \leq \ell, t, b \leq 10.
\end{aligned}
\right\} \quad (4.31)
$$

The terms $\tau(x)$, $\sigma(x)$, $P_c(x)$ and $\delta(x)$ are given below:

$$
\begin{aligned}
\tau(x) &= \sqrt{(\tau'(x))^2 + (\tau''(x))^2 + \ell\tau'(x)\tau''(x)/\sqrt{0.25[\ell^2 + (h+t)^2]}}, \\
\sigma(x) &= \frac{504\,000}{t^2 b},
\end{aligned}
$$

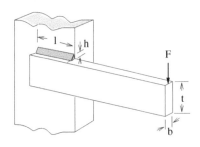

Figure 72   The welded beam design problem.

$$P_c(\mathbf{x}) = 64\,746.022(1 - 0.028\,234\,6t)tb^3,$$

$$\delta(\mathbf{x}) = \frac{2.1952}{t^3 b},$$

where:

$$\tau'(\mathbf{x}) = \frac{6000}{\sqrt{2}h\ell},$$

$$\tau''(\mathbf{x}) = \frac{6000(14 + 0.5\ell)\sqrt{0.25[\ell^2 + (h + t)^2]}}{2\{0.707h\ell[\ell^2/12 + 0.25(h + t)^2]\}}.$$

The optimized solution reported in the literature (Reklaitis et al., 1983) is $h^* = 0.2444$, $\ell^* = 6.2187$, $t^* = 8.2915$, and $b^* = 0.2444$, with a function value equal to $f^* = 2.38116$. Binary GAs are applied to this problem in an earlier study (Deb, 1991) and the solution $\mathbf{x} = (0.2489, 6.1730, 8.1789, 0.2533)$ with $f = 2.43$ (within 2% of the above best solution) was obtained with a population size of 100. However, it was observed that the performance of the GAs largely depended on the chosen penalty parameter values.

In order to obtain further insights into the working of the GAs, we apply binary GAs with the tournament selection and a single-point crossover operator with $p_c = 0.9$ on this problem. Each variable is coded in 10 bits, so that the total string length is 40. A population size of 80 is used and GAs with 50 different initial populations are run until 500 generations have been achieved. All constraints are normalized and a single penalty parameter R is used. Table 7 shows the performance of such binary GAs for different penalty parameter values. For each case, the best, median (the optimized objective function values of 50 runs are arranged in ascending order and the 25th value in the list is called the median optimized function value) and worst values of 50 optimized objective function values are also shown in the table. With $R = 1$, although three out of 50 runs have found a solution within 10% of the best-known solution, 13 GA runs have not been able to find a single feasible solution in 40 080 function evaluations. This happens because with a small R there is not much pressure for the solutions to become feasible. With a large value of R, the pressure

Table 7 The number of runs (out of 50) converged within ϵ% of the best-known solution are shown when using binary GAs with different penalty parameter values on the welded beam design problem.

R	ϵ						Infea-sible	Optimized $f_w(\mathbf{x})$		
	$\leq$ 1%	$\leq$ 2%	$\leq$ 10%	$\leq$ 20%	$\leq$ 50%	$>$ 50%		Best	Median	Worst
10^0	0	1	3	3	12	25	13	2.41	7.62	483.50
10^1	0	0	0	0	12	38	0	3.14	4.33	7.45
10^3	0	0	0	0	1	49	0	3.38	5.97	10.66
10^6	0	0	0	0	0	50	0	3.73	5.88	9.42

for solutions to become feasible is more and all 50 runs then found feasible solutions. However, because of the larger emphasis assigned for solutions to become feasible, a feasible solution has a large selective advantage over an infeasible solution. If new and different feasible solutions are not created, GAs would overemphasize this sole feasible solution and soon prematurely converge near this solution. This is exactly what has happened in GA runs with larger R values, where the best solution obtained, in most cases, is more than 50% away (in terms of function values) from the true constrained optimum. Similar experiences have been reported by other researchers in applying GAs with an exterior penalty function approach to other constrained optimization problems. Thus, if the penalty function method is to be used, the user usually have to take many runs or 'adjust' the penalty parameters to get a solution within an acceptable limit.

There exist a number of *dynamic* penalty methods where the penalty parameter is changed with the generation counter t (Joines and Houck, 1994):

$$F(\mathbf{x}) = f(\mathbf{x}) + (C \cdot t)^\alpha \left(\sum_{j=1}^{J} \langle g_j(\mathbf{x}) \rangle^\beta + \sum_{k=1}^{K} |h_k(\mathbf{x})|^\beta \right), \tag{4.32}$$

where C, α and β are user-defined constants. A value of $\beta = 2$ is suggested. The penalty parameter R can be updated from the last k generations of population statistics and updated at every generation. However, both static and dynamic penalty methods require exogenous parameters, which must be tuned for every new problem.

Methods Based on Feasible Over Infeasible Solutions

In these methods, a clear distinction between feasible and infeasible solutions in the search space is made. In the *behavioral memory method* (Schoenauer and Xanthakis, 1993), all constraints are considered in a sequence. When sufficient feasible solutions satisfying the constraints taken from the top of the sequence are found, the next constraint from the sequence is considered. In this way, each constraint gets its turn to divert the population members in the region of its feasibility.

In another approach (Powell and Skolnick, 1993), an additional generation-dependent term is added to the static penalty term as follows:

$$F(\mathbf{x}) = f(\mathbf{x}) + R \left(\sum_{j=1}^{J} \langle g_j(\mathbf{x}) \rangle + \sum_{k=1}^{K} |h_k(\mathbf{x})| \right) + \lambda(t, \mathbf{x}). \tag{4.33}$$

At every generation, the term $\lambda(t, \mathbf{x})$ is the difference between the best static penalized function value among all infeasible solutions and the worst feasible solution. This would make the best infeasible solution in the population to have the same F() value as that of the worst feasible solution in the population. For a hypothetical single-variable function f(x) having one constraint $g(x) \geq 0$, the left plot of Figure 73 shows the construction of F(x). This figure illustrates the procedure with six population

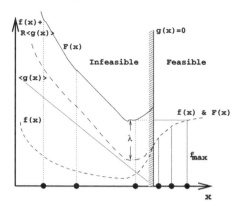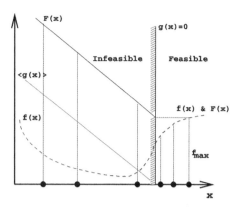

Figure 73   The left figure shows Powell and Skolnick's constraint handling method, while the right figure shows Deb's constraint handling method. Reprinted from Computer Methods in Applied Mechanics and Engineering, Volume 186, K. Deb, 'An efficient constraint handling method for genetic algorithms', Pages 311–338, Copyright 2000, with permission from Elsevier Science.

members, of which three are feasible. Notice that for all feasible solutions, $F(x) = f(x)$. As discussed in an another study (Deb, 2000), this method has at least two difficulties:

1. The need to fix a penalty parameter R still exists.
2. The fitness function $F(\mathbf{x})$, constructed by using the above approach, may have an artificial local optimum.

A recent study (Deb, 2000) suggested a modification, which does not need any penalty parameter:

$$F(\mathbf{x}) = \begin{cases} f(\mathbf{x}), & \text{if } \mathbf{x} \text{ is feasible;} \\ f_{max} + \sum_{j=1}^{J} \langle g_j(\mathbf{x}) \rangle + \sum_{k=1}^{K} |h_k(\mathbf{x})|, & \text{otherwise.} \end{cases} \qquad (4.34)$$

Here, f_{max} is the objective function value of the worst feasible solution in the population. The right figure of Figure 73 shows the construction procedure of $F(x)$ from $f(x)$ and $g(x)$ for a single-variable objective function. One fundamental difference between this approach and the previous approach is that the objective function value is *not* computed for any infeasible solution. Since all feasible solutions have zero constraint violation and all infeasible solutions are evaluated according to their constraint violations only, both the objective function value and constraint violation are not combined in any solution in the population. Thus, there is no need to have any penalty parameter R for this approach. Moreover, the approach is also quite pragmatic. Since infeasible solutions are not to be recommended for use, there is no real reason for one to find the objective function value for an infeasible solution. The method uses a binary tournament selection operator, where two solutions are compared at a time, and the following scenarios are always assured (Goldberg, 1992):

1. Any feasible solution is preferred to any infeasible solution.

2. Among two feasible solutions, the one having a better objective function value is preferred.

3. Among two infeasible solutions, the one having a smaller constraint violation is preferred.

To steer the search towards the feasible region and then towards the optimal solution, the method recommends the use of a niched tournament selection (where two solutions are compared in a tournament only if their Euclidean distance is within a pre-specified limit). This ensures that even if a few isolated solutions are found in the feasible space, they will be propagated from one generation to another for maintaining diversity among the feasible solutions.

Using this penalty parameter-less constraint handling strategy, real-parameter GAs (with an SBX-2 operator), using a niched tournament selection and a population size of 80, has found 50 out of 50 runs within a maximum of 0.1% of the optimal objective function value in the welded beam design problem (Deb, 2000). This means that with the proposed approach, one run is enough to find a satisfactory solution close to the true optimal solution. The best, median and mean objective function values of 50 runs are 2.381 45, 2.382 63, and 2.383 55, respectively.

Methods Based on Decoders

In these approaches, a chromosome stores information about how to fix an infeasible solution. For example, the chromosome may keep information about an ordering of decision variables which may be altered to make a solution feasible. In some instances, a decoder imposes a mapping between a feasible solution and a decoded solution. One such example is the use of a homomorphous mapping between an n-dimensional cube and a feasible search space (Koziel and Michalewicz, 1998). The mapping allows every solution in the cube to be mapped into a feasible solution in the search space. Moreover, the entire feasible region is mapped in the cube. In this way, a search in the cube will result in a search in the feasible region only. In such methods, care must be taken in assuring that the computational complexity of the mapping is not too high. Moreover, the continuity in the decoded space (in the cube) must preserve the continuity of actual solutions in the feasible search region.

Hybrid Methods

Since classical constrained optimization methods have been studied and applied for many years, it may be wise to combine such a method with the existing operators of an evolutionary algorithm. In one application, the Hooke–Jeeves pattern search method is combined with evolutionary algorithms (Waagen et al., 1992). In the coevolutionary approach suggested by Paredis (1994), where a population of decision variable vectors is evolved along with a population of constraints, fitter solutions satisfy more constraints and fitter constraints are violated by fewer solutions. A different twist to this coevolutionary approach, suggested elsewhere (Barbosa, 1996), uses a population of solutions and a population of Lagrange multipliers. Since the constrained optimum solutions are *saddle* points of the corresponding Lagrangian function, the coevolutionary approach helps to find the optimum solution along with optimum Lagrange multipliers. A knowledge of the optimal Lagrange multiplier is extremely helpful in performing post-optimal sensitivity analysis.

4.3   Evolution Strategies

More or less contemporary to the development of GAs, research on a field very similar to such algorithms was in progress in Germany. As a result of this work, P. Bienert, I. Rechenberg and H.-P. Schwefel of the Technical University of Berlin suggested the so-called *evolution strategy* (ES) during the early Sixties. The first applications of ES were experimental and attempted to solve the shape optimization of a bended pipe (Lichtfuss, 1965), the drag minimization of a jointed plate (Rechenberg, 1965), and the shape optimization of a flashing nozzle (Schwefel, 1968), along with various other problems. Since the evaluation of a solution in each of these problems was difficult and time-consuming, a simple two-membered ES was used in all of the early studies. However, Schwefel was the first to simulate a different version of the ES on a computer in 1965. Thereafter, multi-membered ESs, recombinative ESs, and self-adaptive ESs were all suggested. However, the early ES procedure is fundamentally different from binary GAs in mainly two ways: (i) ESs use real parameter values and (ii) early ESs do not use any crossover-like operator. However, an ES's working principle is similar to that of a real-parameter GA used with selection and mutation operators only.

Realizing the similarity between these two surprisingly similar (although practiced in two geographically distant places) procedures, recent ES studies have introduced crossover-like operators. In the following subsections, we will describe these approaches and discuss an important issue which is mostly studied in the context of evolution strategies.

4.3.1   Non-Recombinative Evolution Strategies

In this section, we describe a number of different ESs where no recombination operator is used.

Two-Membered ESs: (1+1)-ES

This is the simplest of all ESs. In each iteration, one parent is used to create one offspring by using a Gaussian mutation operator. The step-by-step procedure is shown below. (The procedure is given for solving minimization problems.)

Two-Membered ES: (1+1)-ES

> **Step 1** Choose an initial solution $\mathbf{x}$ and a mutation strength σ.
>
> **Step 2** Create a mutated solution:

$$\mathbf{y} = \mathbf{x} + \mathbf{N}(0, \sigma), \tag{4.35}$$

where $N(0, \sigma)$ is a vector of instances created using a zero-mean normal distribution and having a standard deviation σ.

Step 3 If $f(\mathbf{y}) < f(\mathbf{x})$, replace $\mathbf{x}$ with $\mathbf{y}$.

Step 4 If a termination criterion is satisfied, stop. Otherwise, go to Step 2.

Here, all decision variables are mutated with a normal distribution having the same mutation strength (a term used for the 'standard deviation' in ES research). It is intuitive that the success of this algorithm in finding a solution close to the true optimum solution largely depends on the chosen σ value. For a few simple problems, the optimal σ values are calculated for obtaining the maximum rate of convergence towards the optimum solution. In order to find the minimum of a linear function $f(\mathbf{x}) = c_0 + c_1 x_1$ in the range $-b/2 \leq x_i \leq b/2$ for $i = 1, 2, \ldots, n$, Rechenberg (1973) calculated the optimal mutation strength:

$$\sigma^* = \sqrt{\frac{\pi}{2} \frac{b}{n}}. \tag{4.36}$$

For a sphere model $f(\mathbf{x}) = \sum_{i=1}^{n} x_i^2$, the optimal mutation strength is found as follows:

$$\sigma^* = 1.224 \frac{r}{n}, \tag{4.37}$$

where r denotes the current Euclidean distance from the true optimum solution. It is interesting to note that for this function the optimal mutation strength must change dynamically. The chosen mutation strength must be inversely proportional to the number of variables and as the solution comes closer and closer to the optimum, the mutation strength should reduce proportionately. This phenomenon is exploited in developing *self-adaptive* ESs, where in addition to the decision variables, mutation strength is also evolved in the ES. We will discuss self-adaptive ESs in Section 4.3.3 below.

Using these two calculations, the probabilities of a successful mutation occurring in both functions are found to be 0.184 and 0.270. Based on these numbers, Rechenberg postulated the famous '1/5 success rule', where the mutation strength σ is modified across generations (Rechenberg, 1973, p. 123):

> The ratio of successful mutations to all mutations should be 1/5. If this ratio is greater than 1/5, increase the mutation strength, while if it is less than 1/5, decrease the mutation strength.

A mutation is defined as successful if the mutated offspring is better than the parent solution. If many successful mutations are observed, this indicates that the solutions are residing in a better region in the search space. Hence, it is time to increase the mutation strength in the hope of finding even better solutions closer to the optimum solution. By counting the number of successful mutations p_s over n trials, Schwefel (1981) suggested a factor $c_d = 0.817$ in the following σ-update rule:

$$\sigma^{t+1} = \begin{cases} c_d \sigma^t, & \text{if } p_s < 1/5, \\ \frac{1}{c_d} \sigma^t, & \text{if } p_s > 1/5, \\ \sigma^t, & \text{if } p_s = 1/5. \end{cases} \tag{4.38}$$

In the presence of constraints, a mutated solution is considered better if it is feasible and is better than the parent in terms of the objective function value. Oyman, Deb, Beyer (1999) suggested a better constraint handling ES along the lines of constraint handling strategies used in genetic algorithms.

Multi-Membered ESs: $(\mu + \lambda)$-ES and (μ, λ)-ES

With the use of optimal mutation strength and the 1/5 success rule, the (1+1)-ES is better than the simple Monte Carlo method, where a solution in the neighborhood is created with an arbitrary user-defined probability distribution. However, the population approach used in multi-membered ESs makes them much better algorithms. There are two ways to introduce multiple members in an ES. We first discuss the 'plus' strategy, where μ (> 1) parent solutions are used to create λ offspring using mutation alone.

Multi-Membered ES: $(\mu + \lambda)$-ES

Step 1 Choose an initial population P of μ solutions $\mathbf{x}^{(i)}$, $i = 1, 2, \ldots, \mu$ and a mutation strength σ.

Step 2 Create λ mutated solutions:

$$\mathbf{y}^{(j)} = \mathbf{x}^{(i)} + \mathbf{N}(0, \sigma). \tag{4.39}$$

To create the j-th offspring, a parent i is chosen at random from μ parent solutions. $\mathbf{N}(0, \sigma)$ is defined as earlier.

Step 3 Modify the parent population P by combining all parent and offspring:

$$P = \left(\cup_{j=1}^{\lambda} \{\mathbf{y}^{(j)}\} \right) \cup \left(\cup_{i=1}^{\mu} \{\mathbf{x}^{(i)}\} \right).$$

From P, the best μ solutions are chosen and the rest are deleted.

Step 4 If a termination criterion is satisfied, stop. Otherwise, go to Step 2.

For many problems, $\lambda/\mu \approx 5$ is suggested. Figure 74 shows a schematic of this procedure. Since in the selection operation both parent and offspring populations are used, the $(\mu + \lambda)$-ES is an elitist algorithm. Although elite-preservation is desired, this procedure does not control the extent of elitism. In certain problems, where new successful mutations are difficult to achieve, the search in a $(\mu + \lambda)$-ES will stagnate. In order to avoid this problem, an alternate (μ, λ)-ES is suggested. The only difference between this and a $(\mu + \lambda)$-ES is in Step 3. Instead of combining the parent and offspring populations before selection, only the offspring population is now used. Thus, the modified population P before selection is $P = \cup_{j=1}^{\lambda} \{\mathbf{y}^{(j)}\}$. This requires that $\lambda \geq \mu$. In this way, the parent solutions are ignored and the success of the (μ, λ)-ES depends on the ability of the algorithm to find better offspring. Figure 75 shows this procedure. The (μ, λ)-ES is a non-elitist procedure. A strategy, where some limited copies of the better parent solutions are chosen and the rest are filled from

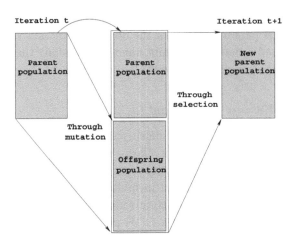

Figure 74 The procedure for a $(\mu + \lambda)$-ES.

the offspring population, would introduce elitism in a controlled manner and may turn out to be a good compromise between a $(\mu + \lambda)$-ES and a (μ, λ)-ES. One way to achieve this is to allow a life span, κ (≥ 1), for each parent (Schwefel and Bäck, 1998). A parent older than κ generations is not considered further in the selection process, thereby allowing new offspring solutions to enter into the new population. Using this notation, the two extreme strategies are (μ, λ)-ES for $\kappa = 1$ and $(\mu + \lambda)$-ES for $\kappa = \infty$.

4.3.2   Recombinative Evolution Strategies

In the recombinative ESs, a set of chosen parents are first recombined to find a new solution. Thereafter, the solution is mutated as before. Instead of choosing two parents or all parents for the recombination, a set of ρ (a value between 1 and μ inclusive) parents are chosen at random. A value of $\rho = 1$ means no recombination. Two types of recombination are mainly in use: (i) intermediate and (ii) discrete. In the intermediate

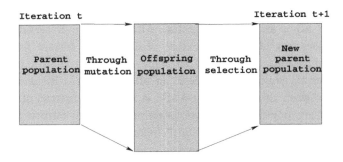

Figure 75 The procedure for a (μ, λ)-ES.

recombination operator, the average ρ chosen solution vectors are calculated as follows:

$$y = \frac{1}{\rho} \sum_{i=1}^{\rho} x^{(i)}. \tag{4.40}$$

It is clear that if all parents are used in each recombination, or $\rho = \mu$, the procedure has a tendency to cluster around the centroid of the current population. In problems, where a population brackets the true optimum, this may be a good strategy.

In the discrete recombination, each decision variable is chosen from one of the ρ parents at random. For example, a discretely recombined solution is shown derived from the following four solutions ($\rho = \mu$):

Parent 1:	$x_1^{(1)}$	$x_2^{(1)}$	$x_3^{(1)}$	$x_4^{(1)}$	$x_5^{(1)}$	$x_6^{(1)}$
Parent 2:	$x_1^{(2)}$	$x_2^{(2)}$	$x_3^{(2)}$	$x_4^{(2)}$	$x_5^{(2)}$	$x_6^{(2)}$
Parent 3:	$x_1^{(3)}$	$x_2^{(3)}$	$x_3^{(3)}$	$x_4^{(3)}$	$x_5^{(3)}$	$x_6^{(3)}$
Parent 4:	$x_1^{(4)}$	$x_2^{(4)}$	$x_3^{(4)}$	$x_4^{(4)}$	$x_5^{(4)}$	$x_6^{(4)}$
Recombinant:	$x_1^{(2)}$	$x_2^{(3)}$	$x_3^{(4)}$	$x_4^{(2)}$	$x_5^{(4)}$	$x_6^{(3)}$

This operator is similar to the uniform crossover procedure used in genetic algorithms. Different combinations of decision variables of the existing solutions can be obtained by this procedure. The step-by-step procedure of the entire algorithm is as follows:

Recombinant ES: $(\mu/\rho + \lambda)$-ES

Step 1 Choose an initial population P of μ solutions $x^{(i)}$, $i = 1, 2, \ldots, \mu$ and a mutation strength vector σ.

Step 2 Create λ mutated solutions, each using ρ parents randomly chosen from μ parents, as follows:

1. Calculate the recombinant solution y either by using an intermediate or a discrete recombination from ρ parents.
2. Mutate the recombinant solution:

$$y = y + N(0, \sigma). \tag{4.41}$$

Step 3 Modify the parent population P by combining all parent and offspring:

$$P = \left(\cup_{j=1}^{\lambda} \{y^{(j)}\} \right) \cup \left(\cup_{i=1}^{\mu} \{x^{(i)}\} \right).$$

From P the best μ solutions are chosen and the rest are deleted.

Step 4 If a termination criterion is satisfied, stop. Otherwise, go to Step 2.

For the $(\mu/\rho, \lambda)$-ES, only the offspring population would be used to create the new population in Step 3 of the above algorithm.

4.3.3 Self-Adaptive Evolution Strategies

Self-adaptation is a phenomenon which makes evolutionary algorithms flexible and closer to natural evolution. Among the evolutionary methods, self-adaptation properties have been explored with evolution strategies (ESs) (Bäck, 1992, 1997; Beyer, 1995a, 1995b; Hansen and Ostermeier, 1996; Rechenberg, 1973; Sarvanan et al., 1995; Schwefel, 1987a) and evolutionary programming (EP) (Fogel et al., 1995), although there exist some studies of self-adaptation in genetic algorithms (GAs) with the mutation operator (Bäck, 1992). When applied to function optimization, there are a number of reasons why evolutionary algorithmists should pay attention to self-adaptation:

1. Knowledge of lower and upper bounds for the optimal solution may not be known a priori.
2. It may be necessary to know the optimal solution with arbitrary precision.
3. The objective function and the optimal solution may change with time.

There are three different ways where self-adaptation is introduced into ESs:

1. A hierarchically organized population-based meta-ES (Herdy, 1992).
2. Adaptation of the covariance matrix (CMA) determining the probability distribution for mutation (Hansen and Ostermeier, 1996).
3. Explicit use of self-adaptive control parameters (Rechenberg, 1973; Schwefel, 1987a).

The meta-ES method of self-adaptation uses two levels of ESs: the top level optimizes the strategy parameters (such as mutation strengths), a solution of which is used to optimize the true objective function in the lower-level ES. The second method (CMA) records the population history for a certain number of iterations to calculate covariance and variance information among the object variables. The subsequent search effort is influenced by these variance values. Here, we discuss the third type of self-adaptive ES, where the strategy parameters (variances σ_i^2 and covariances c_{ij}^2) are explicitly coded along with the decision variables and updated by using predefined update rules in each generation. There are basically three different implementations which are in use.

Isotropic Self-Adaptation

In this self-adaptive ES, a single mutation strength σ is used for all variables. In addition to n object variables, the strategy parameter σ is also used in a population member. The logarithmic update rules for the decision variables and the mutation strength are as follows:

$$\sigma^{(t+1)} \;=\; \sigma^{(t)} \exp(\tau_0 N(0,1)), \tag{4.42}$$

$$x_i^{(t+1)} \;=\; x_i^{(t)} + \sigma^{(t+1)} N_i(0,1), \tag{4.43}$$

where $N(0,1)$ and $N_i(0,1)$ are realizations of a one-dimensional normally distributed random variable with a mean of zero and a standard deviation of one. The parameter

τ_0 is the learning parameter which should be set as $\tau_0 \propto n^{-1/2}$, where n is the dimension of the variable vector (Schwefel, 1987a). Beyer (1995b) has shown that, for the sphere model, the optimal learning parameter for a $(1, \lambda)$-ES is $\tau_0 = c_{1,\lambda}/\sqrt{n}$, where $c_{1,\lambda}$ is the progress coefficient. For multi-membered ESs, an optimal learning parameter value is not determined. However, most studies use $c_{\mu,\lambda}$ or $c_{\mu/\mu,\lambda}$ as the constant of proportionality in the corresponding ES. The above update rule for σ requires an initial value, which may be chosen as $\sigma^{(0)} = (x_i^{(U)} - x_i^{(L)})/\sqrt{12}$, assuming a uniform distribution of solutions within the specified range of x_i values.

Besides the logarithmic update rules described above, there also exist symmetric and generalized two-point update rules (Schwefel, 1995).

Non-Isotropic Self-Adaptation

Here, a different mutation strength σ_i is used for each variable. Thus, this type of self-adaptive ES is capable of learning to adapt to problems where each variable has an unequal contribution to the objective function. In addition to n object variables, n other strategy parameters are included in the decision variable vector. The logarithmic update rules for decision variables and mutation strengths are as follows:

$$\sigma_i^{(t+1)} = \sigma_i^{(t)} \exp\left(\tau' N(0, 1) + \tau N_i(0, 1)\right), \tag{4.44}$$

$$x_i^{(t+1)} = x_i^{(t)} + \sigma_i^{(t+1)} N_i(0, 1), \tag{4.45}$$

where $\tau' \propto (2n)^{-1/2}$ and $\tau \propto (2n^{1/2})^{-1/2}$. Due to lack of any theoretical results on this self-adaptive ES, the progress coefficients for the (μ, λ)-ESs or $(\mu/\mu, \lambda)$-ESs are normally used as the constant of proportionality for τ' and τ. Similar initial values for $\sigma_i^{(0)}$, as discussed for isotropic self-adaptive ESs, may be chosen.

Correlated Self-Adaptation

In correlated self-adaptation, in addition to n mutation strengths, $\binom{n}{2}$ covariances, at most, are included in each individual solution. Thus, there are a total of $\left(\binom{n}{2} + n\right)$ exogenous strategy parameters to be updated for each solution. Thus, this type of self-adaptive ES can adapt to problems where decision variables (x) are correlated. Schwefel (1981) suggested that instead of using covariances, one can replace them with a rotation angle corresponding to each pair of coordinates. In a correlated problem (where decision variables are nonlinearly interacting with each other), the task is to find all pair-wise coordinate rotations and the spread of solutions in each rotated coordinate system so that the objective function is completely uncorrelated in the new coordinate system. Thus, the update rules are suggested for n decision variables, n mutation strengths σ_i and $\binom{n}{2}$ rotation angles α_j as follows:

$$\sigma_i^{(t+1)} = \sigma_i^{(t)} \exp\left(\tau' N(0, 1) + \tau N_i(0, 1)\right), \tag{4.46}$$

$$\alpha_j^{(t+1)} = \alpha_j^{(t)} + \beta_\alpha N_j(0, 1), \tag{4.47}$$

$$x^{(t+1)} = x^{(t)} + N\left(0, C(\sigma^{(t+1)}, \alpha^{(t+1)})\right), \tag{4.48}$$

where $N(0, C(\sigma^{(t+1)}, \alpha^{(t+1)}))$ is a realization of a normally distributed correlated mutation vector with a zero mean vector and a covariance matrix C. The parameter β_α is fixed as 0.0873 (or $5°$) (Schwefel, 1987a). Whenever an angle α_j goes outside the range $[-\pi, \pi]$, it is mapped inside the range, that is, if $|\alpha_j| > \pi$, $\alpha_j = \alpha_j - 2\pi \cdot \alpha_j/|\alpha_j|$. The parameters τ', τ and $\sigma_i^{(0)}$ are set as before. The rotation angles are initialized within zero and 180 degrees at random.

4.3.4   Connection Between Real-Parameter GAs and Self-Adaptive ESs

Without loss of generality, we will now discuss the similarity in the working of a real-parameter GA with an SBX operator and the isotropic self-adaptive ES. Under the latter, the difference (say Δ_i) between the offspring $(x_i^{(t+1)})$ and its parent $(x_i^{(t)})$ can be written from equations (4.42) and (4.43) as follows:

$$\Delta_i = \left[\sigma^{(t)} \exp(\tau_0 N(0, 1))\right] N(0, 1). \tag{4.49}$$

Thus, an instantiation of Δ_i is a normal distribution with zero mean and a variance which depends on $\sigma^{(t)}$, τ_0, and the instantiation of the log-normal distribution. For our argument, there are two aspects of this procedure:

1. For a particular realization of the log-normal distribution, the difference Δ_i is normally distributed with zero mean. That is, offspring closer to the parents are monotonically more likely to be created than offspring away from the parents.
2. The standard deviation of Δ_i is proportional to the mutation strength $\sigma^{(t)}$, which signifies, in some sense, the population diversity.

In Figure 76, we show the effect of a 50% reduction in $\sigma^{(t)}$ on the distribution of Δ_i under a self-adaptive ES.

Under the SBX operator, we write the term Δ_i by using equations (4.12) or (4.13) as follows:

$$\Delta_i = \frac{\delta_{p_i}}{2}(\beta_{q_i} - 1), \tag{4.50}$$

where δ_{p_i} is the absolute difference in the two parent solutions. There is a similarity between equations (4.49) and (4.50). The latter equation suggests that the distribution of Δ_i depends on the distribution of $(\beta_{q_i} - 1)$ for a particular pair of parents. The distribution of β_{q_i} has its mode at $\beta_{q_i} = 1$ (refer to equation (4.11)). Thus, the distribution of $(\beta_{q_i} - 1)$ will have its mode at zero. Although the distribution $(\beta_{q_i} - 1)$ is not a normal distribution, Figure 60 suggests that a small Δ_i has a higher probability to be created than a large Δ_i, and that this distribution is monotonic to the distance from a parent. The variance of this distribution depends on δ_{p_i}, which signifies the population diversity. Figure 77 shows the effect of a 50% reduction in δ_{p_i} on the distribution of Δ_i under the SBX operator. Comparing Figures 76 and 77 we observe that in self-adaptive ES, the diversity in the offspring population is controlled by the explicit mutation strength parameter, while in GAs with SBX operator it is controlled by the parent population diversity. Thus, there is a remarkable similarity in the way

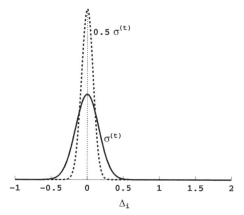
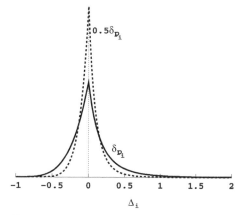

Figure 76 Effect of mutation strength $\sigma^{(t)}$ on the probability distribution of Δ_i under the self-adaptive ES. Reproduced from Deb and Beyer (2001) (© 2001 by the Massachusetts Institute of Technology).

Figure 77 Effect of δ_{p_i} on the probability distribution of Δ_i under the SBX operator (with $\eta_c = 2$). Reproduced from Deb and Beyer (2001) (© 2001 by the Massachusetts Institute of Technology).

that offspring are assigned in both an isotropic self-adaptive ES and in a GA with an SBX operator. In both cases, the offspring closer to the parent solutions are assigned a larger probability to be created than solutions away from parents, and the variance of this probability distribution depends on the current population diversity.

Simulation Results

We consider the following 30-variable ellipsoidal model:

$$\text{ELP}: \begin{cases} \text{Minimize} & \sum_{i=1}^{30} 1.5^{i-1}x_i^2, \\ & -1 \leq x_i \leq 1 \quad \text{for } i = 1, 2, \ldots, 30. \end{cases} \tag{4.51}$$

Since each variable has an unequal contribution to the objective function, we use non-isotropic self-adaptive ESs. A tournament selection operator with size 3 and no mutation operator is used in GAs with the SBX operator (with $\eta_c = 1$). For the recombinative ES, we use the dominant crossover on object variables and intermediate recombination on strategy parameters, as suggested elsewhere (Bäck and Schwefel, 1993). Figure 78 shows the objective function value of the best solution in the population with a GA and with an (10/10,100)-ES. Since the vertical axis is drawn on a logarithmic scale, both approaches proceed towards the true optimum solution exponentially quickly. Since the slopes are more or less the same, the rate of convergence to the optimum is also similar in both approaches.

Figure 79 plots the population standard deviation in x_1, x_{15} and x_{30} variables in the population for the real-parameter GA with the SBX operator. Since they are scaled as 1.0, $1.5^{14} \approx 292$, and $1.5^{29} \approx 127\,834$, respectively, the 30-th variable is likely to

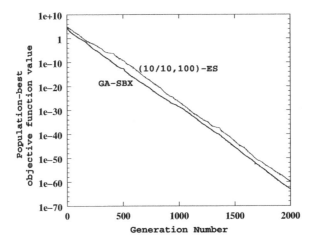

Figure 78   The best objective function value in the population is shown for two evolutionary algorithms – a real-parameter GA with SBX and a self-adaptive (10/10,100)-ES. Reproduced from Deb and Beyer (2001) (© 2001 by the Massachusetts Institute of Technology).

have a smaller population variance than the first variable. The figure shows this fact clearly. Since, the ideal mutation strengths for these variables are also likely to be inversely proportionate to 1.5^{i-1}, we find a similar ordering with the non-isotropic self-adaptive ES as well (Figure 80). Thus, there is a remarkable similarity in how both real-parameter GAs with SBX operator and self-adaptive ESs work. In the former case, the population diversity becomes adapted by the fitness landscape, which in turn helps the SBX operator to create x_i value of offspring proportionately. In the latter case, the population diversity becomes controlled by independent mutation strengths, which become adapted based on the fitness landscape. Comparative simulations on other problems can also be found in the original study (Deb and Beyer, 2001).

4.4   Evolutionary Programming (EP)

Evolutionary programming (EP) is a mutation-based evolutionary algorithm applied to discrete search spaces. David Fogel (Fogel, 1988) extended the initial work of his father Larry Fogel (Fogel, 1962) for applications involving real-parameter optimization problems. Real-parameter EP is similar in principle to evolution strategy (ES), in that normally distributed mutations are performed in both algorithms. Both algorithms encode mutation strength (or variance of the normal distribution) for each decision variable and a self-adapting rule is used to update the mutation strengths. Several variants of EP have been suggested (Fogel, 1992). Here, we will discuss the so-called *standard* EP, for tackling continuous parameter optimization problems.

EP begins its search with a set of solutions initialized randomly in a given bounded space. Thereafter, EP is allowed to search anywhere in the real space, similar to the real-parameter GAs. Each solution is evaluated to calculate its objective function value

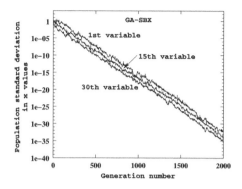

Figure 79 Population standard deviations in the variables x_1, x_{15} and x_{30} shown with a real-parameter GA with the SBX operator for the function ELP. Reproduced from Deb and Beyer (2001) (© 2001 by the Massachusetts Institute of Technology).

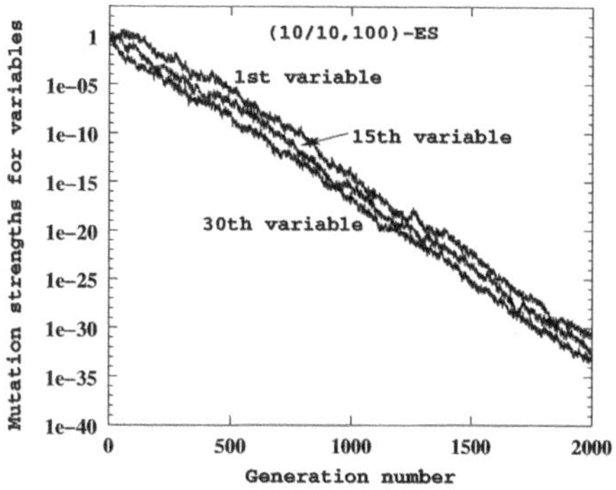

Figure 80 Mutation strengths for the variables x_1, x_{15} and x_{30} shown with a self-adaptive (10/10,100)-ES for the function ELP. Reproduced from Deb and Beyer (2001) (© 2001 by the Massachusetts Institute of Technology).

$f(x)$. Then, a fitness value $F(x)$ is calculated by some user-defined transformation of the objective function. However, the fitness function can also be the same as the objective function.

After each solution is assigned a fitness, it is mutated by using a zero-mean normally distributed probability distribution and a variance dependent on the fitness function, as follows:

$$x_i^{t+1} = x_i^t + \left(\sqrt{\beta_i F(x) + \gamma_i} \right) N_i(0, 1), \qquad (4.52)$$

where the parameters β_i and γ_i must be tuned for a particular problem, such that the term inside the square-root function is positive. If the fitness function $F(x)$ is normalized between zero and one (with $F(x) = 0$ corresponding to the global minimum), the parameters $\gamma_i = 0$ can be used. In such an event, as the solutions approach the global minimum, the mutation strength becomes smaller and smaller. However, for other fitness functions, a proper γ_i must be chosen. To avoid this user-dependent parameter setting, Fogel (1992) suggested a meta-EP, where the mutation strength for each decision variable is also evolved, similar to the self-adaptive evolution strategies. In a meta-EP, problem variables constitute decision variables x and mutation strengths σ. Each is mutated as follows:

$$\begin{aligned} x_i^{t+1} &= x_i^t + \sigma_i^t N_i(0, 1), & (4.53) \\ (\sigma_i^{t+1})^2 &= (\sigma_i^t)^2 + \zeta \sigma_i^t N_i(0, 1), & (4.54) \end{aligned}$$

where ζ is a user-defined exogenous parameter, which must be chosen to make sure that the mutation strength remains positive. It is important to mention here that there also exists a Rmeta-EP, where in addition to the mutation strengths, correlation coefficients are used as decision variables, as in the case of correlated evolution strategies (Fogel et al., 1992).

After the mutation operation, both parent and offspring populations are combined together and the best N solutions are probabilistically selected for the next generation. The selection operator is as follows. For each solution $x^{(i)}$ in the combined population, a set of η members are chosen at random from the combined population. Thereafter, the fitness of $x^{(i)}$ is compared with that of each member of the chosen set and a *score* equal to the number of solutions having the worse fitness is counted. Thereafter, all 2N population members are sorted according to the descending order of their scores and the best N solutions having the better score are selected. It is clear that if $\eta = 2N$ is chosen, the procedure is identical to the deterministic selection of the best N solutions. For $\eta < 2N$, the procedure is probabilistic. Since both parent and offspring populations are combined before selecting N solutions, the selection procedure preserves elitism. Fogel (1992) proved the above algorithm for convergence. Bäck (1996) showed that for the sphere model having a large number of variables, the implicit mutation strength used in the EP is larger than the optimal mutation strength, thereby causing the EP to perform worse than an optimal algorithm. Fogel's correction to equation (4.52) with:

$$\beta_i = \frac{1}{n^2},\qquad(4.55)$$

for $n > 2$ helps make the implicit mutation strength of the same order as the optimal mutation strength. An EP with the above fix-up is expected to perform as good as an evolution strategy with an optimal mutation strength.

4.5   Genetic Programming (GP)

Genetic programming (GP) is a genetic algorithm applied to computer programs to evolve efficient and meaningful programs for solving a task (Koza, 1992; Koza, 1994; Banzhaf et al., 1998). Instead of decision variables, a program (usually a C, C++, or Lisp program) represents a procedure to solve a task. For example, let us consider solving the following differential equation using a GP:

$$\left.\begin{array}{ll} \text{Solve} & \dfrac{dy}{dx} = 2x, \\ \text{with} & y(x = 0) = 2. \end{array}\right\}\qquad(4.56)$$

The objective in this problem is to find a solution $y(x)$ (y as a function of x) which will satisfy the above problem description. Here, a solution in a GP method is nothing but an arbitrary function of x. Some such solutions are $y(x) = 2x^2 - 2x + 1$, $y(x) = \exp(x) + 2$, or $y(x) = 2x + 1$. It is the task of a GP technique to find the function which is the correct solution to the above problem through genetic operations.

In order to create the initial population, a *terminal* set T and a *function* set F must be pre-specified. A terminal set consists of constants and variables, whereas a function set consists of operators and basic functions. For example, for the above problem $T = \{x, 1, 2, -1, -2, \ldots\}$ and $F = \{\div, \times, +, -, \times, \exp, \sin, \cos, \log, \ldots\}$ can be used. By the use of two such sets, valid syntactically correct programs can be developed. Two such programs are shown in Figure 81. Like a GA, a GP process starts with a set of randomly created syntactically correct programs. Each program (or solution) is evaluated by testing the program in a number of S instances in a supervised manner. By comparing the outcome of the program on each of these instances with the actual outcome, a fitness is assigned. Usually, the number of *matched* instances (say I_i) are counted and a fitness equal to I_i/S is assigned to the i-th individual. By maximizing this artificial fitness metric, one would hope to find a solution with a maximum of S matched instances. In the above problem, the fitness of a solution can be calculated in the following manner. First, a set of solutions can be chosen. We illustrate the fitness assignment procedure by choosing a set of 11 instances of x, as shown in Table 8. The true value of the right side of the differential equation can be computed for each solution (column 3 in the table). Thereafter, for each solution, the left side of the differential equation can be computed numerically. In column 4 of the table, we compute the exact value of the left side of the differential equation for the solution $y^{(1)}(x) = 2x^2 - 2x + 1$. Now, an error measure between these two columns (3 and 4) can be calculated as the fitness of the solution. Obviously, such a fitness metric must be minimized in order to get the exact solution of the differential equation. By computing the absolute differences between columns 3 and 4 and summing, we determine the fitness of the solution as 11.0. Since $y^{(1)}(x = 0) = 1$ does not match with the given initial condition, $y(x = 0) = 2$, we add a penalty to the above fitness proportional to this difference and calculate the overall fitness of the solution. With a penalty parameter of one, the overall fitness becomes $11.0 + 1.0$, or 12.0. This way, even if the differential equation is satisfied by a function but the initial condition is not satisfied, then the overall fitness will be non-zero. Only when both are satisfied, will the overall fitness take the lowest possible value.

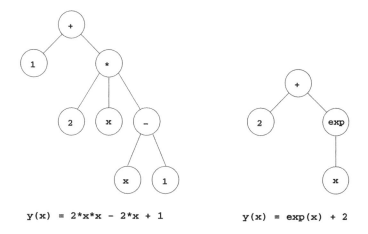

$$y(x) = 2*x*x - 2*x + 1 \qquad\qquad y(x) = exp(x) + 2$$

Figure 81   Two GP solutions.

Table 8 A set of 11 instances and evaluation of a solution $y^{(1)}(x) = 2x^2 - 2x + 1$.

Instances	x	2x	$dy^{(1)}/dx$
1	0.0	0.0	−2.0
2	0.1	0.2	−1.6
3	0.2	0.4	−1.2
4	0.3	0.6	−0.8
5	0.4	0.8	−0.4
6	0.5	1.0	0.0
7	0.6	1.2	0.4
8	0.7	1.4	0.8
9	0.8	1.6	1.2
10	0.9	1.8	1.6
11	1.0	2.0	2.0

The crossover and mutation operators are similar to that in GAs. We illustrate the working of a typical GP crossover operator on two parent solutions in Figure 82. A sub-program to be exchanged between two parents is first chosen at random in each parent. Thereafter, the crossover operator is completed by exchanging the two sub-programs, as shown in Figure 82.

The mutation operator can be applied to a terminal or to a function. If applied to a terminal, the current value is exchanged with another value from the terminal set, as shown in Figure 83. Similarly, if the mutation operation is to be applied to a function (a node), a different function is chosen from the function set. Here, care must be taken to keep the syntax of the chosen function valid.

Like GAs, GP operators are usually run for a pre-specified number of generations, or until a solution with a best fitness value is found. In addition to the above simple GP operators, there exist a number of other advanced operators, such as automatically defined functions, meaningful crossover, and various others, which have been used to solve a wide variety of problems (Koza, 1994).

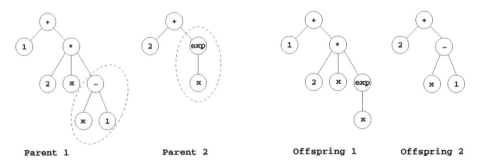

Figure 82 The GP crossover operator.

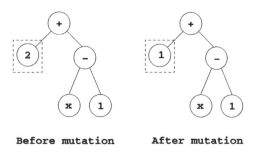

Before mutation   **After mutation**

Figure 83   The GP mutation operator.

4.6   Multi-Modal Function Optimization

As the name suggests, multi-modal functions have multiple optimum solutions, of which many are local optimal solutions. Multi-modality in a search and optimization problem usually causes difficulty to any optimization algorithm in terms of finding the global optimum solutions. This is because in these problems there exist many attractors in which an algorithm can become directed to. Finding the global optimum solution in such a problem becomes a challenge to any optimization algorithm.

However, in this section, we will discuss an optimization task which is much more difficult than just finding the global optimum solution in a multi-modal optimization problem. The objective in a multi-modal function optimization problem is to find multiple optimal solutions having either equal or unequal objective function values. This means that in addition to finding the global optimum solution(s), we are interested in finding a number of other local optimum solutions. It may apparently look puzzling as to why one would be interested in finding local optimum solutions. However, the knowledge of multiple local and global optimal solutions in the search space is particularly useful in obtaining an insight into the function landscape. Such information is also useful to design engineers and practitioners for choosing an alternative optimal solution, as and when required. In practice, because of the changing nature of a problem, new constraints may make a previous optimum solution infeasible to implement. The knowledge of alternate optimum solutions (whether global or local optimum) allows a user to conveniently shift from one solution to another.

Most classical search and optimization methods begin with a single solution and modify the solution iteratively and finally recommend only one solution as the obtained optimum solution (refer to Figure 39). Since our goal is to find multiple optimal solutions, there are at least two difficulties with the classical methods:

1. A classical optimization method has to be applied a number of times, each time starting with a different initial guess solution to hopefully find multiple optimum solutions.
2. Different applications of a classical method from different initial guess solutions do not guarantee finding different optimum solutions. This scenario is particularly true if the chosen initial solutions lie in the basin of attraction of an identical optimum.

Since EAs work with a population of solutions instead of one solution, some change in the basic EA, which is described in the previous section, can be made so that the final EA population consists of multiple optimal solutions. This will allow an EA to find multiple optimal solutions simultaneously in one single simulation run. The basic framework for the working of an EA suggests such a plausibility and makes EAs unique algorithms in terms of finding multiple optimal solutions. In this section, we will discuss different plausible modifications to a basic genetic algorithm for finding and maintaining multiple optimal solutions in multi-modal optimization problems. These ideas are then borrowed in subsequent chapters to find multiple Pareto-optimal solutions in handling multi-objective optimization problems.

4.6.1 Diversity Through Mutation

Finding solutions closer to different optimal solutions in a population and maintaining them over many generations are two different matters. Early on in an EA simulation, solutions are distributed all over the search space. Because of parallel schema processing in multiple schema partitions, solutions closer to multiple optimal solutions get emphasized in early generations. Eventually, when the population contains clusters of good solutions near many optima, competitions among clusters of different optima begins. This continues until the complete population converges to a single optimum. Therefore, although it is possible to discover multiple solutions during early generations, it may not be possible to maintain them automatically in a GA. In order to maintain multiple optimal solutions, an explicit diversity-preserving operator must be used.

The mutation operator is often used as a diversity-preserving operator in an EA. Although, along with selection and crossover operators, it can help find different optimal solutions, it can help little in preserving useful solutions over a large number of generations. The mutation operator has a constructive as well as a destructive effect. As it can create a better solution by perturbing a solution, it can also destroy a good solution. Since we are interested in accepting the constructive effect and since it is computationally expensive to check the worth of every possible mutation for its outcome, the mutation is usually used with a small probability in a GA. This makes the mutation operator insufficient to use as a sole diversity-preserving operator for maintaining multiple optimal solutions.

4.6.2 Preselection

Cavicchio (1970) was the first to introduce an explicit mechanism of maintaining diversity in a GA. Replacing an individual with a *like* individual is the main concept in a preselection operator. Cavicchio introduced preselection in a simple way. When an offspring is created from two parent solutions, the offspring is compared with respect to both parents. If the offspring has a better fitness than the worst parent solution, it replaces that parent. Since an offspring is likely to be similar (in genotypic space and in phenotypic space) to both parents, replacement of a parent with an offspring allows many different solutions to co-exist in the population. This operation is also carried out in all pairs of the mating parents. Thus, the resulting successful offspring replace one of their parents, thereby allowing multiple important solutions to co-exist in the population. Although Cavicchio's study did not bring out the effect of introducing preselection because of the use of many other operators, Mahfoud (1992) later suggested a number of variations, where the preselection method is used to solve a couple of multi-modal optimization problems.

4.6.3 Crowding Model

DeJong (1975) in his doctoral dissertation used a crowding model to introduce diversity among solutions in a GA population. Although his specific implementation turns out to be weak for multi-modal optimization, he laid the foundation for many popular diversity preservation techniques which are in use to date.

As the name suggests, in a crowding model, crowding of solutions anywhere in the search space is discouraged, thereby providing the diversity needed to maintain multiple optimal solutions. In his crowding model, DeJong used an overlapping population concept and a crowding strategy. In his GA, only a proportion G (called the *generation gap*) of the population is permitted to reproduce in each generation. Furthermore, when an offspring is to be introduced in the overlapping population, CF solutions (called the *crowding factor*) from the population are chosen at random. The offspring is compared with these CF solutions and the solution which is most *similar* to the offspring is replaced. In his study, DeJong used $G = 0.1$ and $CF = 2$ or 3. Since a similar string is replaced by the offspring, diversity among different strings can be preserved by this process. Similarity among two strings can be defined in two spaces – phenotypic (the decision variable space) and genotypic space (the Hamming space). DeJong's crowding model has been subsequently used in machine learning applications (Goldberg, 1983) and has also been analyzed (Mahfoud, 1995). The main contribution of DeJong's study was the suggestion of replacing one solution by a similar solution in maintaining multiple optimum solutions in an evolving population.

4.6.4 Sharing Function Model

Goldberg and Richardson (1987) suggested another revolutionary concept, where instead of replacing a solution by a similar solution, the focus was more on degrading the fitness of similar solutions. The investigators viewed the population as a battlefield for multiple optimal solutions, with each trying to survive with the largest possible occupants. Since most subsequent GA studies have used this model in solving multi-modal optimization problems, we will describe it here in somewhat greater detail.

Let us imagine that we are interested in finding q optimal solutions in a multi-modal optimization problem. The only resource available to us is a finite population with N slots. It is assumed here that $q \ll N$, so that each optimum can work with an adequate subpopulation (niche) of solutions. Since the population will need representative solutions of all q optima, somehow the available resource of N population slots must have to be *shared* among all representative solutions. This means that if for each optimum i, there is an expected occupancy m_i in the population, $\sum_{i=1}^{q} m_i = N$. If, in any generation, an optimum i has more than this expected number of representative solutions, this has to come at the expense of another optimum. Thus, the fitness of each of these representative solutions must be degraded so that in an overall competition the selection operator does not choose all representative solutions. On the other hand, solutions of an under-represented optimum must be emphasized by the selection operator. Arguing from a two-armed bandit game-playing scenario, the investigators showed that if such a sharing concept is introduced, the arm with a payoff f gets populated by a number of solutions (m) proportionate to f, or $m \propto f$. For q armed bandits (or q optimal solutions), this means that the ratio of payoff f (optimal objective value) and m (number of subpopulation slots) for each optimum would be identical, or:

$$\frac{f_1}{m_1} = \frac{f_2}{m_2} = \cdots = \frac{f_q}{m_q}. \tag{4.57}$$

This suggests an interesting phenomenon. If a hypothetical fitness function f' (called the 'shared fitness function') is defined as follows:

$$f_i' = \frac{f_i}{m_i}, \tag{4.58}$$

then all optima (whether local or global) would have an identical shared fitness value. If a proportionate selection is now applied with the shared fitness values, all optima will get the same expected number of copies, thereby emphasizing each optimum equally. Since this procedure will be performed in each generation, the maintenance of multiple optimal solutions is also possible.

Although the above principle seems like a reasonable scheme for maintaining multiple optimal solutions in a finite population from generation after generation, there is a practical problem. The precise identification of solutions belonging to each true optimum demands a lot of problem knowledge and in most problems is not available. Realizing this, Goldberg and Richardson (1987) suggested an adaptive strategy, where a *sharing function* is used to obtain an estimate of the number of solutions belonging to each optimum. Although a general class of sharing functions is suggested, they used the following function in their simulation studies:

$$Sh(d) = \begin{cases} 1 - \left(\frac{d}{\sigma_{share}}\right)^{\alpha}, & \text{if } d \leq \sigma_{share}; \\ 0, & \text{otherwise.} \end{cases} \tag{4.59}$$

The parameter d is the distance between any two solutions in the population. The above function takes a value in [0, 1], depending on the values of d and σ_{share}. If d is zero (meaning that two solutions are identical or their distance is zero), $Sh(d) = 1$. This means that a solution has full sharing effect on itself. On the other hand, if $d \geq \sigma_{share}$ (meaning that two solutions are at least a distance of σ_{share} away from each other), $Sh(d) = 0$. This means that two solutions which are a distance of σ_{share} away from each other do not have any sharing effect on each other. Any other distance d between two solutions will have a partial effect on each. If $\alpha = 1$ is used, the effect linearly reduces from one to zero. Thus, it is clear that in a population, a solution may not get any sharing effect from some solutions, may get partial sharing effect from few solutions, and will get a full effect from itself. Figure 84 shows the sharing function $Sh(d)$ as it varies with the normalized distance d/σ_{share} for different values of α. If these sharing function values are calculated with respect to all population members (including itself), are added and a niche count nc_i is calculated for the i-th solution, as follows:

$$nc_i = \sum_{j=1}^{N} Sh(d_{ij}), \tag{4.60}$$

then the niche count provides an estimate of the extent of crowding near a solution. Here, d_{ij} is the distance between the i-th and j-th solutions. It is important to note that nc_i is always greater than or equal to one. This is because the right side includes the term $Sh(d_{ii}) = Sh(0.0) = 1$. The final task is to calculate the shared fitness value as $f_i' = f_i/nc_i$, as suggested above. Since all over-represented optima will have a larger nc_i value, the fitness of all representative solutions of these optima would be degraded by a large amount. All under-represented optima will have a smaller nc_i value and the degradation of the fitness value would not be large, thereby emphasizing the under-represented optima. The following describes the step-by-step procedure for calculating the shared fitness value of a solution i in a population.

Shared Fitness Calculation

Step 1 Calculate the sharing function value $Sh(d_{ij})$ with all population members by using equation (4.59).

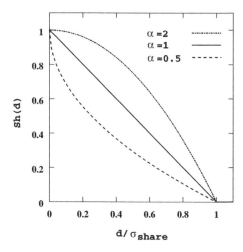

Figure 84   The sharing function plotted with three α values.

Step 2 Add all sharing function values to calculate the niche count nc_i by using equation (4.60).

Step 3 Calculate the shared fitness value as $f'_i = f_i/nc_i$.

After the shared fitness is calculated for all population members, a proportionate selection method is used to find the mating pool. The crossover and mutation operators are applied to the whole population as usual.

Careful thinking will reveal that the shared fitness computation of a population member i is $O(N)$, since this involves finding the distance d_{ij} with every population member. Although some bookkeeping can be used, the computation for calculating the shared fitness value of all population members is $O(N^2)$. However, it is important to highlight that in most real-world problems the calculation of the distance between vectors of decision variables or strings could be negligible compared to expensive objective function computations. Hence, the above computational complexity may not be critical. However, there exists another factor, which has been the source of some criticisms of the sharing function method, a matter we will discuss later (see page 154).

An Example Problem

In order to illustrate the working of the sharing function approach, let us take a bimodal function:

$$\text{Maximize} \quad \left. \begin{array}{l} f(x) = |\sin(\pi x)|, \\ 0 \le x \le 2. \end{array} \right\} \tag{4.61}$$

Let us also consider a population of six solutions, as shown in Table 9. A six-bit coding is used and decoded value of each string is mapped in $[0, 2]$ to calculate the corresponding decision variable value. These six solutions are also shown as circles on a function plot in Figure 85. This figure shows that of the two maxima, the second maximum ($x^* = 1.5$) is populated by more solutions than the first maximum ($x^* = 0.5$). If the sharing approach is not used, this population is most likely to lead a GA to converge to the second maximum alone, thereby not finding the first maximum.

Table 9   Six solutions and corresponding shared fitness values.

Sol. i	String	Decoded value	$x^{(i)}$	f_i	nc_i	f'_i
1	110100	52	1.651	0.890	2.856	0.312
2	101100	44	1.397	0.948	3.160	0.300
3	011101	29	0.921	0.246	1.048	0.235
4	001011	11	0.349	0.890	1.000	0.890
5	110000	48	1.524	0.997	3.364	0.296
6	101110	46	1.460	0.992	3.364	0.295

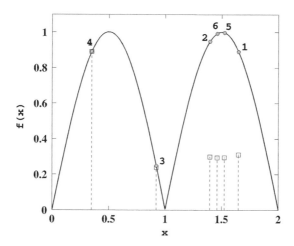

Figure 85 Six solutions and their shared fitness values (shown by dashed lines) on the bimodal function.

We will now illustrate the sharing function approach and calculate the shared fitness values of each population member.

To illustrate this, we use phenotypic sharing and a $\sigma_{share} = 0.5$. In the next subsection, we will show that with two optima and a range of 2.0 for the decision variable, the suggested σ_{share} is indeed 0.5. We also choose $\alpha = 1$. We use the above procedure to calculate the shared fitness of the first solution only.

Step 1 From the first solution, the distances from all population members are as follows:

$$d_{11} = 0, \qquad d_{12} = 0.254, \quad d_{13} = 0.731, \quad d_{14} = 1.302,$$
$$d_{15} = 0.127, \quad d_{16} = 0.191.$$

The corresponding sharing function values are calculated by using equation (4.59):

$$Sh(d_{11}) = 1, \qquad Sh(d_{12}) = 0.492, \quad Sh(d_{13}) = 0, \quad Sh(d_{14}) = 0,$$
$$Sh(d_{15}) = 0.746, \quad Sh(d_{16}) = 0.618.$$

Note that since solutions 3 and 4 are more than a 0.5 unit away from solution 1, their sharing effect is zero.

Step 2 The niche count of the first solution is simply the addition of the above sharing function values, or:

$$nc_1 = 1 + 0.492 + 0 + 0 + 0.746 + 0.618 = 2.856.$$

Table 10 Computation of niche counts for all six solutions.

| | Sharing function values | | | | | | nc_i |
	1	2	3	4	5	6	
1	1	0.492	0	0	0.746	0.618	2.856
2	0.492	1	0.048	0	0.746	0.874	3.160
3	0	0.048	1	0	0	0	1.048
4	0	0	0	1	0	0	1.000
5	0.746	0.746	0	0	1	0.872	3.364
6	0.618	0.874	0	0	0.872	1	3.364

Step 3 The shared fitness value of the first solution is:

$$f_1' = f(x^{(1)})/nc_1 = 0.890/2.856 = 0.312.$$

Table 10 shows the niche count calculations of all six solutions. The right-most column of this table shows the niche count values, which are copied in the sixth column of Table 9. The corresponding shared fitness values are shown in the seventh column of Table 9. Figure 85 also shows these shared fitness values with dashed lines. Since, solution 4 is near-optimal to $x^* = 0.5$ and this is the only solution representing this optimum, its shared fitness value is higher than any other solution in the population. On the other hand, since there are four representatives (solutions 1, 2, 5 and 6) of the second maximum, their shared fitness values are somewhat less. Since the selection operator will be performed with these shared fitness values, more copies of solution 4 are expected in the mating pool. By calculating the average shared fitness of the population, it is found that $f_4'/f_{avg}' = 2.29$, whereas the average of f_i'/f_{avg}' of the four solutions near the second maximum is 0.77. Thus, by the action of a proportionate selection, it is expected that solution 4 would have two or more copies and the four solutions near the second maximum may reduce to three solutions ($4 \times 0.77 = 3.08$). These calculations show how the sharing function approach can restore a balance between the number of solutions in each optimum.

Parameter Setting

The sharing function $Sh(d)$ defined in equation (4.59) introduced two user-defined parameters − α and σ_{share}. Although α does not have too much effect on the performance of the sharing function method (Deb, 1989), the second parameter σ_{share} must be set right in order to define the *niche size* of an optimum. Ideally, the latter is directly related to the basin of attraction of the optimum. Since this information is not available in an arbitrary problem, Deb and Goldberg (1989) suggested different ways to set this parameter. We will first present the suggestions for phenotypic sharing and then discuss the same for genotypic sharing. In most applications, an $\alpha = 1$ or 2 is used.

For phenotypic sharing, the Euclidean distance between two decision variable vectors $x^{(i)}$ and $x^{(j)}$ can be calculated as d_{ij}:

$$d_{ij} = \sqrt{\sum_{k=1}^{n} (x_k^{(i)} - x_k^{(j)})^2}. \tag{4.62}$$

Imagining that the n-dimensional hypersphere of radius r must be divided among q optima equally (each small hypersphere having a radius σ_{share}), this reduces to the following equation:

$$q(\sigma_{share}^n) = cr^n, \quad \text{or}$$

$$\sigma_{share} = \frac{r}{\sqrt[n]{q}}.$$

Here, $r = \frac{1}{2}\sqrt{\sum_{k=1}^{n}(x_k^{(U)} - x_k^{(L)})^2}$. Thus, as estimate of σ_{share} for introducing q equispaced niches in the search space is:

$$\sigma_{share} = \frac{\sqrt{\sum_{k=1}^{n}(x_k^{(U)} - x_k^{(L)})^2}}{2\sqrt[n]{q}}. \tag{4.63}$$

For $n = 1$ (single variable problems), the above equation reduces to $\sigma_{share} = (x_k^{(U)} - x_k^{(L)})/(2q)$. If normalized distance values $d_{ij} = \sqrt{\sum_{k=1}^{n}[(x_k^{(i)} - x_k^{(j)})/(x_k^{(U)} - x_k^{(L)})]^2}$ are used, the following normalized sharing parameter value can be employed:

$$\sigma_{share} = \frac{0.5}{\sqrt[n]{q}}. \tag{4.64}$$

For genotypic sharing, the Hamming distance between two ℓ-bit binary strings $s^{(i)}$ and $s^{(j)}$ is used as d_{ij}. The Hamming distance is defined as the number of positions where bits are different in two strings. This distance value only takes an integer value in the range $[0, \ell]$. Counting the number of strings having different bit differences with a given string in the entire search space, Deb and Goldberg (1989) calculated the number of allowed bit differences K for which all strings having K or less bit differences occupy at least the $1/q$-th portion of the entire search space, or:

$$\sum_{i=0}^{K} \binom{\ell}{i} = \frac{2^\ell}{q}. \tag{4.65}$$

Since there may not exist an integer K satisfying the above equation, we calculate the minimum value of K for which the left side becomes more than the right side. This minimum value of K is assigned as σ_{share}.

Simulation Results

In order to demonstrate the working of the sharing function method, we show one GA simulation run on the following five-peaked function:

$$\text{Maximize} \quad \left.\begin{array}{l} f(x) = 2^{-2((x-0.1)/0.8)^2} \sin^6(5\pi x), \\ 0 \leq x \leq 1. \end{array}\right\} \tag{4.66}$$

We use a population size of 100, the proportionate selection method, and a single-point crossover operator with a probability of 0.9. No mutation is used. The GA is run until 200 generations have taken place, in order to demonstrate that subpopulations in all five optima are possible to maintain even for a large number of generations. Figure 86 shows that without the sharing function approach, a GA converges to the best maximum. However, Figure 87 shows that all five maxima are found in a single simulation run of a GA with the sharing function approach with $\sigma_{\text{share}} = 0.1$ (obtained using equation (4.63) with $n = 1$, $q = 5$, $x^{(U)} = 1$ and $x^{(L)} = 0$).

Difficulties of the Sharing Function Approach

There are two difficulties with the sharing function approach. First, there is a need for choosing a σ_{share}. We have already suggested estimates for choosing an appropriate σ_{share} earlier. Both types of sharing require a user to fix a value of the number of optima q to calculate an estimate of an adequate σ_{share}. If the q chosen to compute σ_{share} is larger than the actual number of optima in the search space, the sharing function approach tends to form more niches than the function can allow. This may lead to finding a number of suboptimal solutions in addition to the optimal solutions. On the other hand, if the chosen q underestimates the actual number of optima in the search space, not all optima may be found by the sharing function approach. In

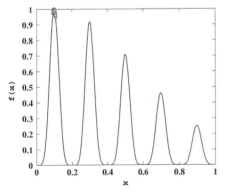

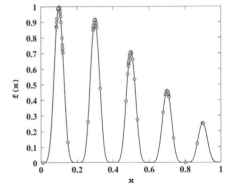

Figure 86   The population obtained without the sharing function approach.

Figure 87   The population obtained with the sharing function approach. Reproduced from Deb (1997b) (© 1997 IOP Publishing Ltd and Oxford University Press).

most applications where an exact number of optima is not known a priori, a q = 5
to 10 may be tried. It is also important to note that if q niches are needed to be
found, a GA needs an adequate population size to support all q subpopulations. We
suggest using N = κq, where κ is a constant and must be calculated as the minimum
population size needed to find one optimum independently.

The computed niche count for each solution provides an idea of the number of
solutions around an optimum and can be used to resize σ_{share} at every generation.
Such a dynamic update of σ_{share} may provide a better performance of an EA than
a fixed value of σ_{share}. More research in this direction must be pursued to evaluate
the efficacy of such techniques. We will discuss one such dynamic update procedure
in Section 5.8.6.

Secondly, there is an inherent scaling problem that shared fitness values can
introduce. Consider the population shown using circles on the two-optimum function
in Figure 88. The shared fitness values (marked by dashed lines in the figure) suggest
that a sub-optimal solution (solution A) has the best shared fitness value. This is
because the solution has a fitness value which is not too much worse than the optimal
solutions, but is not crowded by many solutions. Since the proportionate selection
will emphasize this solution the most (even more than the near-optimal solutions),
the population may contain some sub-optimal solutions. Figure 87 also shows the
existence of some such solutions, even after 200 generations. However, many of the
non-optimal solutions shown in Figure 87 are also the outcome of crossovers between
solutions from two different optima, a matter we will discuss below in Section 4.6.7.

Real-Parameter GAs for Multi-Modal Optimization

In the above simulations, binary-coded GAs are used. Since the implementation of
the sharing function approach affects only the selection operator, it can also be used

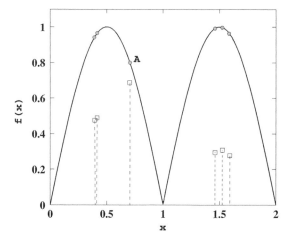

Figure 88   The shared fitness of solution A is maximum here.

in a straightforward manner with real-parameter GAs. However, there is one matter which needs discussing.

The role of any niching method is to emphasize solutions corresponding to poorly represented regions in the population. The sharing function approach achieves this task by artificially degrading the fitness of the crowded solutions in the population. Although the mating pool obtained by such a procedure may contain multiple copies of important solutions, if highly destructive recombination and mutation operators are used, some such solutions may get destroyed. Thus, there is a need to use restricted search operators. Fortunately, in most of the real-parameter crossover operators described earlier, the *search power* can be controlled with a tunable parameter. In the SBX operator, this parameter is η_c, while in the BLX operator, the parameter is α. To achieve a restricted search power, in SBX a large value of η_c must be used, while in BLX a small value of α must be used.

Deb and Kumar (1995) studied the effect of η_c on the performance of a real-parameter GA to maintain a good distribution of solutions among multiple optima on a number of multi-modal functions. We reproduce here the outcome of one such function. This function is the same five-peaked function used in the previous subsections. Instead of showing the population members on the function, Figure 89 presents a performance measure derived from the distribution of solutions in the final population. The actual number of solutions (n_i) near each of the five optima are calculated and a chi-square like deviation measure is evaluated:

$$\iota = \sqrt{\sum_{i=1}^{q+1} \left(\frac{n_i - \bar{n}_i}{\sigma_i} \right)^2}, \tag{4.67}$$

where q is the number of optima, $\bar{n}_i$ is the desired number of solutions in the i-th optimum, and σ_i is the standard deviation in the number of solutions in the i-th optimum. The desired number of solutions is calculated in proportion to the optimal function value of each optimum. For the above problems, the optimal function values are 1.0, 0.917, 0.707, 0.458 and 0.25. Thus, the corresponding desired numbers of solutions in each optimum in a population of size 100 are $\bar{n}_1 = 30$, $\bar{n}_2 = 28$, $\bar{n}_3 = 21$, $\bar{n}_4 = 14$, and $\bar{n}_5 = 7$. The $(q+1)$-th set is considered as all solutions in the population not belonging to any of the optima. For this set, $\bar{n}_{q+1} = 0$. The variances of all groups can be calculated as follows:

$$\sigma_i^2 = \begin{cases} \bar{n}_i \left(1 - \frac{\bar{n}_i}{N} \right) & \text{for } i = 1, 2, \ldots, q, \\ \sum_{j=1}^{q} \sigma_j^2, & \text{for } i = q + 1. \end{cases}$$

Thus, the smaller the above measure ι, then the closer is the distribution to the desired distribution. In Figure 89, we observe that with η_c, the distributing ability (calculated by using equation (4.67)) of real-parameter GAs with the SBX operator gets better in solving a problem having five optima. The figure also shows that at around $\eta_c = 30$, the distributing ability of real-parameter GAs with an SBX is equivalent to that of binary-coded GAs with phenotypic sharing. It is important to realize that with such

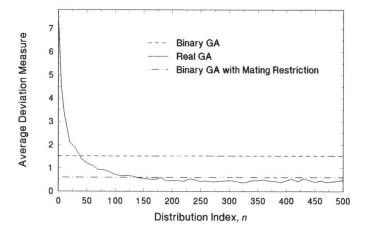

Figure 89 Real-parameter GAs with an SBX operator performs well in distributing solutions across multiple optima (Deb and Kumar, 1995). Reprinted with permission.

a large η_c, on most occasions offspring are created close to the parent solutions. This allows a slow yet steady search for better solutions in a multi-modal problem, instead of wild searches which may lead to lethal solutions.

Effect of Selection Operators

Oei et al. (1991) argued that the sharing function approach will not work well with the tournament or any ranking selection operator. Because of the proportionality law between the number of expected solutions and the optimal function value, only a proportionate selection can maintain subpopulations with a proportionate number of solutions. The investigators suggested a different niching strategy for the tournament selection. In their continuously updating sharing technique, a tournament between two solutions is played by finding the number of solutions in the vicinity of each solution in the new partial population. Instead of calculating the niche count with solutions in the old population, as in the sharing function approach described earlier, the niche count is calculated with solutions of the new population, as and when the population is formed. Between two solutions, the one with a better shared fitness value computed in the new population is chosen.

Instead of using the above continuously updated sharing approach, the investigators have also suggested a few practical implementation procedures. A niche size parameter n^* is defined by the user beforehand. For each of the two solutions participating in a tournament, the number of individuals in a niche with a σ_{share} parameter can be computed. Alternatively, the niche counts can also be computed. If the number of individuals in a niche or niche count of both solutions is less than n^*, the one with the better fitness value wins; otherwise, the one with the smaller number of crowded individuals or smaller niche count wins. Thus, if one solution is overly crowded beyond n^* and the other is not, the second solution is chosen. However, if both solutions competing in a tournament are adequately crowded, the one which is comparatively less crowded is selected. The fitness comparison is used only when both solutions are not adequately crowded. The attractive aspect of this strategy is that the importance to fitness or to crowding can be controlled by using the n^* parameter. Since the tournament selection operator is found to have both a fast *take-over* time and a quick execution time, this shared tournament selection is now attracting a lot of attention.

Instead of using the complete population to compute the niche count or number of crowded individuals, these computations can be performed on a small subpopulation randomly chosen from the complete population. This will reduce the computational complexity at the expense of introducing noise in the niching calculations.

Another simple procedure to introduce niching with a tournament selection would be to seek a maximum of k individuals as a partner for the selection of an individual i. If any of k individuals lie within σ_{share}, that solution competes with the solution i and the one with a better fitness value wins. On the other hand, if none of the k solutions is found to lie within σ_{share} of solution i, it is automatically declared selected. Once again, the selection pressure based on fitness can be controlled by changing k. Such an idea has been implemented in a recent study (Deb, 2000).

4.6.5 Ecological GA

Davidor (1991) suggested an ecological GA to evolve both the artificial niche and species in a population. In order to achieve this, he placed the population members on a two-dimensional grid (formed to make a torus) so that each member is surrounded by eight other members. In a steady-state GA, one member in the grid is chosen at random. The genetic operations are performed within a subpopulation consisting of nine individuals (the chosen individual and eight other surrounding individuals) to create an offspring string. This string contests with parent strings and survives according to the fitness of the string. Since an offspring can only survive by replacing its parents, like population members (individuals in the basin of each optimal solution in a multi-modal problem) are expected to stay close to each other, thereby forming islands of different optimal solutions. Davidor has shown that in solving a multi-modal problem, separate islands can be formed in the torus. However, the ecological model does not have any mechanism for maintaining islands of local peaks for many generations. Thus, local peaks are formed but eventually get overwhelmed by the global optimum solution due its slow diffusion effect.

4.6.6 Other Models

Goldberg and Wang (1998) proposed an alternate niching method (coevolutionary shared niching (CSN)), which is inspired by the economic model of monopolistic competition. In a population of customers, a set of businessmen is needed to be placed in such a way that each businessman, staying at a minimum distance to each other, serves the maximum number of customers. In a coevolutionary context of two populations involving businessmen and customers, each population seeks to maximize their separate interests, thereby creating adequately spaced niches. The usual GA population of solutions is named as the customer population here. The diversity in solutions is maintained by the usual fitness sharing scheme, where the fitness of a customer is calculated by degrading its objective function by the total number of customers its businessman serves. On the other hand, the fitness of a businessman is assigned by summing the objective function value of all of its customers. Initially, all customers and businessmen are picked from the search space at random. A selecto-recombinative GA is applied to the customer population and a selecto-mutation GA is applied to the businessman population. For each customer, the closest businessman is chosen. Thereafter, its shared fitness is calculated and the stochastic universal selection (SUS) operator is applied. The single-point crossover is used in the original study. For a businessman, a mutated solution is created and compared. If the mutated businessman is better than the original solution and is at least a critical distance away from other businessmen, it replaces the original businessman. On two test problems, it was possible to maintain stable subpopulations in all optima.

In solving an immune system problem, Smith et al. (1993) suggested an antigen–antibody matching procedure, which has an implicit fitness sharing strategy. The sharing function approach is not explicitly used here. In order to find a match in an antibody population for each antigen chosen from an antigen population, the procedure showed its ability to maintain a diverse set of antibodies covering a set of antigens. Such a coevolutionary strategy can be implemented in other optimization contexts without resorting to an explicit sharing approach.

Jelasity (1998) proposed the *universal evolutionary global optimization algorithm* (or UEGO) for multi-modal function optimization. UEGO works in different levels, starting from the complete search space. Each level embodies a number of species. Each species is designated with a center and a radius to mark its working domain in the search space. A new species is formed from two solutions of a previously formed species by comparing their objective function values with that of the centroid of the two solutions. If the centroid is worse than both parents, two new species are formed at two parent solutions. This heuristic is derived from the fact that there always lies a worse region between two optimal regions. UEGO also uses a fusion of species based on their nearness to each other, deletion of species to control over-growth of the number of species, and optimization within a species in search of better solutions representing the species. Each of these procedures involves some parameter settings. Several suggestions for fixing most of these parameters are outlined in the study. Although this procedure was used to find only one global optimum solution, multiple optimal solutions (that is, other global or local optimal solutions) are also found as byproducts of this procedure. This is because multiple species are maintained up until the end of a simulation run.

4.6.7 Need for Mating Restriction

In the previous subsections, we have seen that multiple optimal solutions in a multi-modal optimization problem can be found simultaneously by forming artificial niches (subpopulations) in the population. Each niche can be considered to represent an optimum. In order to capture a number of optima simultaneously and maintain these for many generations, a niching method is used. Niching helps to emphasize and maintain solutions around multiple optima, but it is the task of the search operators to find the best point within a niche efficiently. With the usual crossover operator applied uniformly to the whole population, some search effort is wasted in the recombination of solutions from different optima. A *speciation* method restricts mating among solutions in the basin of one optimum and helps find the best or a near-best optimum solution efficiently. If a speciation method is not used, individual optimum or near-optimum solutions may be obtained, but because of the creation of *lethal* solutions (solutions that do not represent any optimum) either a larger population size or more generations are necessary to achieve a specified accuracy in the obtained solutions. In speciation methods, only 'like individuals' are allowed to mate. Thus, the probability of creating lethal individuals is small and the search is mostly confined in the vicinity of each optimum. Thus, in the presence of both niching and speciation, niching maintains subpopulation of solutions around many optima and the speciation allows us to make a parallel search in all optimal basins to find multiple optimal solutions simultaneously.

A number of speciation methods have been suggested and implemented in genetic algorithms. Of the earlier works related to mating restriction in genetic algorithms, Hollstein's inbreeding scheme (Hollstein, 1971), where mating was allowed between close individuals in his simulation of animal husbandry problems, Booker's taxon-exemplar scheme (Booker, 1982) for restrictive mating in his simulation of learning pattern classes, Holland's suggestion of a tag-template scheme (Holland, 1975), Perry's adaptive GA for creating multiple species (Perry, 1984), Sannier and Goodman's restrictive mating in forming separate coherent groups in a population (Sannier and Goodman, 1987), Deb's phenotypic and genotypic mating restriction schemes (Deb, 1989), and Davidor's ecological model (Davidor, 1991) are just a few studies. Here, we will briefly mention the phenotypic and the genotypic mating restriction schemes.

We have developed two mating restriction schemes based on the phenotypic and genotypic distances between mating individuals (Deb, 1989). These mating restriction schemes are straightforward. In order to choose a mate for an individual, their distances (for a phenotypic mating restriction, the Euclidean distance, and for a genotypic mating restriction, the Hamming distance) are computed. If the distance is closer than a parameter σ_{mating}, they are participated in the crossover operation; otherwise, another individual is chosen at random and their distances are computed. This process is continued until a suitable mate is found or all population members are exhausted, in which case a random individual is chosen as a mate. In all simulations of this study, σ_{mating} is kept the same as the value of the parameter σ_{share} chosen for the niching method. Figure 90 shows the population after 200 generations, obtained by using the phenotypic mating restriction scheme with $\sigma_{\text{mating}} = 0.1$ and with the

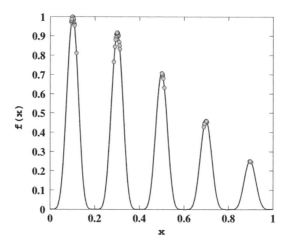

Figure 90   The population obtained after 200 generations using the sharing function and mating restriction approaches. Reproduced from Deb (1997b) (© 1997 IOP Publishing Ltd and Oxford University Press).

sharing function approach. The rest of the parameter settings are the same as before. This figure shows that the GA with niching and mating restrictions finds a better distributed population across the peaks than GAs with the sharing function approach alone (shown in Figure 87).

It is interesting to note in Figure 89 that real-parameter GAs with the SBX operator having $\eta_c \geq 140$ performed better than binary GAs with sharing and mating restriction operations.

4.7   Summary

Before understanding the working principles of multi-objective evolutionary algorithms, it is essential to know the workings of single-objective evolutionary algorithms. In this chapter we have discussed a number of evolutionary algorithms (EAs). Of them, genetic algorithms (GAs) and evolution strategy (ES) are used as baseline algorithms in most of the popular multi-objective evolutionary algorithms. Thus, we have discussed these two algorithms in somewhat more detail. Later, we have also described briefly two other evolutionary algorithms, namely evolutionary programming (EP) and genetic programming (GP). Although there exists a number of different evolutionary algorithms (Bäck et al., 1997), they are all motivated by the natural evolutionary principles. Most of these methods share a number of common properties, as follows:

1. They work with a population of solutions, instead of one solution in each iteration. By starting with a random set of solutions, an EA modifies the current population to a different population in each iteration. Working with a number of solutions

provides an EA with the ability to capture multiple optimal solutions in one single simulation run.

2. They have two distinct operations, namely selection and search. In the selection operation, better solutions in the current population are emphasized by duplicating them in the mating pool. In the search operation, new solutions are created by exchanging partial information (in the decision variable space) among solutions of the mating pool and by perturbing them in their neighborhood.

3. They do not use any gradient information during the above two operations. In addition, their representations are flexible. These properties make EAs flexible enough to be used in a wide variety of problem domains.

4. The operators use stochastic principles. Since deterministic rules are not used, EAs do not assume any particular structure of a problem to be solved.

Much of the research carried out in the area of evolutionary computation is available in a number of well-written books (Bäck, 1996; Fogel, 1995; Fogel et al., 1966; Goldberg, 1989; Gen and Cheng, 1997; Holland, 1975; Michalewicz, 1992; Mitchell, 1996; Koza, 1992, 1994; Schwefel, 1995). In addition, the following regular conference proceedings provide useful source material: The International Conference on Genetic Algorithms (ICGA), a biannual conference started in 1985, The Genetic and Evolutionary Computation Conference (GECCO), an annual conference started in 1999, The International Conference on Evolutionary Computation (ICEC), an annual conference started in 1994, The Congress on Evolutionary Computation (CEC), an annual conference started in 1999, The Evolutionary Programming Conference, an annual conference started in 1992, The Genetic Programming Conference (GP), an annual conference started in 1996, Parallel Problem Solving from Nature (PPSN) Conference, a biannual conference started in 1990, Foundations of Genetic Algorithms (FOGA), a biannual workshop started in 1990, and many other related events. The following journals also regularly publish papers on evolutionary computation: IEEE Transactions on Evolutionary Computation (TEC), published by the IEEE Press, Evolutionary Computation Journal (ECJ), published by the MIT Press, and Genetic Programming and Evolvable Machines published by the Kluwer Academic Publishers, plus several other domain-specific journals, including a couple of on-line publications.

Multi-objective optimization is a mixture of two optimization strategies: (i) minimizing the *distance* of the obtained solutions from the true optimal solutions, and (ii) maximizing the spread of solutions in the obtained set of solutions. The former task is similar to the usual task in a single-objective optimization, while the latter task is similar to finding multiple optimal solutions in a multi-modal optimization. In this chapter we have also discussed a number of multi-modal optimization techniques using evolutionary algorithms. The ability to capture multiple solutions in an EA population makes multi-modal and multi-objective evolutionary algorithms elegant and useful. With such descriptions of both of these optimization tasks, we have now set a platform from which we can now make our journey into the world of multi-objective evolutionary optimization algorithms.

Exercise Problems

1. In the mutation operator used in $(1 + 1)$-ES, a neighboring point is created with the following probability distribution:

$$p(x, \sigma) = \frac{a}{\sigma^2 + x^2}.$$

What must be the value of a, in order to have a valid probability distribution? The current point is $(x, \sigma) \equiv (10.0, 2.0)$. In order to perform mutation, a random number 0.723 is used. What is the new point x'?

2. In a three-variable problem, the following variable bounds are specified.

$$-5 \leq x \leq 10,$$

$$0.001 \leq y \leq 0.005,$$

$$1(10^3) \leq z \leq 1(10^4).$$

What should be the minimum string length of any point $(x, y, z)^\mathsf{T}$ coded in binary string to achieve the minimum accuracy in the solution as given below:

(a) $\Delta x = 0.01$, $\Delta y = 1(10^{-5})$, and $\Delta z = 10$.
(b) $\Delta x = 0.1$, $\Delta y = 1(10^{-6})$, and $\Delta z = 1$.

3. We would like to use genetic algorithms to solve the following NLP problem:

$$\begin{aligned}
\text{Minimize} \quad & (x_1 - 1.5)^2 + (x_2 - 4)^2 \\
\text{subject to} \quad & 4.5x_1 + x_2^2 - 18 \leq 0, \\
& 2x_1 - x_2 - 1 \geq 0, \\
& 0 \leq x_1, x_2 \leq 4.
\end{aligned}$$

We decide to have three and two decimal places of accuracy for variables x_1 and x_2, respectively.

(a) How many bits are required for coding the variables?
(b) Write down the fitness function which you would be using in the selection procedure.
(c) Which schemata represent the following regions according to your coding:
    (a) $0 \leq x_1 \leq 2$, (b) $1 \leq x_1 \leq 2$, and $x_2 > 2$.

4. In order to solve the following problem:

$$\text{Maximize} \quad f(x) = (2x + 1)^5$$

in the range $0 \leq x \leq 1$ using GAs, we are interested in the growth of the schema $(1\ 1\ 0\ *\ *\ *\ *\ *)$ under the following selection operators:

(i) Proportionate selection.

(ii) Binary tournament selection.

Assume random initial population. What proportions of the population will be occupied by the above schema after three iterations due to the above selection operators?

5. (a) Write at least four fundamental differences of GAs with most traditional search and optimization methods.

   (b) Do you support the following remark?

   Selection and mutation are the main operators in genetic algorithms.

   Justify your argument through examples/sketches.

6. For a maximization problem, an initial population of binary strings and their fitness values are shown below:

String	Fitness
∩U∩UU∩	15
UU∩∩∩U	5
∩∩UU∩∩	14
U∩U∩UU	8
UU∩∩UU	7
∩UUU∩U	10

Using crossover and mutation probabilities $p_c = 0.8$ and $p_m = 0.1$, estimate using schema theorem how many strings representing the schema (∩∗ ∗ U∗ ∗) would be produced after one iteration of GAs with roulette-wheel reproduction, single-point crossover, and bit-wise mutation operators. Which of the two schemata (∩∗ ∗ ∗ ∗ ∗) and (U∗ ∗ ∗ ∗ ∗) is likely to take over the population in a GA run with above parameter setting? Give reasons.

7. A cantilever beam of circular cross-section (diameter d) has to be designed for minimizing the cost of the beam. The beam is of length $l = 0.30$ m and carries a maximum of $F = 1$ kN force at the free end. The beam material has $E = 100$ GPa, $S_y = 100$ MPa, and density $\rho = 7866$ Kg/m^3. The material costs Rs. 20 per Kg. There are two constraints: (i) Maximum stress in the beam ($\sigma = 32Fl/(\pi d^3)$) must not be more than the allowable strength S_y, (ii) Maximum deflection of the beam must not be more than 1 mm. The deflection at the free end due to the load F is $\delta = 64Fl^3/(3\pi Ed^4)$. The volume of the beam is $\pi d^2 l/4$. The initial population contains the following solutions: $d = 0.06, 0.1, 0.035, 0.04, 0.12,$ and 0.02 m.

   (a) For each of six solutions, calculate cost in rupees and determine its feasibility in a tabular format.

   (b) Determine the fitness of each solution using Deb's penalty parameter-less constraint handling strategy. Which is the best solution in the above population?

8. (i) Using the simulated binary crossover (SBX) with $\eta_c = 2$, find two offspring from parent solutions $x^{(1)} = 10.53$ and $x^{(2)} = -0.68$. Use the random number 0.723.

   (ii) Show that, on an average, the polynomial probability distribution used in SBX operator produces diverging offspring (solutions which are outside the range of parent solutions) for any η_c.

(iii) How would you modify the polynomial probability distribution of the SBX operator so that the expected range of offspring is the same as that of the parent solutions?

9. Using real-coded GAs with simulated binary crossover (SBX) having $\eta = 2$, find the probability of creating children solutions in the range $0 \le x \le 1$ with two parents $x^{(1)} = 0.5$ and $x^{(2)} = 3.0$. Recall that for SBX operator, the children solutions are created using the following probability distribution:

$$P(\beta) = \begin{cases} 0.5(\eta + 1)\beta^{\eta}, & \text{if } \beta \le 1; \\ 0.5(\eta + 1)/\beta^{\eta+2}, & \text{if } \beta > 1. \end{cases}$$

where $\beta = |(c_2 - c_1)/(p_2 - p_1)|$ and c_1 and c_2 are children solutions created from parent solutions p_1 and p_2.

10. What are the differences between (μ, λ)-ES and $(\mu + \lambda)$-ES?
11. For the two parents

$$x^{(1)} = (2.0, 5.0)^{\mathsf{T}}, \quad x^{(2)} = (4.0, 3.0)^{\mathsf{T}},$$

find the probability of finding the solution $(1.0, 1.0)^{\mathsf{T}}$ using

(a) intermediate recombination operator of ES,
(b) discrete recombination operator of ES.

Use mutation strength $\sigma_i = 1.0$ for $i = 1, 2$.

12. Two parents are shown below:

$$x^{(1)} = (10.0, 3.0)^{\mathsf{T}}, \quad x^{(2)} = (5.0, 5.0)^{\mathsf{T}}.$$

Calculate the probability of finding a child in the range $x_i \in [0.0, 3.0]$ for $i = 1, 2$ using

(a) Simulated binary crossover (SBX) operator with $\eta_c = 2$,
(b) Blend crossover (BLX) operator with $\alpha = 0.67$,
(c) Fuzzy recombination (FR) operator with $d = 1.0$.

13. Apply the polynomial mutation operator to create a mutated child of the solution $x^{(t)} = 5.0$ using a random number 0.675.
14. Why is a mating restriction scheme used in a GA?
15. Consider the following population of five strings, having three choices in each place:

Strings	Fitness
♡♣♢♣♣♡	6
♢♢♢♡♣♣	1
♡♢♢♣♢♡	8
♣♡♣♢♢♢	2
♢♣♡♣♢♡	6

(a) How many copies would ($\heartsuit * * * * *$) have after one iteration of a proportionate selection operation, single-point crossover operator with $p_c = 0.8$, and bit-wise mutation operator with $p_m = 0.1$?

(b) What proportion of the entire search space is represented by the schema ($\clubsuit \diamondsuit * * * *$)?

(c) Assuming that the above representation maps the integer space, how many discrete regions are represented by the schema ($\clubsuit * * \diamondsuit \heartsuit *$)?

16. Consider the following non-linear programming problem:

$$\text{Minimize} \quad f(\mathbf{x}) = x_1^2 - x_1 x_3 + x_2^2,$$
$$\text{subject to} \quad g_1(\mathbf{x}) \equiv x_1 - 2 \geq 0,$$
$$g_2(\mathbf{x}) \equiv x_1 x_2 x_3 - 10 \geq 0.$$

Calculate the fitness of each solution of the population shown below using Powell and Skolnick's constraint-handling approach and Deb's constraint-handling approach:

$$\mathbf{x}^{(1)} = (1, 2, 5)^\mathsf{T}, \quad \mathbf{x}^{(2)} = (6, 1, 2)^\mathsf{T}, \quad \mathbf{x}^{(3)} = (3, 4, 1)^\mathsf{T},$$
$$\mathbf{x}^{(4)} = (0, 2, 1)^\mathsf{T}, \quad \mathbf{x}^{(5)} = (5, 1, 1)^\mathsf{T}, \quad \mathbf{x}^{(6)} = (1, 2, 1)^\mathsf{T}.$$

Normalize the constraints and use $R = 1$.

17. We would like to find all four maxima of the following multimodal function defined in the range $x_1, x_2 \in [0, 1]$ using GAs with sharing functions:

$$F(x_1, x_2) = (1 - x_1 - x_2) \sum_{i=1}^{2} f_i(x_i),$$

where $f_i(x_i) = \sin^2(2\pi x_i)$. Assume optima lie at $x_1, x_2 = 0.25$ and 0.75. To avoid the stochastic sampling error, we require at least 5 copies at each optimum. What should be the minimum population size?

18. For the following maximization problem

$$\text{Maximize} \quad f(x) = |\sin(\pi x)|$$
$$0 \leq x \leq 2,$$

the following six strings are used.

(1) 110110   (2) 101100   (3) 011101
(4) 001011   (5) 110000   (6) 101110

Using $\sigma_{\text{share}} = 0.5$ and $\alpha = 1$, calculate the niche count and shared fitness value of each solution. Present your results in a tabular format.

Determine how many copies are expected in the population by the action of a proportionate selection alone in the left half and right half of the search space for the following two cases:

(a) Without sharing approach,
(b) With sharing approach.

19. In a binary decision problem with expected payoff $f_1 = 10$ for decision one and $f_0 = 5$ for decision zero, calculate the expected number of decision one in the next generation under the following selection schemes:

(i) Roulette-wheel selection only.

(ii) Roulette-wheel selection with genotypic sharing ($\sigma_{share} = 1$, $\alpha = 1$).

(iii) Roulette-wheel selection with genotypic sharing ($\sigma_{share} = 2$, $\alpha = 1$).

Assume the population currently contains 70 ones and 30 zeros.

20. For the bimodal maximization problem

$$f(x) = x \sin^2 \pi x, \qquad 0 \le x \le 2,$$

following four solutions are used in a population:

$$x^{(1)} = 0.4, \quad x^{(2)} = 1.6, \quad x^{(3)} = 1.2, \quad x^{(4)} = 0.9.$$

Calculate the shared fitness value of each of the above solutions using $\sigma_{share} = 0.4$.

5

Non-Elitist Multi-Objective Evolutionary Algorithms

A striking difference between a classical search and optimization method and an evolutionary algorithm (EA) is that in the latter a population of solutions is processed in every iteration (or generation). This feature alone gives an EA a tremendous advantage for its use in solving multi-objective optimization problems (MOOPs). Recall from Section 2.2 that one of the goals of an ideal multi-objective optimization procedure is to find as many Pareto-optimal solutions as possible. Since an EA works with a population of solutions, in theory we should be able to make some changes to the basic EA described in the previous chapter so that a population of Pareto-optimal solutions can be captured in one single simulation run of an EA. If we can achieve this, it will eliminate repetitive use of a single-objective optimization method for finding one different Pareto-optimal solution in each run. This will also eliminate the need of any parameters such as weight vectors, ϵ vectors, target vectors, and various others. These parameters are needed in a classical multi-objective optimization algorithm in order to transform the MOOP into a single-objective optimization problem favoring a single Pareto-optimal solution. This is because each setting of these vectors is associated with a particular Pareto-optimal solution in some problems. Using classical methods to find multiple Pareto-optimal solutions, one then chooses a different weight vector and solves the resulting single-objective optimization problem again.

On the other hand, the EA's population approach can be exploited to emphasize all non-dominated solutions in a population equally and to simultaneously preserve a diverse set of multiple non-dominated solutions using a niche-preserving operator, described in the previous chapter. In this way multiple good solutions can be found and maintained in a population. After some generations, this process may lead the population to converge close to the Pareto-optimal front and with a good spread.

In this chapter, we will describe such modifications to a basic EA for finding multiple Pareto-optimal solutions simultaneously. First, we concentrate on algorithms which do not use any elite-preserving operator. Once we have a clear idea of these algorithms, we shall present multi-objective EAs which use an elite-preserving operator in the next chapter.

5.1   Motivation for Finding Multiple Pareto-Optimal Solutions

Before we discuss the multi-objective EAs, let us go one step ahead and ponder for a while the advantages of finding multiple Pareto-optimal solutions. Let us say that we use a multi-objective EA and finally manage to find a number of Pareto-optimal solutions in one single simulation run (for example, see Figures 37 and 38 earlier). We recall that each of these solutions would correspond to the optimum (or a near-optimum) solution of a composite problem trading-off different objective functions. Thus, each solution is important with respect to some trade-off relationship among the objectives. Now the obvious question is: 'What should we do with such multiple Pareto-optimal solutions?' After all, from a practical standpoint, we only need one solution to implement. So the real question is which one of these multiple solutions are we interested in? This may not be a difficult question to answer in the presence of many obtained trade-off solutions, but is difficult to answer in the absence of any such knowledge. This is precisely the difficulty a user faces while using a classical method. If a user knows the exact trade-off among objective functions (in other words, the user knows a weight vector that (s)he is interested in), there is no need to find multiple solutions. A weight-based classical method would be good enough to find the corresponding optimal solution. We have termed this approach a preference-based approach in Chapter 1, while the classical literature calls this approach an 'a priori' approach.

However, unfortunately, a user is usually not sure of an *exact* trade-off relationship among objectives. In this scenario, we argue that from a practical standpoint, it is better to find a set of Pareto-optimal solutions first and then choose one solution from the set by using some other higher-level information or considerations. We have termed this approach an *ideal* approach in Chapter 1. After a non-dominated set of solutions are found, each solution can be associated with an artificial weight vector estimated from the location of the solutions in the non-dominated set. Thereafter, if some rough preference knowledge about the trade-off among objectives is known, it can be used to choose one of the non-dominated solutions, which will closely match the information conveyed by the artificial weight vector. This method, at least, gives a user an overall perspective of other possible optimal solutions that the underlying MOOP offers before choosing a solution and also allows a user to choose a solution according to some desired degree of importance to each objective.

One can argue that in order to get a perspective of different Pareto-optimal solutions that the underlying MOOP offers, one can still use the 'a priori' approach of classical methods in the following way. In this so-called 'posteriori' approach, first a set of weight vectors (or ϵ vectors) are chosen. Thereafter, for each weight vector, we construct a single-objective optimization problem and find the corresponding optimum solution. When all such optimizations are performed, we get a set of Pareto-optimal solutions. Figure 91 illustrates this posteriori approach and the ideal approach on a two-objective hypothetical minimization problem. For each approach, the resulting Pareto-optimal solutions are also illustrated. In the repeated application of the a priori approach shown in the left figure, nine different equi-spaced weight vectors are chosen.

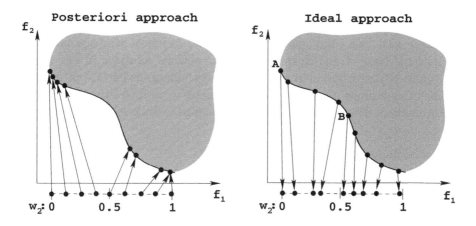

Figure 91 Classical 'posteriori' approach and the proposed ideal approach.

The figure shows w_2 only, but it is assumed that $w_1 = 1 - w_2$. The direction of the arrows in the left figure indicates that the weight vectors are chosen first. Thereafter, Pareto-optimal solutions are found by using each weight vector. In the absence of any additional knowledge, it is wise to use a uniform-spaced set of weight vectors. In a sufficiently nonlinear problem, it is likely that the optimal solutions resulting from a uniformly spaced set of weight vectors may not result in a uniformly spaced set of Pareto-optimal solutions, as depicted in the figure.

The right portion of Figure 91 shows the working principle of the proposed ideal approach. Here, no knowledge of any weight vector is assumed at first and a population-based optimization algorithm is used to find multiple Pareto-optimal solutions. Since the original MOOP is not converted to a single-objective optimization problem and since usually the concept of domination along with a diversity-preserving mechanism are to be used to drive the search process, this approach may find a good spread of Pareto-optimal solutions. Here, an arrow pointing towards a weight value indicates that a weight vector is calculated *after* the Pareto-optimal solutions are found. By using the individual objective function value of each of these solutions, an estimate of a trade-off among different objectives can then be made for each solution. For example, the left-most Pareto-optimal solution (marked 'A' on the figure) will correspond to a 100% preference of f_1 and a 0% preference of f_2. This is because this solution corresponds to the minimum f_1 among all obtained Pareto-optimal solutions. We can relate these preference values loosely with the weights. The solution 'A' may be assigned a weight vector $(w_1, w_2) = (1, 0)$. In this way, all obtained solutions can be assigned a pseudo-weight vector (refer to equation (8.1) later for a way to achieve a pseudo-weight vector). The user can now compare different solutions with their associated weight vector and choose a solution most desirable to the user. For example, if the user is interested in finding a solution with more or less equal trade-off between the two objectives, (s)he should choose the solution closer to a weight vector

$w_1 = w_2 = 0.5$. For the solutions illustrated in the figure, the user's choice would be solution 'B'.

We argue that this ideal approach is a better practical strategy, although the computational issue of finding multiple Pareto-optimal solutions in one simulation using a multi-objective optimizer versus iteratively finding one Pareto-optimal solution at a time using a single-objective optimizer remains as an interesting future research. Most EA methods described in this book are assumed to work with the ideal approach.

5.2 Early Suggestions

As in the classical optimization, early researchers of evolutionary algorithms have also realized the need of handling multiple objectives. G. E. P. Box (1957) hand-simulated an *evolutionary operation* for multiple objectives. He, being an experimentalist, clearly appreciated the necessity of using more than one objectives in a design (Box, 1957, p. 84):

> ...in practice we are interested not only in the productivity of the process, but also in the physical properties of the product which is manufactured.

Furthermore, in discussing how to handle multiple objectives using an evolutionary operation, he clearly mentioned the difficulties of using a weighted approach (Box, 1957, p. 96):

> Although it is theoretically possible to equate all responses to a single criterion such as profitability, this usually presents great practical difficulties. As a general rule it is best to represent the problem as one of improving a *principal* response (for example the cost per lb. of product) subject to satisfying certain conditions on a number of *auxiliary* responses. These auxiliary responses usually measure the quality and important physical properties of the product.
>
> Very careful thought in the selection of the principal response is essential. The vital question to ask is: 'If this response is improved will it mean necessarily that the processes improved?'

It is interesting to note the suggestion of an ϵ-constraint-based strategy for use with the evolutionary operation. This suggestion came much earlier than a formal introduction of an ϵ-constraint strategy by Haimes, Lasdon, and Wismer (1971). Box was also concerned with the need for choosing an appropriate criterion as the objective function of the converted single-objective optimization problem.

Fogel et al. (1966) suggested and simulated on a computer a weighted approach in handling multiple objectives (or goals) in two different scenarios. With the task of finding a logic for transforming a sequence of input symbols to a sequence of output symbols, the investigators suggested a way to evaluate a solution for predicting the n-th symbol into the future (where $n > 0$) with an evaluation function f_n. In order to solve a combined problem of predicting two or more symbols in the future, each corresponding evaluation function can be combined with a weight directly related to the importance of predicting that symbol. In an effort to control the growth of the finite-state machines (FSMs) by evolutionary programming, they were faced with two different objectives – maximize the performance of prediction and minimize the machine complexity. Once again, they suggested adding a penalty term (related to the second objective) to the first objective. This procedure is similar to the use of a weighted approach, with the constant in the penalty term acting as the relative weight associated with the second objective of minimizing complexity. On a computer simulation, they showed that by increasing the constant in the penalty term, small-sized FSMs can be evolved. Although the non-domination concept in handling multi-objective optimization problems is not used, this application was one of the earliest evolutionary algorithms used to handle multiple objectives.

Rosenberg (1967) in his dissertation suggested, but did not simulate, a genetic search method for finding the chemistry of a population of single-celled organisms with multiple properties or objectives. He referred to a set of chemical concentrations as a *property*. Although his formulations included minimizing the deviations of multiple properties from desired properties, in all of his simulations he considered only one property, thereby missing the opportunity to investigate the effect of multiple objectives in a GA.

The first real implementation of a multi-objective evolutionary algorithm (vector evaluated GA or VEGA) was suggested by David Schaffer in 1984. Schaffer (1984) modified the simple tripartite genetic algorithm GAs with selection, crossover and mutation by performing independent selection cycles according to each objective. We will describe the VEGA in detail in a subsequent subsection. The algorithm worked efficiently for some generations but in some cases suffered from its bias towards individual objective champions. Although the first goal of multi-objective optimization was fulfilled, the second task of maintaining a good spread of solutions in the obtained front was not entirely accomplished by the proposed VEGA.

Ironically, no significant study was performed for almost a decade after the pioneering work of Schaffer. Goldberg (1989), while discussing the VEGA in his book, realized a better implementation of domination principle in an EA and suggested a revolutionary 10-line sketch of a new non-dominated sorting procedure. Since an EA needs one fitness function for reproduction, the trick was to find a single metric from a number of objective functions. Goldberg's suggestion was to use the concept of domination to assign more copies to non-dominated individuals in a population. Since diversity is another concern, he also suggested the use of a *niching* strategy among solutions of a non-dominated class. Realizing the potential of a good multi-objective evolutionary algorithm which can be derived from his suggestions, at least three independent groups of researchers have developed different versions of multi-objective evolutionary algorithms (multi-objective GAs (MOGAs), niched Pareto GAs (NPGAs) and non-dominated sorting GAs (NSGAs)). These algorithms differ in the way a fitness is assigned to each individual. We will discuss each of these implementations in detail in subsequent subsections. A comprehensive summary of some of these algorithms can also be found elsewhere (Coello, 1999; Veldhuizen, 1999).

   Besides the algorithms described in this chapter, many other non-elitist MOEAs have also been suggested. We certainly do not make any attempt to include all such algorithms in this present text. Instead, we have included a few which are most commonly used.

5.3   Example Problems

Most of the MOEAs described in this book are illustrated by showing a hand simulation of the algorithm on a simple example problem. Since some algorithms are specifically designed to handle minimization problems and some are designed to handle maximization problems, we have chosen two different example problems.

5.3.1   Minimization Example Problem: Min-Ex

We consider the following two-objective, two-variable minimization problem for illustrating the working of most of the algorithms presented in this and the next chapter:

$$\text{Min-Ex:} \quad \left\{ \begin{array}{ll} \text{Minimize} & f_1(\mathbf{x}) = x_1, \\ \text{Minimize} & f_2(\mathbf{x}) = \dfrac{1 + x_2}{x_1}, \\ \text{subject to} & 0.1 \leq x_1 \leq 1, \\ & 0 \leq x_2 \leq 5. \end{array} \right. \tag{5.1}$$

Although this problem looks simple, it produces conflicting scenarios between both objectives, resulting in a set of Pareto-optimal solutions. A straightforward manipulation of the above two functions will allow us to find the following relationship between both objectives:

$$f_2 = \frac{1 + x_2}{f_1}. \tag{5.2}$$

The function f_1 is a function of x_1 only. Thus, the boundaries of the search space correspond to the minimum and maximum values of the numerator of the right side of the above equation. This happens for $x_2^* = 0$ and $x_2^* = 5$. Figure 92 shows the partial search space with two boundaries. Since both functions are to be minimized, all solutions in the lower boundary (with $x_2^* = 0$) are members of the Pareto-optimal set. Thus, for all Pareto-optimal solutions, two objective functions are related as $f_2 = 1/f_1$, thereby constituting the trade-off among the Pareto-optimal solutions. The figure also shows the complete Pareto-optimal front, which comprises the solutions $0.1 \leq x_1^* \leq 1$ and $x_2^* = 0$. It is important to highlight that the resulting Pareto-optimal region (shown by a bold line) is convex.

   We have also chosen six random solutions in the search space for illustrating the working principle of algorithms described in this chapter. These solutions are also marked in Figure 92 and are tabulated in Table 11.

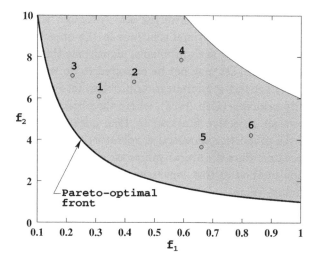

Figure 92 The example problem Min-Ex with six randomly chosen solutions.

5.3.2 Maximization Example Problem: Max-Ex

On the other extreme of the Pareto-optimal front in the above problem, there lies a search boundary corresponding to $f_2 = 6/f_1$. Incidentally, all solutions in this boundary belong to the Pareto-optimal set of the problem where the same objective functions are being maximized:

$$\left.\begin{array}{rl} \text{Maximize} & f_1(\mathbf{x}) = x_1, \\ \text{Maximize} & f_2(\mathbf{x}) = \dfrac{1 + x_2}{x_1}, \\ \text{subject to} & 0.1 \le x_1 \le 1, \\ & 0 \le x_2 \le 5. \end{array}\right\} \quad (5.3)$$

However, besides the basic difference in the type of optimization, there is a subtle difference between Min-Ex and the above problem. In the former case (minimization of objective functions) the resulting Pareto-optimal set is a convex set in the objective function space, whereas in the latter case (maximization of objective functions), the

Table 11 Six randomly chosen solutions in the objective space of Min-Ex.

Solution	x_1	x_2	f_1	f_2
1	0.31	0.89	0.31	6.10
2	0.43	1.92	0.43	6.79
3	0.22	0.56	0.22	7.09
4	0.59	3.63	0.59	7.85
5	0.66	1.41	0.66	3.65
6	0.83	2.51	0.83	4.23

resulting Pareto-optimal set is a nonconvex set. Although the variable space (the x_1–x_2 space) is convex, Pareto-optimal solutions in the objective space behave in a nonconvex manner. In tackling such problems, algorithms that somehow exploit the convexity of Pareto-optimal set may not be expected to do well.

One of the ways to convert a minimization problem into a maximization problem is to use the duality principle (Deb, 1995), where the objective function is multiplied by -1. This allows an objective function to take negative values. Some selection operators (such as the proportionate selection operator) cannot process negative fitness values since in these operators the fitness values are converted to probability measures directly. A common method of this conversion is to subtract the objective function from a large maximum value:

$$f_i(\mathbf{x}) \leftarrow f_i^{\max} - f_i(\mathbf{x}), \tag{5.4}$$

where $f_i^{\max}$ is the maximum function value of f_i in the entire search space. In the problem stated in equation (5.1), the maximum objective function values are 1 and 60, respectively. Thus, we modify the problem in equation (5.3) as follows:

$$\text{Max-Ex:} \quad \left\{ \begin{array}{ll} \text{Maximize} & f_1(\mathbf{x}) = 1.1 - x_1, \\ \text{Maximize} & f_2(\mathbf{x}) = 60 - \dfrac{1 + x_2}{x_1}, \\ \text{subject to} & 0.1 \leq x_1 \leq 1, \\ & 0 \leq x_2 \leq 5. \end{array} \right. \tag{5.5}$$

Since x_1 varies between 0.1 and 1.0, we have used a constant 1.1 instead of 1.0 to avoid any divide-by-zero error in computing f_2. This modified function has a Pareto-optimal front at $x_2^* = 0$ and $0.1 \leq x_1^* \leq 1.0$ (the same as in the Min-Ex problem), although the objective functions vary as $f_2(\mathbf{x}) = 60 - 1/[1.1 - f_1(\mathbf{x})]$. Figure 93 shows the partial search space and the complete Pareto-optimal front (with a bold line). The Pareto-optimal set is now a convex set in the objective space. A randomly created initial population of six members is shown in the objective space in Figure 93. The decision parameter values and corresponding objective function values are also shown in Table 12. These solutions will be used to illustrate the fitness assignment procedure for the different multi-objective evolutionary algorithms discussed in subsequent sections.

Table 12   Six randomly chosen solutions in the objective space of Max-Ex.

Solution	x_1	x_2	f_1	f_2
1	0.31	0.89	0.79	53.90
2	0.43	1.92	0.67	53.21
3	0.22	0.56	0.88	52.91
4	0.59	3.63	0.51	52.15
5	0.66	1.41	0.44	56.35
6	0.83	2.51	0.27	55.77

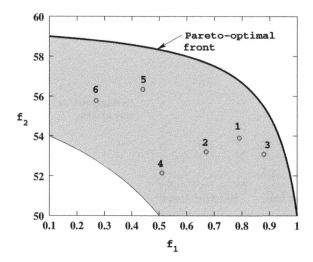

Figure 93   The maximization problem Max-Ex with six randomly chosen solutions.

5.4   Vector Evaluated Genetic Algorithm

Schaffer (1984) implemented the first multi-objective GA to find a set of non-dominated solutions. Subsequently, he compared his GA with an adaptive random search technique (Schaffer, 1985) and observed a better performance obtained by his algorithm. He called his GA the *vector evaluated GA*, or VEGA. The name is appropriate for multi-objective optimization, because his GA evaluated an objective vector (instead of a scalar objective function), with each element of the vector representing each objective function.

VEGA is the simplest possible multi-objective GA and is a straightforward extension of a single-objective GA for multi-objective optimization. Since a number of objectives (say M) have to be handled, Schaffer (1984) thought of dividing the GA population at every generation into M equal subpopulations randomly. Each subpopulation is assigned a fitness based on a different objective function. In this way, each of the M objective functions is used to evaluate some members in the population.

We illustrate this fitness evaluation scheme in Figure 94 for five objective functions. The population at any generation is divided into five equal divisions. Each individual in the first subpopulation is assigned a fitness based on the first objective function only, while each individual in the second subpopulation is assigned a fitness based on the second objective function only, and so on. In order to reduce the positional bias in the population, it is better to *shuffle* the population before it is partitioned into equal subpopulations. After each solution is assigned a fitness, the selection operator, restricted among solutions of each subpopulation, is applied until the complete subpopulation is filled. This is particularly useful in handling problems where objective functions take values of different orders of magnitude. Since all members in a subpopulation are assigned a fitness based on a particular objective

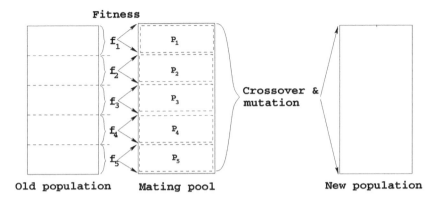

Figure 94 Schematic of a VEGA.

function, restricting the selection operator only within a subpopulation emphasizes good solutions corresponding to that particular objective function. Moreover, since no two solutions are compared for different objective functions, disparity in the ranges of different objective functions does not create any difficulty either. Schaffer used the proportionate selection operator (Goldberg, 1989).

Vector Evaluated Genetic Algorithm Procedure

Step 1 Set an objective function counter $i = 1$ and define $q = N/M$.

Step 2 For all solutions $j = 1 + (i - 1) * q$ to $j = i * q$, assign fitness as:

$$F(\mathbf{x}^{(j)}) = f_i(\mathbf{x}^{(j)}).$$

Step 3 Perform proportionate selection on all q solutions to create a mating pool P_i.

Step 4 If $i = M$, go to Step 5. Otherwise, increment i by one and go to Step 2.

Step 5 Combine all mating pools together: $P = \cup_{i=1}^{M} P_i$. Perform crossover and mutation on P to create a new population.

A little thought will reveal that the above algorithm emphasizes solutions which are good for individual objective functions. In order to find intermediate trade-off solutions, Schaffer allowed crossover between any two solutions in the entire population. In this way, Schaffer thought that a crossover between two good solutions, each corresponding to a different objective, may find offspring which are good compromised solutions between the two objectives. The mutation operator is applied on each individual as usual.

5.4.1 Hand Calculations

We consider the two-objective, two-variable optimization problem Max-Ex to better illustrate the working of the VEGA. Since a VEGA uses the proportionate selection operator, which in principle maximizes the fitness function, we have chosen the Max-Ex problem. The randomly created initial population of six members is shown in the objective space in Figure 93. The Pareto-optimal front comprises solutions $0.1 \leq x_1^* \leq 1$ and $x_2^* = 0$. It is important to highlight that the resulting Pareto-optimal region (shown by the bold curve in Figure 93) is convex. Now, we follow the step-by-step procedure of the VEGA.

Step 1 We set the counter $i = 1$ and parameter $q = 6/2 = 3$. Since there are two objectives, we partition the population into two subpopulations. Let us say that partition 1 comprises solutions 1, 2 and 5, and partition 2 comprises solutions 3, 4 and 6. Figure 95 shows this partitioning in the objective space.

Step 2 Now, all solutions of the first partition will be assigned a fitness based on the first objective function. Table 13 presents the assigned fitness to each of these three solutions of the first subpopulation. The best solutions in each partition is shown in italic script.

Step 3 It can be seen that solution 1 has the best fitness among all solutions in the first partition (Figure 96). Under a proportionate selection operator with scaling, solutions 1 and 2 are expected to have two copies and one copy, respectively, in the first partition. It is likely that the worst solution (solution 5) will get eliminated after the selection operation.

Step 4 Next we move to Step 2 for the second objective function ($i = 2$).

Step 2 Solutions 3, 4 and 6 belong to the second partition and are assigned fitness values based on the second objective function. These fitness values are presented in Table 13.

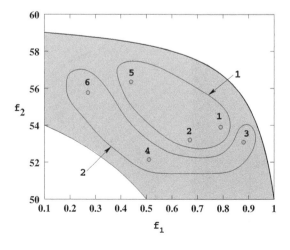

Figure 95 Six solutions are partitioned into two groups in VEGA.

Table 13 Fitness assignment procedure under a VEGA.

Solution	x_1	x_2	f_1	f_2	Partition	Assigned fitness
1	0.31	0.89	0.79	53.90	1	0.79
2	0.43	1.92	0.67	53.21	1	0.67
3	0.22	0.56	0.88	52.91	2	52.91
4	0.59	3.63	0.51	52.15	2	52.15
5	0.66	1.41	0.44	56.35	1	0.44
6	0.83	2.51	0.27	55.77	2	55.77

Step 3 The solution 6 has the best fitness among all solutions in the second partition (Figure 96). During the selection operation in the second partition, two copies of solution 6 and one copy of solution 3 are expected.

Step 4 Now, we move to Step 5, where crossover and mutation operations will be performed on the complete population obtained after both selection operations. This completes one iteration of the VEGA.

Figure 96 shows that there is a bias for solutions towards individual champions.

5.4.2 Computational Complexity

Since a stochastic remainder proportionate selection operator takes $O(N/M)$ computations (Goldberg and Deb, 1991) for each of M objectives, a VEGA has the same computational complexity as that of single-objective GAs.

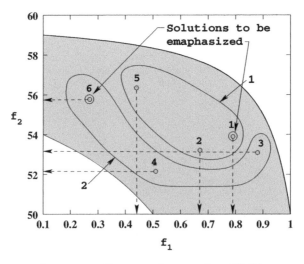

Figure 96 Fitness assignment in a VEGA.

5.4.3   Advantages

The main advantage of a VEGA is that it uses a simple idea and is easy to implement. Only minor changes are required to be made in a simple GA to convert it to a multi-objective GA and this does not incur any additional computational complexity. This is a definite advantage of the VEGA over all other multi-objective GAs.

We have discussed earlier that a VEGA has a tendency to find solutions near the individual best solution of each objective. In problems where such solutions are desired, the VEGA may be found to be a good alternative. One such application of VEGA on constrained function optimization has been nicely demonstrated elsewhere (Coello, 2000).

Although the VEGA is simple to implement, there exist some limitations in its working principle. We will describe these in the next section.

5.4.4   Disadvantages

Recall that each solution in a VEGA is evaluated only with one objective function. Thus, every solution is not tested for other $(M - 1)$ objective functions, all of which are also important in the context of multi-objective optimization. During a simulation run of a VEGA, it is likely that solutions near the optimum of an individual objective function would be preferred by the selection operator in a subpopulation. Such preferences take place in parallel with other objective functions in different subpopulations. In the original study (Schaffer, 1984), it was assumed that the crossover operator would combine these individual champion solutions together to find other trade-off solutions in the Pareto-optimal region. However, as observed by Schaffer, even in convex search space problems, the crossover between individual champion solutions could not find diverse solutions in the population. Eventually, the VEGA converges to individual champion solutions only.

5.4.5   Simulation Results

Figure 97 shows the population obtained by using a VEGA on the two-objective optimization problem (Max-Ex). In this experiment, we have used the following GA parameter settings:

Population Size	40,
String length	24 for x_1 and 26 for x_2,
Selection	Stochastic universal selection without any scaling,
Crossover	Single-point crossover with $p_c = 0.9$,
Mutation	Not used.

Although a few solutions close to the Pareto-optimal solutions were found during 30 to 50 generations (not shown here), the VEGA cannot sustain the intermediate solutions. If all GA parameters are chosen right, the VEGA may come very close to the Pareto-optimal front, but eventually it is destined to get attracted by the individual

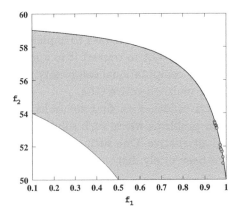

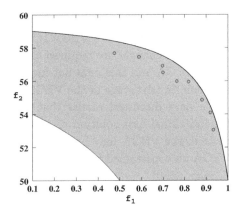

Figure 97 Non-dominated solutions after 500 generations of the VEGA without the mutation operator.

Figure 98 Non-dominated solutions at the end of 500 generations of the VEGA with the mutation operator.

champion solutions, thereby losing the intermediate solutions in the Pareto-optimal front. Figure 97 shows all non-dominated solutions obtained after 500 generations. It is clear that after a substantial number of generations the VEGA fails to sustain its diversity along the Pareto-optimal front and converges near one of the individual champion solutions.

When mutation is used with a probability of 0.02 (equivalent to 1/(string-length)), the VEGA seems to find a good spread of non-dominated solutions (Figure 98), but because of the absence of an elite-preserving operator and being stochasticity involved in the bit-wise mutation operator, the VEGA cannot maintain its solutions on the Pareto-optimal front. Recall that in order to remain on the Pareto-optimal front, the second decision variable x_2 must be zero (the lower bound coded by the binary string of all zeros). With the presence of mutation (on an average one bit per string), it is difficult to maintain solutions on the Pareto-optimal front. However, the same mutation when applied to the substring coding the first decision variable (x_1) has provided some diversity in the non-dominated solutions.

Although the mutation operator has provided some diversity, it has been achieved at the expense of convergence near the Pareto-optimal front. In fact, Schaffer's VEGA did not include any other operator for maintaining diversity explicitly. However, his later studies included two modifications, which we will discuss next.

5.4.6 Non-Dominated Selection Heuristic

Schaffer realized that one of the ways to maintain diverse solutions in the Pareto-optimal region is to emphasize non-dominated solutions in a population. One of the ways to give importance to a solution in a GA is to assign a large selection probability. In the proportionate selection operator, a solution with a fitness f_i is selected with a probability $f_i / \sum_{j=1}^{N} f_j$. Thus, the solution i can be emphasized by adding an extra term ϵ_i to this probability, but care must be taken to make sure that this extra amount ϵ_i is subtracted from some other solutions in the population in order to preserve the overall probability of selecting the complete population as one.

In this method, the non-dominated solutions in the current population are first found. Thereafter, an amount ϵ is subtracted from each dominated solution, thereby reducing the importance of the dominated solutions. For a subpopulation of N' population members having ρ' non-dominated solutions, this amounts to a total reduction of $(N' - \rho')\epsilon$ from all $(N' - \rho')$ dominated solutions. This amount is then redistributed among ρ' non-dominated solutions by adding an amount $\frac{N'-\rho'}{\rho'}\epsilon$ to each non-dominated solution. This method of redistribution of selection probabilities has two important implications:

1. Non-dominated solutions are emphasized over dominated solutions.
2. Additional emphasis to each non-dominated solution is given equally.

We shall see throughout this book that these two features are important ingredients of most multi-objective optimization algorithms that exist to date. However, since the fitness assignment is still performed based on individual objective functions on members of separate subpopulations, all non-dominated solutions do not have the same absolute selection probability.

The rest of the algorithm is the same as the VEGA. Each member in a subpopulation is selected with its modified selection probability. Crossover and mutation are performed as usual on the whole population. We apply this algorithm to the same two-objective test problem and observe that the ability of this modified VEGA is not very different from the original VEGA. We use $\epsilon = 0.02$. After 500 generations, we observe in Figure 99 that the modified VEGA cannot improve the diversity to a large extent in the obtained Pareto-optimal solutions. In fact, as Schaffer also pointed out, if there exist only a few non-dominated solutions in the population (ρ' is small), they get over-emphasized by the selection operator. This may cause the algorithm to lose its population diversity.

When a mutation operator is invoked with a bit-wise probability of 0.02 (equivalent to the $1/(\text{string-length})$ rule) and the modified VEGA is run up until 500 generations, there seems to be a good spread of solutions on the Pareto-optimal front (Figure 100). The modified VEGA seems to have found a better spread of solutions, compared to the original VEGA, in the presence of the bit-wise mutation operator. The emphasis of non-dominated solutions provided by the assignment of an extra selection probability matches well with the mutation's search power to find and maintain solutions on the Pareto-optimal front, even in the absence of an explicit elite-preserving operator.

5.4.7 Mate Selection Heuristic

With the selection operation used in the original VEGA, it is expected that specialist solutions of each objective will populate a GA population. Schaffer realized that the random mating (crossover) between any two solutions may not guarantee that solutions of different subpopulations are used in the crossover operation. For example,

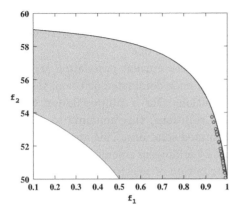

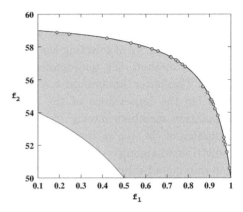

Figure 99 Non-dominated solutions after 500 generations obtained with a VEGA using the non-domination selection heuristic and without the mutation operator.

Figure 100 Non-dominated solutions after 500 generations obtained with a VEGA using the non-domination selection heuristic and mutation.

with two objective functions, there is only a 50% chance that the two mating solutions are from two different subpopulations. In order to create utopian solutions (intermediate Pareto-optimal solutions), Schaffer thought that cross-breeding of these specialist solutions is necessary. He proposed two techniques. In the first approach, one parent is chosen at random from the mating pool. To choose its mate, a solution having the maximum Euclidean distance in the objective space is chosen from the population. This mate would most likely be a solution from the specialists of an objective different from that in the first parent. This will allow cross-breeding of two different specialist solutions and intermediate solutions are expected to be created. Unfortunately, Schaffer reported that this strategy did not work well compared to the original random mating strategy. The reason for its poor performance is that since the first parent is randomly chosen, it may not be a specialist. Thus, a solution which is widely different from it may not be a specialist as well. Moreover, we argue that the basic idea of cross-breeding two specialist solutions to create an intermediate utopian solution may not occur in many problems, particularly in those with a nonconvex search region.

Schaffer's second approach to restrict mating was to use an *improvement distance*, instead of the Euclidean distance, to choose a mate. In the distance metric computation, only positive (improved) distances are summed over all dimensions. A solution with the maximum improvement distance is chosen as a mate. His conclusion was that although this heuristic worked better than the first heuristic, it is worse than the random mating (the crossover operator used in the original VEGA). Since none of these methods performed better than the original VEGA in the original study, we do not present any simulation results with these methods on our test problem. There exists a number of other MOEAs which use a fitness assignment scheme similar to that in the VEGA. Next, we will describe a few of these algorithms.

Fourman (1985), in his symbolic layout design study, suggested two ways to handle multiple criteria in a GA. The first algorithm is a straightforward weighted approach. By changing the weight associated with each objective, different solutions have been found using a simple GA. The second approach is similar to that of a VEGA. However, the selection operator was similar to the binary tournament selection. When two solutions are picked for comparison, one of the M objective functions is chosen at random and is used to compare two solutions. The algorithm also introduced a way to vary the frequency of choosing an objective. In this way, it is possible to introduce problem knowledge, if available, into his GA. On layout design problems, he observed that this strategy was able to find a 'greater variability' in resulting populations than the first approach.

The extreme of assigning a relative selection frequency for objectives is to use lexicographic ordering of objectives (Ben-Tal, 1980). By making one of the objectives infinitely more important than the rest of the objectives, that objective is first optimized alone. After the first level of optimization with the most-preferred objective, if there exist multiple optimal solutions, the second-best objective is tried. A constraint limiting the first objective to take no worse than the obtained optimal value in the previous case is added. This makes sure that previously found better solutions are always retained. The algorithm terminates when there exists only one optimum solution to an optimization problem.

Bentley and Wakefield (1997) suggested a number of ranking schemes as plausible fitness functions in a GA. The weighted average ranking (WAR) worked well in many multi-objective optimization problems. In this method, population members are ranked separately according to each objective function. A fitness equal to the sum of the ranks in each objective is assigned. In this way, individual champion solutions and intermediate solutions will have comparable fitness values, thereby emphasizing a variety of solutions to co-exist in the GA population.

Lis and Eiben (1996) proposed a multi-sexual GA (MSGA) for multi-objective optimization, which is similar in principle to the VEGA. Each solution is marked with an additional integer variable, representing the objective function for which it has to be evaluated. Investigators decided to remain closer to nature by referring to this additional variable as the sex variable. All solutions marked with a sex variable value k are evaluated by the k-th objective function only. Instead of using the proportionate selection, a ranking selection is used to choose a solution for mating. Crossover is allowed only between solutions having a different sex variable value. In a uniform crossover operation, the offspring inherits the sex variable of that parent from which it inherited the maximum number of genes (bits). Mutation is performed as well, but no mutation is allowed to the sex variable gene. The MSGA also maintains an external set of non-dominated solutions and updates it with current non-dominated solutions at every generation. Since the elite set of solutions is not used in the genetic operations (rather it is used for bookkeeping purposes), we do not call this strategy an elitist MOEA. However, on a couple of test problems, a widely distributed set of non-dominated solutions were found. It is interesting to note that the algorithm does not use any explicit niche-preserving mechanism, except the use of a sex marker and a restricted mating which ensures crossovers among good solutions of different objectives. It is a matter of future research to investigate how such a strategy without an explicit niche-preserving operator is able to find a well-distributed Pareto-optimal set in more complex problems.

5.5   Vector-Optimized Evolution Strategy

A few years after the VEGA study, but well before the study of other multi-objective EAs, Frank Kursawe (1990) suggested a vector-optimized evolution strategy (VOES) for multi-objective optimization. In this approach, the basic self-adaptive evolution strategy for single-objective optimization is modified to handle multi-objective optimization problems. Although the fitness assignment scheme was similar to the VEGA, Kursawe's algorithm used a number of other aspects from nature.

First, a solution is represented by using a diploid chromosome, each having a dominant and a recessive string. Two different solution vectors (each with a decision variable vector x and the corresponding strategy parameter vector σ) are used as an individual in a population. One of the two solutions is marked as 'dominant' and the other is marked as 'recessive'. Thus, an individual i results in two objective vector evaluations: (i) $\mathbf{f}_d$, calculated by using the decision variables of the dominant genotype and (ii) $\mathbf{f}_r$, calculated by using the decision variables of the recessive genotype. The overall fitness evaluation and the selection procedure are as follows.

The selection process is completed in M steps (where M is the number of objective functions). For each step, a user-defined probability vector is used to choose an objective. This vector can either be kept fixed or varied with generations. If the m-th objective function is chosen, the fitness of an individual i is computed as a weighted average of its dominant objective value $(\mathbf{f}_d)_m^{(i)}$ and recessive objective value $(\mathbf{f}_r)_m^{(i)}$ in a 2 : 1 ratio:

$$F^{(i)} = \frac{2}{3}(\mathbf{f}_d)_m^{(i)} + \frac{1}{3}(\mathbf{f}_r)_m^{(i)}. \tag{5.6}$$

From μ parent solutions, λ mutated solutions are created, as usual. The update of strategy parameters and decision variables are also performed, as described earlier in Section 4.3.3. An additional *swap* operator exchanging decision variables of recessive and dominant solutions is used with a variable-wise probability of 1/3. This allows mixing of dominant genes with recessive genes. This operator is similar to the uniform crossover operator performed between dominant and recessive chromosomes, but, on an average, one-third of the decision variables are swapped.

For each selection step, the population is sorted according to each objective function and the best $(M-1)/M$-th portion of the population is chosen as parents. This procedure is repeated M times, every time using the survived population from the previous sorting. Thus, the relationship between λ and μ is as follows:

$$\mu = \left(\frac{M-1}{M}\right)^M \lambda. \tag{5.7}$$

For two objectives, $\mu = 0.25\lambda$. If the probability vectors are chosen properly, good solutions corresponding to each objective will be chosen for the mating pool.

All new μ solutions are copied into an external set, which stores all non-dominated solutions found since the beginning of the simulation run. After new solutions are inserted into the external set, a non-domination check is made on the whole external set and only new non-dominated solutions are retained. If the size of the external set exceeds a certain limit, a niching mechanism is used to eliminate closely packed solutions. This allows maintenance of a diversity in the solutions of the external set. Although the preservation of non-dominated solutions in the external set is an elite-preserving strategy, these solutions are not used in further genetic processing. Hence, we do not call this approach an elitist approach.

5.5.1 Advantages and Disadvantages

We observe that this algorithm performs a domination check to retain non-dominated solutions and a niching mechanism to eliminate crowded solutions. We have argued that both of these features are essential in a good MOEA. Unfortunately, Kursawe showed simulation results on a single problem and no further work has been pursued since this study. It remains a challenging task to investigate how this algorithm will fare in more complex problems.

Although some variations to this algorithm are possible and the necessity of diploid chromosomes is questionable, in Section 6.6 we shall discuss a simpler and a well-studied multi-objective ES. Because of its complications and the fact that it is not much used by current researchers, we do not show any hand simulation of the VOES. Interested readers may refer to the original study (Kursawe, 1990) for more details.

5.6 Weight-Based Genetic Algorithm

Hajela and Lin (1993) proposed a weight-based genetic algorithm (we call this WBGA) for multi-criterion optimization in 1993. As the name suggests, each objective function f_i is multiplied by a weight w_i. A GA string represents all decision variables x_i as well as their associated weights w_i. The weighted objective function values are then added together to calculate the fitness of a solution. However, unlike the weighted sum approach used in the classical multi-criterion optimization literature, the investigators suggested a technique which exploits the population-approach of GAs. Each individual in a GA population is assigned a different weight vector. Thus, instead of finding one Pareto-optimal solution corresponding to a particular weight vector, a GA population simultaneously maintains multiple weight vectors, thereby finding multiple Pareto-optimal solutions in one simulation run. Thus, the key issue in WBGAs is to maintain diversity in the weight vectors among the population members. In WBGAs, the diversity in the weight vectors is maintained in two ways. In the first approach, a niching method is used only on the substring representing the weight vector, while in the second approach, carefully chosen subpopulations are evaluated for different pre-defined weight vectors, an approach similar to that of the VEGA. Before we discuss these approaches, we would like to reiterate that any weight-based approach for multi-objective optimization will, in principle, fail to find solutions in a nonconvex Pareto-optimal region. We have discussed this matter in Chapter 2. Fortunately, not many real-world multi-objective optimization problems have been found to have a nonconvex Pareto-optimal region. This is the reason why the classical weighted sum approaches are still popularly used in practice. To those problems, both of the following approaches may be found useful.

5.6.1 Sharing Function Approach

In addition to n decision variables, the user codes a weight vector of size M in a string. The weights must be chosen in a way so as to make the sum of all weights in every string one. One of the ways to achieve this is to normalize each weight by the sum of all weights:

$$w_i \leftarrow \frac{w_i}{\sum_{j=1}^{M} w_j}.$$

On the other hand, if only a few Pareto-optimal solutions are desired, a fixed set of weight vectors may be chosen beforehand and an integer variable x_w can be used to represent one of the weight vectors. We will use this latter approach here.

The integer variable x_w varies between 1 and a maximum number K, which denotes the maximum number of weight vectors that one is interested in. Since, in most problems, each weight vector will result in one Pareto-optimal solution, the choice of K will be governed by the maximum number of desired Pareto-optimal solutions. The choice of weight combinations is entirely up to the user. With each value of $x_w = k$, the user must associate a particular weight vector w^k. For example, in a two-objective optimization problem, nine (K = 9) different weight vectors can be represented by the integer variable x_w varying between 1 to 9, as shown in the following table.

x_w	1	2	3	$\cdots$	9
Weight vector	(0.1,0.9)	(0.2,0.8)	(0.3,0.7)	$\cdots$	(0.9,0.1)

A solution $x^{(i)}$ is then assigned a fitness equal to the weighted sum of the normalized objective function values computed as follows:

$$F(x^{(i)}) = \sum_{j=1}^{M} w_j^{x_w^{(i)}} \frac{f_j(x^{(i)}) - f_j^{min}}{f_j^{max} - f_j^{min}}. \tag{5.8}$$

The normalization of the objective function values as shown above are essential, simply because each objective function may take different range of values.

In order to preserve diversity in the weight vectors, a sharing strategy is proposed. Since the weights are directly mapped in the x_w variable, the sharing is performed with x_w variable only. That is, in the computation of the distance metric d_{ij} between two solutions i and j, only $x_w^{(i)}$ and $x_w^{(j)}$ values are considered and no other decision variables are used:

$$d_{ij} = |x_w^{(i)} - x_w^{(j)}|. \tag{5.9}$$

The sharing function $Sh(d_{ij})$ is computed with a niching parameter σ_{share}. Recall that x_w is an integer variable which takes values in steps of one. Thus, a σ_{share} value of 1 would mean that solutions with identical x_w values are only shared. A $\sigma_{share} = 2$ will allow contributions from neighboring weight vectors, but this may reduce the diversity needed to find all K different Pareto-optimal solutions, each corresponding to a weight combination.

For each solution i in the population, the sharing function values are added together for all solutions including itself and the niche count nc_i is computed. Thereafter, the fitness (calculated using equation (5.8)) is degraded by the niche count nc_i to calculate the shared fitness $F'_i = F(x^{(i)})/nc_i$.

Since fitness is degraded when using the sharing function concept, the proportionate selection method needs to be used. The crossover and mutation operators are applied on the whole population as usual. Investigators have also suggested the use of a mating restriction strategy (Deb, 1989) for better convergence of the algorithm. In the distance metric computation for deciding which two solutions should mate with each other, only the integer variable x_w is used. If the computed distance is within a threshold σ_{mating}, two solutions are mated; otherwise, another solution is tried.

Weight-Based GA Fitness Assignment

Step 1 For each objective function j, set upper and lower bounds as f_j^{max} and f_j^{min}.

Step 2 For each solution $i = 1, 2, \ldots, N$, calculate the distance $d_{ik} = |x_w^{(i)} - x_w^{(k)}|$ with all solutions $k = 1, 2, \ldots, N$. Then calculate the sharing function value as follows:

$$Sh(d_{ik}) = \begin{cases} 1 - \dfrac{d_{ik}}{\sigma_{share}}, & \text{if } d_{ik} \leq \sigma_{share}; \\ 0, & \text{otherwise.} \end{cases}$$

Thereafter, calculate the niche count of the solution i as $nc_i = \sum_{k=1}^{N} Sh(d_{ik})$.

Step 3 For each solution $i = 1, 2, \ldots, N$, follow the entire procedure below. Corresponding to the $x_w^{(i)}$ value, identify the weight vector $w^{(i)}$ from the user-defined mapping between the integer variable $x_w^{(i)}$ and the weight vector $w^{(i)}$. Assign fitness F_i according to equation (5.8). Calculate the shared fitness as $F'_i = F_i/nc_i$.

When all population members are assigned a fitness F', the proportionate selection is applied to create the mating pool. Thereafter, crossover and mutation operators are applied on the entire string, including the substring representing x_w.

Hand Calculations

WBGAs use the proportionate selection operator which is ideal for solving maximization problems. Thus, we choose the two-objective optimization problem Max-Ex to illustrate the fitness assignment procedure. We will use an identical population of six solutions as used in the VEGA, but include a third integer variable x_w. The following mapping between x_w and the corresponding five weight vectors are used:

x_w	1	2	3	4	5
Weight vector	(0.1,0.9)	(0.3,0.7)	(0.5,0.5)	(0.7,0.3)	(0.9,0.1)

Thus, an additional integer value between 1 and 5 accompanies each solution. This value suggests a weight vector (from the above table) by which the objective functions are weighted for a particular solution. Next, we follow the step-by-step procedure of the fitness assignment procedure of WBGAs, as outlined above.

Step 1 By observing the lower and upper bounds[1] parameter values, we set $f_1^{max} = 1.0$, $f_1^{min} = 0.1$, $f_2^{max} = 59.0$ and $f_2^{min} = 0.0$.

Step 2 Now, we calculate the niche count of each of the six solutions, only using the x_w variable (Table 14). We first compute the distance values:

$$d_{11} = 0, \quad d_{12} = |2 - 5| = 3, \quad d_{13} = 0,$$
$$d_{14} = 1, \quad d_{15} = 2, \quad d_{16} = 3.$$

We use $\sigma_{share} = 1$ and calculate the sharing function values as follows:

$$Sh(d_{11}) = 1, \quad Sh(d_{12}) = 0, \quad Sh(d_{13}) = 1,$$
$$Sh(d_{14}) = 0, \quad Sh(d_{14}) = 0, \quad Sh(d_{16}) = 0.$$

The niche count of the 1st solution is the sum of the above sharing function values, or $nc_1 = 2$. Similarly, we find the niche count of the other solutions in the following:

$$nc_2 = 2, \quad nc_3 = 2, \quad nc_4 = 1, \quad nc_5 = 1, \quad nc_6 = 2.$$

Step 3 We now assign the fitness to each solution. We first calculate the fitness of the 1st solution. The variable x_w takes the value 2 for this solution. With respect to the mapping (from the table above), we observe that the corresponding weight

---

[1] For an arbitrary problem, a guess of the upper and lower bounds may be used. One way to get an idea of these bounds is to calculate a sample of random solutions and then use the sample maximum and minimum values as the bounds. If in any generation, bounds outside of these ranges are found, the sample maximum and minimum values can be replaced with the new bounds. Another approach would be to first find the individual best solution for each objective. Thereafter, at each individual best solution, all other objective values are computed. Finally, for each objective, the minimum and maximum values of all such solutions are noted.

vector is $(w_1^{(1)}, w_2^{(1)})^T = (0.3, 0.7)^T$. Thus,

$$
\begin{aligned}
F_1 &= w_1^{(1)} \frac{f_1^{(1)} - f_1^{min}}{f_1^{max} - f_1^{min}} + w_2^{(1)} \frac{f_2^{(1)} - f_2^{min}}{f_2^{max} - f_2^{min}}, \\
&= 0.3 \frac{0.79 - 0.1}{1 - 0.1} + 0.7 \frac{53.90 - 0}{59 - 0}, \\
&= 0.869.
\end{aligned}
$$

The shared fitness of this solution is $F_1' = F_1/nc_1 = 0.869/2 = 0.435$.

Similarly, we calculate the fitness of other solutions as follows:

$$
F_2 = 0.660, \quad F_3 = 0.888, \quad F_4 = 0.841, \quad F_5 = 0.551, \quad F_6 = 0.265.
$$

The corresponding shared fitness values are as follows:

$$
F_2' = 0.330, \quad F_3' = 0.444, \quad F_4' = 0.841, \quad F_5' = 0.551, \quad F_6' = 0.133.
$$

These values are tabulated in Table 14. Figure 101 shows the solutions and their weighted contour lines. Column 8 of Table 14 reveals that the solution 3 has the highest weighted fitness, but not the highest shared fitness.

During the selection operation, comparison of solutions with different weight vectors does not make much sense, and must not be allowed. For example, a comparison of solutions 1 and 2 does not make sense, because solution 1 is evaluated with the first weight vector (emphasizing f_2), while solution 2 is evaluated with the fifth weight vector (emphasizing f_1). However, the original WBGA does not suggest any restriction in the selection operation. The proportionate selection operator is used on the entire population. However, for solutions having an identical x_w, the selection prefers a solution closer to the Pareto-optimal set. In the above example, solutions 1 and 3 have the same weight vector and the table reveals that solution 3 is better compared to solution 1. When sharing is performed on the weight variable, solution 3 would be emphasized over solution 1. This is because they are both competing for the same

Table 14   Fitness assignment under a WBGA with a sharing function approach.

Solution	x_1	x_2	x_w	f_1	f_2	Niche count	Fitness	Shared fitness
1	0.31	0.89	2	0.79	53.90	2	0.869	0.435
2	0.43	1.92	5	0.67	53.21	2	0.660	0.330
3	0.22	0.56	2	0.88	52.91	2	0.888	0.444
4	0.59	3.63	1	0.51	52.15	1	0.841	0.841
5	0.66	1.41	4	0.44	56.35	1	0.551	0.551
6	0.83	2.51	5	0.27	55.77	2	0.265	0.133

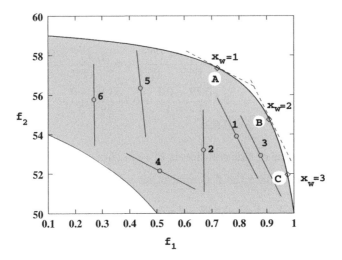

Figure 101 Weighted direction for each solution.

weight vector and the figure shows that solution 3 is closer to the Pareto-optimal front than solution 1. A similar conclusion can be made for solutions 2 and 6 having the same weight vector. Here, solution 2 is better. Solutions 4 and 5 are only representatives of the weight vectors $(0.1, 0.9)^T$ and $(0.7, 0.3)^T$, respectively. It is interesting to note how these solutions are emphasized by means of their shared fitness values. In this way, the sharing procedure will maintain different weight vectors in the population and the fitness assignment procedure will emphasize the correct solution from the niche of solutions having an identical weight vector. In this problem, the true Pareto-optimal solution $(x_1^*, x_2^*)^T$ corresponding to any given weight vector $(w_1, w_2)^T$ can be calculated by differentiating the weighted normalized objective function and is found to be as follows:

$$x_1^* = \sqrt{\frac{0.9}{59.0} \frac{w_2}{w_1}}, \tag{5.10}$$

$$x_2^* = 0. \tag{5.11}$$

Figure 101 also shows three Pareto-optimal solutions (A, B and C) corresponding to $x_w = 1$, 2 and 3, respectively. The exact location of these solutions are calculated by using equations (5.10) and (5.11). The other two Pareto-optimal solutions (corresponding to $x_w = 4$ and 5) lie outside the range $0.1 \le x_1 \le 1$. It is also interesting to note that a uniformly spread set of weight vectors may not correspond to a uniformly spread set of Pareto-optimal solutions in the objective space. In the following table, we show the Pareto-optimal function values for different weight vectors:

Weight vector $(f_1^*, f_2^*)^T$	$(0.02, 0.98)^T$ $(0.24, 58.84)^T$	$(0.1, 0.9)^T$ $(0.73, 57.30)^T$	$(0.3, 0.7)^T$ $(0.91, 54.70)^T$	$(0.5, 0.5)^T$ $(0.98, 51.90)^T$
Weight vector $(f_1^*, f_2^*)^T$	$(0.7, 0.3)^T$ $(1.02, 47.63)^T$	$(0.9, 0.1)^T$ $(1.06, 35.71)^T$	$(0.98, 0.02)^T$ $(1.08, 3.32)^T$	

Computational Complexity

Step 2 of a WBGA computes the distance metric between all pairs of solutions, thereby requiring $O(N^2)$ computations. Other computations are smaller than this computation. Thus, the overall complexity of the fitness assignment procedure of a WBGA is $O(N^2)$. However, if a weight vector is coded, $O(MN^2)$ computations are necessary.

It is important to highlight here that if a stochastic sharing function approach is used (Deb and Kumar, 1995; Goldberg and Richardson, 1987), this complexity can be reduced. Instead of calculating the distance between one solution with all $(N-1)$ other solutions, the niche count can be calculated with a sample (of size $\eta \ll N$) of solutions. Although the niche count computation may not be accurate, the overall complexity can be reduced to $O(\eta N)$. Elsewhere (Deb and Kumar, 1995), it is shown that a subpopulation of size of about one-tenth of the population size is enough to maintain adequate diversity in solutions in a number of multi-modal problems.

Advantages

Since a WBGA uses a single-objective GA, not much change is needed to convert a simple GA implementation into a WBGA one. Moreover, the complexity of the algorithm (with a single extra variable x_w) is smaller than other multi-objective evolutionary algorithms.

Disadvantages

The WBGA uses a proportionate selection procedure on the shared fitness values. Thus, in principle, the WBGA will work in a straightforward manner in finding solutions for maximization problems. However, if objective functions are to be minimized, they are required to be converted into a maximization problem. Moreover, for mixed types of objective functions (some are to be minimized and some are to be maximized), complications may arise in trying to construct a fitness function.

Weighted sum approaches share a common difficulty in finding Pareto-optimal solutions in problems having nonconvex Pareto-optimal region. Thus, a WBGA may also face difficulty in solving such problems.

We have discussed above that usually a uniformly distributed set of weight vectors may not result in a set of uniformly distributed Pareto-optimal solutions. Thus, if a uniform spread of non-dominated solutions is desired, it is difficult to choose an appropriate set of weight vectors.

Simulation Results

The same GA parameter settings that were used in the previous algorithm are used here. We use $\sigma_{\text{share}} = 1$. Figure 102 shows the non-dominated solutions found after 500 generations of the WBGA. The bit-wise mutation operator is used here. The figure

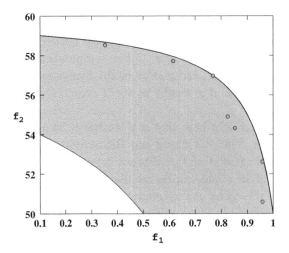

Figure 102   Non-dominated solutions obtained after 500 generations of the WBGA with the sharing approach.

also shows that only a few solutions on or near the Pareto-optimal front are found.

5.6.2   Vector Evaluated Approach

The vector evaluated approach is similar to the VEGA. First, a set of K different weight vectors $\mathbf{w}^{(k)}$ (where $k = 1, 2, \ldots, K$) are chosen. Thereafter, each weight vector $\mathbf{w}^{(k)}$ is used to compute the weighted normalized fitness for all N population members. Among them, the best N/K members (decided on the basis of the weighted normalized fitness values) are grouped together into a subpopulation. This subpopulation is then associated with the weight vector $\mathbf{w}^{(k)}$ for subsequent genetic processing. Selection, crossover and mutation are all restricted to solutions of each subpopulation. The subpopulation size N/K for each weight vector is maintained so that at the end of these operations the total population size remains the same as N. In the next generation, once again each weight vector is used to find the best subpopulation from the new population and the generations continue.

   It is clear from the above procedure that there is a total of K subpopulations, one for each chosen weight vector. In addition, one population member can be associated with more than one weight vector. In particular, the middle solutions will be included in more than one subpopulations. Hajela and Lin (1993) also tested a scheme where one population member is strictly allowed to be included in exactly one subpopulation. However, finally they concluded that this scheme is not as good as the former one. Therefore, we will only use the former scheme here.

   In this algorithm, multiple solutions are found by explicitly preserving K weight vectors. Thus, there is no need for introducing any additional niching operator. One subpopulation is processed for each weight vector and genetic operations search for the best possible solution for that weight vector. In this way, diversity among weight vectors is maintained explicitly and convergence to the Pareto-optimal solutions is attempted using a genetic algorithm procedure.

In the following, we present one generation of this approach in a step-by-step manner. It is assumed that a set of K weight vectors are known.

Vector Evaluated Approach

Step 1 Set weight vector counter $k = 1$.

Step 2 Find the fitness F_j of each solution j with its weight vector $w^{(k)}$ using equation (5.8). Choose the best N/K solutions according to fitness F. Copy these solutions in the subpopulation P_k.

Step 3 Perform selection, crossover and mutation on P_k to create a new population of size N/K.

Step 4 If $k < K$, increment k by one and go to Step 2. Otherwise, combine all subpopulations to create the new population $P = \cup_{k=1}^{K} P_k$. If $|P| < N$, add randomly created solutions to make the population size equal to N.

Step 2 evaluates N solutions and is performed K times, thereby requiring a total of KN evaluations. If K is very small compared to N, the complexity involved in each generation of this algorithm is $O(N)$.

We now apply the above procedure without Step 3 on the two-variable example problem Max-Ex. Let us assume that we consider the following two weight vectors:

x_w	1	2
Weight vector	(0.1,0.9)	(0.5,0.5)

Step 1 We set $k = 1$. We also use the same lower and upper bounds of the objective values as used before.

Step 2 For the first weight vector $w^{(1)} = (0.1, 0.9)^T$, we evaluate all six population members in Table 15 (see below). Since $K = 2$ and $N = 6$, we need to choose only

Table 15 Fitness assignment under a WBGA with the vector evaluated approach.

Solution	x_1	x_2	f_1	f_2	Fitness with $w^{(1)}$	Fitness with $w^{(2)}$
1	0.31	0.89	0.79	53.90	0.899	0.840
2	0.43	1.92	0.67	53.21	0.875	0.768
3	0.22	0.56	0.88	52.91	0.894	0.881
4	0.59	3.63	0.51	52.15	0.841	0.669
5	0.66	1.41	0.44	56.35	0.897	0.666
6	0.83	2.51	0.27	55.77	0.870	0.567

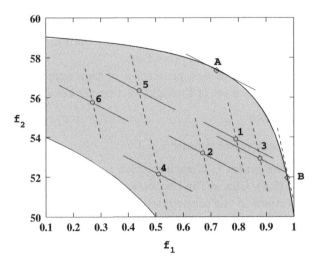

Figure 103   Weighted directions in the vector evaluated approach.

the three best solutions for each weight vector. It turns out that solutions 1, 3, and 5 belong to the first subpopulation P_1. Figure 103 also shows that among all six solutions, these three solutions are closer to solution A, which is the Pareto-optimal solution corresponding to the first weight vector. The continuous lines are used to show the weighted directions.

Step 3 Ideally, we should now use these three solutions and perform an iteration of selection, crossover and mutation to create a new subpopulation of size 3. Since we are interested in illustrating the fitness assignment scheme here, we ignore this step.

Step 4 Next, we increment k to 2 and move to Step 2.

Step 2 With the second weight vector $\mathbf{w}^{(2)} = (0.3, 0.7)^{\mathsf{T}}$, we find that solutions 1, 2, and 3 belong to the second subpopulation P_2 (refer to the seventh column in Table 15). Figure 103 also shows the weighted directions with dashed lines. These solutions are closer to solution B, which is the Pareto-optimal solution corresponding to the second weight vector.

Step 3 As before, three new solutions should be created with three solutions (1, 2 and 3). However, we will again ignore this step here.

Step 4 The population at the end of this iteration will be the union set of two new subpopulations (P_1 and P_2) which would have been created in Step 3 for k = 1 and 2. If there exist some common members between these two subpopulations, randomly created solutions can be added to fill up the population.

Advantages and Disadvantages of the Vector Evaluated Approach

As in any weight-based method, a knowledge of the weight vectors is also essential in this approach. This method is better in complexity than the sharing function approach, because no pair-wise distance metric is required here. There is also no need for keeping any additional variable for the weight vectors. However, since GA operators are applied to each subpopulation independently, an adequate subpopulation size is required for finding the optimal solution corresponding to each weight vector. This requires a large population size.

By applying both methods in a number of problems, investigators concluded that the vector evaluated approach is better than the sharing function approach.

Simulation Results

We have attempted to solve the test problem Max-Ex by using the vector evaluated approach. Figure 104 shows the non-dominated solutions after 500 generations. Comparing Figures 102 and 104, we observe that a WBGA with the vector evaluated approach is able to find a better spread of population members on the Pareto-optimal front. Moreover, the convergence to the Pareto-optimal front is also better with the vector evaluated approach.

Coello and Christiansen (1999) used a similar technique in their multi-objective optimization tool for engineering design (MOSES) by executing independent GAs with user-supplied weight vectors. For each weight vector, one solution is found. When all GAs found this optimized solutions, they are declared as the set of non-dominated solutions. Unlike in the above WBGA method, weights are not coded, but a fixed user-defined weight vector is used in each GA run.

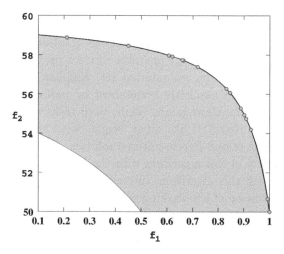

Figure 104 Non-dominated solutions after 500 generations using a WBGA and the vector evaluated approach.

5.7   Random Weighted GA

Murata and Ishibuchi (1995) suggested a random weighted GA (RWGA) similar to the above WBGA, except that a random normalized weight vector $\mathbf{w}^{(i)} = (w_1, w_2, \ldots, w_M)^T_{(i)}$ is assigned to the i-th solution in the population. The fitness of the solution is assigned as the weighted sum of the objectives with $\sum_{j=1}^{M} w_j = 1$:

$$f(\mathbf{x}^{(i)}) = \sum_{j=1}^{M} w_j^{(i)} f_j(\mathbf{x}^{(i)}). \qquad (5.12)$$

After all population members are assigned a fitness, a proportionate selection operator, a crossover operator, and a mutation operator are applied to form a new population. Before accepting the new population, a random portion of the population is replaced with an equal number of solutions chosen from an external population (created at random). Although the original study referred to this process as an elite-preserving operator, it is not clear how randomly picked solutions can qualify as elites in subsequent generations. In a later version of the RWGA (Ishibuchi and Murata, 1998a), investigators have clearly demonstrated the use of maintaining an external population containing non-dominated solutions only. There, this external population gets modified with current non-dominated solutions, as performed in other elitist MOEAs (described in the next chapter). Nevertheless, the original study and subsequent studies of the investigators have shown the applicability of the RWGA in many optimization problems such as multi-objective flow-shop scheduling and fuzzy rule selection problems. In comparison to the VEGA, they have shown better convergence results in these problems. We discuss a modified RWGA later in Section 8.11.

In the RWGA, the diversity in the non-dominated solutions is maintained in two ways: (i) by using a random weight vector to evaluate each solution, thereby emphasizing solutions which may lead to different solutions in the Pareto-optimal region, and (ii) by using an exchange operator where a proportion of the population is replaced with solutions from an external set. Emphasis for solutions closer to the Pareto-optimal front is explicitly maintained by performing the proportionate selection with weighted function values. Thus, both tasks needed in an MOEA are present in the RWGA. However, like other weight-based approaches, the RWGA is also likely to be unable to find Pareto-optimal solutions in nonconvex problems.

Due to the similarity of this algorithm with the WBGA, we will not perform any simulation studies with this algorithm. Interested readers may refer to the original report for a demonstration of the working of this algorithm.

5.8   Multiple Objective Genetic Algorithm

Fonseca and Fleming (1993) first introduced a multi-objective GA (called MOGA) which used the non-dominated classification of a GA population. The investigators were the first to suggest a multi-objective GA which explicitly caters to emphasize non-dominated solutions and simultaneously maintains diversity in the non-dominated solutions.

The MOGA differs from a standard tripartite GA in the way fitness is assigned to each solution in the population. The rest of the algorithm (the stochastic universal selection (SUS), a single-point crossover, and a bit-wise mutation) is the same as that in a classical GA. We describe the MOGA in the following.

First, each solution is checked for its domination in the population. To a solution i, a rank equal to one plus the number of solutions n_i that dominate solution i is assigned:

$$r_i = 1 + n_i. \tag{5.13}$$

In this way, non-dominated solutions are assigned a rank equal to 1, since no solution would dominate a non-dominated solution in a population. A little thought reveals that in any population, there must be at least one solution with rank equal to one and the maximum rank of any population member cannot be more than N (the population size). Figure 105 shows a two-objective minimization problem having 10 solutions, while Figure 106 shows the rank of each solution. The shaded region represents the feasible search space. It is clear that the ranking procedure may not assign all possible ranks (between 1 and N) to any population. For example, ranks 7, 9 and 10 are missing in the population used in the figure. Once the ranking is performed, a raw fitness to a solution is assigned based on its rank. To perform this, first the ranks are sorted in ascending order of magnitude. Then, a raw fitness is assigned to each solution by using a linear (or any other) mapping function. Usually, the mapping function is chosen so as to assign fitness between N (for the best-rank solution) and 1 (for the worst-rank solution). Thereafter, solutions of each rank are considered at a time and their raw fitnesses are averaged. This average fitness is now called the assigned fitness to each solution of the rank. In this way, the total allocated raw fitness and total assigned fitness to each rank remains identical. Moreover, the mapping and averaging procedure ensures that the better ranked solutions have a higher assigned fitness. In this way, non-dominated solutions are emphasized in a population.

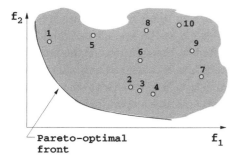

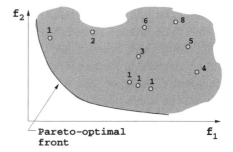

Figure 105 A two-objective search space and the corresponding Pareto-optimal front.

Figure 106 Ranking procedure in the MOGA.

Lee and Esbensen (1997) also used the above ranking procedure to assign copies in a proportionate selection. Although the idea of converting multiple objective values to a ranked metric by using the non-domination principle is a step towards an efficient multi-objective optimization, the second goal of maintaining a diverse set of solutions cannot be guaranteed by this simple procedure.

In order to maintain diversity among non-dominated solutions, Fonseca and Fleming (1993) have introduced niching among solutions of each rank. The niche count with σ_{share} is found as described in the previous chapter. The sharing function with $\alpha = 1$ is used, but the distance metric is computed with the objective function values, instead of the parameter values. Thus, the normalized distance between any two solutions i and j in a rank is calculated as follows:

$$d_{ij} = \sqrt{\sum_{k=1}^{M} \left(\frac{f_k^{(i)} - f_k^{(j)}}{f_k^{max} - f_k^{min}} \right)^2},$$ (5.14)

where f_k^{max} and f_k^{min} are the maximum and minimum objective function value of the k-th objective. For the solution i, d_{ij} is computed for each solution j (including i) having the same rank. Equation (4.59) (see above) is used with $\alpha = 1$ to compute the sharing function value. Thereafter, the niche count is calculated by summing the sharing function values:

$$nc_i = \sum_{j=1}^{\mu(r_i)} Sh(d_{ij}),$$ (5.15)

where $\mu(r_i)$ is the number of solutions in rank r_i. In an MOGA, the sharing function approach, as described earlier in Section 4.6.4, is followed and a shared fitness value is calculated by dividing the fitness of a solution by its niche count. Although all solutions of any particular rank have the identical fitness, the shared fitness value of a solution residing in a less-crowded region has a better shared fitness. This produces a large selection pressure for poorly represented solutions in any rank.

Dividing the assigned fitness values by the niche count (always equal to or greater than one) reduces the fitness of each solution. In order to keep the average fitness of the solutions in a rank the same as that before sharing, these fitness values are scaled so that their average shared fitness value is the same as the average assigned fitness value. After these calculations, the focus is shifted to the solutions of the next rank and an identical procedure is executed. This procedure is continued until all ranks are processed. Thereafter, the stochastic universal selection (SUS) (with shared fitness values), the single-point crossover, and the bit-wise mutation operators are applied to create a new population.

MOGA Fitness Assignment Procedure

> **Step 1** Choose a σ_{share} (a dynamically updated procedure for fixing σ_{share} is described later). Initialize $\mu(j) = 0$ for all possible ranks $j = 1, \ldots, N$. Set solution counter $i = 1$.

Step 2 Calculate the number of solutions (n_i) that dominates solution i. Compute the rank of the i-th solution as $r_i = 1 + n_i$. Increment the count for the number of solutions in rank r_i by one, that is, $\mu(r_i) = \mu(r_i) + 1$.

Step 3 If $i < N$, increment i by one and go to Step 1. Otherwise, go to Step 4.

Step 4 Identify the maximum rank r^* by checking the largest r_i which has $\mu(r_i) > 0$. The sorting according to rank and fitness-averaging yields the following assignment of the average fitness to any solution $i = 1, \ldots, N$:

$$F_i = N - \sum_{k=1}^{r_i-1} \mu(k) - 0.5(\mu(r_i) - 1). \tag{5.16}$$

To each solution i with rank $r_i = 1$, the above equation assigns a fitness equal to $F_i = N - 0.5(\mu(1) - 1)$, which is the average value of $\mu(1)$ consecutive integers from N to $N - \mu(1) + 1$. Set a rank counter $r = 1$.

Step 5 For each solution i in rank r, calculate the niche count nc_i with other solutions ($\mu(r)$ of them) of the same rank by using equation (5.15). Calculate the shared fitness using $F'_j = F_j/nc_j$. To preserve the same average fitness, scale the shared fitness as follows:

$$F'_j \leftarrow \frac{F_j \mu(r)}{\sum_{k=1}^{\mu(r)} F'_k} F'_j.$$

Step 6 If $r < r^*$, increment r by one and go to Step 5. Otherwise, the process is complete.

We now illustrate the working of the MOGA on the same two-objective optimization problem Min-Ex.

5.8.1 Hand Calculations

We begin with the six-membered population used earlier. To illustrate the fitness assignment procedure of the MOGA better, we follow the above step-by-step procedure.

Step 1 We choose $\sigma_{share} = 0.5$, and $\mu(j) = 0$ for all $j = 1$ to 6 (since there are $N = 6$ population members).

Steps 2 and 3 For solution 1, we find that $n_1 = 0$, and thus, $r_1 = 1 + 0 = 1$. Similarly, we find r_i for all other solutions and list them in Table 16. We observe that there are three solutions with rank 1 (thus, $\mu(1) = 3$), two solutions with rank 2 ($\mu(2) = 2$) and one solution with rank 4 ($\mu(4) = 1$). Therefore, there is no solution in rank 3, 5 and 6, or $\mu(3) = \mu(5) = \mu(6) = 0$.

Step 4 The maximum rank r^* assigned to any solution in the population is 4. Sorting of the population according to rank yields the following sequence (Table 16):

$$(1,3,5)\ (2,6)\ (\ )\ (4).$$

Table 16 Fitness assignment procedure of the MOGA.

Solution	x_1	x_2	f_1	f_2	Rank	Average fitness	Niche count	Shared fitness	Scaled fitness
1	0.31	0.89	0.31	6.10	1	5.0	1.752	2.854	4.056
2	0.43	1.92	0.43	6.79	2	2.5	1.000	2.500	2.500
3	0.22	0.56	0.22	7.09	1	5.0	1.702	2.938	4.176
4	0.59	3.63	0.59	7.85	4	1.0	1.000	1.000	1.000
5	0.66	1.41	0.66	3.65	1	5.0	1.050	4.762	6.768
6	0.83	2.51	0.83	4.23	2	2.5	1.000	2.500	2.500

These ranks are also shown against solutions on Figure 107. We now assign a raw fitness of 6.00 to solution 1, a raw fitness of 5.00 to solution 3, and so on. Finally, solution 4 gets a raw fitness equal to 1. Then, we average the raw fitnesses of all solutions in each rank and reassign that value as the average fitness of each solution having the same rank. Alternatively, the above procedure can be substituted using equation (5.16), as follows:

$$F_1 = N - 0.5(\mu(1) - 1) = 6 - 0.5(3 - 1) = 5.0,$$
$$F_2 = N - \mu(1) - 0.5(\mu(2) - 1) = 6 - 3 - 0.5(2 - 1) = 2.5,$$
$$F_3 = N - 0.5(\mu(1) - 1) = 6 - 0.5(3 - 1) = 5.0,$$
$$F_4 = N - \mu(1) - \mu(2) - \mu(3) - 0.5(\mu(4) - 1)$$
$$= 6 - 3 - 2 - 0 - 0.5(1 - 1) = 1.0,$$

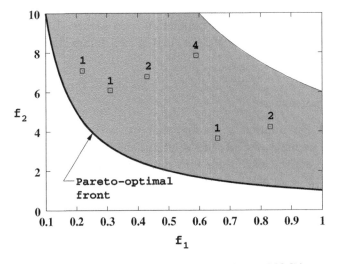

Figure 107 Ranking of solutions under an MOGA.

$$F_5 = N - 0.5(\mu(1) - 1) = 6 - 0.5(3 - 1) = 5.0,$$
$$F_6 = N - \mu(1) - 0.5(\mu(2) - 1) = 6 - 3 - 0.5(2 - 1) = 2.5.$$

It is seen that all three solutions of the first rank has the same average fitness of 5.0.

Step 5 For three solutions of the first rank, we will now calculate the niche count in the objective function space. Assuming, $f_1^{min} = 0.1$, $f_1^{max} = 1.0$, $f_2^{min} = 1$, and $f_2^{max} = 10$, we have the following normalized distances:

$$d_{13} = 0.149, \quad d_{15} = 0.475, \quad d_{35} = 0.621.$$

Using these values and $\sigma_{share} = 0.5$, we calculate the corresponding sharing function values (with $\alpha = 1$):

$$Sh(d_{13}) = 0.702, \quad Sh(d_{15}) = 0.050, \quad Sh(d_{35}) = 0.$$

Of course, $Sh(d_{11}) = Sh(d_{33}) = Sh(d_{55}) = 1$. Thus, the niche counts for solutions 1, 3 and 5 are as follows:

$$
\begin{aligned}
nc_1 &= 1 + 0.702 + 0.050 = 1.752, \\
nc_3 &= 1 + 0.702 + 0 = 1.702, \\
nc_5 &= 1 + 0.050 + 0 = 1.050.
\end{aligned}
$$

Now, dividing the average fitness values by the corresponding niche count, we calculate the shared fitness F_j' (column 9 in Table 16).

Now, we scale these fitness values so that their average is the same as the original average fitness value. The scaling factor is $F_1\mu(1)/(F_1' + F_3' + F_5')$ or $5 \times 3/(2.854 + 2.938 + 4.762)$ or 1.421. We multiply column 9 in Table 16 by 1.421 to calculate the scaled fitness values of solutions 1, 3 and 5.

Step 6 We now move to Step 5 with solutions of rank 2.

Step 5 We compute $d_{26} = 0.528$. Thus, $Sh(d_{26}) = 0$ (solutions 2 and 6 have no effect on each other) and niche counts $nc_2 = nc_6 = 1$. Thus, the shared fitness values are the same as the average fitness values and the scaling factor is one. Table 16 shows the corresponding scaled fitness values.

Step 6 Now, we move to Step 5 again for rank 3 solutions. Since there are none, we move to Step 5 for rank 4 solutions. Since there is only one solution in rank 4, its scaled fitness value is the same as its average fitness value. Since the maximum rank is 4 in the population, we stop here. This completes the fitness assignment procedure of the MOGA.

5.8.2   Computational Complexity

It is clear from the algorithm that Steps 2 and 3 are executed N times, each time comparing $(N - 1)$ solutions for domination. Although the procedure can be made faster by using proper bookkeeping, this procedure requires $O(MN^2)$ comparisons. Since there could be a maximum of N ranks, the sorting procedure in Step 4, niche count computation in Step 5, and the scaling of shared fitness values will not require worse than $O(N^2)$ computations. Thus, the overall complexity of the MOGA is $O(MN^2)$.

5.8.3   Advantages

The fitness assignment scheme is simple in MOGA. Since niching is performed in the objective space, the MOGA can be easily applied to other optimization problems, such as combinatorial optimization problems. If a spread of Pareto-optimal solutions is required on the objective space, the MOGA is the suitable choice.

5.8.4   Disadvantages

Although the concept of domination is used to assign fitness, all solutions in a particular non-dominated front (except the first one), need not have the same assigned fitness. This may introduce an unwanted bias towards some solutions in the search region. In particular, this algorithm may be sensitive to the shape of the Pareto-optimal front and to the density of solutions in the search space.

The shared fitness computation procedure does not make sure that a solution in a poorer rank will always have a worse scaled fitness F' than every solution in a better rank. This may happen particularly if there exist many crowded solutions with a better rank. The niche count for these solutions would be large and the resulting shared fitness may be small. If this happens, adequate selection pressure may not exist to all solutions in a better rank, thereby leading to a slow convergence or inability to find a good spread in the Pareto-optimal front. In an NSGA, this problem is avoided by assigning fitness values front-wise. Each solution in a better front is assigned a larger shared fitness value.

Like other GAs which use niche counts to maintain diversity, the σ_{share} parameter needs fixing. However, developers of MOGAs have suggested a dynamic update of σ_{share}, which does not require any pre-fixing of this parameter. We will describe this procedure in a later subsection.

5.8.5   Simulation Results

We apply an MOGA on the two-objective test problem using a fixed $\sigma_{share} = 0.158$ and other parameter settings as before (population size 40, crossover probability 0.9 and binary coding with (24+26) or 50 bits). The MOGA without the mutation operator cannot find a good spread, as shown in Figure 108. At the end of 50 generations,

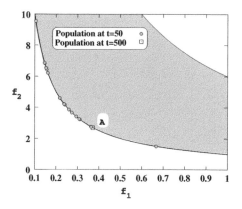

Figure 108   Population after 50 and 500 generations obtained with an MOGA without mutation. At 500 generation, all solutions converge to the solution A.

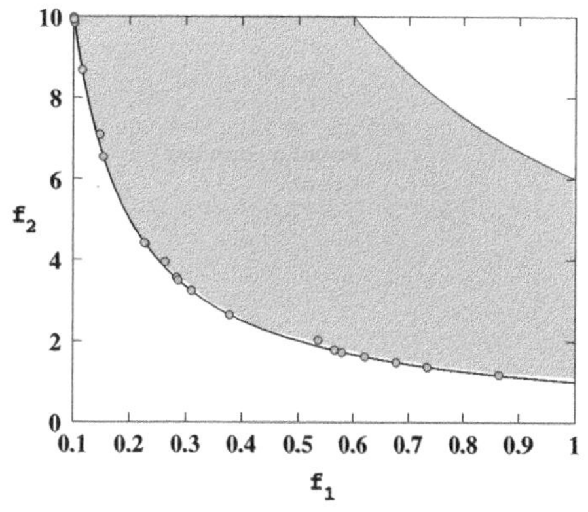

Figure 109   Population after 500 generations obtained with an MOGA with mutation.

many distinct Pareto-optimal solutions are found; however, as the MOGA is run for a large number of generations (500 generations), the population converges to only one Pareto-optimal solution (marked 'A' in the figure). In the absence of any other diversity preserving mechanism (such as mutation), the niching method adopted in an MOGA does not seem to maintain the needed diversity among the Pareto-optimal solutions.

When MOGA is applied to Min-Ex with the bit-wise mutation operator with a mutation probability of 0.02, a different scenario emerges. The algorithm can now find more solutions near the front, as shown in Figure 109. Even after 500 generations, a good spread is maintained by the help of the mutation operator.

5.8.6   Dynamic Update of the Sharing Parameter

The sharing parameter σ_{share} signifies the maximum distance (in phenotype or genotype space) between any two solutions before they can be considered to be in the same niche. In the context of multi-objective optimization, we should choose a value of the σ_{share} parameter in such a way that we have a uniform distribution of solutions in the Pareto-optimal front. If we are following the objective function space niching, as in an MOGA, we compute distance as the Euclidean distance between any two solutions in the objective function space. Thus, if the Pareto-optimal front is known, we would like to have all N (population size) solutions uniformly spaced on the Pareto-optimal front. This is depicted in Figure 110. Knowing the perimeter (L) of the Pareto-optimal front, we would calculate $\sigma_{share} = L/N$. In this way, each niche will contain one solution, thereby having a near-uniform distribution of Pareto-optimal solutions. However, the difficulty of the above simple approach is that the Pareto-optimal region is not known a priori.

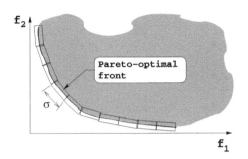

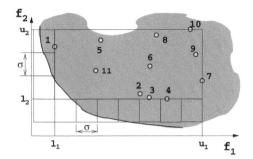

Figure 110 The Pareto-optimal region is divided equally to find a suitable σ_{share}.

Figure 111 Fonseca and Fleming's method of calculating σ_{share}.

In the absence of any knowledge of the Pareto-optimal front, Fonseca and Fleming (1993) suggested a simple procedure for calculating σ_{share} in each generation. From the population members, first find the minimum and maximum bounds of each of the objective function values (l_i and u_i). Now, it can be argued that all solutions in the current population lie in a hypervolume $V = \Pi_{i=1}^{M}(u_i - l_i)$. On each objective, we now increment the range by adding σ_{share} and find the new hypervolume $V' = \Pi_{i=1}^{M}(u_i - l_i + \sigma_{\text{share}})$. The difference in these two hypervolumes is $\Delta V = V' - V$ and all N solutions should ideally lie in this differential hypervolume. Figure 111 shows a population of 11 solutions in the objective function space at any particular generation and how the hypervolume is divided into 11 equal squares. Since each niche occupies a hypervolume of $\sigma_{\text{share}}^{M}$ and there are N such hypervolumes in ΔV, we use the following equation to calculate σ_{share}:

$$\Delta V = \Pi_{i=1}^{M}(u_i - l_i + \sigma_{\text{share}}) - \Pi_{i=1}^{M}(u_i - l_i) = N(\sigma_{\text{share}})^{M}. \qquad (5.17)$$

For more than three objective functions ($M > 2$), a numerical root finding method (Press et al., 1988) can be used to find σ_{share}. For two objective functions ($M = 2$), the above equation reduces to

$$\sigma_{\text{share}} = \frac{u_2 - l_2 + u_1 - l_1}{N - 1}. \qquad (5.18)$$

In order to use the above equations, care must be taken to normalize each objective function so that an identical σ_{share} can be used for each of the objective functions. If normalized objective function values are used, equation (5.17) can be written for the normalized niching parameter σ'_{share} as follows:

$$(1 + \sigma'_{\text{share}})^{M} - 1 = N(\sigma'_{\text{share}})^{M}. \qquad (5.19)$$

For $M = 2$, the above equation yields $\sigma'_{\text{share}} = 2/(N - 1)$. Table 17 lists these normalized niching parameter values for a few cases by finding σ'_{share} from the above equation. It is intuitive that as the population size increases, the σ'_{share} value reduces in order to accommodate more niches. It is also interesting to note that as the number

Table 17 Normalized σ'_{share} values for several cases of M and N.

N	Number of objectives, M			
	2	3	4	5
30	0.069	0.377	0.707	1.014
50	0.041	0.280	0.560	0.827
70	0.029	0.231	0.484	0.730
100	0.020	0.190	0.417	0.643
150	0.013	0.152	0.354	0.560
200	0.010	0.131	0.316	0.510
300	0.007	0.105	0.270	0.447
400	0.005	0.090	0.242	0.408
500	0.004	0.081	0.223	0.381

of objective functions increases, the σ'_{share} value also increases. This is due to the increase in the dimensionality of the search space.

The above equation allows a static normalized σ'_{share} to be used, thereby eliminating the need for evaluating σ_{share} in every generation, as proposed by the investigators. However, instead of computing the Euclidean distance metric, we now have to compute the normalized Euclidean distance, as follows:

$$d'_{ij} = \sqrt{\sum_{k=1}^{M} \left(\frac{f_k^{(i)} - f_k^{(j)}}{u_k - l_k} \right)^2}. \tag{5.20}$$

Note that the above equation for computing the distance metric is different from that shown earlier in equation (5.14) (see page 202). Here, the bounds u_k and l_k for the objective functions change with generations, while in the latter, the bounds are fixed over the entire GA run.

When equation (5.18) is used to update σ_{share} in each iteration of MOGA, similar performance of the algorithm is observed in the two-objective test problem Min-Ex. The obtained Pareto-optimal solutions without and with the mutation operator are shown in Figures 112 and 113, respectively. Besides these studies, more experimental studies are needed to investigate if the added complexity in dynamic σ_{share} computation is worthwhile over a reasonable static (fixed) value (Deb and Goldberg, 1989).

5.9 Non-Dominated Sorting Genetic Algorithm

Goldberg's idea of using the non-dominated sorting concept in GAs (Goldberg, 1989) was more directly implemented by Srinivas and Deb in 1994. Once again, the dual objectives in a multi-objective optimization algorithm are maintained by using a fitness assignment scheme which prefers non-dominated solutions and by using a

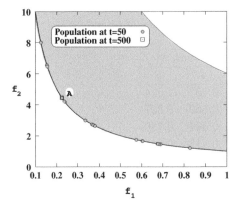

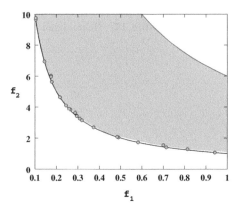

Figure 112 Populations after 50 and 500 generations obtained with an MOGA without mutation. At generation 500, all solutions converge to the solution A.

Figure 113 Population after 500 generations obtained with an MOGA with mutation.

sharing strategy which preserves diversity among solutions of each non-dominated front. In the following paragraphs, we will discuss the non-dominated sorting GA (or NSGA) in some detail.

The first step of an NSGA is to sort the population P according to non-domination. Different algorithms suitable for this task are described in Section 2.4 above. This classifies the population into a number of mutually exclusive equivalent classes (or non-dominated sets) P_j:

$$P = \cup_{j=1}^{\rho} P_j.$$

It is important to reiterate that any two members from the same class cannot be said to be better than one another with respect to all objectives. The total number of classes (or fronts), denoted as ρ in the above equation, depends on the population P and the underlying problem. As we have discussed before, the procedure of completely classifying the population according to non-domination is at most $O(MN^2)$. Let us consider the population shown in Figure 105. After the non-dominated sorting, the population members are classified into four distinct non-dominated sets, as shown in Figure 114.

Once the classification task is over, it is clear that all solutions in the first set, that is, all $i \in P_1$, belong to the best non-dominated set in the population. The second best solutions in the population are those that belong to the second set, or all members of P_2, and so on. Obviously, the worst solutions are those belonging to the final set, or all members of P_ρ, where ρ is the number of different non-dominated sets in the population. By examining Figure 114 carefully, it is clear that solutions of the first front are best in terms of their closeness to the true Pareto-optimal front in the population. Thus, it makes sense to assign the highest fitness to solutions of the best non-dominated front and then assign a progressively worse fitness to solutions

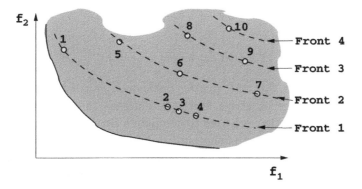

Figure 114 Ten solutions of Figure 105 are classified into different non-domination fronts.

of higher fronts. This is because the best non-dominated solutions in a population are nearest to the true Pareto-optimal front compared to any other solution in the population. In the following, we will describe the exact NSGA fitness assignment procedure which was adopted by Srinivas and Deb (1994).

The fitness assignment procedure begins from the first non-dominated set and successively proceeds to dominated sets. Any solution i of the first (or best) non-dominated set is assigned a fitness equal to $F_i = N$ (the population size). This specific value of N is used for a particular purpose. Since all solutions in the first non-dominated set are equally important in terms of their closeness to the Pareto-optimal front relative to the current population, we assign the same fitness to all of them. With respect to Figure 114, solutions 1 to 4 are all assigned a fitness equal to 10 (the population size).

Assigning more fitness to solutions belonging to a better non-dominated set ensures a selection pressure towards the Pareto-optimal front. However, in order to achieve the second goal, diversity among solutions in a front must also be maintained. Unless an explicit diversity-preserving mechanism is used, GAs (or any EAs) are not likely to ensure diversity. In an NSGA, the diversity is maintained by degrading the assigned fitness based on the number of neighboring solutions. In order to explain the need of diversity preservation, let us refer to the solutions in the first front in Figure 114. It is clear that the front is not represented adequately with the four solutions 1 to 4. Three solutions (2 to 4) are crowded in one portion of the front, whereas there is only one solution representing the entire top-left portion of the front. If the solution 1 is not emphasized adequately and if it gets lost in subsequent genetic operations, the GA has to rediscover this part of the Pareto-optimal front. In order to avoid this problem, we have to make sure that less crowded regions in a front are adequately emphasized. In an NSGA, the sharing function method (refer to Section 4.6.4) is used for this purpose.

The sharing function method is used front-wise. That is, for each solution i in the front P_1, the normalized Euclidean distance d_{ij} from another solution j in the same front is calculated, as follows:

$$d_{ij} = \sqrt{\sum_{k=1}^{|P_1|} \left(\frac{x_k^{(i)} - x_k^{(j)}}{x_k^{max} - x_k^{min}} \right)^2}. \tag{5.21}$$

It is important to note that the sharing distance is calculated with the decision variables, instead of the objective function values as in an MOGA. Once these distances are calculated, they are used to compute a sharing function value by using equation (4.59) with $\alpha = 2$. The sharing function takes a value between zero and one, depending on the distance d_{ij}. Any solution j which has a distance greater than σ_{share} from the i-th solution contributes nothing to the sharing function value. Although a dynamic update procedure of σ_{share} is suggested (Fonseca and Fleming, 1995), described above in Section 5.8.6, an NSGA uses a fixed value, employing equations (4.63) or (4.64). After all $|P_1|$ sharing function values are calculated, they are added together to calculate the niche count nc_i of the i-th solution. The niche count, in some sense, denotes the number of solutions in the neighborhood of the i-th solution, including the latter. If there exists no other solution within a radius of σ_{share} from a solution, the niche count for that solution would be one. On the other hand, if all $|P_1|$ solutions in the front are very close to each other compared to σ_{share}, the niche count of any solution in the group would be close to $|P_1|$. For example, if a proper σ_{share} is chosen, the niche count for solution 1 in Figure 114 would be one, whereas the niche count of the other three solutions (2, 3 or 4) in the first front would be close to three.

The final task is to reduce the fitness of the i-th solution by its niche count and obtain the shared fitness value, $F_i' = F_i/nc_i$. This process of degrading fitness of a solution which is crowded by many solutions helps emphasize the solutions residing in less crowded regions. This is precisely the purpose of the diversity preserving mechanism. In the context of Figure 114, the shared fitness of solution 1 will be $F_1' = 10$, whereas that of the other three solutions would be close to $10/3$ or 3.33. We shall illustrate the exact calculation of the shared fitness values in the next subsection.

This procedure completes the fitness assignment procedure of all solutions in the first front. In order to proceed to the second front, we note the minimum shared fitness in the first front and then assign a fitness value slightly smaller than this minimum shared fitness value. This makes sure that no solution in the first front has a shared fitness worse than the assigned fitness of any solution in the second front. Once again, the sharing function method is applied among the solutions of the second front and the corresponding shared fitness values are computed. This procedure is continued until all solutions are assigned a shared fitness.

Since the sharing function method works with the proportionate selection operator, as discussed above in Section 4.6.4, the NSGA used the less-noisy stochastic remainder roulette-wheel operator (Goldberg, 1989). This assigns copies in the mating pool proportional to the shared fitness. In this way, each solution in the first front has a better chance of surviving in the mating pool than that in the second front, and so on. On the other hand, the sharing function approach among solutions in each front makes sure that solutions in the less crowded region get more copies in the mating pool. In this way, the dual goals of multi-objective optimization are achieved by the above fitness assignment procedure. As mentioned above, the crossover and mutation operators are applied as usual to the whole population.

The NSGA fitness assignment procedure is described in a step-by-step algorithm in the following.

NSGA Fitness Assignment

> **Step 1** Choose sharing parameter σ_{share} and a small positive number ϵ and initialize $F_{min} = N + \epsilon$. Set front counter $j = 1$.

> **Step 2** Classify population P according to non-domination:
> $(P_1, P_2, \ldots, P_\rho) = \text{Sort}(P, \preceq)$.

> **Step 3** For each $q \in P_j$

>> **Step 3a** Assign fitness $F_j^{(q)} = F_{min} - \epsilon$.

>> **Step 3b** Calculate niche count nc_q using equation (4.60) among solutions of P_j only.

>> **Step 3c** Calculate shared fitness $F'^{(q)}_j = F_j^{(q)}/nc_q$.

> **Step 4** $F_{min} = \min(F'^{(q)}_j : q \in P_j)$ and set $j = j + 1$.

> **Step 5** If $j \leq \rho$, go to Step 3. Otherwise, the process is complete.

This fitness assignment procedure can be embedded in a single-objective GA. The proportionate selection operator must be used. However, any crossover and mutation operator can be employed. In the following, we illustrate the above fitness assignment procedure on the Min-Ex problem.

5.9.1   Hand Calculations

A randomly created initial population of six members was shown in the objective space in Figure 92. This figure also marked the Pareto-optimal front, which comprises solutions $0.1 \leq x_1^* \leq 1$ and $x_2^* = 0$. Table 18 lists these six solutions and their objective function values.

Step 1 We choose $\sigma_{share} = 0.158$ using equation (4.59). We have used $\epsilon = 0.22$, in particular to round-off the fitness value of the solutions in the second front. We set $F_{min} = 6.22$ and $j = 1$.

Step 2 Next, we classify the population into different non-domination sets. It is observed that there are three sets, as marked in Figure 115. Solutions 1, 3 and 5 belong to the first non-domination set P_1. Solutions 2 and 6 belong to the second non-domination set P_2, while the solution 4 belongs to the third set P_3 of non-domination.

Table 18   Fitness assignment under an NSGA.

Solution	x_1	x_2	f_1	f_2	Front Id	Assigned fitness	Shared fitness
1	0.31	0.89	0.31	6.10	1	6.00	4.22
2	0.43	1.92	0.43	6.79	2	4.00	4.00
3	0.22	0.56	0.22	7.09	1	6.00	4.22
4	0.59	3.63	0.59	7.85	3	3.78	3.78
5	0.66	1.41	0.66	3.65	1	6.00	6.00
6	0.83	2.51	0.83	4.23	2	4.00	4.00

Step 3 The next task is to begin from the solutions of the first non-dominated set and assign a fitness equal to the population size. In this example, there are six solutions. Hence, we assign a fitness of 6.00 to all solutions of the first front. Now, we modify these fitness values to emphasize diversity among the solutions. For this purpose, we calculate the normalized Euclidean distance between the solutions by using equation (5.21):

$$d_{13} = 0.120, \quad d_{15} = 0.403, \quad d_{35} = 0.518.$$

We now calculate the sharing function values by using equation (4.59):

$$Sh(d_{13}) = 0.423, \quad Sh(d_{15}) = 0, \quad Sh(d_{35}) = 0.$$

Using equation (4.60) (but replacing N with $|P_1| = 3$), we calculate the niche

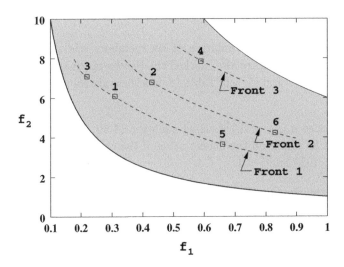

Figure 115   The population is classified into three non-dominated fronts.

count for all three solutions:

$$nc_1 = Sh(d_{11}) + Sh(d_{13}) + Sh(d_{15})$$
$$= 1 + 0.423 + 0$$
$$= 1.423,$$

$$nc_3 = Sh(d_{31}) + Sh(d_{33}) + Sh(d_{35})$$
$$= 0.423 + 1 + 0$$
$$= 1.423,$$

$$nc_5 = Sh(d_{51}) + Sh(d_{53}) + Sh(d_{55})$$
$$= 0 + 0 + 1$$
$$= 1.000.$$

Finally, we compute the shared fitness by dividing the assigned fitness with the above niche count values. The corresponding shared fitness values are shown in Table 18 and Figure 116.

Step 4 Now, $F_{min} = 4.22$. We set $j = 2$.

Step 5 We continue the above procedure for assigning fitness to solutions in the second and third fronts. Corresponding shared fitness values are shown in Table 18 and in Figure 116.

From the above fitness assignment, the following two observations can be made:

1. No solution in a front has a worse shared fitness value than a solution in a worse front. For example, any solution in front 1 has a better fitness than any solution in fronts 2 or 3.

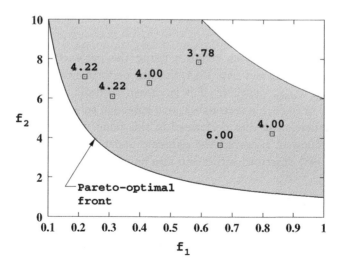

Figure 116 Shared fitness values of six solutions.

2. Within a front, less dense solutions (in the parameter space) have better fitness
   values.

The first property provides a selection pressure towards the non-dominated solutions,
a process which would hopefully lead NSGA towards the Pareto-optimal solutions.
The second property ensures that diversity is maintained among the non-dominated
solutions.

5.9.2   Computational Complexity

The computational complexity of the above fitness assignment procedure is
mainly governed by the non-dominated sorting procedure and the sharing function
implementation. The first task requires at most $O(MN^2)$ complexity (refer to
Section 2.4 above). The sharing function implementation requires every solution in a
front to be compared with every other solution in the same front, thereby incurring
a total of $\sum_{j=1}^{P} |P_j|^2$ distance computations. Each distance computation requires
evaluation of n differences between parameter values. In the worst case (when all
population members belong to one set, $\rho = 1$ and $|P_1| = N$), this computation is
$O(nN^2)$. Thus, the overall complexity of the NSGA is the larger of $O(MN^2)$ and
$O(nN^2)$.

5.9.3   Advantages

The main advantage of an NSGA is the assignment of fitness according to non-
dominated sets. Since better non-dominated sets are emphasized systematically, an
NSGA progresses towards the Pareto-optimal region front-wise. Moreover, performing
sharing in the parameter space allows phenotypically diverse solutions to emerge when
using NSGAs. If desired, the sharing can also be performed in the objective space.

5.9.4   Disadvantages

The sharing function approach requires fixing the parameter σ_{share}. It has been
observed earlier (Srinivas and Deb, 1994) that the performance of an NSGA is sensitive
to the parameter σ_{share}; however, if it is chosen according to the suggestions (see
equations (4.63) or (4.64)), a reasonably good spread of solutions can be obtained. For
most of the simulation studies performed in this book using NSGAs, the suggested
guidelines are followed. However, the dynamic update procedure of σ_{share} (Fonseca
and Fleming, 1993) described earlier can also be used.

5.9.5   Simulation Results

We now apply the NSGA on the two-objective optimization problem with GA
parameter settings identical to that used for other algorithms. Two variables are coded
in binary strings of lengths 24 and 26, respectively. First, we use the NSGA without

a mutation operator. We use a fixed $\sigma_{share} = 0.158$. Figure 117 shows that the NSGA without a mutation operator is able to find a good spread of solutions as early as in 50-th generation. The NSGA is started with 40 random solutions in the search space. The combined action of emphasis of non-dominated solutions and niching operator allows the NSGA to come close to the Pareto-optimal front and also maintain a good diversity.

When the NSGA is allowed to run up to 500 generations, the spread of solutions near the Pareto-optimal region is maintained and the distribution becomes better (Figure 118). This shows the ability of the NSGA to maintain diversity of solutions even without the use of a mutation operator.

We now turn on the bit-wise mutation operator with a probability of 0.02. Figure 119 shows the non-dominated solutions after 50 generations. Since no elite-preserving operator is used in any of these NSGA runs, mutation with a 1/(string-length) probability seems to be destructive. The number of solutions near the Pareto-optimal region is less than that in Figure 117 (without mutation). A similar behavior of the NSGA is observed at 500 generations (Figure 120). These results suggest that NSGAs (or any other multi-objective GAs) must use some sort of elitism so that previously discovered solutions near the Pareto-optimal region are not lost in subsequent generations. One can then rely on the mutation operator to try and find better solutions than those that have been found already. If better solutions are not found, such mutations are a waste, but if better solutions are found, they will not be lost easily due to the preservation of elitism. We shall discuss more about elitist multi-objective GAs in the next chapter.

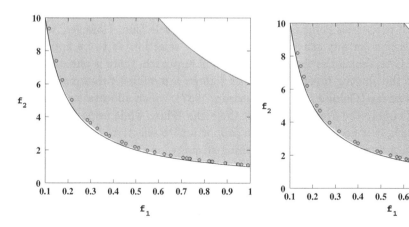

Figure 117 The NSGA distributes solutions near the Pareto-optimal region at the 50-th generation.

Figure 118 The NSGA distributes solutions near the Pareto-optimal region at the 500-th generation.

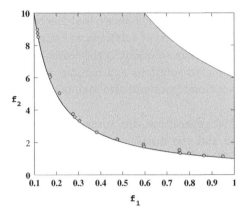

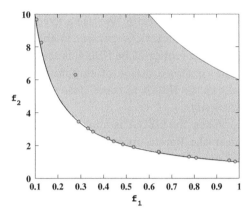

Figure 119 Population at the 50-th generation with an NSGA having a mutation operator.

Figure 120 Population at the 500-th generation with an NSGA having a mutation operator.

5.10 Niched-Pareto Genetic Algorithm

Horn et al. (1994) have proposed a multi-objective GA based on the non-domination concept. This method differs from the previous methods in the selection operator. The niched Pareto genetic algorithm (NPGA) uses the binary tournament selection, unlike the proportionate selection method used in VEGAs, NSGAs and MOGAs. Their choice of tournament selection over proportionate selection is motivated by the theoretical studies (Goldberg and Deb, 1991) on the selection operators used in single-objective GAs. It has been shown that the tournament selection has better growth and convergence properties compared to proportionate selection. However, when a tournament selection (or any other non-proportionate selection) is to be applied to multi-objective GAs, which use the sharing function approach, there is one difficulty. The working of the sharing function approach follows a modified two-armed bandit game playing strategy (Goldberg and Richardson, 1987), which allocates solutions in a niche proportional to the average fitness of the niche. Thus, while using the sharing function approach, it is necessary to use the proportionate selection operator.

In the NPGA, a dynamically updated niching strategy, originally suggested in Oei et al. (1991), is used. During the tournament selection, two solutions i and j are picked at random from the parent population P. They are compared by first picking a subpopulation T of size $t_{dom} (\ll N)$ solutions from the population. Thereafter, each solution i and j is compared with every solution of the subpopulation for domination. Of the two solutions, if one dominates all solutions of the subpopulation but the other is dominated by at least one solution from the subpopulation, the former is chosen. We call this scenario 1. However, if both solutions i and j are either dominated by at least one solution in the subpopulation or are not dominated by any solution in the subpopulation, both are checked with the current (partially filled) offspring population Q. We call this scenario 2. Each solution is placed in the offspring population and its niche count is calculated. The solution with the smaller niche count wins the tournament. In the following, we will first outline the NPGA tournament selection procedure and then present one generation of the algorithm. The implementation of this algorithm is different from that of a standard GA. In the beginning when the offspring population is empty, a slight deviation from the above procedure is adopted. If scenario 2 is faced, one of the two solutions is chosen as a parent at random. The same procedure is followed for another pair of solutions to choose the second parent. These two parents are then mated and mutated to create two offspring. From the third tournament onwards, the original procedure is adopted.

For any two solutions i and j, and the current offspring population Q, the following dynamically updated binary tournament selection procedure is used.

NPGA Tournament Selection Procedure

      `winner = NPGA-tournament(i, j, Q)`

    Step T1 Pick a subpopulation T_{ij} of size t_{dom} from the parent population P.

    Step T2 Find α_i as the number of solutions in T_{ij} that dominates i. Calculate α_j as the number of solutions in T_{ij} that dominates j.

    Step T3 If $\alpha_i = 0$ and $\alpha_j > 0$, then i is the winner. The process is complete.

    Step T4 Otherwise, if $\alpha_i > 0$ and $\alpha_j = 0$, then j is the winner. The process is complete.

    Step T5 Otherwise, if $|Q| < 2$, i or j is chosen as a winner with probability 0.5. The process is complete. Alternatively, the niche counts nc_i and nc_j are calculated by placing i and j in the current offspring population Q, independently. With the niching parameter σ_{share}, nc_i is calculated as the number of offspring ($k \in Q$) within a σ_{share} distance d_{ik} from i. The distance d_{ik} is the Euclidean distance between solutions i and k in the objective space:

$$d_{ik} = \sqrt{\sum_{m=1}^{M} \left(\frac{f_m^{(i)} - f_m^{(k)}}{f_m^{max} - f_m^{min}} \right)^2}. \tag{5.22}$$

    Note that, f_m^{max} and f_m^{min} are the maximum and minimum bounds of the m-th objective function in the search space.

    Step T6 If $nc_i \leq nc_j$, solution i is the winner. Otherwise, solution j is the winner.

One attractive aspect of an NPGA is that there is no need for specifying any particular fitness value to each solution. The tournament selection prefers non-dominated solutions in a stochastic manner (domination-check is performed only within a subpopulation of solutions) and whenever a decision about non-domination cannot be established, the parents which reside in less crowded regions in the offspring population are chosen. The former feature provides a selection pressure towards Pareto-optimal solutions, while the latter feature allows diverse solutions to be created. As pointed out earlier, these two features are essential for a multi-objective optimization algorithm and an NPGA, like NSGAs and MOGAs, also possesses the above two essential features.

In the following, we outline how the above tournament selection is used in one cycle of the NPGA algorithm.

NPGA Procedure

Step 1 Shuffle P, set $i = 1$, and set $Q = \emptyset$.

Step 2 Perform the above tournament selection and find the first parent, $p_1 =$ NPGA-tournament$(i, i+1, Q)$.

Step 3 Set $i = i+2$ and find the second parent, $p_2 =$ NPGA-tournament$(i, i+1, Q)$.

Step 4 Perform crossover with p_1 and p_2 and create offspring c_1 and c_2. Perform mutation on c_1 and c_2.

Step 5 Update offspring population $Q = Q \cup \{c_1, c_2\}$.

Step 6 Set $i = i+1$. If $i < N$ Go to Step 2. Otherwise, if $|Q| = N/2$, shuffle P, set $i = 1$, and go to Step 2. Otherwise, the process is complete.

In addition to the σ_{share} parameter, the NPGA is also sensitive to the t_{dom} parameter. Although the proposers of this algorithm did not suggest any guidelines for fixing t_{dom}, the numerical experiments performed by the proposers reveal that t_{dom} should be an order of magnitude smaller than the population size. If the chosen t_{dom} value is too small, the non-domination check would be noisy, which may not amply emphasize the true non-dominated solutions in a population. On the other hand, if the chosen t_{dom} is too large, true non-dominated solutions would be emphasized, but the computational complexity of the NPGA would be greater. Ideally, this parameter should also depend on the number of objectives being optimized (refer to Section 8.8 below for further discussions on this aspect).

5.10.1 *Hand Calculations*

Let us choose the Min-Ex problem and show a hand-calculation with an initial population of six members, as shown in Table 11. We follow the NPGA in a step-by-step manner.

Step 1 Set $i = 1$ and $Q = \emptyset$. For ease of demonstration, we use eight bits for coding x_1 and 10 bits for coding x_2.

Step 2 We move to the NPGA tournament selection procedure with solutions 1 and 2.

   Step T1 Let us choose a subpopulation (with $t_{\text{dom}} = 2$) with solutions 3 and 5.

Step T2 Since neither solutions 3 nor 5 dominate solutions 1 or 2, we have $\alpha_1 = \alpha_2 = 0$.

Steps T3 and T4 Skipped.

Step T5 Since there is no solution in the offspring population yet, we choose one at random. Say we choose solution 1.

Thus, $p_1 = 1$.

Step 3 Now, we play another NPGA tournament with solutions 3 and 4.

Step T1 Let us say we choose another subpopulation with solutions 1 and 6.

Step T2 We observe that both solutions 1 and 6 do not dominate solution 3, but solution 1 dominates solution 4.

Step T3 Thus, $\alpha_3 = 0$ and $\alpha_4 = 1$. So, solution 3 is the winner.

Thus, $p_2 = 3$.

Step 4 We now perform a single-point crossover independently on each variable (site 4 in x_1 and site 3 in x_2) between solutions 1 and 3 and assume no mutation here.

Parent		Offspring	
00111011	0010110110	00110010	0011110011
(0.31)	(0.89)	(0.28)	(1.19)
00100010	0001110011	00101011	0000110110
(0.22)	(0.56)	(0.25)	(0.26)

with a $\Rightarrow$ between the Parent and Offspring columns.

Step 5 Thus, the resulting offspring population Q has the following two solutions:

Solution	x_1	x_2	f_1	f_1
c1	0.28	1.19	0.28	7.82
c2	0.25	0.26	0.25	5.04

Step 6 Now, we set $i = 4$ and move to Step 2.

Step 2 We play a tournament with solutions 5 and 6.

Step T1 We choose a subpopulation with solutions 2 and 4.

Step T2 Neither solutions 2 nor 4 dominate solutions 5 or 6. Thus, we move to Step T5.

Step T5 We now calculate the niche counts nc_5 and nc_6 in the objective space. We calculate the following distances:

$$d_{5(c1)} = \sqrt{\left(\frac{0.66 - 0.28}{1 - 0.1}\right)^2 + \left(\frac{3.65 - 7.82}{10 - 1}\right)^2} = 0.627,$$

$$d_{5(c2)} = 0.481,$$

$$d_{6(c1)} = 0.730,$$

$$d_{6(c2)} = 0.651.$$

Using a $\sigma_{share} = 0.5$, we observe that only one solution $c2$ is within the σ_{share} distance from solution 5, whereas neither $c1$ nor $c2$ are within the σ_{share} distance from solution 6. Thus, $nc_5 = 1$ and $nc_6 = 0$. So, we choose solution 5 as the winner of the tournament. Note that solution 5 dominates solution 6. However, since both offspring are located closer to solution 5, it makes a selective disadvantage of solution 5 over solution 6.

Step 3 In order to create the second parent, we shuffle the population and follow the above procedure.

5.10.2 Computational Complexity

We perform a worst case analysis, where every tournament requires both scenarios 1 and 2 to be performed. In each tournament, scenario 1 requires a domination check of two solutions with a set of t_{dom} solutions, thereby requiring $2Mt_{dom}$ comparisons. For all population members, this amounts to a total of $2MNt_{dom}$ comparisons. In scenario 2, the niche count calculation requires $2|Q|$ distance computations for each tournament. Since the size of the offspring population Q linearly varies from 2 to $(N - 2)$, a total of $O(N^2)$ distance computations are needed. Thus, the overall complexity of the algorithm is $O(N^2)$. If the size of t_{dom} is of the same order as N, the overall complexity is $O(MN^2)$.

5.10.3 Advantages

As mentioned above, one of the main advantages of the NPGA is that no explicit fitness assignment is needed. While VEGAs, NSGAs and MOGAs are not free from their subjectivity in the fitness assignment procedure, the NPGA does not have this problem. Another advantage of the NPGA is that this is the first proposed multi-objective evolutionary algorithm which uses the tournament selection operator. However, the dynamically updated tournament selection somewhat reduces the elegance with which tournament selection is applied in single-objective optimization problems. If the subpopulation size t_{dom} is kept much smaller than N, the complexity of the NPGA does not depend much on the number of objectives (M), as is evident from the above complexity analysis. Thus, the NPGA may be found to be computationally efficient in solving problems having many objectives.

5.10.4 Disadvantages

The NPGA requires fixing two important parameters: σ_{share} and t_{dom}. The parameter σ_{share} has more effect on an NPGA than an NSGA. This is because the niche count nc_i of a solution i is calculated as the number of population members present within a distance σ_{share} from solution i. Whether a solution resides very near to solution i or a solution is situated at a distance almost equal to σ_{share} is not a concern here. They all are counted equally in the niche count calculation. Whereas in the NSGA niche count calculation, although all solutions within σ_{share} distance are counted, the contribution of a solution away from solution i is less when compared to a solution close to i. Thus, there is a greater need to use an appropriate value for σ_{share} in an NPGA.

The performance of the NPGA depends on the choice of t_{dom}, as described above.

5.10.5   Simulation Results

We apply the NPGA to the Min-Ex problem for 500 generations. Identical GA parameter settings to those used in the NSGA study are used here. In addition, $t_{dom} = 6$ and $\sigma_{share} = 0.158$ are used. The NPGA without mutation is used first. Although there exist some solutions near the Pareto-optimal front after 30 generations, the NPGA's dynamically updated niching mechanism cannot maintain them until 500 generations (Figure 121). All solutions at 500 generations converge to one point in the Pareto-optimal front (marked 'A' on the figure).

However, when mutation is introduced with a bit-wise probability of $p_m = 0.02$, a good spread is observed. Figure 122 shows the obtained solutions after 500 generations. Although a reasonable diversity is obtained as early as 50 generations, the figure shows that even after 500 generations, the mutation operator is able to provide a good diversity among the optimized solutions.

5.11   Predator–Prey Evolution Strategy

Laumanns et al. (1998) suggested a multi-objective evolution strategy (ES) which is radically different from any of the methods discussed so far. This algorithm does

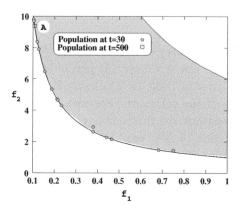

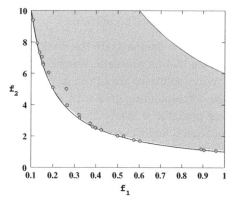

Figure 121   The population after 30 and 500 generations obtained with an NPGA without mutation. At the 500-th generation, all solutions converge to the solution A.

Figure 122   The population after 500 generations obtained with an NPGA with mutation.

not use a domination check to assign fitness to a solution. Instead, the concept of a predator–prey model is used. Preys represent a set of solutions ($x^{(i)}$, $i = 1, 2, \ldots, N$), which are placed on the vertices of a undirected connected graph. First, each predator is randomly placed on any vertex of the graph. Each predator is associated with a particular objective function. Thus, at least M predators are initialized in the graph. Secondly, staying in a vertex, a predator looks around for preys in its neighboring vertices. The predator 'catches' a prey having the worst value of its associated objective function. When a prey $x^{(i)}$ is 'caught', it is erased from the vertex and a new solution is obtained by mutating (and recombining) a random prey in the neighborhood of $x^{(i)}$. The new solution is then kept in the vertex. After this event is over, the predator takes a random walk to any of its neighboring vertices. The above procedure continues in parallel (or sequentially) with all predators until a pre-specified number of iterations have elapsed. The simultaneous presence of predators favoring each objective allows trade-off solutions to co-exist in the graph.

The proposers of this method have suggested using a two-dimensional torus as a graph, which ensures that the maximum expected number of iterations required by a predator to visit every vertex is bounded between $N(\log N)^2$ and $2(N - 1)^2$, where N is the number of vertices in the torus. On a couple of test problems, they have also observed that using multiple predators per objective is a good strategy. Moreover, instead of using a constant mutation strength for the entire EA run, a decreasing schedule of mutation strength is better. Starting from a large mutation strength, this is decreased by a fixed factor every time a prey produces an offspring. In particular, the investigators used $\sigma_{k+1} = 0.99\sigma_k$, where in every iteration the mutation strength is reduced by 1%. Since the mutation strength is reduced over iterations, preys progressively create solutions in a closer neighborhood to themselves.

5.11.1 Hand Calculations

In order to demonstrate the working of the predator–prey evolution strategy (PPES) on the Min-Ex problem, we use a toroidal grid of size 4×4 (or 16) preys. These 16 random solutions are shown in Figure 123 and listed in Table 19. The figure also shows the corresponding grid with marked locations of the solutions. For each objective, one predator is used. Thus, there are two predators: the first one is placed on grid location 6 and the second one is placed on location 15. The neighbors of vertex 6 are vertices 3, 5, 6, 7 and 11. However, the neighbors of vertex 15 are vertices 2, 10, 14, 15 and 16.

Table 19 shows that prey 7 has the worst value of the first objective function among all neighboring preys of the first predator. Thus, this solution must be replaced by a new solution. The new solution is calculated by first choosing a random solution among the neighbors of this worst solution (solution 7). Thus, we choose one solution at random from solutions 2, 6, 8 and 10. Say, we choose solution 8. We mutate this solution by using a normal distribution with a chosen mutation strength 0.05. We find a mutated solution $(0.79, 3.96)^\top$. The corresponding function values are $(0.79, 6.28)^\top$.

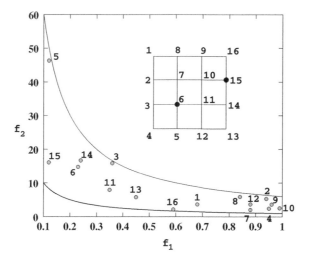

Figure 123   A population of 16 preys and 2 predators in a 4 × 4 grid.

Thus, vertex 7 gets this new solution.

Figure 124 shows that solution 7 is worse than the new solution in terms of f_1. By comparing these two solutions, it can also be seen that the new solution is not as close to the Pareto-optimal front as the original solution 7. However, the PPES compared both of these solutions by using only the first objective function, thereby allowing such an unwanted replacement.

We now play the same game with the second predator, which resides in vertex 15. We observe that prey 14 has the worst value of the second objective function among the neighbors of the second predator. We choose vertex 11 at random and mutate the solution at this vertex. We obtain solution $(0.36, 1.90)^{\top}$. The corresponding function values are $(0.36, 8.06)^{\top}$. Thus, vertex 14 gets this new solution. The modified set of 16 solutions are shown in the table. Figure 125 shows the location of the new solution,

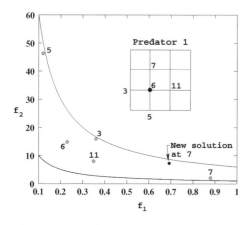

Figure 124   Predator 1 deletes prey 7.

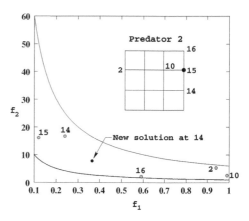

Figure 125   Predator 2 deletes prey 14.

Table 19   The predator–prey algorithm is illustrated in a random population. The worst objective function value for a predator is shown in bold.

Point	Current population				Neighbors		New population	
	x_1	x_2	f_1	f_2	f_1	f_2	x_1	x_2
1	0.68	1.49	0.68	3.63	—	—	0.68	1.49
2	0.94	4.04	0.94	5.34	—	5.34	0.94	4.04
3	0.36	4.65	0.36	15.90	0.36	—	0.36	4.65
4	0.95	1.30	0.95	2.43	—	—	0.95	1.30
5	0.12	4.47	0.12	46.31	0.12	—	0.12	4.47
6	0.23	2.41	0.23	14.79	0.23	—	0.23	2.41
7	0.88	0.80	0.88	2.05	**0.88**		0.79	3.96
8	0.84	3.96	0.84	5.92	—	—	0.84	3.96
9	0.96	2.48	0.96	3.64	—	—	0.96	2.48
10	0.99	1.53	0.99	2.55	—	2.55	0.99	1.53
11	0.35	1.82	0.35	7.97	0.35		0.35	1.82
12	0.88	2.23	0.88	3.69	—	—	0.88	2.23
13	0.45	1.58	0.45	5.77	—	—	0.45	1.58
14	0.24	3.03	0.24	16.69	—	**16.69**	0.36	1.90
15	0.12	0.87	0.12	16.12	—	16.12	0.12	0.87
16	0.59	0.28	0.59	2.18	—	2.18	0.59	0.28

which will replace solution 14.

5.11.2   Advantages

The main advantage of this method is its simplicity. Random walks and replacing worst solutions of individual objectives with mutated solutions are all simple operations, which are easy to implement.

Another advantage of this algorithm is that it does not emphasize non-dominated solutions directly. As we have previously mentioned in Chapter 2, one of the difficulties with the domination-based algorithms is that all solutions in a non-dominated set need not be members of the true Pareto-optimal front. Since the domination principle is not used, at least this algorithm is free from this difficulty.

5.11.3   Disadvantages

No explicit operator is used to maintain a spread of solutions in the obtained non-dominated set. Instead, each predator is assigned the task of eliminating the worst neighboring solution with respect to a different objective. Although this procedure is different from preferring individual champion solutions (as it is in VEGA), there is no special care taken to maintain the intermediate solutions. Thus, this algorithm may also have difficulties in finding a good spread of solutions in the Pareto-optimal set.

In the experiments shown in the original study, it was clear that the algorithm is sensitive to the choice of mutation strength parameter and the ratio of predator and prey used in a simulation study. If a low predator-to-prey ratio is used, individual optima are emphasized, thereby making the algorithm difficult to find the intermediate Pareto-optimal solutions. On the other hand, if a high predator-to-prey ratio is chosen, the preys are compared with respect to all objectives more often. In an EA, this produces an effect similar to that of emphasizing intermediate non-dominated solutions. This raises an important issue of finding the optimal predator-to-prey ratio for a problem. The original study used only two test problems, both having two objective functions. How such a simple strategy without any diversity-preserving operator can scale up in higher-dimensional objective space would be an interesting study for future research.

In the following subsection, we present simulation results of this predator-prey approach on a test problem.

5.11.4 Simulation Results

We now apply the algorithm on the following test problem for a maximum of 10 000 iterations per objective:

$$
\left.
\begin{array}{ll}
\text{Minimize} & f_1(\mathbf{x}) = x_1^2, \\
\text{Minimize} & f_2(\mathbf{x}) = \dfrac{1 + x_2^2}{x_1^2}, \\
\text{subject to} & \sqrt{0.1} \leq x_1 \leq 1, \\
& 0 \leq x_2 \leq \sqrt{5}.
\end{array}
\right\}
\tag{5.23}
$$

This problem is different from the Min-Ex problem. The modification is used in order to keep the decision variables corresponding to the Pareto-optimal solutions well inside the search region. In the original problem, the optimum value corresponding to x_2 was zero (the lower bound on x_2). Since mutation is used as the only search operator here and the variable is coded directly instead of a binary representation, maintaining the variable boundaries becomes a difficult task.

We use a 10×10 grid with 100 preys. First, we use one predator for each objective. These are randomly placed on any vertex. We also use a normally distributed probability distribution for mutation with a fixed mutation strength of 0.05. Figure 126 shows all preys after 1000 iterations per objective. The figure shows that preys crowd near the Pareto-optimal region, with a tendency to have more solutions in the intermediate region of the Pareto optimal set. This is in accordance with the original study. Interestingly, when the algorithm is allowed to run for further iterations, the diversity in the solutions reduce and solutions tend to come closer together. Figure 127 show the preys after 10 000 iterations per objective. It is clear from the figures how preys lose diversity with iterations. Since no explicit diversity-preserving mechanism is used, the presence of two predators in trying to enhance each objective's niche cannot maintain enough diversity, particularly when each predator is allowed to visit

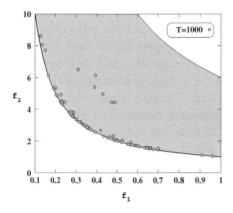

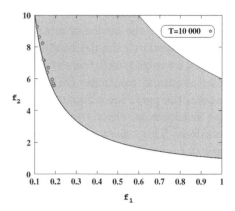

Figure 126 All 100 preys after 1000 iterations per objective with a PPES algorithm.

Figure 127 All 100 preys after 10 000 iterations of the fixed mutation scheme.

every vertex a reasonable number of times. We argue that the remaining diversity in the population is due to the fixed mutation strength of 0.05. If a larger mutation strength were used, we would have more diversity but the solutions would have been away from the Pareto-optimal set as well. Later, we will suggest an improvement to this algorithm with an explicit diversity-preserving mechanism in the hope of keeping better diversity of solutions in the population with iterations.

We will now try a decreasing mutation schedule, $\sigma_{k+1} = 0.9999\sigma_k$, where k is incremented every time an offspring is created. To start with, we use $\sigma_0 = 0.5$. We use 10 predators per objective. All 20 predators are placed on random vertices in the initial population. Figure 128 shows the prey population after 8000 iterations per objective (thereby allowing each predator to have 800 iterations). Although we have a good spread of solutions at this stage, when the algorithm is run for a longer period, all preys converge to the one solution on the Pareto-optimal front, as shown in Figure 129 after 500 000 iterations.

5.11.5   A Modified Predator–Prey Evolution Strategy

In the above studies, we have observed that all preys tend to become identical when a decreasing mutation scheme is used. As discussed above, the original predator–prey algorithm does not explicitly implement any diversity-preserving operator. Thus, if the predators are allowed to visit each vertex many times, eventually the competition among predators will prevail and preys will tend to become similar. There could be several ways to introduce diversity to the above algorithm. We suggest one approach here.

Instead of associating one objective to each predator, we associate a different weight vector with each predator. In our implementation for two objectives, we use nine

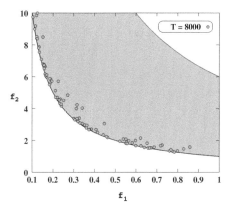 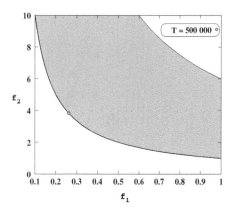

Figure 128   All 100 preys after 8000 iterations per objective with a PPES algorithm.

Figure 129   All preys converge to one solution after 500 000 iterations of the decreasing mutation scheme.

predators each corresponding to a different weight vector. For example, the first predator takes a weight vector $(w_1 = 1, w_2 = 0)$, the second predator takes a weight vector $(w_1 = 0.875, w_2 = 0.125)$, and so on, until the ninth predator takes the weight vector $(w_1 = 0, w_2 = 1)$. Each predator evaluates its neighboring prey with respect to the weighted sum of the objectives. The prey with the worst weighted objective value must be deleted. This modification is a general version of the PPES algorithm, where only two weight vectors $(w_1 = 1, w_2 = 0)$ and $(w_1 = 0, w_2 = 1)$ were allowed. Since each predator emphasizes solutions corresponding to each weight vector, the modified PPES algorithm may be able to maintain diversity in obtained solutions explicitly (at least for problems having a convex Pareto-optimal region). In addition, we suggest two further modifications.

1. To motivate faster convergence, we replace the worst vertex by a solution obtained by mutating the best neighboring vertex, instead of mutating a random neighboring vertex.

2. To allow multiple solutions to co-exist in the grid, we also restrict the predator's movement in a special way. Instead of a random walk, a predator makes a move to one of its neighbors (including its current position) based on the weighted objective value of each neighboring vertex. Vertices with better weighted objective have a larger probability to attract the predator. This mechanism also allows predators to form their own niches in the grid.

We first tried this modified PPES on the problem shown in equation (5.23). Since objectives f_1 and f_2 are unequally scaled, we use the following weighted normalized function: $w_1 f_1 + w_2 f_2 / 50$, for a predator with the weight vector $(w_1, w_2)^\mathsf{T}$. With a fixed mutation scheme ($\sigma = 0.05$), we use a 10×10 grid. Figure 130 shows all 100 preys after 20 000 iterations per predator. Although solutions on the front were found much earlier, we show the preys after 20 000 iterations to demonstrate that the modified

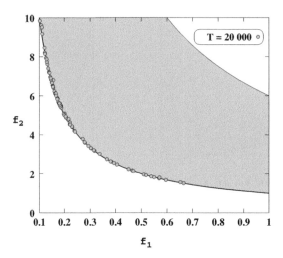

Figure 130   All 100 preys after 20 000 iterations per predator.

algorithm can maintain a good diversity in the Pareto-optimal solutions for a large number of generations. However, like all other weighted approaches, the algorithm is sensitive to the normalization procedure and to the convexity of the Pareto-optimal region.

Next, we will attempt to solve one of the two problems which were used in the original study:

$$\left. \begin{aligned} \text{Minimize} \quad & f_1(\mathbf{x}) = x_1^2 + x_2^2, \\ \text{Minimize} \quad & f_2(\mathbf{x}) = (x_1 + 2)^2 + x_2^2, \\ & -50 \leq x_1 \leq 50, \quad -50 \leq x_2 \leq 50. \end{aligned} \right\} \tag{5.24}$$

We use a 30×30 (or 900) preys (the same as that used in the original PPES study) and a fixed mutation scheme with a mutation strength equal to 0.05. Figure 131 shows that a good distribution is already formed on and near the true Pareto-optimal front as early as 500 iterations per predator. This means that at this stage a total of 4500 solutions have been evaluated. Figure 132 shows all 100 preys after 50 000 iterations per predator. The algorithm is run long enough to show that even after many iterations the diversity in the preys close to the Pareto-optimal region is maintained. Since each predator tries to maintain its own niche, multiple solutions co-exist in the grid.

5.12   Other Methods

In this section, we will briefly describe a number of other approaches to multi-objective optimization which do not explicitly use an elite-preserving operator. Besides them, a number of other algorithms are also suggested and summarized elsewhere (Coello, 1999; Horn, 1997).

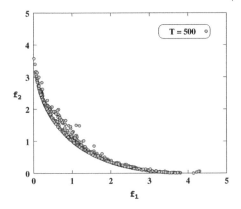

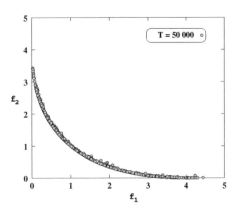

Figure 131 All preys after 500 iterations per predator of the modified algorithm.

Figure 132 All preys after 50 000 iterations per predator of the modified algorithm.

5.12.1   Distributed Sharing GA

Hiroyasgu et al. (1999) suggested a distributed sharing GA (DSGA) for multi-objective optimization. In this approach, the distributed island model (see Section 4.6.5 above) is used to maintain diversity among non-dominated solutions. The GA population is divided into a number of subpopulations (or islands) and independent genetic operations are performed to each island. The occasional migration of solutions from each island to other islands is allowed. Subpopulations from all islands are collected together and the non-dominated solutions are recorded. When the size of the non-dominated set is more than a desired number, sharing with a suitable σ_{share} is performed to choose the desired number of solutions. These chosen solutions are again divided into a number of islands and the process continues until a convergence criterion is satisfied. On a four-objective optimization problem, the investigators reported better distributed solutions than a canonical GA applied many times. Besides the sharing parameter, the DSGA requires a number of other parameters (such as number of islands, the number of desired non-dominated solutions, convergence criterion, migration rate, etc.). However, this study is different from other studies in that a different niching technique is used in a distributed environment.

5.12.2   Distributed Reinforcement Learning Approach

Mariano and Morales (2000) suggested a distributed reinforcement learning approach, where a family of agents are assigned to different objective functions. Each agent proposes a solution to optimize its objective function. All such solutions are combined and a non-dominated compromised set of solutions are identified. Each non-dominated solution is rewarded. In the context of solving continuous search space problems, an agent considers solutions in a particular search direction from its current location. The solution is evaluated by the agent's corresponding objective function. The rewarding mechanism provides a direction for the algorithm to move towards the Pareto-optimal region and the non-domination check maintains a diverse set of solutions, while simultaneous creation of multiple solutions by a directional search method helps find new solutions in the search space. This algorithm involves a number of parameter fixings. On four test problems (SRN, ZDT3, VNT and TNK, discussed below in Section 8.3), they used different parameters and showed better performance measures (error ratio and spacing, discussed later in Section 8.2) than MOMGA – an elitist MOEA which will be described in the next chapter.

5.12.3 Neighborhood Constrained GA

Loughlin and Ranjithan (1997) suggested a neighborhood constrained GA (NCGA), which uses a population approach to the ϵ-constraint strategy discussed earlier in Chapter 3. All $(M-1)$ objectives are converted into constraints and only one objective is kept for optimization. Instead of using different ϵ vectors in different simulation runs, as in the case of the classical ϵ-constraint method, each population member is assigned an index in the population. According to the index value, a different ϵ vector is assigned to each member. A linear mapping for the ϵ vector within user-defined lower and upper constraint levels is used for this purpose. Since solutions in the population are gradually made more constrained, a restricted mating confining the crossover to take place between neighboring solutions is adopted. The offspring solutions are placed consecutively in the new population. The selection operator is also performed within the neighboring solutions to promote diversity in the population. The investigators demonstrated the working of this algorithm by applying it to an air pollution problem. Compared to a single-objective GA and a Pareto-based GA, their NCGA has found better diversity in the obtained solutions. It is interesting to note that such diversity in these solutions is ensured by the explicit assignment of different ϵ vectors for different population members. Even without any explicit non-dominated sorting procedure the approach, with an added penalty for constraint violations is able to provide enough selection pressure for the algorithm to progress towards the Pareto-optimal region. Although a number of other fix-ups are suggested, the algorithm requires a number of parameter settings such as neighborhood distance, penalty parameter, lower and upper limits on each constraint value, etc.

5.12.4 Modified NESSY Algorithm

Köppen and Rudlof (1997) suggested a neural evolutional strategy system (NESSY) for single-objective optimization. The idea is very different from the standard evolutionary algorithms. The evolutionary search takes place in a neural network paradigm. There are three layers in this neural network. The first layer is called the solution layer, where each neuron represents a decision variable vector x. Thus, the set of neurons in the solution layer constitutes a population of solutions. The second layer is the generation layer, where each neuron represents a separate gene (or decision variable). Thus, if there are n decision variables, there are exactly n neurons in the generation layer. All neurons in the solution layer are connected to each neuron in the generation layer. The third layer contains one neuron representing the lower bound of the objective function. Figure 133 shows a schematic of such a neural network.

Each neuron j in the solution layer is assigned a fitness $F^{(j)}$, calculated from the objective function value of $x^{(j)}$. Each neuron in the generation layer i chooses exactly one neuron j in the solution layer with a probability proportional to the connection weight w_{ij}. The higher the weight, then the better is the probability of establishing an association between the corresponding decision variable and the solution. All weights are initialized randomly in $[0, 1]$. After the association is made, the fitness $g^{(i)}$ of a neuron i in the generation layer is assigned the same as the fitness of the solution-neuron j associated with the i-th neuron. Thereafter, the connection weights are updated using the following rule:

$$w_{ij}^t = w_{ij}^{t-1} - \alpha \frac{F^{(j)} - g^{(i)}}{O}, \tag{5.25}$$

where α is the learning rate and O represents the state of the output neuron used to normalize the difference in fitness values. It could be taken as the best fitness of solution-neurons at generation t. It is important to note that if the fitness of a solution-neuron is more (worse) than that of the generational-neuron, this means that the connection between the corresponding gene and solution vectors is weak. The above update rule reduces the weight in such a case. On the other hand, when the converse happens, the weight is enhanced. For the currently associated neurons, the weight is unchanged.

Besides this update, each solution-neuron is modified by comparing its fitness with that of an arbitrary neuron in the generational layer. If the latter fitness is better, the corresponding gene (decision variable value) is put in the solution-neuron. The investigators termed this operator the *transduction* operator. Each decision

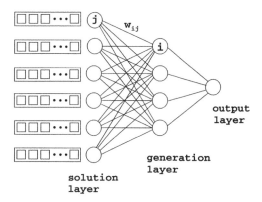

Figure 133   A schematic of the single-objective NESSY algorithm.

variable value is also mutated with a zero-mean normally distributed probability distribution. The overall user-defined parameters of the NESSY algorithm are the learning rate α, mutation probability p_m, and the standard deviation of the normal probability distribution used in the mutation operation. Ideally, there must be n standard deviations used, each for a different decision variable.

For the use of the NESSY algorithm in multi-objective optimization (MONESSY), the investigators have also suggested multiple neurons in the third layer, one corresponding to each objective function. The fitness of a neuron $g^{(i)}$ in the generational layer is suggested to be computed by using a randomly chosen objective function. The association with a solution-neuron is also made with the same objective function. Weight updates and the transduction operations are performed with the same objective function chosen for the generational-neuron. However, the choice of an objective function for each generational-neuron is made independently in each generation. Since all objectives are used, solutions corresponding to individual optimum are expected to survive in the population. On the test problem KUR (discussed on page 341), the investigators have shown that the MONESSY algorithm is able to distribute more solutions near the Pareto-optimal front than anywhere in the search space. Since no elite-preserving operator is used, although solutions close to the Pareto-optimal front are found, they may get lost after a few generations. Simulation results with the MONESSY algorithm showed that it is able to rediscover solutions near the Pareto-optimal regions periodically. Moreover, the lack of any explicit elite-preserving operator may cause difficulty in allowing a diverse of set of solutions to co-survive in the solution layer.

5.12.5   Nash GA

Sefrioui and Periaux (2000) suggested a Nash GA for multi-objective optimization. Motivated by a game-theoretic approach, they allowed one player to get associated with each objective function. A player tries to optimize its objective function while keeping other objective functions unchanged. In a periodic sequences of operations, the Nash GA is terminated when no more improvement is recorded. At this steady-state scenario, the resulting solution is a Nash-equilibrium solution and is a candidate Pareto-optimal solution. Although the investigators claim better convergence properties of this GA compared to the NSGA, it is clear that an explicit niche-forming operator must be used to maintain multiple Pareto-optimal solutions.

5.13   Summary

The attractive feature of multi-objective evolutionary algorithms is their ability to find a wide range of non-dominated solutions close to the true Pareto-optimal solutions. Because of this advantage, the philosophy of solving multi-objective optimization problems can be revolutionized. Instead of pre-fixing a weight vector and finding the corresponding Pareto-optimal solution, we have suggested an ideal approach where a first attempt is made to find a number of trade-off solutions in one single simulation run. We have argued that it becomes less subjective to choose one solution from such a set of multiple trade-off solutions, instead of finding a solution by using a predefined trade-off among objectives.

Evolutionary algorithms process a population of solutions in each iteration, thereby making them ideal candidates for finding multiple trade-off solutions in one single simulation run. In this chapter, we have presented a number of early multi-objective evolutionary algorithms, which can be used to find multiple non-dominated solutions close to the Pareto-optimal solutions. The common aspect of these algorithms is that none of them have used an elite-preserving operator, thereby making the algorithms easy to understand and implement. Even without an elite-preserving operator, some of these algorithms have found well converged and well distributed trade-off solutions in an example problem here and in many problems in their original studies.

As early as 1957, researchers have suggested a weighted approach for handling multi-objective optimization problems. However, the first real MOEA was suggested in 1984 by David Schaffer (Schaffer, 1984), where three different implementations, including a domination-based strategy, were tried. Since no explicit diversity-preserving mechanism, other than the mutation operator, was used in the VEGA, the population members eventually had a tendency to crowd near the individual optimum solutions. This difficulty was eliminated in a number of implementations about a decade later by carefully introducing the non-domination concept and an explicit diversity-preserving operator. These implementations – MOGAs, NPGAs and NSGAs – followed a suggestion of David E. Goldberg (Goldberg, 1989) and have found well-converged and well-distributed sets of trade-off solutions in test problems, as well as in real-world problems (refer to Chapter 9). At about the same time, two other implementations, using a diploidy approach and a weighted sum approach, were also suggested. Along with a few other algorithms, all of these MOEAs have clearly demonstrated their use as candidate algorithms for the ideal multi-objective optimization procedure, as described in Chapter 1.

In the next chapter, we will discuss elitist multi-objective evolutionary algorithms which are supposedly faster and better than the algorithms described in this chapter. However, the algorithms of this chapter provide a basic understanding of the working principles of a multi-objective evolutionary algorithm, a matter which will be essential to comprehend the algorithms of the next chapter.

Exercise Problems

1. Instead of dividing a population (of size N) into M subpopulations and reproducing each subpopulation using a different objective function used in VEGA, a modified approach is proposed. All individuals are evaluated for each objective function and only N/M individuals are selected using the proportionate selection with each objective function. The rest of the procedure is the same as that in VEGA. Discuss the advantages and disadvantages of the modified approach compared to

the original VEGA.

2. For the population shown in Table 13, use the non-dominated selection heuristic method to find the selection probability of each individual using $\epsilon = 0.05$.

3. Consider the following two-objective optimization problem:

$$\text{Maximize} \quad f_1(x) = x_1,$$
$$\text{Maximize} \quad f_2(x) = 1 + x_2 - x_1^2,$$
$$\text{subject to} \quad 0 \leq x_1 \leq 1,$$
$$0 \leq x_2 \leq 3.$$

If we desire to find the Pareto-optimal solutions in the step of 0.1 in f_1, which weight vectors are to be used in the weight-based GA?

4. The following population is used in MOGA to solve the problem stated above:

$$x^{(1)} = (0.2, 2.0)^T, \quad x^{(2)} = (0.7, 1.0)^T, \quad x^{(3)} = (0.8, 2.5)^T,$$
$$x^{(4)} = (0.4, 1.2)^T, \quad x^{(5)} = (0.1, 0.5)^T, \quad x^{(6)} = (0.3, 0.9)^T.$$

Use $\sigma_{\text{share}} = 0.5$ to calculate the shared fitness of each individual.

5. A new σ_{share} update strategy is proposed in the adjoining figure for minimization problems with a convex search space.

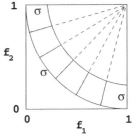

All objectives are normalized to have values between zero and one. For $M = 2$ objectives, calculate σ_{share} for various values of population size N:

$$N = 50, 70, 100, 200, 500.$$

Can you extend the idea for three objectives?

6. In problem 4, calculate the shared fitness value of each individual using NSGA. Choose

(a) $\sigma_{\text{share}} = 0.5$

(b) $\sigma_{\text{share}} = 1.5$

(c) $\sigma_{\text{share}} = 0.5$ but sharing is performed in the objective space.

7. Consider the following problem:

$$\text{Minimize} \quad f_1(x) = x_1^2 + x_2^2,$$
$$\text{Minimize} \quad f_2(x) = x_2^2 - 4x_1,$$
$$-2 \leq x_1, x_2 \leq 2,$$

and the following solutions:

$$x^{(1)} = (0, 1)^T, \quad x^{(2)} = (1, 0)^T, \quad x^{(3)} = (1, 2)^T,$$
$$x^{(4)} = (1, 1)^T, \quad x^{(5)} = (-1, 1)^T, \quad x^{(6)} = (-1, -2)^T.$$

Find the winner of the NPGA tournament selection operation applied between the following pairs of solutions:

(a) $x^{(2)}$ and $x^{(4)}$

(b) $x^{(1)}$ and $x^{(3)}$

(c) $x^{(2)}$ and $x^{(6)}$

In each case, use the rest of the population members as t_{dom}. Use $\sigma_{share} = 0.3$.

8. Consider the problem:

$$\begin{aligned}
\text{Minimize} \quad & f_1(x) = x_1^2 + x_2^2, \\
\text{Minimize} \quad & f_2(x) = (x_1 + 2)^2 + x_2^2, \\
\text{subject to} \quad & -10 \le x_1, x_2 \le 10.
\end{aligned}$$

Sixteen solutions $x_j^{(i)} = -10 + 1.25i$ for $j = 1, 2$ and $i = 1, 2, \ldots, 16$ are used in a toroidal grid (nodes are numbered using the top-to-bottom and left-to-right scheme) for the predator-prey approach. Choose 7-th and 13-th nodes as the predators. Find the new set of 16 solutions using a mutation strength of 0.1 (on each variable). Use random numbers in the sequence given below

$$0.123, 0.678, 0.562, 0.818, 0.054, 0.773$$

for the Gaussian mutation operator.

9. Consider the following problem of two objectives: (i) maximize number of 1s from the left of the string and (ii) maximize the number of 0s from the right of the string. Consider the following population:

Soln.	String
1	100100
2	110000
3	001100
4	111010
5	001001
6	111000

First three solutions are evaluated using the first objective and the last three solutions are evaluated using the second objective.

(a) Calculate the number of copies the schema 1****0 obtained after the selection operator in VEGA.

(b) If a two-point crossover operator and a bit-wise mutation operator are used, what is the survival probability of this schema in the next generation? Use $p_c = 0.8$ and $p_m = 0.1$.

6

Elitist Multi-Objective Evolutionary Algorithms

In the previous chapter, we have presented a number of multi-objective EAs which do not use any *elite-preserving* operator. As the name suggests, an elite-preserving operator favors the elites of a population by giving them an opportunity to be directly carried over to the next generation. In single-objective optimization problems, elitism is introduced into a GA in various ways. In a stead-state EA, the elitism can be introduced in a simple manner. After two offspring are created using the crossover and mutation operators, they are compared with both of their parents. Thereafter, among the four parent–offspring solutions, the best two are selected, thereby allowing elite parents to compete with their offspring for a slot in the next generation. Elitism can also be introduced globally in a generational sense. Once the offspring population is created, both parent and offspring population can be combined together. Thereafter, the best N members may be chosen to form the population of the next generation. In this way too, parents get a chance to compete with the offspring population for their survival in the next generation.

No matter how elitism is introduced, it makes sure that the fitness of the population-best solution does not deteriorate. In this way, a good solution found early on in the run will never be lost unless a better solution is discovered. The absence of elitism does not guarantee this aspect. In fact, Rudolph (1996) has proved that GAs converge to the global optimal solution of some functions in the presence of elitism. Moreover, the presence of elites enhances the probability of creating better offspring.

Based on the above discussion, it is clear that elitism, in some form, is important in an evolutionary algorithm. Although elitism is important, it is not clear to what extent the elitism must be introduced into an MOEA. Elitism can be implemented to different degrees. For example, one can simply keep track of the best solution in a population and update it if a better solution is discovered at any generation, but not use the elite solutions in any genetic operations. In this way, no evolutionary advantage is exploited from the elites. On the other hand, in another extreme implementation, all elites present in the current population can be carried over to the new population. In this way, not many new solutions get a chance to enter the new population and the search does not progress anywhere. The interesting implementations are when an intermediate degree of elitism is introduced. In most single-objective EA implementations, the best α solutions of the population are used as the elites. However, the choice of an appropriate α value becomes important in the successful working of the algorithm. The parameter α is directly related to the *selection pressure* associated with the elites. Since they are directly carried over to the next generation and also participate in genetic operations, they tend to influence the population to crowd around themselves. If a large α is used, this influence is more and the population loses its diversity. However, if a too small an α is used, the adequate advantage of the elites will not be exploited. Ideally, this procedure introduces the parameter α to experiment with in any problem, but many prefer to use a fixed value of α. Values of $\alpha = 1$ to $\alpha = 0.1N$ (10% of the population size) are common.

In single-objective optimization, elites are easy to identify. They are the better solutions (with higher objective function values) in a population. The best elite is the solution with the best objective function value. However, which solutions are elites in the context of multi-objective optimization? Since there are many objective functions, it is not as straightforward as in the single-objective case to identify the elites. In such situations, the non-domination ranking comes to our rescue. A solution can be evaluated as good or bad based on its non-domination rank in the population. As we have seen earlier, there usually exists more than one solution in the best non-dominated set in a population at any generation. Since all non-dominated solutions are treated equally well, all such best non-dominated solutions become elites of identical importance. Thus, the choice of a predefined α does not work in multi-objective optimization. Introducing elitism with $\alpha = 1$ means that all best non-dominated solutions are called elites. With $\alpha = 2$, all solutions in the first two non-dominated fronts (with rank 1 and 2) may be called elites, and so on. Since the number of non-dominated solutions in any front is not controllable, it becomes difficult to introduce elitism in a controlled manner in multi-objective optimization. Moreover, we will discuss later in Section 8.8 that the number of solutions in the best non-dominated front is usually large, particularly with a large number of objectives.

Thus, even if we would like to introduce elitism with $\alpha = 1$, we have a good proportion of elite solutions coming from the previous population. As discussed earlier, this allows the population to lose its diversity soon. The presence of elitism should improve the performance of a multi-objective EA, but care must be taken to control the effective degree of elitism introduced in the process. In the following sections, we will present a number of algorithms that attempt to achieve a controlled elitism in multi-objective evolutionary optimization.

6.1   Rudolph's Elitist Multi-Objective Evolutionary Algorithm

Rudolph (2001) suggested, but did not simulate, a multi-objective evolutionary algorithm which uses elitism. This algorithm cannot be implemented as suggested and requires the introduction of a diversity preservation mechanism. Nevertheless, the algorithm provides a useful plan for introducing elitism into a multi-objective EA.

In its general format, μ parents are used to create λ offspring using genetic operators. Now there are two populations: the parent population P_t and the offspring population Q_t. The algorithm works in three phases. In the first phase, the best solutions of Q_t are first identified. Since non-dominated solutions are best solutions in a population, the non-dominated solutions of Q_t are found. These are deleted from Q_t and placed in an elite population Q^*. Thus, all solutions of the modified Q_t are worse (or dominated) than at least one solution in Q^*.

In the second phase, each solution q of Q^* is compared with each solution of the parent population P_t. If q dominates any solution of P_t, that solution cannot be an elite solution in accordance with Q^* and is thus deleted from P_t. Since this offspring is special (at least it has dominated one parent solution), we take it out of Q^* and put it in a set P'. On the other hand, if q does not dominate any solution of P_t and q does not get dominated by any solution in P_t, then q belongs to the same non-dominated set as all solutions of P_t. Such a solution q is also special to a lower degree than the elements of P' in the sense that at least q does not get dominated by any parent solution of P_t. We take it out of Q^* and put it into another set, Q'. This ends the second phase. It is important to note that during this phase the original parent population P_t gets modified.

In the third phase, all of the above sets are arranged in a special order of preference to fill up the next generation parent population. First, the modified parent solution set P and the set P' are combined together. If they together do not fill up all μ population slots, we take solutions from different sets in the following order: Q', Q^* and Q_t. Solutions are accepted until all population slots are filled. This algorithm systematically prefers the best non-dominated solutions from both parent and offspring populations and reduces $(\mu + \lambda)$ solutions to μ.

Rudolph's Elitist EA

> **Step 1** Create the parent population P_t to create an offspring population Q_t. Find non-dominated solutions of Q_t and put them in Q^*. Delete Q^* from Q_t or $Q_t = Q_t \backslash Q^*$. Set $P' = Q' = \emptyset$.
>
> **Step 2** For each $q \in Q^*$, find all solutions $p \in P_t$ that are dominated by q. Put these solutions of P_t in a set $D(q)$. If $|D(q)| > 0$, carry out the following:

$$
\begin{aligned}
P_t &= P_t \backslash D(q), \\
P' &= P' \cup \{q\}, \\
Q^* &= Q^* \backslash \{q\}.
\end{aligned}
$$

> Otherwise, if $D(q) = \emptyset$ and q is non-dominated with all existing members of P_t, then perform the following:

$$
Q' = Q' \cup \{q\}, \quad Q^* = Q^* \backslash \{q\}.
$$

> **Step 3** Form a combined population $P_{t+1} = P_t \cup P'$. If $|P_{t+1}| < \mu$, then fill P_{t+1} with elements from the following sets in the order mentioned below

until $|P_{t+1}| = \mu$:

$$(Q', Q^*, Q_t).$$

6.1.1 Hand Calculations

We now consider the same two-objective Min-Ex problem (defined above in equation (5.1)) to illustrate the above elite-preservation algorithm. We assume an offspring population of six solutions, as shown in Table 20 and in Figure 134. Here, we show a serial implementation of Step 2, whereas a parallel implementation can also be done.

Step 1 The first task is to find the best non-dominated front of the offspring population Q_t. The resulting front consists of solutions a, b and e. Thus, $Q^* = \{a, b, e\}$. Figure 135 shows that these three belong to the non-dominated front of the offspring population. We eliminate these solutions from Q_t, or $Q_t = \{c, d, f\}$. For further processing, we set $P' = Q' = \emptyset$.

Step 2 We find that $q = a$ dominates the four solutions 1 to 4 of the parent population. Thus, $D(a) = \{1, 2, 3, 4\}$. We delete these from the parent population: $P_t = \{5, 6\}$ and put the offspring a in $P' = \{a\}$. We also eliminate solution a from Q^* and have $Q^* = \{b, e\}$.

Next, we consider solution b. We find that it dominates solution 6. Thus, $D(b) = \{6\}$. Thus, we delete solution 6 from P_t, or $P_t = \{5\}$. We also put this solution in P', or $P' = \{a, b\}$. Finally, we delete this member from Q^*. The modified set is $Q^* = \{e\}$. We also observe that solution 5 of the parent population dominates offspring b. However, the above algorithm does not take account of this aspect.

We now consider the last element of Q^*. We observe that solution e is non-dominated with solution 5 (the remaining member of P_t). Thus, we keep solution e in Q', or $Q' = \{e\}$. We also delete this element from Q^*. Thus, the modified $Q^* = \emptyset$.

Table 20 Parent and offspring with their objective function values.

| Parent population, P_t | | | | | Offspring population, Q_t | | | | |
Solution	x_1	x_2	f_1	f_2	Solution	x_1	x_2	f_1	f_2
1	0.31	0.89	0.31	6.10	a	0.21	0.24	0.21	5.90
2	0.43	1.92	0.43	6.79	b	0.79	2.14	0.79	3.97
3	0.22	0.56	0.22	7.09	c	0.51	2.32	0.51	6.51
4	0.59	3.63	0.59	7.85	d	0.27	0.87	0.27	6.93
5	0.66	1.41	0.66	3.65	e	0.58	1.62	0.58	4.52
6	0.83	2.51	0.83	4.23	f	0.24	1.05	0.24	8.54

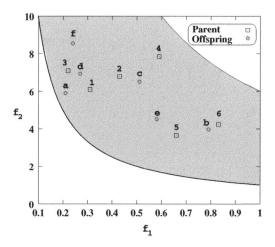

Figure 134   Parent (numerals) and offspring (alphabets).

Step 3 The resulting sets are as follows:

$$P_t = \{5\}, \quad P' = \{a, b\}, \qquad Q' = \{e\},$$
$$Q^* = \emptyset, \quad Q_t = \{c, d, f\}.$$

We first form $P_t \cup P' = \{5, a, b\}$. Since three more elements are needed to fill the population up to its size, we first include the only member of Q', and any two members from Q_t. Let us say that we choose solution c and d. Thus, the new population is $P_{t+1} = \{5, a, b, c, d, e\}$. Figure 136 shows the resulting solutions, joined by dashed lines. This population becomes the parent population of the next generation.

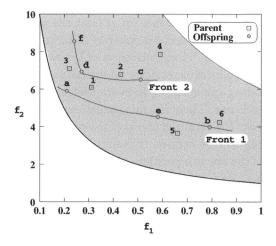

Figure 135   Offspring are classified into two non-dominated fronts.

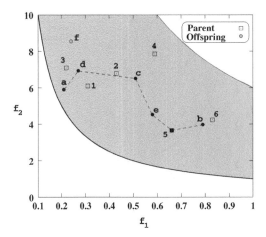

Figure 136 The parent population of the next generation consists of solutions joined by dashed lines.

6.1.2 Computational Complexity

In Step 1, the best non-dominated solutions are found. This requires $O(M\lambda^2)$ computations. In Step 2, for each non-dominated solution in the offspring population, all solutions in the parent population are checked, thereby requiring the worst case $O(\mu\lambda)$ comparisons. If $\mu = \lambda = N$ is used, the worst case complexity is $O(MN^2)$. However, it is important to remember that this algorithm does not attempt to maintain any diversity in the obtained solutions.

6.1.3 Advantages

Rudolph (2001) has proven that the above algorithm with a 'positive variation kernel' of its search operators allows convergence (of at least one solution in the population) to the Pareto-optimal front in a finite number of trials (refer to Section 8.9 below) in finite search space problems. The positiveness of the variation kernel (representing the joint transition probabilities of crossover and mutation operators) makes sure that the probability of creating any offspring from an arbitrary set of parent solutions in a finite number of trials is one. Thus, in any population, if no Pareto-optimal solution exists, the positiveness of the variation kernel of the combined search operators ensures that one such member will be created in a finite number of trials. With the elite-preserving strategy of the above algorithm, this member cannot ever be deleted from the population. This is because this solution will not be dominated by any other solution in the search space. Therefore, this solution will always lie in P_t.

6.1.4 Disadvantages

However, what has not been proved is a guaranteed maintenance of diversity among the Pareto-optimal solutions. If a population-based evolutionary algorithm converges to a single Pareto-optimal solution, it cannot be justified for its use over other classical algorithms. What makes evolutionary algorithms attractive is their ability to find multiple Pareto-optimal solutions in one single run.

Whenever all μ parent solutions are Pareto-optimal, no new Pareto-optimal solutions can be accommodated in the new population. This shortcoming does not allow the above algorithm to find a more diverse set of Pareto-optimal solutions. Although a niche formation scheme can be used in choosing the elements from Q', Q^* and Q_t to fill up the population with a diverse set of solutions from these sets, the algorithm stagnates as soon as an arbitrary set of μ Pareto-optimal solutions are found as parents.

The set P' is given priority over Q' in forming the new population P_{t+1}, although all members of Q' are non-dominated with all members of P_t. Since a member of P_t can dominate some members of P', the resulting set P_{t+1} may not contain all non-dominated solutions.

Since this algorithm does not ensure any diversity among the obtained solutions, we have not used it for testing on the example problem.

6.2 Elitist Non-Dominated Sorting Genetic Algorithm

This author and his students suggested an elitist non-dominated sorting GA (termed NSGA-II) in 2000 (Deb et al., 2000a, 2000b). Unlike the above method of using only an elite-preservation strategy as in the previous algorithm, NSGA-II also uses an explicit diversity-preserving mechanism. In most aspects, this algorithm does not have much similarity with the original NSGA, but the authors kept the name NSGA-II to highlight its genesis and place of origin.

Like Rudolph's algorithm, in NSGA-II, the offspring population Q_t is first created by using the parent population P_t. However, instead of finding the non-dominated front of Q_t only, first the two populations are combined together to form R_t of size 2N. Then, a non-dominated sorting is used to classify the entire population R_t. Although this requires more effort compared to performing a non-dominated sorting on Q_t alone, it allows a global non-domination check among the offspring and parent solutions. Once the non-dominated sorting is over, the new population is filled by solutions of different non-dominated fronts, one at a time. The filling starts with the best non-dominated front and continues with solutions of the second non-dominated front, followed by the third non-dominated front, and so on. Since the overall population size of R_t is 2N, not all fronts may be accommodated in N slots available in the new population. All fronts which could not be accommodated are simply deleted. When the last allowed front is being considered, there may exist more solutions in the last front than the remaining slots in the new population. This scenario is illustrated in Figure 137. Instead of arbitrarily discarding some members from the last front, it would be wise to use a niching strategy to choose the members of the last front, which reside in the least crowded region in that front.

A strategy like the above does not affect the proceedings of the algorithm much

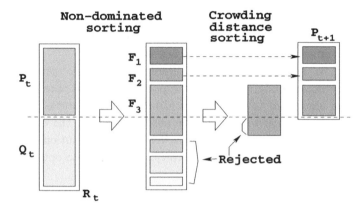

Figure 137   Schematic of the NSGA-II procedure.

in the early stages of evolution. This is because, early on, there exist many fronts in the combined population. It is likely that solutions of many good non-dominated fronts are already included in the new population, before they add up to N. It then hardly matters which solution is included to fill up the population. However, during the latter stages of the simulation, it is likely that most solutions in the population lie in the best non-dominated front. It is also likely that in the combined population R_t of size 2N, the number of solutions in the first non-dominated front exceeds N. The above algorithm then ensures that niching will choose a diverse set of solutions from this set. When the entire population converges to the Pareto-optimal front, the continuation of this algorithm will ensure a better spread among the solutions.

In the following, we outline the algorithm in a step-by-step format. Initially, a random population P_0 is created. The population is sorted into different non-domination levels. Each solution is assigned a fitness equal to its non-domination level (1 is the best level). Thus, minimization of the fitness is assumed. Binary tournament selection (with a crowded tournament operator described later), recombination and mutation operators are used to create a offspring population Q_0 of size N. The NSGA-II procedure is outlined in the following.

NSGA-II

> **Step 1** Combine parent and offspring populations and create $R_t = P_t \cup Q_t$. Perform a non-dominated sorting to R_t and identify different fronts: $\mathcal{F}_i$, $i = 1, 2, \ldots$, etc.
>
> **Step 2** Set new population $P_{t+1} = \emptyset$. Set a counter $i = 1$. Until $|P_{t+1}| + |\mathcal{F}_i| < N$, perform $P_{t+1} = P_{t+1} \cup \mathcal{F}_i$ and $i = i + 1$.
>
> **Step 3** Perform the Crowding-sort$(\mathcal{F}_i, <_c)$ procedure (described below on page 248) and include the most widely spread $(N - |P_{t+1}|)$ solutions by using the crowding distance values in the sorted $\mathcal{F}_i$ to P_{t+1}.
>
> **Step 4** Create offspring population Q_{t+1} from P_{t+1} by using the crowded tournament selection, crossover and mutation operators.

In Step 3, the crowding-sorting of the solutions of front i (the last front which could not be accommodated fully) is performed by using a *crowding distance metric*, which we will describe a little later. The population is arranged in descending order of magnitude of the crowding distance values. In Step 4, a crowding tournament selection operator, which also uses the crowding distance, is used.

It is important to mention here that the non-dominated sorting in Step 1 and filling up population P_{t+1} can be performed together. In this way, every time a non-dominated front is found, its size can be used to check if it can be included in the new population. If this is not possible, no more sorting is needed. This will help in reducing the run-time of the algorithm.

6.2.1   Crowded Tournament Selection Operator

The crowded comparison operator ($<_c$) compares two solutions and returns the winner of the tournament. It assumes that every solution i has two attributes:

1. A non-domination rank r_i in the population.
2. A local crowding distance (d_i) (which we define in the next subsection) in the population.

The crowding distance d_i of a solution i is a measure of the search space around i which is not occupied by any other solution in the population. Based on these two attributes, we can define the crowded tournament selection operator as follows.

Definition 6.1. *Crowded Tournament Selection Operator: A solution i wins a tournament with another solution j if any of the following conditions are true:*

1. *If solution i has a better rank, that is, $r_i < r_j$.*
2. *If they have the same rank but solution i has a better crowding distance than solution j, that is, $r_i = r_j$ and $d_i > d_j$.*

The first condition makes sure that chosen solution lies on a better non-dominated front. The second condition resolves the tie of both solutions being on the same non-dominated front by deciding on their crowded distance. The one residing in a less crowded area (with a larger crowding distance d_i) wins. The crowding distance d_i can be computed in various ways. Of these, the niche count metric (nc_i) and the head count metric (hc_i) are common (refer to Section 8.2 later). Both of these metrics require $O(MN^2)$ complexity. Although they can be used, they have to be used in a reverse sense. A solution having a smaller niche count or head count means a lesser dense solution and has to be preferred. Thus, in the second condition, one should choose solution i if $r_i = r_j$ and either $nc_i < nc_j$ or, in terms of head counts, $hc_i < hc_j$. However, in NSGA-II, we use a crowding distance metric, which requires $O(MN \log N)$ computations.

Crowding Distance

To get an estimate of the density of solutions surrounding a particular solution i in the population, we take the average distance of two solutions on either side of solution i along each of the objectives. This quantity d_i serves as an estimate of the perimeter of the cuboid formed by using the nearest neighbors as the vertices (we call this the *crowding distance*). In Figure 138, the crowding distance of the i-th solution in its front (marked with solid circles) is the average side-length of the cuboid (shown by a dashed box). The following algorithm is used to calculate the crowding distance of each point in the set $\mathcal{F}$.

Crowding Distance Assignment Procedure: Crowding-sort$(\mathcal{F}, <_c)$

    Step C1 Call the number of solutions in $\mathcal{F}$ as $l = |\mathcal{F}|$. For each i in the set, first assign $d_i = 0$.

    Step C2 For each objective function $m = 1, 2, \ldots, M$, sort the set in worse order of f_m or, find the sorted indices vector: $I^m = \text{sort}(f_m, >)$.

    Step C3 For $m = 1, 2, \ldots, M$, assign a large distance to the boundary solutions, or $d_{I_1^m} = d_{I_l^m} = \infty$, and for all other solutions $j = 2$ to $(l-1)$, assign:

$$d_{I_j^m} = d_{I_j^m} + \frac{f_m^{(I_{j+1}^m)} - f_m^{(I_{j-1}^m)}}{f_m^{max} - f_m^{min}}.$$

The index I_j denotes the solution index of the j-th member in the sorted list. Thus, for any objective, I_1 and I_l denote the lowest and highest objective function values, respectively. The second term on the right side of the last equation is the difference in objective function values between two neighboring solutions on either side of solution I_j. Thus, this metric denotes half of the perimeter of the enclosing cuboid with the nearest neighboring solutions placed on the vertices of the cuboid (Figure 138). It is interesting to note that for any solution i the same two solutions $(i+1)$ and $(i-1)$ need not be neighbors in all objectives, particularly for $M \geq 3$. The parameters f_m^{max} and f_m^{min} can be set as the population-maximum and population-minimum values of the m-th objective function.

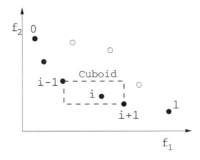

Figure 138 The crowding distance calculation. This is a reprint of Figure 1 from Deb et al. (2000b) (© Springer-Verlag Berlin Heidelberg 2000).

The above metric requires M sorting calculations in Step C2, each requiring $O(N \log N)$ computations. Step C3 requires N computations. Thus, the complexity of the above distance metric computation is $O(MN \log N)$. For large N, this complexity is smaller than $O(MN^2)$, which denotes the computational complexity required in other niching methods. Now, we illustrate the procedure while hand-simulating NSGA-II to the example problem Min-Ex.

6.2.2 Hand Calculations

We use the same parent and offspring populations as used in the previous section (Table 20).

Step 1 We first combine the populations P_t and Q_t and form $R_t = \{1, 2, 3, 4, 5, 6, a, b, c, d, e, f\}$. Next, we perform a non-dominated sorting on R_t. We obtain the following non-dominated fronts:

$$
\begin{aligned}
\mathcal{F}_1 &= \{5, a, e\}, \\
\mathcal{F}_2 &= \{1, 3, b, d\}, \\
\mathcal{F}_3 &= \{2, 6, c, f\}, \\
\mathcal{F}_4 &= \{4\}.
\end{aligned}
$$

These fronts are shown in Figure 139.

Step 2 We set $P_{t+1} = \emptyset$ and $i = 1$. Next, we observe that $|P_{t+1}| + |\mathcal{F}_1| = 0 + 3 = 3$. Since this is less than the population size N $(= 6)$, we include this front in P_{t+1}. We set $P_{t+1} = \{5, a, e\}$. With these three solutions, we now need three more solutions to fill up the new parent population.

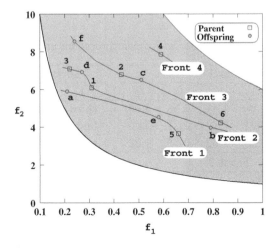

Figure 139   Four non-dominated fronts of the combined population R_t.

Now, with the inclusion of the second front, the size of $|P_{t+1}| + |\mathcal{F}_2|$ is $(3+4)$ or 7. Since this is greater than 6, we stop including any more fronts into the population. Note that if fronts 3 and 4 had not been classified earlier, we could have saved these computations.

Step 3 Next, we consider solutions of the second front only and observe that three (of the four) solutions must be chosen to fill up three remaining slots in the new population. This requires that we first sort this subpopulation (solutions 1, 3, b, and d) by using the $<_c$ operator (defined above in Section 6.2.1). We calculate the crowded distance values of these solutions in the front by using the step-by-step procedure.

Step C1 We notice that $l = 4$ and set $d_1 = d_3 = d_b = d_d = 0$. We also set $f_1^{max} = 1$, $f_1^{min} = 0.1$, $f_2^{max} = 60$ and $f_2^{min} = 0$.

Step C2 For the first objective function, the sorting of these solutions is shown in Table 21 and is as follows: $I^1 = \{3, d, 1, b\}$.

Step C3 Since solutions 3 and b are boundary solutions for f_1, we set $d_3 = d_b = \infty$. For the other two solutions, we obtain:

$$d_d = 0 + \frac{f_1^{(1)} - f_1^{(3)}}{f_1^{max} - f_1^{min}} = 0 + \frac{0.31 - 0.22}{1 - 0.1} = 0.10.$$

$$d_1 = 0 + \frac{f_1^{(b)} - f_1^{(d)}}{f_1^{max} - f_1^{min}} = 0 + \frac{0.79 - 0.27}{1 - 0.1} = 0.58.$$

Now, we turn to the second objective function and update the above distances. First, the sorting on this objective yields $I^2 = \{b, 1, d, 3\}$. Thus,

Table 21   The fitness assignment procedure under NSGA-II.

Solution	Front 1				Sorting in		Distance
	x_1	x_2	f_1	f_2	f_1	f_2	
5	0.66	1.41	0.66	3.65	third	first	∞
a	0.21	0.24	0.21	5.90	first	third	∞
e	0.58	1.62	0.58	4.52	second	second	0.54

Solution	Front 2				Sorting in		Distance
	x_1	x_2	f_1	f_2	f_1	f_2	
1	0.31	0.89	0.31	6.10	third	second	0.63
3	0.22	0.56	0.22	7.09	first	fourth	∞
b	0.79	2.14	0.79	3.97	fourth	first	∞
d	0.27	0.87	0.27	6.93	second	third	0.12

$d_b = d_3 = \infty$ and the other two distances are as follows:

$$d_1 = d_1 + \frac{f_2^{(d)} - f_2^{(b)}}{f_2^{max} - f_2^{min}} = 0.58 + \frac{6.93 - 3.97}{60 - 0} = 0.63.$$

$$d_d = d_d + \frac{f_1^{(3)} - f_2^{(1)}}{f_2^{max} - f_2^{min}} = 0.10 + \frac{7.09 - 6.10}{60 - 0} = 0.12.$$

The overall crowded distances of the four solutions are:

$$d_1 = 0.63, \quad d_3 = \infty, \quad d_b = \infty, \quad d_d = 0.12.$$

The cuboids (rectangles here) for these solutions are schematically shown in Figure 140. Solution d has the smallest perimeter of the hypercube around it than any other solution in the set $\mathcal{F}_2$, as evident from the figure. Now, we move to the main algorithm.

Step 3 A sorting according to the descending order of these crowding distance values yields the sorted set $\{3, b, 1, d\}$. We choose the first three solutions.

Step 4 The new population is $P_{t+1} = \{5, a, e, 3, b, 1\}$. These population members are shown in Figure 141 by joining them with dashed lines. It is important to note that this population is formed by choosing solutions from the better non-dominated fronts.

The offspring population Q_{t+1} has to be created next by using this parent population. We realize that the exact offspring population will depend on the chosen pair of solutions participating in a tournament and the chosen crossover and mutation operators. Let us say that we pair solutions $(5, e)$, $(a, 3)$, $(1, b)$, $(a, 1)$, (e, b) and $(3, 5)$, so that each solution participates in exactly two

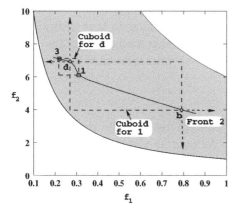

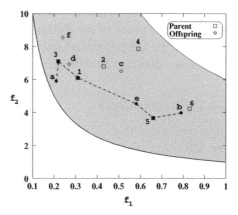

Figure 140   The cuboids of solutions 1 and d. The cuboids for solutions 3 and b extend to infinity.

Figure 141   The new parent population P_{t+1} joined by dashed lines.

tournaments. The crowding distance values for solutions of the first front are also listed in Table 21. In the first tournament, we observe that solutions 5 and e belong to the same front ($r_5 = r_e = 1$). Thus, we choose the one with the larger crowding distance value. We find that solution 5 is the winner.

In the next tournament between solutions a and 3, solution a wins, since it belongs to a better front. Performing other tournaments, we obtain the mating pool: $\{5, a, a, b, b, e\}$. Now, these solutions can be mated pair-wise and mutated to create Q_{t+1}. This completes one generation of the NSGA-II.

6.2.3   Computational Complexity

Step 1 of the NSGA-II requires a non-dominated sorting of a population of size 2N. This requires at most $O(MN^2)$ computations. Sorting in Step 3 requires crowding distance computation of all members of front i. However, the crowded tournament selection in Step 4 requires the crowding distance computation of the complete population P_{t+1} of size N. This requires $O(MN \log N)$ computations. Thus, the overall complexity of the NSGA-II is at most $O(MN^2)$.

6.2.4   Advantages

The diversity among non-dominated solutions is introduced by using the crowding comparison procedure which is used with the tournament selection and during the population reduction phase. Since solutions compete with their crowding distances (a measure of the density of solutions in the neighborhood), no extra niching parameter (such as σ_{share} needed in the MOGAs, NSGAs or NPGAs) is required here. Although the crowding distance is calculated in the objective function space, it can also be implemented in the parameter space, if so desired (Deb, 1999c).

In the absence of the crowded comparison operator, this algorithm also exhibits a convergence proof to the Pareto-optimal solution set similar to that in Rudolph's algorithm, but the population size would grow with the generation counter. The elitism mechanism does not allow an already found Pareto-optimal solution to be deleted.

6.2.5   Disadvantages

However, when the crowded comparison is used to restrict the population size, the algorithm loses its convergence property. As long as the size of the first non-dominated set is not larger than the population size, this algorithm preserves all of them. However, in latter generations, when more than N members belong to the first non-dominated set in the combined parent–offspring population, some closely-packed Pareto-optimal solutions may give their places to other non-dominated yet non-Pareto-optimal solutions. Although these latter solutions may get dominated by other Pareto-optimal solutions in a later generation, the algorithm can resort into this cycle of generating Pareto-optimal and non-Pareto-optimal solutions before finally

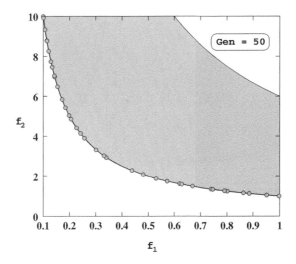

Figure 142 The population after 50 generations with the NSGA-II without mutation.

converging to a well-distributed set of Pareto-optimal solutions.

As mentioned earlier, the non-dominated sorting needs to be performed on a population of size 2N, instead of a population of size N required in most algorithms.

6.2.6 Simulation Results

We apply the above algorithm to the Min-Ex problem. The GA parameter settings are also identical to that used before. A population size of 40 and a crossover probability of 0.9 are used. The decision variables 1 and 2 are coded in 24 and 26 bits, respectively. Figure 142 shows the population of 40 solutions at the end of 50 generations without a mutation operator. This figure shows that even without the mutation operator, the NSGA-II can make the population reach the true Pareto-optimal front with a good spread of solutions. The population at the end of 500 generations, shown in Figure 143, confirms that the NSGA-II can sustain the spread of solutions without mutation even after a large number of generations are elapsed. However, the distribution of solutions gets better with generations.

The next figure shows the NSGA-II population obtained using a bit-wise mutation operator. Figure 144 shows the population after 500 generations with a mutation probability of 0.02. With the mutation operator, the exploration power of the NSGA-II gets enhanced and a better distribution of solutions is observed.

6.3   Distance-Based Pareto Genetic Algorithm

Osyczka and Kundu (1995) suggested an elitist GA, a distance-based Pareto genetic algorithm (DPGA), which attempts to emphasize the progress towards the Pareto-optimal front and the diversity along the obtained front by using one fitness measure. This algorithm maintains two populations: one standard GA population P_t where

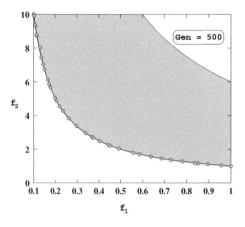

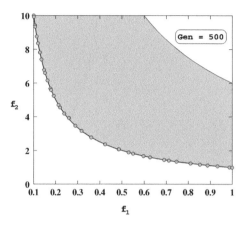

Figure 143 The population after 500 generations with the NSGA-II without mutation.

Figure 144 The population after 500 generations with the NSGA-II with mutation.

genetic operations are performed, and another elite population E_t containing all non-dominated solutions found thus far.

The initial population P_0 of size N is created at random. The first population member is assigned a positive random fitness F_1 (chosen arbitrarily) and is automatically added to the elite set E_0. Thereafter, each solution is assigned a fitness based on its distance from the elite set, $E_t = \{e^{(k)} : k = 1, 2, \ldots, K\}$, where K is the number of solutions in the elite set. Each elite solution $e^{(k)}$ has M function values, or $e^{(k)} = \left(e_1^{(k)}, e_2^{(k)}, \ldots, e_M^{(k)}\right)^T$. The distance of a solution $\mathbf{x}$ from the elite set is calculated as follows:

$$d^{(k)}(\mathbf{x}) = \sqrt{\sum_{m=1}^{M} \left(\frac{e_m^{(k)} - f_m(\mathbf{x})}{e_m^{(k)}}\right)^2}. \tag{6.1}$$

For the solution $\mathbf{x}$, the minimum $d^{(k)}(\mathbf{x})$ of all $k = 1, 2, \ldots, K$ is found as follows:

$$d^{\min} = \min_{k=1}^{K} d^{(k)}(\mathbf{x})$$

and the index k^* for the minimum distance is also recorded. Thereafter, if the solution $\mathbf{x}$ is a non-dominated solution with respect to the existing elite set, it is accepted in the elite set and its fitness is calculated by adding the fitness of the elite member with minimum distance from it and its distance from the minimum member:

$$F(\mathbf{x}) = F(e^{(k^*)}) + d^{\min}. \tag{6.2}$$

The elite set is updated by deleting all elite solutions dominated by $\mathbf{x}$, if any. On the other hand, if the solution $\mathbf{x}$ is dominated by any elite solution, it is not accepted in the elite set and its fitness is calculated as follows:

$$F(\mathbf{x}) = \max\left[0, (F(e^{(k^*)}) - d^{\min})\right]. \tag{6.3}$$

In this way, as population members are evaluated for their fitness, the elite set is constantly updated. At the end of the generation (when all N population members are evaluated), the maximum fitness F_{max} among the existing elite solutions is calculated and all existing elite solutions are assigned a fitness equal to F_{max}. At the end of a generation, selection, crossover and mutation operators are used to create a new population.

It is interesting to note that a non-dominated solution lying a large distance away from the existing elite set gets a large (better) fitness. This helps in two ways. First, if the new solution dominates a few members of the elite set, the fitness assignment procedure helps in emphasizing solutions closer to the Pareto-optimal set. A distant solution here means a solution distant from the existing elite set but closer to the Pareto-optimal set. Assigning a large fitness to such a solution helps to progress towards the Pareto-optimal front. On the other hand, if the new solution lies in the same non-dominated front along with the elite solutions, the fitness assignment procedure helps in maintaining diversity among them. A distant solution here means an isolated solution on the same front. Assigning a large fitness to an isolated solution helps to maintain diversity among obtained non-dominated solutions.

Distance-Based Pareto GA (DPGA)

Step 1 Create an initial random population P_0 of size N and set the fitness of the first solution as F_1. Set generation counter $t = 0$.

Step 2 If $t = 0$, insert the first element of P_0 in an elite set $E_0 = \{1\}$. For each population member $j \geq 2$ for $t = 0$ and $j \geq 1$ for $t > 0$, perform the following steps to assign a fitness value.

Step 2a Calculate the distance $d_j^{(k)}$ with each elite member k (with fitness $e_m^{(k)}, m = 1, 2, \ldots, M$) as follows:

$$d_j^{(k)} = \sqrt{\sum_{m=1}^{M} \left(\frac{e_m^{(k)} - f_m^{(j)}}{e_m^{(k)}} \right)^2}.$$

Step 2b Find the minimum distance and the index of the elite member closest to solution j:

$$d_j^{min} = \min_{k=1}^{|E_t|} d_j^{(k)},$$

$$k_j^* = \{k : d_j^{(k)} = d_j^{min}\}.$$

Step 2c If any elite member dominates solution j, the fitness of j is:

$$F_j = \max \left[0, F(e^{(k_j^*)}) - d^{min} \right].$$

Otherwise, the fitness of j is:

$$F_j = F(e^{(k_j^*)}) + d^{min}$$

and j is included in E_t by eliminating all elite members that are dominated by j.

Step 3 Find the maximum fitness value of all elite members:

$$F_{max} = \max_{k=1}^{|E_t|} F_i.$$

All elite solutions are assigned a fitness F_{max}.

Step 4 If $t < t_{max}$ or any other termination criterion is satisfied, the process is complete. Otherwise, go to Step 5.

Step 5 Perform selection, crossover and mutation on P_t and create a new population P_{t+1}. Set $t = t + 1$ and go to Step 2.

Note that no separate genetic processing (like reproduction, crossover or mutation operation) is performed on the elite population E_t explicitly. The fitness F_{max} of elite members is used in the assignment of fitness of solutions of P_t.

6.3.1 Hand Calculations

We use the two-variable Min-Ex problem. Let us also use the same parent population as used in the previous section (Table 20).

Step 1 We arbitrarily choose the fitness of the first population member as $F_1 = 10.0$. We also set $t = 0$.

Step 2 Since $t = 0$, we include the first population member in the elite set $E_0 = \{1\}$. Now, for the second individual, we run through Steps 2a – 2c as follows:

Step 2a The distance d_2^1 is:

$$d_2^1 = \sqrt{\left(\frac{0.31 - 0.43}{0.31}\right)^2 + \left(\frac{6.10 - 6.79}{6.10}\right)^2},$$
$$= 0.40.$$

Step 2b Since there is only one member in E_0, the minimum distance is $d_2^{min} = 0.40$ and $k_2^* = 1$.

Step 2c Since solution 1 dominates solution 2, the fitness of solution 2 is:

$$F_2 = \max[0, (10.0 - 0.40)]$$
$$= 9.60.$$

The elite set remains unchanged here.

We go back to Step 2a to show the fitness computation for solution 3. The distance $d_3^1 = 0.33$. Since solution 3 is non-dominated with solution 1, the fitness of solution 3 is $F_3 = 10.00 + 0.33 = 10.33$. Solution 3 is also included in the elite set. Thus, $E_0 = \{1, 3\}$.

For solution 4, the distances $d_4^1 = 0.95$ and $d_4^3 = 1.69$. Thus, $d_4^{\min} = 0.95$ and the fitness of solution 4 is $F_4 = 10.00 - 0.95 = 9.05$. The elite set is unchanged here.

Similarly, the fitness values of solutions 5 and 6 are also calculated. The corresponding minimum distance values and elite sets are shown in Table 22.

Step 3 There are three individuals (solutions 1, 3 and 5) in the elite set. The maximum fitness of these two solutions is 11.20. Thus, these three solutions are assigned a fitness 11.20 as an elite member. However, their fitness values in the GA population remain as 10.00, 10.33 and 11.20, respectively.

Step 4 As no termination criterion is satisfied, we proceed to the next step.

Step 5 The population is now operated by a selection operation based on the fitness values shown in the table, followed by crossover and mutation operations. However, the elite set $E_1 = \{1, 3, 5\}$ passes on to the next population and the fitness of the new population members will be computed with respect to this elite set. In the process, the elite set may also get modified.

Figure 145 shows the six population members and the elite set obtained at the end of the first generation. Arrows on this figure connect population members (ends of arrows) with elite members (starts of arrows) having minimum distances in the objective space. A continuous arrow marks an improvement in fitness, while a dashed arrow shows a decrement in fitness of the population member from the fitness of the elite member. A natural hierarchy of decreasing fitness for increasingly dominated solutions evolves with this fitness assignment scheme. Since non-dominated solutions are assigned more fitness, the selection operator will emphasize these solutions. The crossover and mutation operators will then help create solutions near these non-dominated solutions, thereby helping the search to proceed towards the Pareto-optimal

Table 22   Fitness computations with the DPGA.

Solution	GA population						
(j)	x_1	x_2	f_1	f_2	$d_j^{\min}$	Fitness	E_0
1	0.31	0.89	0.31	6.10	—	10.00	$\{1\}$
2	0.43	1.92	0.43	6.79	0.40	9.60	$\{1\}$
3	0.22	0.56	0.22	7.09	0.33	10.33	$\{1,3\}$
4	0.59	3.63	0.59	7.85	0.95	9.05	$\{1,3\}$
5	0.66	1.41	0.66	3.65	1.20	11.20	$\{1,3,5\}$
6	0.83	2.51	0.83	4.23	0.30	10.90	$\{1,3,5\}$

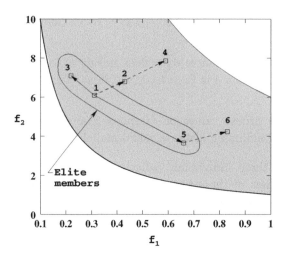

Figure 145 The DPGA fitness assignment procedure.

region. Notice how the fitness assignment scheme takes care of the crowding of solutions in the above example case. Since solution 3 is located close to solution 1 in the objective space and since there exists no solution close to solution 5, the fitness of solution 3 is smaller than that of solution 5. Thus, an emphasis of non-dominated solutions and a maintenance of diversity are both obtained by the above fitness assignment scheme. However, it is important to note that this may not always be true, a matter we discuss later in Section 6.3.4.

6.3.2   Computational Complexity

The distance computation and dominance test require $O(M\eta^2)$ computations, where η is the current size of the elite set E_t. Since the algorithm does not restrict the size of the elite set, it is likely that η will increase with generation. Thus, the complexity of the algorithm increases with the number of generations.

6.3.3   Advantages

For some sequence of fitness evaluations, both goals of progressing towards the Pareto-optimal front and maintaining diversity among solutions are achieved without any explicit niching method. However, as we shall see in the next subsection, an appropriate niching may not form in certain sequence of fitness assignments.

6.3.4   Disadvantages

As mentioned above, the elite size is not restricted and is allowed to grow to any size. This will increase the computational complexity of the algorithm with generations. However, this difficulty can be eliminated by restricting the elite size in Step 2c. A population member is allowed to be included in the elite set only when the elite size does not exceed a user-specified limit. If there is no room left in the elite population, the new solution can be used to replace an existing elite member. Several criteria can be used to decide the member that must be deleted. One approach would be to find the crowded distance (see the NSGA-II discussion in Section 6.2 above). If the crowded distance of the new solution in the elite set is more than the crowded distance of any elite member, the new solution replaces that elite member. In this way, the elite set is always restricted to a specified size and better diversity can be obtained.

The original study suggested the use of a proportionate selection operator. In such a case, the choice of the initial F_1 is important. However, for tournament selection or ranking selection methods, the algorithm is not sensitive to F_1.

The DPGA fitness assignment scheme is sensitive to the ordering of individuals in a population. In the above example (Table 22), if solution 1 and 5 are interchanged in the population, the fitnesses of solutions 3 and 5 are more than that of solution 1 (Figure 146). However, solution 1 now resides in a least crowded area. This fitness assignment is likely to have an adverse effect on the diversity of the obtained solutions.

6.3.5   *Simulation Results*

We apply the DPGA to the Min-Test problem. As in the other algorithms, we use a population of size 40 and run the DPGA up until 500 generations. As suggested by the investigators, we will use the proportionate selection scheme. (We have used the stochastic universal selection scheme here.) The single-point crossover with $p_c = 0.9$ and the bit-wise mutation operator with $p_m = 0.02$ are used. Two variables are coded in 24 and 26 bits. No restriction on the size of the elite set is kept. Figures 147 and 148 show the elite populations after 50 and 500 generations, respectively. It is

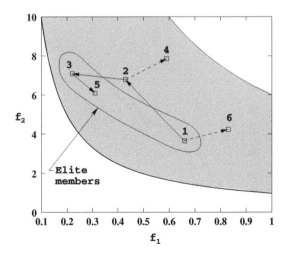

Figure 146   The DPGA fitness assignment fails to emphasize the least crowded solution.

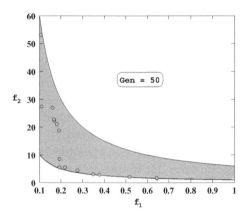

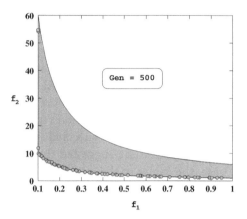

Figure 147   The population after 50 generations with the DPGA.

Figure 148   The population after 500 generations with the DPGA.

interesting to note that the DPGA cannot find many non-dominated solutions after 50 generations. However, after 500 generations, the algorithm can find as many as 81 solutions near the Pareto-optimal front. Moreover, the population has maintained a good distribution.

In order to support our argument on the gradual increase in the size of the elite population, we have counted the number of non-dominated solutions in the elite set at the end of each generation and plotted these data in Figure 149. It is clear that the size increases with generation, except for occasional drops in the number. When a better non-dominated solution is discovered, a few existing elite population members may become dominated and hence the size of the elite population reduces. Since the elite

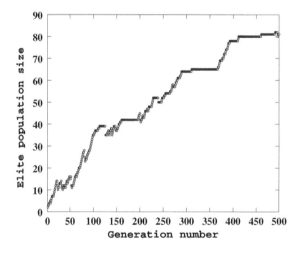

Figure 149   Growth of the elite population size in the DPGA.

population did not entirely converge on the exact Pareto-optimal front even after 500 generations, this sudden drop followed by increase in the size of the elite population continues up until 500 generations. When all elite population members converge to the true Pareto-optimal front, the elite size will monotonically increase or remain constant with the generation counter.

6.4 Strength Pareto Evolutionary Algorithm

Zitzler and Thiele (1998a) proposed an elitist evolutionary algorithm, which they called the strength Pareto EA (SPEA). This algorithm introduces elitism by explicitly maintaining an external population $\overline{P}$. This population stores a fixed number of the non-dominated solutions that are found until the beginning of a simulation. At every generation, newly found non-dominated solutions are compared with the existing external population and the resulting non-dominated solutions are preserved. The SPEA does more than just preserving the elites; it also uses these elites to participate in the genetic operations along with the current population in the hope of influencing the population to steer towards good regions in the search space.

This algorithm begins with a randomly created population P_0 of size N and an empty external population $\overline{P}_0$ with a maximum capacity of $\overline{N}$. In any generation t, the best non-dominated solutions (belonging to the first non-dominated front) of the population P_t are copied to the external population $\overline{P}_t$. Thereafter, the dominated solutions in the modified external population are found and deleted from the external population. In this way, previously found elites which are now dominated by a new elite solution get deleted from the external population. What remains in the external population are the best non-dominated solutions of a combined population containing old and new elites. If this process is continued over many generations, there is a danger of the external population being overcrowded with non-dominated solutions, like that in the DPGA. In order to restrict the population to over-grow, the size of the external population is bounded to a limit $(\overline{N})$. That is, when the size of the external population is less than $\overline{N}$, all elites are kept in the population. However, when the size exceeds $\overline{N}$, not all elites can be accommodated in the external population. This is where the investigators of the SPEA considered satisfying the second goal of multi-objective optimization. Elites which are less crowded in the non-dominated front are kept. Only that many solutions which are needed to maintain the fixed population size $\overline{N}$ are retained. The investigators suggested a clustering method to achieve this task. We shall describe this clustering method a little later.

Once the new elites are preserved for the next generation, the algorithm then turns to the current population and uses genetic operators to find a new population. The first step is to assign a fitness to each solution in the population. It was mentioned earlier that the SPEA also uses the external population $\overline{P}_t$ in its genetic operations. In addition to the assigning of fitness to the current population members, fitness is also assigned to external population members. In fact, the SPEA assigns a fitness (called the *strength*) S_i to each member i of the external population first. The strength S_i

is proportional to the number (n_i) of current population members that an external
solution i dominates:

$$S_i = \frac{n_i}{N+1}. \tag{6.4}$$

In other words, the above equation assigns more strength to an elite which dominates
more solutions in the current population. Division by $(N + 1)$ ensures that the
maximum value of the strength of any external population member is never one or
more. In addition, a non-dominated solution dominating a fewer solutions has a smaller
(or better) fitness.

Thereafter, the fitness of a current population member j is assigned as one more
than the sum of the strength values of all external population members which weakly
dominate j:

$$F_j = 1 + \sum_{i \in \overline{P}_t \wedge i \preceq j} S_i. \tag{6.5}$$

The addition of one makes the fitness of any current population member P_t to be
more than the fitness of any external population member $\overline{P}_t$. This method of fitness
assignment suggests that a solution with a smaller fitness is better. Figure 150 shows
this fitness assignment scheme for both external (shown by circles) and EA population
members (shown by squares) on a two-objective minimization problem. The fitness
values are also marked on this figure. The latter shows that the external population
members get smaller fitness values than the EA population members. Since there are
six population members in the EA population, the denominator in the fitness values
is seven. EA population members dominated by many external members get large
fitness values. With these fitness values, a binary tournament selection procedure is
applied to the combined $(\overline{P}_t \cup P_t)$ population to choose solutions with smaller fitness
values. Thus, it is likely that external elites will be emphasized during this tournament
procedure. As usual, crossover and mutation operators are applied to the mating pool
and a new population P_{t+1} of size N is created. In the following, we will describe one

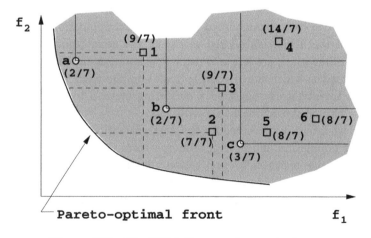

Figure 150 The SPEA fitness assignment scheme.

iteration of the algorithm in a step-by-step format. Initially, $\bar{P}_t = \emptyset$ is assumed.

Strength Pareto Evolutionary Algorithm (SPEA)

Step 1 Find the best non-dominated set $\mathcal{F}_1(P_t)$ of P_t. Copy these solutions to $\bar{P}_t$, or perform $\bar{P}_t = \bar{P}_t \cup \mathcal{F}_1(P_t)$.

Step 2 Find the best non-dominated solutions $\mathcal{F}_1(\bar{P}_t)$ of the modified population $\bar{P}_t$ and delete all dominated solutions, or perform $\bar{P}_t = \mathcal{F}_1(\bar{P}_t)$.

Step 3 If $|\bar{P}_t| > \bar{N}$, use a clustering technique to reduce the size to $\bar{N}$. Otherwise, keep $\bar{P}_t$ unchanged. The resulting population is the external population $\bar{P}_{t+1}$ of the next generation.

Step 4 Assign fitness to each elite solution $i \in \bar{P}_{t+1}$ by using equation (6.4). Then, assign fitness to each population member $j \in P_t$ by using equation (6.5).

Step 5 Apply a binary tournament selection with these fitness values (in a minimization sense), a crossover and a mutation operator to create the new population P_{t+1} of size N from the combined population $(\bar{P}_{t+1} \cup P_t)$ of size $(\bar{N} + N)$.

Steps 3 and 5 result in the new external and current populations, which are then processed in the next generation. This algorithm continues until a stopping criterion is satisfied.

6.4.1 Clustering Algorithm

Let us now describe the clustering algorithm which reduces the size of the external population $\bar{P}_t$ of size $\bar{N}'$ to $\bar{N}$ (where $\bar{N}' > \bar{N}$).

At first, each solution in $\bar{P}_t$ is considered to reside in a separate cluster. Thus, initially there are $\bar{N}'$ clusters. Thereafter, the *cluster-distances* between all pairs of clusters are calculated. In general, the distance d_{12} between two clusters C_1 and C_2 is defined as the average Euclidean distance of all pairs of solutions ($i \in C_1$ and $j \in C_2$), or mathematically:

$$d_{12} = \frac{1}{|C_1||C_2|} \sum_{i \in C_1, j \in C_2} d(i, j). \qquad (6.6)$$

The distance $d(i, j)$ can be computed in the decision variable space or in the objective space. The proposers of the SPEA preferred to use the latter. Once all $\binom{\bar{N}'}{2}$ cluster-distances are computed, the two clusters with the minimum cluster-distance are combined together to form one bigger cluster. Thereafter, the cluster-distances are recalculated for all pairs of clusters and the two closest clusters are merged together. This process of merging crowded clusters is continued until the number of clusters in the external population is reduced to $\bar{N}$. Thereafter, in each cluster, the solution with the minimum average distance from other solutions in the cluster is retained and all other solutions are deleted. The following is the algorithm in a step-by-step format.

Clustering Algorithm

> **Step C1** Initially, each solution belongs to a distinct cluster or $C_i = \{i\}$, so that $C = \{C_1, C_2, \ldots, C_{\overline{N}'}\}$.
>
> **Step C2** If $|C| \leq \overline{N}$, go to Step C5. Otherwise, go to Step C3.
>
> **Step C3** For each pair of clusters (there are $\binom{|C|}{2}$ of them), calculate the cluster-distance by using equation (6.6). Find the pair (i_1, i_2) which corresponds to the minimum cluster-distance.
>
> **Step C4** Merge the two clusters C_{i_1} and C_{i_2} together. This reduces the size of C by one. Go to Step C2.
>
> **Step C5** Choose only one solution from each cluster and remove the others from the clusters. The solution having the minimum average distance from other solutions in the cluster can be chosen as the representative solution of a cluster.

Since in Step C5, all but one representative solution is kept, the resulting set has at most $\overline{N}$ solutions.

Complexity of Clustering Algorithm

The way the above algorithm is presented makes the understanding of the clustering algorithm easier, but an identical implementation demands a large computational burden: $O(M\overline{N}'^3)$. However, the clustering algorithm can be performed with proper bookkeeping and the computations can then be reduced. Since solutions are not removed until Step C5 is performed, all distance computations can in practice be performed once in the beginning. This requires $O(M\overline{N}'^2)$ computations. Thereafter, the calculation of the average distances of clusters, the minimum distance calculation of all pairs of clusters (with a complicated bookkeeping procedure), and the merging of clusters can all be done with special bookkeeping[1] in a linear time at each call of Steps C2 to C4. Since they are called $(\overline{N}' - \overline{N})$ times, the overall complexity of the cluster-updates is $O(\overline{N}'^2)$. The final removal of extra solutions with intra-cluster distance computation is linear to $\overline{N}'$. Thus, if implemented with care, the complexity of the above clustering algorithm can be reduced to $O(M\overline{N}'^2)$.

6.4.2   Hand Calculations

We consider the Min-Ex problem again. Table 23 shows the EA and the external population members used to illustrate the working of the SPEA. We have $P_t =$

---

[1] The investigators of the SPEA did not suggest any such bookkeeping methods. However, by using information about which clusters were the neighbors of each other cluster and updating this information as clusters merge together, a linear complexity algorithm with a large storage requirement is possible to achieve.

Table 23 Current EA and external populations with their objective function values.

	EA population P_t				External population $\overline{P}_t$				
Solution	x_1	x_2	f_1	f_2	Solution	x_1	x_2	f_1	f_2
1	0.31	0.89	0.31	6.10	a	0.27	0.87	0.27	6.93
2	0.43	1.92	0.43	6.79	b	0.79	2.14	0.79	3.97
3	0.22	0.56	0.22	7.09	c	0.58	1.62	0.58	4.52
4	0.59	3.63	0.59	7.85					
5	0.66	1.41	0.66	3.65					
6	0.83	2.51	0.83	4.23					

$\{1,2,3,4,5,6\}$ and $\overline{P}_t = \{a,b,c\}$. Figure 151 shows all of these solutions. Note that here $N = 6$ and $\overline{N} = 3$.

Step 1 First, we find the non-dominated solutions of P_t. We observe from Figure 151 that they are $\mathcal{F}_1(P_t) = \{1,3,5\}$. We now include these solutions in $\overline{P}_t$. Thus, $\overline{P}_t = \{a,b,c,1,3,5\}$.

Step 2 We now calculate the non-dominated solutions of this modified population $\overline{P}_t$ and observe $\mathcal{F}_1(\overline{P}_t) = \{a,c,1,3,5\}$. We set this population as the new external population.

Step 3 Since the size of $\overline{P}_t$ is 5, which is greater than the external population size $(\overline{N} = 3)$, we need to use the clustering algorithm to find which three will remain in the external population.

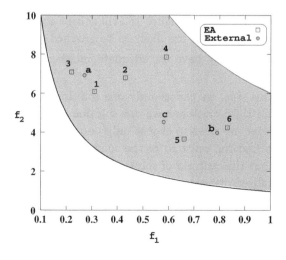

Figure 151 EA and external populations.

Step C1 Initially, all five solutions belong to a separate cluster:

$$C_1 = \{a\}, \quad C_2 = \{c\}, \quad C_3 = \{1\}, \quad C_4 = \{3\}, \quad C_5 = \{5\}.$$

Step C2 Since there are 5 clusters, we move to Step C3.

Step C3 We use $f_1^{max} = 1$, $f_1^{min} = 0.1$, $f_2^{max} = 60$ and $f_2^{min} = 1$. All $\binom{5}{2}$ or 10 cluster-distances are as follows:

$$d_{12} = 0.35, \quad d_{13} = 0.05, \quad d_{14} = 0.06, \quad d_{15} = 0.44, \quad d_{23} = 0.30,$$
$$d_{24} = 0.40, \quad d_{25} = 0.09, \quad d_{34} = 0.09, \quad d_{35} = 0.39, \quad d_{45} = 0.49.$$

We observe that the minimum cluster-distance occurs between the first and the third clusters.

Step C4 We merge these clusters together and have only four clusters:

$$C_1 = \{a, 1\}, \quad C_2 = \{c\}, \quad C_3 = \{3\}, \quad C_4 = \{5\}.$$

Step C2 Since there are four clusters, we move to Step C3 to reduce one more cluster.

Step C3 Now, the average distance between the first and second cluster is the average distance between the two pairs of solutions (a,c) and (1,c). The distance between solutions a and c is 0.35 and that between solutions 1 and c is 0.30. Thus, the average distance d_{12} is 0.325. Similarly, we can find other distances of all $\binom{4}{2}$ or 6 pairs of clusters:

$$d_{12} = 0.325, \quad d_{13} = 0.075, \quad d_{14} = 0.415,$$
$$d_{23} = 0.400, \quad d_{24} = 0.090, \quad d_{34} = 0.490.$$

The minimum distance occurs between clusters 1 and 3.

Step C4 Thus, we merge clusters 1 and 3 and have the following three clusters:

$$C_1 = \{a, 1, 3\}, \quad C_2 = \{c\}, \quad C_3 = \{5\}.$$

It is also intuitive from Figure 152 that three solutions a, 1, and 3 reside close to each other and are likely to be grouped into one cluster.

Step C2 Since we now have an adequate number of clusters, we move to Step C5.

Step C5 In this step, we choose only one solution in every cluster. Since the second and third clusters have one solution each, we accept them as they are. However, we need to choose only one solution for cluster 1. The first step is to find the centroid of the solutions belonging to the cluster. We observe that the centroid of solutions a, 1 and 3 is $c_1 = (0.27, 6.71)$. Now the normalized distance of each solution from this centroid is as follows:

$$d(a, c_1) = 0.005, \quad d(1, c_1) = 0.049, \quad d(3, c_1) = 0.052.$$

We observe that solution a is closest to the centroid c_1 (this fact is also clear from Figure 152). Thus, we choose solution a and delete solutions 1 and 3 from this cluster. Therefore, the new external population is $\overline{P}_{t+1} = \{a, c, 5\}$.

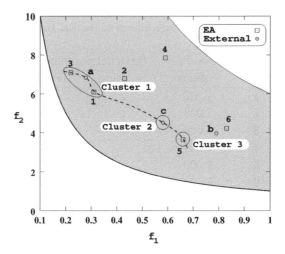

Figure 152   Illustration of the three clusters.

Step 4 Now, we assign fitness values to the solutions of populations P_t and $\overline{P}_{t+1}$. Note that solution 5 is a member of both P_t and $\overline{P}_{t+1}$ and is treated as two different solutions. First, we concentrate on the external population. We observe that solution a dominates only one solution (solution 4) in P_t, Thus, its fitness (or strength) is assigned as:

$$F_a = \frac{n_a}{N+1} = \frac{1}{6+1} = 0.143.$$

Similarly, we find $n_c = n_5 = 1$, and their fitness values are also $F_c = F_5 = 0.143$.

Next, we calculate the fitness values of the solutions of P_t. Solution 1 is dominated by no solution in the external population. Thus, its fitness is $F_1 = 1.0$. Similarly, solutions 2 and 3 are not dominated by any external population members and hence their fitness values are also 1.0. However, solution 4 is dominated by two external members (solutions a and c) and hence its fitness is $f_4 = 1 + 0.143 + 0.143 = 1.286$. We also find that $F_5 = 1.0$ and $F_6 = 1.143$. These fitness values are listed in Table 24.

Step 5 Now, using the above fitness values we would perform six tournaments by randomly picking solutions from the combined population of size nine (in effect, there are only eight distinct solutions) and form the mating pool. Thereafter, crossover and mutation operators will be applied to the mating pool to create the new population P_{t+1} of size six.

This completes one generation of the SPEA. These hand calculations illustrate how elitism is introduced in such an algorithm.

Table 24 The fitness assignment procedures of the EA and external populations.

EA population P_t		External population $\overline{P}_{t+1}$	
Solution	Fitness	Solution	Fitness
1	1.000	a	0.143
2	1.000	c	0.143
3	1.000	5	0.143
4	1.286		
5	1.000		
6	1.143		

6.4.3 Computational Complexity

Step 1 requires the computation of the best non-domination front of P_t. This needs at most $O(MN^2)$ computations. If N and $\overline{N}$ are of the same order, Step 2 also has a similar computational complexity. Step 3 invokes the clustering procedure, requiring at most $O(MN^2)$ computations. All other steps in the algorithm do not have computational complexities more than the above computation. Thus, the overall complexity needed in each generation of the SPEA is $O(MN^2)$.

6.4.4 Advantages

It is easy to realize that once a solution in the Pareto-optimal front is found, it immediately gets stored in the external population. The only way it gets eliminated is when another Pareto-optimal solution, which leads to a better spread in the Pareto-optimal solutions, is discovered. Thus, in the absence of the clustering algorithm, the SPEA has a similar convergence proof to that of Rudolph's algorithm. Clustering ensures that a better spread is achieved among the obtained non-dominated solutions.

This clustering algorithm is parameter-less, thereby making it attractive to use. The fitness assignment procedure in the SPEA is more or less similar to that of Fonseca and Fleming's MOGA (Fonseca and Fleming, 1993) and is easy to calculate. To make the clustering approach find the most diverse set of solutions, each extreme solution can be forced to remain in an independent cluster.

6.4.5 Disadvantages

The SPEA introduces an extra parameter – $\overline{N}$, the size of the external population. A balance between the regular population size N and this external population size $\overline{N}$ is important in the successful working of the SPEA. If a large (comparable to N) external population is used, the selection pressure for the elites will be large and the SPEA may not be able to converge to the Pareto-optimal front. On the other hand, if a small external population is used, the effect of elitism will be lost. Moreover, many solutions in the population will not be dominated by any external population members and their derived fitnesses will be identical. The investigators of the SPEA used a 1:4 ratio between the external population and the EA population sizes.

The NSGA-II uses a crowding strategy which is $O(MN \log N)$; however, the clustering algorithm used in an SPEA has an $O(MN^2)$ complexity. Thus, an SPEA's niche-preservation operator can be made faster by using the crowding strategy used in the NSGA-II.

The SPEA shares a common problem with the MOGA. Since non-dominated sorting of the whole population is not used for assigning fitness, the fitness values do not favor all non-dominated solutions of the same rank equally. This bias in fitness assignment in the solutions of the same front is dependent on the exact population and densities of solutions in the search space.

Moreover, in the SPEA fitness assignment, an external solution which dominates more solutions gets a worse fitness. This assignment is justified when all dominated solutions are concentrated near the dominating solution. Since in most cases this is not true, the crowding effect should come only from the clustering procedure. Otherwise, this fitness assignment may provide a wrong selection pressure for the non-dominated solutions.

6.4.6   Simulation Results

To illustrate the working of the SPEA, we apply this algorithm to the Min-Ex problem. In addition to the standard parameter settings used in previous sections, we will also use the following settings: a population size of 32 and an elite population size of eight. Figure 153 shows the non-dominated solutions of the combined population ($P_t \cup \overline{P}_{t+1}$)

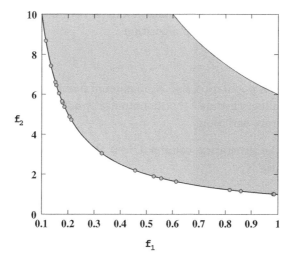

Figure 153   The non-dominated solutions of the combined EA and external population in a SPEA run after 500 generations.

after 500 generations. This figure shows that the SPEA with a mutation operator is able to find a widely distributed set of solutions on the Pareto-optimal front.

6.5 Thermodynamical Genetic Algorithm

Kita et al. (1996) suggested a fitness function which allows a convergence near the Pareto-optimal front and a diversity-preservation among obtained solutions. The fitness function is motivated from the thermodynamic equilibrium condition, which corresponds to the minimum of the Gibb's free energy F, defined as follows:

$$F = \langle E \rangle - HT, \tag{6.7}$$

where $\langle E \rangle$ is the mean energy, H is the entropy, and T is the temperature of the system. The relevance of the above terms in multi-objective optimization is as follows. The mean energy is considered as the fitness measure. In the study, the non-domination rank of a solution in the population is used as the fitness function. In this way, the best non-dominated solutions have a rank (or fitness) equal to 1. Obviously, minimization of the fitness function is assumed here. The second term on the right side of the above equation controls the diversity existent in the population members. In this way, minimization of F means minimization of $\langle E \rangle$ and maximization of HT. In the parlance of multi-objective optimization, this means minimization of the objective function (convergence towards the Pareto-optimal set) and the maximization of diversity in obtained solutions. Since these two goals are precisely the focus of an MOEA, the analogy between the principle of finding the minimum free energy state in a thermodynamic system and the working principle of an MOEA is appropriate.

Although the analogy is ideal, the exact definitions of mean energy and entropy in the context of multi-objective optimization must be made. The investigators have proposed the following step-by-step algorithm.

Thermodynamic GA (TDGA)

> **Step 1** Select a population size N, maximum number of generations t_{max}, and an annealing schedule $T(t)$ (monotonically non-increasing function) for the temperature variation.
>
> **Step 2** Set the generation counter $t = 0$ and initialize a random population $P(t)$ of size N.
>
> **Step 3** By using a crossover and a mutation operator, create the offspring population $Q(t)$ of size N from $P(t)$. Combine the parent and offspring populations together and call this $R(t) = P(t) \cup Q(t)$.
>
> **Step 4** Create an empty set $P(t+1)$ and set the individual counter i of $P(t+1)$ to one.
>
> **Step 5** For every solution $j \in R(t)$, construct $P'(t,j) = P(t+1) \cup \{j\}$ and calculate

the free energy (fitness) of the j-th member as follows:

$$F(j) = \langle E(j) \rangle - T(t) \sum_{k=1}^{\ell} H_k(j), \tag{6.8}$$

where $\langle E(j) \rangle = \sum_{k=1}^{|P'(t,j)|} E_k / |P'(t,j)|$ and the entropy is defined as

$$H_k(j) = - \sum_{\substack{l \in \{0,1\} \\ P'(t,j)}} p_l^k \log p_l^k. \tag{6.9}$$

The term p_l^k is the proportion of bit l on the locus k of the population $P'(t,j)$.

After all 2N population members are assigned a fitness, find the solution $j^* \in R(t)$ which will minimize F given above. Include j^* in $P(t+1)$.

Step 6 If $i < N$, increment i by one and go to Step 5.

Step 7 If $t < t_{max}$, increment t by one and go to Step 3.

The term E_k can be used as the non-dominated rank of the k-th solution in the population $P'(t,j)$. Although in every iteration of Step 5, the same $R(t)$ population members are used in order to choose the solution with the minimum fitness, it is important to realize that the fitness computation is dependent on the temperature term $T(t)$, which varies from generation to generation. The fitness assignment also depends on the $H_k(j)$ term, which depends on the current population $P'(t,j)$. Thus, one member which was minimum in one iteration need not remain minimum in another iteration. Since non-dominated solutions are emphasized by the $\langle E(j) \rangle$ term and the genotypic diversity is maintained through the $H_k(j)$ term, both tasks of a multi-objective optimization are served by this algorithm.

It is interesting to note that since the optimization in Step 5 is performed for all 2N parent and offspring for every member selection for $P(t+1)$, multiple copies of an individual may exist in the population $P(t+1)$.

6.5.1 Computational Complexity

Step 5 requires a non-dominated sorting of 2N population members, thereby requiring $O(MN^2)$ computations. If an exhaustive optimization method is used, each $k \in P'(t,j)$ requires $O(|P'(t,j)|)$ or $O(i)$ complexity in evaluating the corresponding free energy. Since k varies from 1 to 2N, this requires $O(iN)$ computations. In order to find a complete population $P(t+1)$ of size N, i varies from 1 to N, thereby making a total of $O(N^3)$ computational complexity. Thus, the overall complexity of one generation of the TDGA is $O(N^3)$, which is larger than that in the NSGA-II, PAES (to be discussed in the next section) or SPEA.

6.5.2   Advantages and Disadvantages

Since both parent and offspring populations are combined and the best set of N solutions are selected from the diversity-preservation point of view, the algorithm is an elitist algorithm.

The TDGA requires a predefined annealing schedule $T(t)$. It is interesting to note that this temperature parameter acts like a normalization parameter for the entropy term needed to balance between the mean energy minimization and entropy maximization. Since this requires a crucial balancing act for finding a good convergence and spread of solutions, an arbitrary $T(t)$ distribution may not work well in most problems.

Investigators have demonstrated the working of the TDGA on a simple two-dimensional, two-objective optimization problem. Because of the need for choosing a temperature variation $T(t)$ and a local minimization technique for every solution, we do not perform a hand calculation, nor do we present any simulation study here.

6.6   Pareto-Archived Evolution Strategy

Knowles and Corne (2000) suggested a multi-objective evolutionary algorithm which uses an evolution strategy. In its simplest form, the Pareto-archived ES (PAES) uses a (1+1)-ES. The main motivation for using an ES came from their experience in solving real-world telecommunications network design problem. In the single-objective version of the network design problem, they observed that a local search strategy (such as a hill-climber, tabu search method, or simulated annealing) worked better than a population-based approach. Motivated by this fact, they investigated whether a multi-objective evolutionary algorithm with a local search strategy can be developed to solve the multi-objective version of the telecommunications network design problems. Since a (1+1)-ES uses only mutation on a single parent to create a single offspring, this is a local search strategy, and thus the investigators developed their first multi-objective evolutionary algorithm using the (1+1)-ES.

At first, a random solution x_0 (call this a parent) is chosen. It is then mutated by using a normally distributed probability function with zero-mean and with a fixed mutation strength. Let us say that the mutated offspring is c_0. Now, both of these solutions are compared and the winner becomes the parent of the next generation. The main crux of the PAES lies in the way that a winner is chosen in the midst of multiple objectives. We will describe this feature next.

At any generation t, in addition to the parent (p_t) and the offspring (c_t), the PAES maintains an archive of the best solutions found so far. Initially, this archive is empty and as the generations proceed, good solutions are added to the archive and updated. However, a maximum size of the archive is always maintained. First, the parent p_t and the offspring c_t are compared for domination. This will result in three scenarios.

If p_t dominates c_t, the offspring c_t is not accepted and a new mutated solution is created for further processing. On the other hand, if c_t dominates p_t, the offspring is better than the parent. Then, solution c_t is accepted as a parent of the next generation and a copy of it is kept in the archive. This is how the archive gets populated by non-dominated solutions.

The intricacies arise when both p_t and c_t are non-dominated to each other. In such a case, the offspring is compared with the current archive, which contains the set of non-dominated solutions found so far. Three cases are possible here. Figure 154 marks these cases as 1, 2, and 3. In the first case, the offspring is dominated by a member of the archive. This means that the offspring is not worth including in the archive, The offspring is then rejected and the parent p_t is mutated again to find a new offspring for further processing. In the second case, the offspring dominates a member of the archive. This means that the offspring is better than some member(s) of the archive. The dominated members of the archive are deleted and the offspring is accepted without any condition. The offspring then becomes the parent of the next generation. Note that this process does not increase the size of the archive, mainly because the offspring enters the archive by eliminating at least one existing dominated archive member. In the third case, the offspring is not dominated by any member of the archive and it also does not dominate any member of the archive. In this case, it belongs to that non-dominated front in which the archive solutions belong. In such a case, the offspring is added to the archive only if there is a slot available in the latter. Recall that a maximum archive size is always maintained in the PAES. The acceptance of the offspring in the achieve does not qualify it to become the parent of the next generation. This is because as for the offspring c_t, the current parent p_t is also a member of the archive. To decide who qualifies as a parent of the next generation, the density of solutions in their neighborhood is checked. The one residing in the least crowded area in the search space qualifies as the parent. If the archive is full, the above density-based comparison is performed between the parent and the offspring to decide who remains in the archive.

The density calculation is different from the other methods we have discussed so

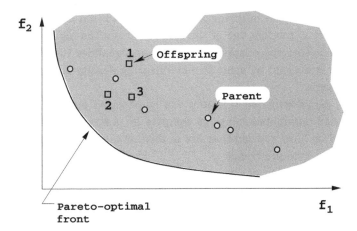

Figure 154   Three cases of an offspring non-dominated with the parent solution. The archive is shown by circles, while offspring are shown by boxes.

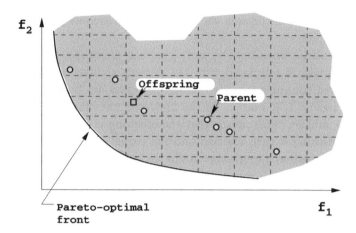

Figure 155 The parent resides in a more dense hypercube than the offspring.

far. A more direct approach is used here. Each objective is divided into 2^d equal divisions,[2] where d is a user-defined depth parameter. In this way, the entire search space is divided into $(2^d)^M$ unique, equal-sized M-dimensional hypercubes. The archived solutions are placed in these hypercubes according to their locations in the objective space. Thereafter, the number of solutions in each hypercube is counted. If the offspring resides in a less crowded hypercube than the parent, the offspring becomes the parent of the next generation. Otherwise, the parent solution continues to be the parent of the next generation. This situation is illustrated by the two-objective minimization problem shown in Figure 155. For the situation in this figure, the offspring becomes the new parent. This is because the offspring is located in a grid with two solutions and the parent is located in a grid with three solutions.

If the archive is already full with non-dominated solutions, obviously the offspring cannot be included automatically. First, the hypercube with the highest number of solutions is identified. If the offspring does not belong to that hypercube, it is included in the archive and at random one of the solutions from the highest-count hypercube is eliminated. Whether the offspring or the parent qualifies to be the parent of the next generation is decided by same parent–offspring density count mentioned in the previous paragraph.

In this way, non-dominated solutions are emphasized and placed in an archive. As and when non-dominated solutions compete for a space in the archive, they are evaluated based on how crowded they are in the search space. The one residing in the least crowded area gets preference.

We outline one iteration of the (1+1)-PAES algorithm in a step-by-step manner with parent p_t, offspring c_t (created by mutating p_t) and an archive A_t. The archive is initialized with the initial random parent solution. The following procedure updates

---

[2] To allow recursive divisions of the search space, 2^d divisions are used. The locating and placing of a solution in a grid becomes computationally effective this way.

the archive A_t and determines the new parent solution p_{t+1}.

Archive and Parent Update in PAES

Step 1 If c_t is dominated by any member of A_t, set $p_{t+1} = p_t$ (A_t is not updated). Process is complete. Otherwise, if c_t dominates a set of members from A_t: $D(c_t) = \{i : i \in A_t \wedge c_t \preceq i\}$, perform the following steps:

$$
\begin{aligned}
A_t &= A_t \backslash D(c_t), \\
A_t &= A_t \cup \{c_t\}, \\
p_{t+1} &= c_t.
\end{aligned}
$$

Process is complete. Otherwise, go to Step 2.

Step 2 Count the number of archived solutions in each hypercube. The parent p_t belongs to a hypercube having n_p solutions, while the offspring belongs to a hypercube having n_c solutions. The highest-count hypercube contains the maximum number of archived solutions.

If $|A_t| < N$, include the offspring in the archive, or $A_t = A_t \cup \{c_t\}$ and $p_{t+1} = \text{Winner}(c_t, p_t)$, and return. Otherwise (that is, if $|A_t| = N$), check if c_t belongs to the highest-count hypercube. If yes, reject c_t, set $p_{t+1} = p_t$, and return. Otherwise, replace a random solution r from the highest-count hypercube with c_t:

$$
\begin{aligned}
A_t &= A_t \backslash \{r\}, \\
A_t &= A_t \cup \{c_t\}, \\
p_{t+1} &= \text{Winner}(c_t, p_t).
\end{aligned}
$$

The process is complete.

The $\text{Winner}(c_t, p_t)$ chooses c_t, if $n_c < n_p$. Otherwise, it chooses p_t. It is important to note that in any parent–offspring scenario, only one of the above two steps will be invoked.

6.6.1 *Hand Calculations*

We now illustrate the working of the (1+1)-PAES on the example problem Min-Ex. We consider an archive of size five ($N = 5$), with these five archive members, i.e. a, b, c, d, and e being marked by circles in Figure 156. To illustrate different scenarios, we present three independent parent-offspring pairs shown within dashed ellipses. Let us consider them one by one. First, we consider $p_t = e$ and $c_t = 3$.

Step 1 Solution e dominates the offspring (solution 3). Thus, we do not accept this offspring. Parent e remains as the parent of the next generation and the archive also remains unchanged.

Note that only Step 1 is executed here. Next, we consider the '$p_t = b$ and $c_t = 2$' scenario.

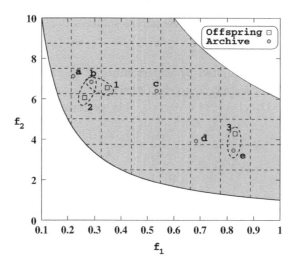

Figure 156 The archive and three offspring considered in the hand calculation.

Step 1 Here, solution 2 dominates solution b and c. Thus, both of these solutions will be replaced by the offspring in the archive and the offspring (solution 2) will become the new parent. The updated archive is $A_t = \{2, a, d, e\}$.

Here also, only Step 1 is executed. The new parent is solution 2. Finally, we consider the case with solution b being the parent ($p_t = b$) and solution 1 being the mutated offspring ($c_t = 1$). We observe that these two solutions do not dominate each other.

Step 1 No member of the archive dominates solution 1, and the latter does not dominate any solution in the archive either.

Step 2 First, we divide the space in each objective into eight divisions (objective 1 within [0.1,1] and objective 2 in [0,10]). From the grids shown in the figure, we find that $n_p = 2$ and $n_c = 1$. That is, there is one additional archive member in the hypercube containing the parent p_t and no other archive members exist in the hypercube containing the offspring c_t. The highest-count archive here is the one containing the parent. Since the archive is full, we check if solution 1 belongs to the highest-count hypercube. Since it does not, we include solution 1 in the archive and delete one of the solutions a or b at random. If we delete solution b, the new five-member archive is $\{a, 1, c, d, e\}$. The new parent is solution 1.

Note that only Step 2 is executed in this case.

6.6.2 Computational Complexity

In one generation of the (1+1)-PAES, Step 1 requires $O(MN)$ comparisons to find if the offspring dominates any solution in the archive (of size N) or if any archive solution dominates the offspring.

When Step 2 is called, each archive member's hypercube location is first identified. This requires $O(Md)$ comparisons for each solution, thereby requiring a total of $O(NMd)$ comparisons for the entire archive of N solutions. Thus, in the worst case (when Step 2 is executed), the complexity of one generation of the PAES is $O(MNd)$.

In order to compare the complexity of the PAES with other generational EAs (such as the NSGA-II or SPEA), we can argue that N generations of the PAES are equivalent to one generation of the NSGA-II or SPEA in terms of the number of total new function evaluations. Thus, the complexity of the (1+1)-PAES is $O(MN^2d)$.

6.6.3   Advantages

The PAES has a direct control on the diversity that can be achieved in the Pareto-optimal solutions. Step 2 of the algorithm emphasizes the less populated the hypercubes to survive, thereby ensuring diversity. The size of these hypercubes can be controlled by choosing an appropriate value of d. If a small d is chosen, hypercubes are large and adequate diversity may not be achieved. However, if a large d is chosen there exist many hypercubes and the computational burden will increase.

Since equal-sized hypercubes are chosen, the PAES should perform better when compared to other methods in handling problems having a search space with non-uniformly dense solutions (Deb, 1999c). In such problems, there can be a non-uniform distribution of solutions in the search space. This naturally favors some regions in the Pareto-optimal front, thus making it difficult for an algorithm to find a uniform distribution of solutions. By keeping the hypercube size fixed, Step 2 of the algorithm will attempt to emphasize solutions in as many hypercubes as possible.

6.6.4   Disadvantages

In addition to choosing an appropriate archive size (N), the depth parameter d, which directly controls the hypercube size, is also an important parameter. We mentioned earlier that the number of hypercubes in the search space varies with d as 2^{dM} or $(2^M)^d$. For a three-objective function, this means 8^d. A change of d changes the number of hypercubes exponentially, thereby making it difficult to arbitrarily control the spread of solutions. Moreover, the side-length of a hypercube in each dimension requires the knowledge of minimum and maximum possible objective function values. These values are often difficult to choose in a problem.

There is another difficulty with the PAES. Since the sizing of the hypercubes is performed with the minimum and maximum bounds of the entire search space, when solutions converge near the Pareto-optimal front, the hypercubes are comparatively large. One alternative method would be to use the population minimum and maximum objective function values to find the hypercube sizes on the fly. However, this has a serious drawback. If for any reason, the population members concentrate in a narrow region, either because of premature convergence or because of the nature of the problem, the resulting hypercube sizes can become smaller than desired.

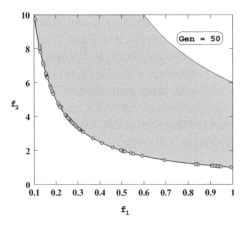

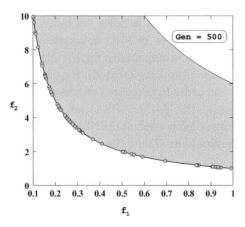

Figure 157   The PAES Archive after 50 generations.

Figure 158   The PAES Archive after 500 generations.

6.6.5   Simulation Results

Next, we apply the PAES to the Min-Ex problem with parameter settings identical to that used for the NSGA-II. Here, we use a depth factor of 4. Figure 157 shows the PAES-archive at the end of 50 generations. It is clear that the PAES has no difficulty in converging to the Pareto-optimal front. However, the distribution of obtained solutions is not as good as that found for the NSGA-II. When the PAES is run up until 500 generations (Figure 158), the distribution becomes somewhat better. The performance of the PAES depends on the depth parameter, which controls the distribution in the obtained solutions.

6.6.6   Multi-Membered PAES

In order to give the PAES a global perspective, the concept of multi-membered ES is introduced (Knowles and Corne, 2000) into an MOEA. At first, we discuss the $(1+\lambda)$-PAES, where one parent solution is mutated λ times, each time creating an offspring. With multiple offspring around, it is then necessary to decide which one of the offspring competes with the parent to fight for the parenthood of the next generation. To facilitate this decision-making, a fitness is assigned to each offspring based on a comparison with the archive and its hypercube location in the search space. The best-fit solution among the parent and offspring becomes the parent of the new population.

In the (μ, λ)-PAES, each of the μ parents and λ offspring are compared with the current archive. Every offspring is checked for its eligibility to become a member of the archive, as before. If it is, then the change in the spread of archive members in the phenotypic space is noted. If the spread has changed by more than a *threshold* amount, the offspring is accepted and the archive is updated. In order to create the new parent population of μ solutions, a fitness is assigned to all population members (μ parents and λ offspring) based on a *dominance score*, which is calculated as follows. If a parent or an offspring dominates any member of the archive, it is assigned a score of 1. If it gets dominated by any member of the archive, its score is -1. If it is non-dominated, it gets a score of 0. Although it is not clear exactly how a fitness is assigned from the scores, a solution with a higher dominance score is always assigned a better fitness. However, for solutions with the same dominance score, a higher fitness is assigned to the solution having a lower population count in its hypercube location. Using the fitness values, a binary tournament selection is applied to the parent–offspring population to choose μ new parent solutions. Each of these μ solutions is then mutated to create a total of λ offspring for their processing in the next generation. In this way, both parent and offspring directly affect the archive and also compete against each other for survival in the next generation.

Based on simulation results on a number of test problems, the proposers of the PAES observed that multi-membered PAESs do not generally perform as well as the (1+1)-PAES. Since offspring are not compared against each other and only compared with the archive, this method does not guarantee that the best non-dominated solutions among the offspring are emphasized enough. For simulations limiting the number of total trails, the (1+1)-ES does well on the test problems chosen in the study. Investigators have also suggested an improved version of the PAES (Corne et al., 2000).

6.7 Multi-Objective Messy Genetic Algorithm

Veldhuizen (1999), in his doctoral dissertation, proposed an entirely different approach to multi-objective optimization. He modified the single-objective messy GAs of Goldberg et al. (1989) to find multiple Pareto-optimal solutions in multi-objective optimization. Since a proper understanding of this approach requires an understanding of the single-objective messy GA approach, we first provide an outline here of the messy GA. Interested readers may refer to the original and other studies on messy GAs (Goldberg et al., 1989, 1990, 1993a; Veldhuizen and Lamont, 2000) for details.

6.7.1 Original Single-Objective Messy GAs

The main motivation in the study of messy GAs (mGAs) was to sequentially solve the two main issues of identifying salient building blocks in a problem and then combining the building blocks together to form the optimal or a near-optimal solution. Messy GAs were successful in solving problems which can be decomposed into a number of overlapping or non-overlapping building blocks (partial solutions corresponding to the true optimal solution). Since the genic combination is an important matter to be discovered in a problem, mGAs use both genic and allelic information in a string. The following is a four-variable ($\ell = 4$) messy chromosome:

$$((2\ 0)\ (1\ 1)\ (2\ 1)\ (3\ 1))$$

There are four members in the chromosome. The first value in each member denotes the gene (or position) number, while the second value is the allele value of that position. Although it is a four-variable chromosome, the fourth gene is missing and the second gene appears twice. A messy chromosome is allowed to have duplicate genes and unspecified genes. By using a left-to-right scanning and a first-come-first-served selection, the above chromosome codes to the binary string: (1 0 1 _). The unspecified problem is solved by filling the unspecified bits from a fixed template string of length ℓ.

The above two tasks of identification and combination of building blocks are achieved in two phases, (i) the primordial phase and (ii) the juxtapositional phase. In the primordial phase, the main focus is to identify and maintain salient building blocks of a certain maximum order k. Since this is the most important task in a problem, the developers of mGAs did not at first take any risk of missing out in any salient building blocks. Thus, they made sure that the initial population size in the primordial phase is large enough to accommodate all order-k genic and allelic combinations. A simple calculation suggests that the required population size is $N_p = \binom{\ell}{k}2^k$. For small k compared to ℓ, this size is polynomial to the order k, or $O(\ell^k)$. Since the length of these initial chromosomes is smaller than the problem length ℓ, the remaining bits are filled from a fixed template string. The investigators argued extensively that in order to have the minimum noise in evaluating partial strings, it is essential that a *locally optimal* solution is used as a template string. We will discuss the issue of finding such a locally optimal string a little later. After the partial string is embedded in a template and is evaluated, a binary tournament selection with a niching approach is used to emphasize salient partial strings in the primordial phase. Since the salient building blocks constitute only a small fraction of all primordial strings, a systematic reduction in population sizing is also used. After a fair number of repetitive applications of tournament selection followed by a population reduction, the primordial phase is terminated.

In the juxtapositional phase, the salient building blocks are allowed to combine together with the help of a cut-and-splice and a mutation operator. The cut-and-splice operator is similar in principle to the single-point crossover, except that the cross sites in both parents need not fall at the same place. This causes variable-length chromosomes to exist in the population. The purpose of tournament selection with thresholding, cut-and-splice, and the mutation operator is to find better and bigger building blocks, eventually leading to the true optimal solution.

The issue of using a locally optimal template string is a crucial one. Since the knowledge of such a string is not known beforehand, proposers have suggested a level-wise mGA, where the mGAs are applied in different eras, starting from the simplest possible era. In the level-1 era of an mGA, all order one (k = 1) substrings are initialized and evaluated with a template string created at random. At the end of this era, the obtained best solution is saved and used as a template for the level-2 era. In this era, all order two (k = 2) substrings are initialized. This process is continued until a specified number of era has elapsed or a specific number of function evaluations has been exceeded.

The above mGA procedure has successfully solved a number of complex optimization problems, which were otherwise difficult to solve by using a simple GA.

6.7.2   Modification for Multi-Objective Optimization

In multi-objective optimization, we are dealing with a number of objectives. Thus, the evaluation and comparison of chromosomes become difficult issues. Each chromosome now has M different objective function values. Although a non-domination comparison can be performed, there is an additional problem. In the primordial phase and in the early generations of the juxtapositional phase, the chromosomes are small and a complete string is formed by filling missing bits from a template string. If one fixed template string is used for all solutions in an mGA era, this will produce a bias towards finding solutions in a particular region in the search space. Since the important task in a multi-objective optimization is to find multiple diverse set of solutions near the Pareto-optimal region, Veldhuizen (1999) suggested the use of M different template strings in an era. Similar to the single-objective mGA, a locally optimal template string for each objective function is found by using the level-wise MOMGA. In the level-1 MOMGA, each partial string is filled from M template strings chosen randomly before the era has begun. Each filled string is evaluated with a different objective function. The objective vector obtained in this process is used in the selection operator, as described in the next paragraph. At the end of an era, the best solution corresponding to each objective function is identified and is assigned as the template string corresponding to that objective function. Since different partial strings are filled with different template strings, each corresponding to a locally optimal solution of an objective, salient building blocks corresponding to all objective functions co-survive in the population.

The primordial phase begins with a population identical to the single-objective mGAs. However, a different tournament selection procedure is used. The niched tournament selection operator is exactly the same as that in the NPGA approach. In order to compare two solutions in a population, a set of t_{dom} solutions are chosen from the latter. If one solution is non-dominated with respect to the chosen set and the other is not, the former solution is selected. On the other hand, if neither of them or both of them are dominated in the chosen set, a niching method is used to find which of the two solutions resides in a least crowded area. For both solutions, the *niche count* is calculated by using the sharing function approach discussed above in Section 4.6 with a pre-defined σ_{share} value. The solution with a smaller niche count is selected. The preference of non-dominated and less-crowded solutions using the above tournament selection helps to provide a selection advantage towards the non-dominated solutions and simultaneously maintains a diverse set of solutions. In all simulations, the investigator used the phenotypic sharing approach. The population reduction procedure is the same as that used in the single-objective mGAs.

The juxtapositional phase is also similar to that in the original mGA study, except that the above-mentioned niched tournament selection procedure is used. At the end of the juxtapositional phase, the current population is combined with an external population which stores a specified number of non-dominated solutions found thus far. Before ending the era, the external population is replaced with non-dominated solutions of the combined population. This dynamically updated procedure introduces elitism, which is an important property in any evolutionary algorithm. At the end of all MOMGA eras, the external set is reported as the obtained non-dominated set of solutions. The investigator has demonstrated the efficacy of the MOMGA in a number of different test problems.

Veldhuizen (1999) also proposed a concurrent MOMGA (or CMOMGA) by suggesting parallel applications of the MOMGA with different initial random templates. In order to maintain diversity and to increase efficiency of the procedure, the investigator also suggested occasional interchanges of obtained building blocks among different MOMGAs. At the completion of all MOMGAs, the obtained external sets of non-dominated solutions are all combined together and the best non-dominated set is reported as the obtained non-dominated set of solutions of the CMOMGA. Although the solutions obtained by this procedure do not indicate the robustness associated with an independent run of an MOMGA, this parallel approach may be desirable in practical problem solving. However, it is important to highlight that the CMOMGA and the original MOMGA requires $O(M\ell^{k_{max}})$ computational effort, which may be much larger than the usual $O(MN^2)$ estimate in most MOEAs. A recent study using a probabilistically complete initialization of MOMGA population to reduce the computational burden is an improvement over the past studies (Zydallis et al., 2001).

6.8 Other Elitist Multi-Objective Evolutionary Algorithms

In this section, we will briefly describe a number of other elitist MOEA implementations. A few other implementations can also be found elsewhere (Coello, 1999; Zitzler et al., 2001).

6.8.1 Non-Dominated Sorting in Annealing GA

Leung et al. (1998) modified the NSGA by introducing a population transition acceptance criterion. This non-dominated sorting in annealing GA (NSAGA) also uses a simulated annealing-like temperature reduction concept along with the Metropolis criterion. After the offspring population Q_t is created, it is first compared with the parent population P_t by using a two-stage probability computation. The first-stage probability calculation is along the lines of finding the transition probability of creating the offspring population from the parent population. The motivation for this $O(N^3)$ complexity probability calculation is hard to understand, although it has facilitated authors to line up a simulated annealing-like proof of convergence to the Pareto-optimal solutions. The second probability calculation is based on the Metroplois criterion, which uses an energy function related to the number of non-dominated solutions in a population. In an elitist sense, an offspring population is accepted only when the probability of creating such a population and accepting it with the Metropolis criterion with an updated temperature concept is adequate. Clearly, the goal of this work is to modify the NSGA procedure with a simulated annealing-like acceptance criterion, so that a proof of convergence can be achieved. In the process, the algorithm is more computationally complex (with respect to population size) than all of the other algorithms we have discussed so far.

6.8.2   Pareto Converging GA

Kumar and Rockett (in press) suggested a criterion for terminating an MOEA simulation based on the growth in proportion of non-dominated solutions in two consecutive generations. In their Pareto converging GA (PCGA), every pair of offspring is tested for non-domination along with the complete parent population. The combined $(N + 2)$ population members are sorted according to non-domination levels. The worst two solutions are eliminated to restore the population size to N. Since a usual generation requires $N/2$ such non-dominated sorting, the PCGA has $O(MN^3)$ computational complexity for generating N new solutions. Moreover, the PCGA does not use any additional niche-formation operator. However, the interesting aspect of this study is the definition of a 'rank histogram', which is calculated as the ratio of the number of solutions in a given non-domination level in the current population to that in a combined population of the current and immediately preceding generation. This requires non-dominated sorting of the current population $P(t)$ and the combined population $P(t) \cup P(t-1)$. If these populations are sorted as follows:

$$
\begin{aligned}
P(t) &= \{P_1(t), P_2(t), \ldots\}, \\
P(t) \cup P(t-1) &= \{P_1'(t), P_2'(t), \ldots\},
\end{aligned}
$$

the rank histogram is calculated as a vector

$$
RH(t) = \left\{ \frac{|P_1(t)|}{|P_1'(t)|}, \frac{|P_2(t)|}{|P_2'(t)|}, \ldots \right\}. \tag{6.10}
$$

The vector has a size equal to the number of non-dominated sorted fronts found in $P(t)$. By observing the vector $RH(t)$ over the generations, the convergence properties of an algorithm can be obtained. For example, if $RH(t)$ has only one element and its value is 0.5, the algorithm has completely converged to a particular front. However, it is important to note that a value of 0.5 does not necessarily mean a complete convergence to the true Pareto-optimal front, nor does it signify anything about the spread of the obtained solutions. A decreasing value of the histogram tail signifies a progress towards the convergence (see Figure 159). However, the first element in $RH(t)$ is an important parameter to track. The reduction or increase of this value with generation indicates the amount of shuffling of the best non-dominated solutions in successive generations. Although the investigators did not suggest this possibility, the $RH(t)$ vector information at each generation t can be used to adaptively change the

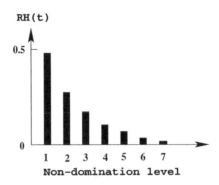

Figure 159   The rank histogram with a decreasing tail.

extent of exploration versus exploitation in an MOEA. A recent study on controlled elitism (described later in Section 8.10) is an effort in this direction, where the extent of elitism is controlled by using a predefined decreasing rank histogram (Deb and Goel, 2001a).

6.8.3   Multi-Objective Micro-GA

For handling problems requiring a large computational overhead in evaluating a solution, Krishnakumar (1989) proposed a micro-GA with a small population size for single-objective optimization. This investigator proposed using four to five population members, participating in selection, crossover and block replacement. Coello and Toscano (2000) suggested a multi-objective micro-GA (MµGA), which maintains two populations. The GA population (of size four) is operated in a similar way to that of the single-objective micro-GA, whereas the elite population stores the non-dominated solutions obtained by the GA. The elite archive is updated with new solutions in a similar way to that achieved in the PAES. The search space is divided into a number of grid cells. Depending on the crowding in each grid with non-dominated solutions, a new solution is accepted or rejected in the archive. In difficult test problems, the proof-of-principle results are encouraging.

6.8.4   Elitist MOEA with Coevolutionary Sharing

Neef et al. (1999) proposed an elitist recombinative multi-objective GA (ERMOCS) based on Goldberg and Wang's coevolutionary sharing concept (1998) (described earlier in Section 4.6.6). For maintaining diversity among non-dominated solutions, the coevolutionary shared niching (CSN) method is used. The elite-preservation is introduced by using a preselection scheme where a better offspring replaces a worse parent solution in the recombination procedure. In the coevolutionary model, the customer and businessman populations interact in the same way as in the CSN model, except an additional imprint operator is used for emphasizing non-dominated solutions. After both customer and businessman populations are updated, each businessman is compared with a random set of customers. If any customer dominates the competing businessman and the latter is at least a critical distance d_{min} away from other businessmen, it replaces the competing businessman. In this way non-dominated solutions from the customer population get filtered and find their place in the businessman population. On a scheduling problem, ERMOCS is able to find well-distributed customer as well as businessman populations after a few generations. The investigators has also observed that the performance of the algorithm is somewhat sensitive to the d_{min} parameter and that there exists a relationship between the parameter d_{min} and the businessman population size.

6.9 Summary

In the context of single-objective optimization, it has been shown that the presence of an elite-preserving operator makes an evolutionary algorithm better convergent. In this chapter we have presented a number of multi-objective evolutionary algorithms, where an elite-preserving operator is included to make the algorithms better convergent to the Pareto-optimal solutions. The wide range of algorithms presented in this chapter demonstrates the different ways that elitism can be introduced into an MOEA.

Rudolph's elitist GA compares non-dominated solutions of both parent and offspring populations and a hierarchy of rules emphasizing the elite solutions is used to form the population of the next generation. Although no simulation is performed, the drawback of this algorithm is the lack of an operator ensuring the second task of maintaining diversity among the non-dominated solutions.

The NSGA-II carries out a non-dominated sorting of a combined parent and offspring population. Thereafter, starting from the best non-dominated solutions, each front is accepted until all population slots are filled. This makes the algorithm an elitist type. For the solutions of the last allowed front, a crowded distance-based niching strategy is used to resolve which solutions are carried over to the new population. This study has also introduced a crowded distance metric, which makes the algorithm fast and scalable to more than two objectives.

The distance-based multi-objective GA (DPGA) maintains an external elite population which is used to assign the fitness of the GA population members based on their proximity to an elite set. Non-dominated solutions are assigned more fitness than dominated solutions. Although the resulting set of solutions depend on the sequence in which population members are assigned a fitness, this is one of the few algorithms which use only one metric for achieving both tasks of multi-objective optimization.

The strength Pareto EA (SPEA) also maintains a separate elite population which contains the fixed number of non-dominated solutions found till the current generation. The elite population participates in the genetic operations and influences the fitness assignment procedure. A clustering technique is used to control the size of the elite set, thereby indirectly maintaining diversity among the elite population members.

The thermodynamical GA (TDGA) minimizes the Gibb's free energy term constructed by using a mean energy term representing a fitness function and an entropy term representing the diversity term needed in a multi-objective optimization problem. The fitness function is used as the non-domination rank of the solution obtained from the objective function values. By carefully choosing solutions from a combined parent and offspring population, the algorithm attempts to achieve both tasks of an ideal multi-objective optimization.

The Pareto-archived evolution strategy (PAES) uses evolution strategy (ES) as the baseline algorithm. Using a (1+1)-ES, the offspring is compared with the parent solution for a place in an elite population. Diversity in the elite population is achieved by deterministically dividing the search space into a number of grids and restricting the maximum number of occupants in each grid. The 'plus' strategy and the continuous update of an external archive (or elite population) with better solutions ensures elitism. The investigators have later implemented the above concept in a multi-membered ES paradigm.

The multi-objective messy GA (MOMGA) extends the original messy GA to solve multi-objective optimization problems. Although the basic structures of the original messy GAs have been retained, multiple template solutions are used, instead of one template solution. Each template is chosen as the best solution corresponding to each objective in the previous era. The evaluation through multiple different templates and assignment of fitness using the domination principle help maintain trade-off solutions in the population.

In addition, we have briefly discussed a number of other elitist MOEAs. Algorithms presented in this chapter have shown different ways elitism can be introduced in MOEAs.

Exercise Problems

1. Consider the parent and offspring populations for a minimization problem:

Soln.	P_t		Soln.	Q_t	
	f_1	f_2		f_1	f_2
1	2.0	3.0	a	0.5	5.0
2	4.0	5.0	b	2.5	0.5
3	5.0	3.0	c	4.6	4.0
4	4.5	1.0	d	3.0	4.0

Find P_{t+1} using Rudolph's method.

2. Calculate P_{t+1} using NSGA-II of the population stated in problem 1.

3. Consider the following parent and offspring populations for a problem of minimizing the first objective and maximizing the second objective:

Soln.	P_t			Soln.	Q_t	
	f_1	f_2			f_1	f_2
1	5.0	2.5		a	3.5	0.0
2	1.0	1.0		b	2.2	0.5
3	1.5	0.0		c	5.0	2.0
4	4.5	1.0		d	3.0	3.0
5	3.5	2.0		e	3.0	1.5

Create P_{t+1} using NSGA-II.

4. Consider the combined population R_t (where all objectives are minimized):

Soln.	f_1	f_2	f_3
1	5.5	4.0	4.5
2	1.0	8.0	0.0
3	3.5	4.0	5.5
4	3.0	2.0	1.0
5	5.0	1.0	4.0
6	2.5	3.0	3.0
7	3.0	6.0	2.5
8	6.0	0.0	0.5

Find the winner of the constrained tournament selection operation in the following cases:

(a) Solutions 1 and 4
(b) Solutions 4 and 5
(c) Solutions 6 and 5
(d) Solutions 3 and 7

5. Using the population given in problem 4 as P_0, find the elite population at the end of one iteration DPGA with $F_1 = 20.0$.

6. Consider the GA and elite populations for a problem where the first objective is maximized and the second objective is minimized:

Soln.	P_t	
	f_1	f_2
1	1.0	1.0
2	3.0	4.0
3	1.0	3.8
4	6.0	0.0
5	5.0	2.0
6	4.0	3.5

Soln.	$\overline{P}_t$	
	f_1	f_2
a	7.0	3.0
b	2.0	5.0
c	8.0	1.0

Calculate the fitness of each individual in P_t and $\overline{P}_t$ using SPEA fitness assignment scheme.

7. Apply the SPEA clustering technique to find three members of the elite population for the next generation in problem 6.

8. We need to minimize f_1 (all values lying in $[0,6]$) and maximize f_2 (all values lying in $[0,5]$). The following archive population is used:

Soln.	f_1	f_2
1	3.1	3.2
2	2.2	2.8
3	4.5	4.3
4	0.8	0.7
5	3.5	3.5

The search space is divided equally in each objective with a step of 1.0. The current parent is the first solution. What is the new archive and parent for each of the following offspring:

(a) $c_t = (5.5, 1.5)^T$
(b) $c_t = (1.5, 1.5)^T$
(c) $c_t = (0.3, 2.4)^T$

9. In the two-objective minimization problem:

$$\text{Minimize} \quad f_1(\mathbf{x}) = x_1^2,$$
$$\text{Minimize} \quad f_2(\mathbf{x}) = 4 - x_1^2 + x_2^2,$$
$$0 \leq x_1, x_2 \leq 2,$$

the following four Pareto-optimal solutions are obtained:

Soln.	x_1	x_2
1	0.7	0.0
2	1.2	0.0
3	1.9	0.0
4	1.3	0.0

Solution 4 is the current parent solution. What is the probability of creating an offspring which would be the parent in the next iteration under PAES? Choose a constant probability for creating a solution within an Euclidean distance of 2.0 units from a parent for mutation. The objective space is discretized at a step of 1.0 unit in each objective.

7

Constrained Multi-Objective Evolutionary Algorithms

All algorithms described in the previous two chapters assumed that the underlying optimization problem is free from any constraint. However, this is hardly the case when it comes to solving real-world optimization problems. Typically, a constrained multi-objective optimization problem can be written as follows:

$$\left. \begin{array}{ll} \text{Minimize/Maximize} & f_m(\mathbf{x}), & m = 1, 2, \ldots, M; \\ \text{subject to} & g_j(\mathbf{x}) \geq 0, & j = 1, 2, \ldots, J; \\ & h_k(\mathbf{x}) = 0, & k = 1, 2, \ldots, K; \\ & x_i^{(L)} \leq x_i \leq x_i^{(U)}, & i = 1, 2, \ldots, n. \end{array} \right\} \quad (7.1)$$

Constraints divide the search space into two divisions – feasible and infeasible regions. Like in the single-objective optimization, all Pareto-optimal solutions must also be feasible. Constraints can be of two types: equality and inequality constraints. Equation (7.1) shows that there are J inequality and K equality constraints in the formulation. Constraints can be *hard* or *soft*. A constraint is considered hard if it *must* be satisfied in order to make a solution acceptable. A soft constraint, on the other hand, can be relaxed to some extent in order to accept a solution. Hard equality constraints are difficult to satisfy, particularly if the constraint is nonlinear in decision variables. Such hard equality constraints may be possible to relax (or made soft) by converting them into an inequality constraint with some loss of accuracy (Deb, 1995). In all of the constraint handling strategies discussed here, we assume greater-than-equal-to type inequality constraints only. Thus, in the nonlinear programming (NLP) formulation of the optimization problem shown in equation (7.1), the equality constraints h_k will be absent in further discussions in this present chapter. It is important to reiterate that this relaxation does not mean that the algorithms cannot handle equality constraints. Instead, it suggests that equality constraints should be handled by converting them into relaxed inequality constraints.

The above formulation also accommodates the smaller-than-equal-to inequality constraint $(g_j(\mathbf{x}) \leq 0)$, if any. When such constraints appear, they can be converted to the greater-than form by multiplying the left side by -1. Thus, if a solution $\mathbf{x}^{(i)}$ makes all $g_j(\mathbf{x}^{(i)}) \geq 0$ for $j = 1, 2, \ldots, J$, the solution is feasible and the constraint violation is zero. On the other hand, if any constraint $g_j(x^{(i)}) < 0$, then the solution is infeasible and the constraint j is violated. The amount of constraint violation in this case is $|g_j(x^{(i)})|$.

7.1   An Example Problem

As before, we construct a simple two-variable, two-objective constraint optimization problem to illustrate the working of the algorithms suggested in this chapter. The objective functions are the same as that used in Min-Ex before. We add two constraints:

$$\text{Constr-Ex:} \begin{cases} \text{Minimize} & f_1(x) = x_1, \\ \text{Minimize} & f_2(x) = \dfrac{1 + x_2}{x_1}, \\ \text{subject to} & g_1(x) \equiv x_2 + 9x_1 \geq 6, \\ & g_2(x) \equiv -x_2 + 9x_1 \geq 1, \\ & 0.1 \leq x_1 \leq 1, \\ & 0 \leq x_2 \leq 5. \end{cases} \qquad (7.2)$$

Recall that the unconstrained problem has the Pareto-optimal solutions: $0.1 \leq x_1^* \leq 1$ and $x_2^* = 0$. This region in the decision variable space is the entire x_1-axis of Figure 160. The corresponding Pareto-optimal region in the objective function space is also shown in Figure 161 (a thin hyperbolic curve). Constraints divide the search space into two regions as shown in both figures. With constraints, a part of the original Pareto-optimal region is not feasible and a new Pareto-optimal region emerges. The combined Pareto-optimal set is as follows:

$$\begin{array}{lll} \text{For } 0.39 \leq x_1^* \leq 0.67: & x_2^* = 6 - 9x_1^* & \text{(Region A)} \\ \text{For } 0.67 \leq x_1^* \leq 1.00: & x_2^* = 0 & \text{(Region B)} \end{array} \qquad (7.3)$$

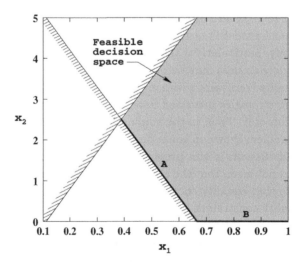

Figure 160   Feasible search region for Constr-Ex in the decision variable space.

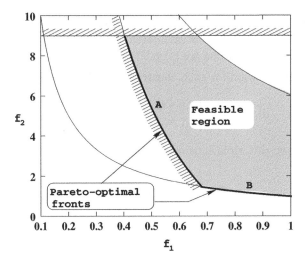

Figure 161   Feasible search region for Constr-Ex in the objective space.

The entire Pareto-optimal region is still convex. The interesting aspect here is that in order to find a diverse set of Pareto-optimal solutions in Region A, both variables x_1^* and x_2^* must be correlated in a way so as to satisfy $x_2^* + 9x_1^* = 6$.

Now, we will discuss different MOEAs which are particularly designed to handle constraints.

7.2   Ignoring Infeasible Solutions

A common and simple way to handle constraints is to ignore any solution that violates any of the assigned constraints (Coello and Christiansen, 1999). Although this approach is simple to implement, in most real-world problems finding a feasible solution (which will satisfy all constraints) is a major problem. In such cases, this naive approach will have difficulty in finding even one feasible solution, let alone finding a set of Pareto-optimal solutions. In order to proceed towards the feasible region, infeasible solutions, as and when created, must be evaluated and compared among themselves and with feasible solutions. One criterion often used in the context of constrained single-objective optimization using EAs is a measure of an infeasible solution's overall constraint violation (Deb, 2000; Michalewicz and Janikow, 1991). In this way, an EA may assign more selection pressure to solutions with less-violated constraints, thereby providing the EA with a direction for reaching the feasible region. Once solutions reach the feasible region, a regular MOEA approach may be used to guide the search towards the Pareto-optimal region.

7.3   Penalty Function Approach

This is a popular constraint handling strategy. Minimization of all objective functions is assumed here. However, a maximization function can be handled by converting it into a minimization function by using the duality principle.

Before the constraint violation is calculated, all constraints are normalized. Thus, the resulting constraint functions are $\underline{g}_j(\mathbf{x}^{(i)}) \geq 0$ for $j = 1, 2, \ldots, J$. For each solution $\mathbf{x}^{(i)}$, the constraint violation for each constraint is calculated as follows:

$$\omega_j(\mathbf{x}^{(i)}) = \begin{cases} |\underline{g}_j(\mathbf{x}^{(i)})|, & \text{if } \underline{g}_j(\mathbf{x}^{(i)}) < 0; \\ 0, & \text{otherwise.} \end{cases} \tag{7.4}$$

Thereafter, all constraint violations are added together to get the overall constraint violation:

$$\Omega(\mathbf{x}^{(i)}) = \sum_{j=1}^{J} \omega_j(\mathbf{x}^{(i)}). \tag{7.5}$$

This constraint violation is then multiplied with a penalty parameter R_m and the product is added to each of the objective function values:

$$F_m(\mathbf{x}^{(i)}) = f_m(\mathbf{x}^{(i)}) + R_m\Omega(\mathbf{x}^{(i)}). \tag{7.6}$$

The functions F_m takes into account the constraint violations. For a feasible solution, the corresponding Ω term is zero and F_m becomes equal to the original objective function f_m. However, for an infeasible solution, $F_m > f_m$, thereby adding a penalty corresponding to total constraint violation. The penalty parameter R_m is used to make both of the terms on the right side of the above equation to have the same order of magnitude. Since the original objective functions could be of different magnitudes, the penalty parameter must also vary from one objective function to another. A number of static and dynamic strategies to update the penalty parameter are suggested in the single-objective GA literature (Michalewicz, 1992; Michalewicz and Schoenauer, 1996; Homaifar et al., 1994). Any of these techniques can be used here as usual. However, most studies in multi-objective evolutionary optimization use carefully chosen static values of R_m (Srinivas and Deb, 1994; Deb, 1999a). Once the penalized function (equation (7.6)) is formed, any of the unconstrained multi-objective optimization methods discussed in the previous chapter can be used with F_m. Since all penalized functions are to be minimized, GAs should move into the feasible region and finally approach the Pareto-optimal set.

Let us consider six solutions shown in Table 25. Figure 162 demonstrates that the first three solutions are not feasible, whereas the other three solutions are feasible. Let us now calculate the penalized function values of all six solutions. The variable values and their unconstrained function values are shown in Table 25.

First of all, the constraints are normalized as follows:

$$\underline{g}_1(\mathbf{x}) = \frac{9x_1 + x_2}{6} - 1 \geq 0, \tag{7.7}$$

$$\underline{g}_2(\mathbf{x}) = \frac{9x_1 - x_2}{1} - 1 \geq 0. \tag{7.8}$$

The first solution is infeasible, because the constraint value is $\underline{g}_1(\mathbf{x}^{(1)}) = [(9 \times 0.31 + 0.89)/6] - 1$ or -0.39, a negative quantity. Thus, the quantity $\omega_1 = 0.39$. However, the

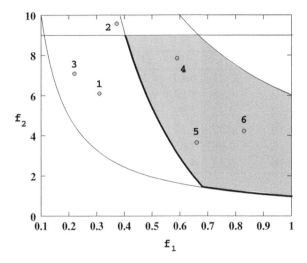

Figure 162 Six solutions are shown in the objective space of Constr-Ex.

second constraint is not violated at this solution $(g_2(x^{(1)}) = 0.90$, which is greater than equal to zero), meaning $w_2 = 0$. Thus, the constraint violation is $\Omega = w_1 + w_2 = 0.39$. The constraint violations of all other solutions are shown in Table 25.

We observe that the f_2 values are an order of magnitude larger than those of f_1, although the constraint violations are of the same order as f_1. Thus, we set $R_1 = 2$ and $R_2 = 20$. The penalized function value of the first solution is:

$$F_1 = f_1 + R_1\Omega,$$
$$= 0.31 + 2 \times 0.39,$$
$$= 1.09.$$
$$F_2 = f_2 + R_2\Omega,$$
$$= 6.10 + 20 \times 0.39,$$
$$= 13.90.$$

Similarly, we compute the penalized fitness of the other five solutions in Table 26.

Table 25 Fitness assignment using the penalty function approach.

Solution	x_1	x_2	f_1	f_2	w_1	w_2	Ω
1	0.31	0.89	0.31	6.10	0.39	0.00	0.39
2	0.38	2.73	0.38	9.82	0.03	0.31	0.34
3	0.22	0.56	0.22	7.09	0.58	0.00	0.58
4	0.59	3.63	0.59	7.85	0.00	0.00	0.00
5	0.66	1.41	0.66	3.65	0.00	0.00	0.00
6	0.83	2.51	0.83	4.23	0.00	0.00	0.00

Table 26 Penalized function values of all six solutions.

Solution	f_1	f_2	Ω	F_1	F_2	Front
1	0.31	6.10	0.39	1.09	13.90	3
2	0.38	9.82	0.34	1.06	16.62	3
3	0.22	7.09	0.58	1.38	18.69	4
4	0.59	7.85	0.00	0.59	7.85	1
5	0.66	3.65	0.00	0.66	3.65	1
6	0.83	4.23	0.00	0.83	4.23	2

Using the F_1 and F_2 values, we find that the first non-dominated set contains
solutions 4 and 5, the second non-dominated set contains solution 6, the third non-
dominated set contains solutions 1 and 2, and the fourth non-dominated set contains
solution 3. This is how infeasible solutions get de-emphasized due to the addition
of penalty terms. Of course, the exact classification depends on the chosen penalty
parameters. It is interesting to note that by this procedure some infeasible solutions
can be on the same front with a feasible solution, particularly if the infeasible solution
resides close to the constraint boundary.

Recall that the non-dominated sorting on the unconstrained objective functions
(based on columns 2 and 3 of Table 26) would reveal the following classification:
$[(1,3,5),(2,4,6)]$. With penalized function values (columns 5 and 6 of Table 26), the
classification is different: $[(4,5),(6),(1,2),(3)]$. Figure 163 shows the corresponding
fronts in the constrained problem. It is interesting to note how this penalty function
approach can change the objective functions so that feasible solutions close to the
Pareto-optimal front are allocated in the best non-dominated front. Among the

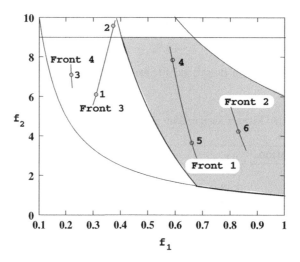

Figure 163 Non-dominated fronts with the penalized function values.

infeasible solutions, the ones close to the constraint boundary are allocated better fronts. However, the classification largely depends on the exact value of the chosen penalty parameters. The above procedure is one of the desired ways to assign fitness in a constraint-handling MOEA.

7.3.1 Simulation Results

To illustrate the working of this naive penalty function approach, we apply the NSGA (see Section 5.9 above) to the two-objective constrained optimization problem, Constr-Ex. The following GA parameters are used:

Population size	40,
Crossover probability	0.9,
Mutation probability	0,
Niching parameter, σ_{share}	0.158.

Niching is performed in the parameter space with normalized parameter values. The constraints are normalized according to equations (7.7) and (7.8) and the bracket operator (similar to equation (7.5), but each ω_j is squared) is used to calculate the penalty terms. It is mentioned above that the success of this naive approach depends on the proper choice of the penalty parameters, R_1 and R_2. To show this effect, we choose $R_1 = R$ and $R_2 = 10R$ and use three different values of R. Figures 164, 165 and 166 show the complete population after 500 generations of the NSGA for different values of R. The reason for continuing simulations for so long is purely to make sure that a stable population is obtained. The first figure shows that a small penalty parameter $R = 0.1$ cannot find all feasible solutions even after 500 generations. Since penalty terms are added to each objective function, the resulting penalized objective functions may form a Pareto-optimal front different from the true Pareto-optimal front, particularly if the chosen penalty parameter values are not adequate. For this

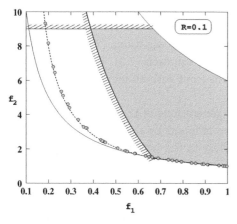
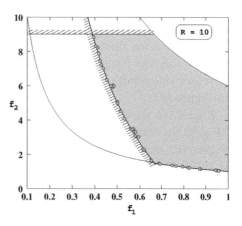

Figure 164   A small R results in many infeasible solutions.

Figure 165   The population with $R = 10$.

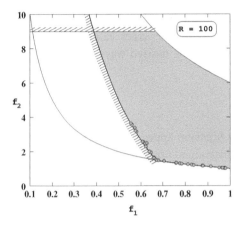

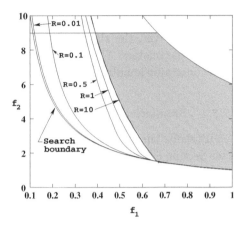

Figure 166   The population with a large R = 100 shows a poor spread of solutions.

Figure 167   Pseudo Pareto-optimal fronts seem to approach the true front with increasing values of R.

particular example problem, this pseudo Pareto-optimal front can be determined by determining the penalized function values:

$$F_1 \;=\; x_1 + R\langle\frac{9x_1 + x_2}{6} - 1\rangle^2 + R\langle 9x_1 - x_2 - 1\rangle^2, \tag{7.9}$$

$$F_2 \;=\; \frac{1 + x_2}{x_1} + 10R\langle\frac{9x_1 + x_2}{6} - 1\rangle^2 + 10R\langle 9x_1 - x_2 - 1\rangle^2. \tag{7.10}$$

With R = 0.1, we randomly create many solutions using the above two functions and find the best non-dominated front. This front is plotted with a dashed line in Figure 164. Since the penalty parameter is small, this front resides in the infeasible region. This phenomenon is also common in single-objective constrained optimization. If a smaller than adequate penalty parameter value is chosen, the penalty effect is less and the resulting optimal solution may be infeasible (Deb, 2000).

However, when the penalty parameter is increased to R = 10, the resulting Pareto-optimal front for the two penalized functions given in equations (7.9) and (7.10) is close to the true Pareto-optimal solution given in equation (7.3). It is interesting to note how this resulting front approaches the true front with an increase of R, as shown in Figure 167.

Figure 165 shows the population after 500 generations of the NSGA with R = 10. Most of the constrained Pareto-optimal region is found by the NSGA. As seen from Figure 167, with R = 10 the pseudo front is close to the true front. Thus, the NSGA is able to locate the true front with R = 10.

It is interesting to note that when we increase the penalty parameter to R = 100, the spread of obtained solutions is not as good as that with R = 10. Although the pseudo front with R = 100 will be even closer to the true Pareto-optimal front compared to that with R = 10, the constraints will be over-emphasized in initial generations. This causes the NSGA to converge near a portion of the Pareto-optimal front.

These results show the importance of the penalty parameter with the naive approach. If an appropriate R is chosen, MOEAs will work well. However, if the choice of R is not adequate, either a set of infeasible solutions or a poor distribution of solutions is likely.

7.4 Jiménez–Verdegay–Goméz-Skarmeta's Method

Jiménez, Verdegay and Goméz-Skarmeta (1999) suggested a systematic constraint handling procedure for the multi-objective optimization. Only inequality constraints of the lesser-than-equal-to type are considered in their study, whereas any other constraints can also be handled by using the procedure. Although most earlier algorithms used the simple penalty function approach, this work suggested a careful consideration of feasible and infeasible solutions and the use of niching to maintain diversity in the obtained Pareto-optimal solutions. The algorithm uses the binary tournament selection in its core. As in the single-objective constraint handling strategies using tournament selection (Deb, 2000; Deb and Agrawal, 1999b), three different cases arise in a two-player tournament. Both solutions may be feasible, or both may be infeasible, or one is feasible and the other is infeasible. The investigators carefully considered each case and suggested the following strategies.

Case 1: Both solutions are feasible. Since infeasibility is not a concern here, the investigators suggested following a procedure similar to the niched Pareto GA proposed by Horn et al. (1994) (see Section 5.10 above). First, a random set of feasible solutions called a comparison set is picked from the current population. Secondly, two solutions chosen for a tournament are compared with this comparison set. If one solution is non-dominated in the comparison set and the other is dominated by at least one solution from the comparison set, the former solution is chosen as the winner of the tournament. Otherwise, if both solutions are non-dominated or both solutions are dominated in the comparison set, both are equally good or bad in terms of their domination with respect to the comparison set. Thus, the neighborhood of each solution is checked to find its niche count. The solution with a smaller niche count means that the solution is situated in a least crowded area and is chosen as the winner of the tournament.

The niche count nc is calculated by using the phenotypic distance measure, that is, problem variables are used to compute the Euclidean distance d between two solutions. Then, this distance is compared with a given threshold σ_{share}. If the distance is smaller than σ_{share}, a sharing function value is computed by using equation (4.59) with an exponent value equal to $\alpha = 2$. Thereafter, the niche count is calculated by summing the sharing function values calculated with each population member. This procedure is the same as that used in the NSGA.

Case 2: One solution is feasible and other is not. The choice is clear here. The feasible solution is declared as the winner.

Case 3: Both solutions are infeasible. As in the first case, a random set of infeasible solutions is first chosen from the population. Secondly, both solutions participating in the tournament are compared with this comparison set. If one is better than the best infeasible solution and the other is worse than the best infeasible solution present in the comparison set, the former solution is chosen. A suitable criterion can be used for this comparison. For example, the constraint violation or nearness to a constraint boundary can be used. If both solutions are either better or worse than the best infeasible solution, both of them are considered as being equally good or bad. A decision with the help of a niche count calculation is made, as before. The niche count is calculated with the phenotypic distance measure (with x_i values) and with respect to the entire population, as described in the first case. The one with the smaller niche count wins the tournament.

This is a more systematic approach than the simple penalty function approach. As long as decisions can be made with the help of feasibility and dominance of solutions, they are followed. However, when both solutions enter a tie with respect to feasibility and dominance considerations, the algorithm attempts to satisfy the second task of multi-objective optimization. The algorithm uses a niching concept to encourage a less-crowded solution.

7.4.1   Hand Calculations

We now illustrate this tournament selection procedure on the six solutions of our example problem. Let us consider that solutions 1 and 2 participate in the first tournament. Since both of them are infeasible, we follow Case 3. We select a subpopulation of infeasible solutions from the population. Since we have considered a small population size for illustration, we choose solution 3 as the only member in the comparison set. Let us also use the constraint violation metric for evaluating an infeasible solution. Thus, we observe that both solutions 1 and 2 are better than solution 3 (the best or only solution of the comparison set). In order to resolve the tie, we need to compute the niche count for each of these two solutions with respect to the entire population. We show the computation for solution 1 only here. However, the niche count values for other solutions in the population are listed in Table 27. The normalized phenotypic distance values calculated are as follows:

$$d_{12} = 0.376, \quad d_{13} = 0.120, \quad d_{14} = 0.630, \quad d_{15} = 0.403, \quad d_{16} = 0.662.$$

We use a sharing parameter $\sigma_{share} = 0.4$ and $\alpha = 2$. The corresponding sharing function values are as follows:

$$Sh(d_{12}) = 0.116, \quad Sh(d_{13}) = 0.910, \quad Sh(d_{14}) = 0, \quad Sh(d_{15}) = 0, \quad Sh(d_{16}) = 0.$$

Thus, the niche count is the sum of all of these sharing function values or $nc_1 = 2.026$. Similarly, we calculate $nc_2 = 1.572$. Since $nc_2 < nc_1$, we choose solution 2 as the winner of the first tournament.

Table 27 The niche count for each solution.

Solution	x_1	x_2	f_1	f_2	Niche count
1	0.31	0.89	0.31	6.10	2.026
2	0.38	2.73	0.38	9.82	1.572
3	0.22	0.56	0.22	7.09	1.910
4	0.59	3.63	0.59	7.85	1.699
5	0.66	1.41	0.66	3.65	1.474
6	0.83	2.51	0.83	4.23	1.717

Next, we take solutions 3 and 4 for the second tournament. Figure 162 reveals that solution 3 is infeasible, but solution 4 is feasible. Case 2 suggests that the winner is solution 4.

Next, we consider solutions 5 and 6. Since both are feasible, we must follow Case 1. Let us choose solution 4 as the only member of the chosen comparison set. We put solution 5 into the comparison set and observe that it is non-dominated. The same is true with solution 6. So, we calculate their niche count in the entire population. The niche count values are listed in Table 27. Since solution 5 has a smaller niche count than that of solution 6, we choose solution 5. Notice that although solution 6 is dominated by solution 5, they both resorted in a tie in the tournament. The difficulty arises because each solution is independently checked for domination with the comparison set. The fact that solution 5 could dominate solution 6 is never checked in the procedure. In the next section, we will present a different tournament-based selection, which does not have this difficulty.

At the end of three tournaments, we have used all solutions exactly once and obtained solutions 2, 4 and 6. Next, we shuffle the population members and perform three more tournaments to fill up six population slots. Let us say that after shuffling the following sequence occurs: $(3, 2, 6, 1, 5, 4)$. Thus, we pair them up as $(3, 2)$, $(6, 1)$ and $(5, 4)$ and play tournaments between them. The first tournament, in the presence of solution 1 as the comparison set, favors solution 2. This is because solution 2 has a smaller constraint violation compared to solution 1, but solution 3 has a larger constraint violation compared to solution 1. The constraint violations are calculated in Table 25 on page 293. Between solutions 6 and 1, solution 6 wins. For the third tournament, we choose solution 6 as the comparison set. Now, both solutions 4 and 5 are non-dominated in the comparison set. Thus, we make our decision based on their niche counts. We find that solution 5 is the winner.

Thus, one application of the tournament selection on the entire original population makes the following population as the mating pool: $(2, 2, 4, 5, 5, 6)$. Although two of the three infeasible solutions are absent in the mating pool, the emphasis of feasible solutions is not particularly ideal.

7.4.2 Advantages

The algorithm requires a niche count computation for each solution. This requires $O(N^2)$ computations, which is comparable to that of other evolutionary algorithms. Another advantage of this method is that the tournament selection is used. As shown elsewhere (Goldberg and Deb, 1991), the tournament selection operator has better convergence properties compared to the proportionate selection operator.

7.4.3 Disadvantages

Since the niche count is calculated with all population members, one wonders why investigators suggested the use of a subpopulation to check domination and constraint violation. The entire set of feasible solutions in the population can also be used as the comparison set without increasing the computational complexity of the algorithm. The flip side of this argument is relevant. It may be better to restrict the niche count computations among the members of the chosen comparison set, instead of the entire population. This would reduce the complexity of the computation.

It is not intuitive why one would be interested in maintaining diversity among infeasible solutions. As in single-objective constrained optimization problems, the goal here is also to move towards the feasible region. By preserving diversity among infeasible solutions explicitly, the progress towards the feasible region may be sacrificed.

There exist a couple of additional parameters (σ_{share} and the size of the comparison set) which a user must set right. In order to make the non-domination check less stochastic, a large comparison set is needed. Furthermore, it was mentioned earlier that the algorithm does not explicitly check the domination of participating solutions in a tournament.

7.4.4 Simulation Results

We consider the two-variable constrained optimization problem given in equation (7.2), using the same GA parameters as before. We would like to mention here that investigators of this algorithm used a different real-parameter crossover and mutation operator to that used here. In the original study, a uniform real-parameter crossover and a dynamically varying mutation operators were used. We use the SBX and the polynomial mutation operator throughout this chapter, including this algorithm.

The initial population and the non-dominated solutions at the end of 500 generations are shown in Figures 168 and 169, respectively. The first figure shows that the initial population contains some feasible and some infeasible solutions. Although the algorithm approached the Pareto-optimal front over the generations, the progress is slow and even after 500 generations, solutions on the constrained Pareto-optimal region are not found.

It is important to note in particular that this algorithm *could not find solutions* on the constrained Pareto-optimal region. In order to be on the constrained Pareto-

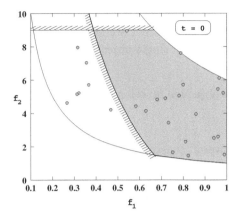
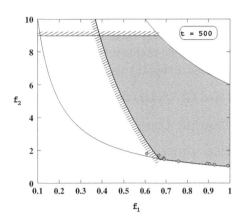

Figure 168 Initial population of 40 solutions used in the Jiménez–Verdegay–Goméz-Skarmeta's algorithm.

Figure 169 Non-dominated solutions after 500 generations.

optimal region, the parameters x_1 and x_2 must be related as $x_2 = 6 - 9x_1$. Since it becomes difficult for the variable-wise crossover and mutation operator (used in the above simulation) to find correlated solutions, satisfying the above equality constraint becomes difficult, thereby making it hard to maintain solutions on the Pareto-optimal front. Once such solutions are found, they have to be adequately emphasized by the selection procedure and the constraint handling technique for their sustained survival. This so-called linkage problem with generic variable-wise (or bit-wise) operators is important in the study of evolutionary algorithms and has motivated many researchers to develop sophisticated EAs (Goldberg et al., 1989; Harik and Goldberg, 1996; Kargupta, 1997; Mühlenbein and Mahnig, 1999; Pelican et al., 1999). The bottom-right portion of the Pareto-optimal region (see Figure 169) is represented by $x_2^* = 0$ and $x_1^* \in [7/9, 1]$. Since the spread in this part of the front can be achieved by keeping diversity in only the variable x_1 and by maintaining $x_2 = 0$, this portion of the Pareto-optimal region is comparatively easier to obtain. This algorithm seems to have found some representative members on this part of the front.

The next strategy uses a much improved tournament selection operator, which is easier in concept and eliminates most of the difficulties of this strategy.

7.5 Constrained Tournament Method

Recently, this author and his students have implemented a penalty-parameter-less constraint handling approach for single-objective optimization. We have outlined this constraint handling technique above in Section 4.2.3. A similar method can also be introduced in the context of multi-objective optimization.

The constraint handling method discussed here uses the binary tournament selection, where two solutions are picked from the population and the better solution is chosen. In the presence of constraints, each solution can be either feasible or infeasible. Thus, there may be at most three situations: (i) both solutions are feasible, (ii) one is feasible and other is not, and (iii) both are infeasible. For single-objective optimization, we used a simple rule for each case:

Case (i) Choose the solution with the better objective function value.

Case (ii) Choose the feasible solution.

Case (iii) Choose the solution with smaller overall constraint violation.

In the context of multi-objective optimization, the latter two cases can be used as they are presented, but the difficulty arises with the first case. This is because now we have multiple objective functions, and we are in a dilemma as to which objective function to consider. The concept of domination can come to our rescue here. When both solutions are feasible, we can check if they belong to separate non-dominated fronts. In such an event, choose the one that belongs to the better non-dominated front. If they belong to the same non-dominated front, we can use the diversity preservation task to resolve the tie. Since maintaining diversity is another goal in multi-objective optimization, we can choose the one which belongs to the least crowded region in that non-dominated set.

We define the following *constrain-domination* condition for any two solutions $x^{(i)}$ and $x^{(j)}$.

Definition 7.1. *A solution $x^{(i)}$ is said to 'constrain-dominate' a solution $x^{(j)}$ (or $x^{(i)} \preceq_c x^{(j)}$), if any of the following conditions are true:*

1. *Solution $x^{(i)}$ is feasible and solution $x^{(j)}$ is not.*
2. *Solutions $x^{(i)}$ and $x^{(j)}$ are both infeasible, but solution $x^{(i)}$ has a smaller constraint violation.*
3. *Solutions $x^{(i)}$ and $x^{(j)}$ are feasible and solution $x^{(i)}$ dominates solution $x^{(j)}$ in the usual sense (see Definition 2.5 above).*

We can use the same non-dominated classification procedure described earlier in Section 2.4.6 to classify a population into different levels of non-domination in the presence of constraints. The only change required is that the domination definition (presented in Section 2.4.2) has to be replaced with the above definition. Among a population of solutions, the set of *non-constrain-dominated* are those that are not constrain-dominated by any member of the population. Let us use Approach 1 (page 34) to classify six solutions in the example problem Constr-Ex into different non-constrain-dominated sets.

Step 1 We set $i = 1$ and set $P' = \emptyset$.

Step 2 We now check if solution 2 constrain-dominates solution 1. Both of these solutions are infeasible. Thus, we use their constraint violation values to decide who wins. The violation values are listed in Table 25. We observe that solution 2 has a smaller constraint violation than solution 1. Thus, solution 2 constrain-dominates solution 1. We now move to Step 4 of Approach 1. This means that solution 1 cannot be in the best non-constrain-dominated front.

Step 4 Increment i to 2 and move to Step 2.

Steps 2 and 3 Solution 1 does not constrain-dominate solution 2. We now check whether solution 3 constrain-dominates solution 1. Both solutions are infeasible and solution 3 has a larger constraint violation. Thus, solution 3 does not constrain-dominate solution 1.

Next, we check solution 4 with solution 2. Solution 4 is feasible and solution 2 is not. Thus, solution 4 constrain-dominates solution 2. Thus, solution 2 cannot be in the best non-constrain-dominated front.

Step 4 Next, we check solution 3.

Steps 2 and 3 Solution 1 constrain-dominates solution 3.

Step 4 Next, we check solution 4.

Step 2 Solutions 1, 2 and 3 do not constrain-dominate solution 4. Next, we check whether solution 5 constrain-dominates solution 4. Both are feasible. We observe that solution 5 does not dominate solution 4.

Step 3 Next, we try with solution 6. This solution also does not dominate solution 4. Since there is no more members left in the population, solution 4 belongs to the first non-constrain-dominated front, or $P' = \{4\}$.

Step 4 We now have to check solution 5.

Step 2 Solutions 1, 2 and 3 do not constrain-dominate solution 5. Solutions 4 and 6 do not dominate solution 5.

Step 3 Thus, solution 5 is also a member of P'.

Step 4 We now check the final solution 6 for its inclusion in the best non-constrain-dominated front.

Step 2 Solutions 1 to 3 do not constrain-dominate this solution. Solution 4 does not dominate solution 6, but solution 5 dominates this solution. Thus, solution 6 cannot be included in the best set.

Step 4 Thus, the first non-constrain-dominated set has two solutions $P_1 = \{4, 5\}$. Figure 170 shows this set.

To obtain the second non-constrain-dominated front, we discount these two solutions (4 and 5) from the population and repeat the above procedure. We shall find $P_2 = \{6\}$.

With the remaining three solutions (1, 2 and 3), we continue the above procedure to find the third non-constrain-dominated front. We shall obtain $P_3 = \{2\}$. This is

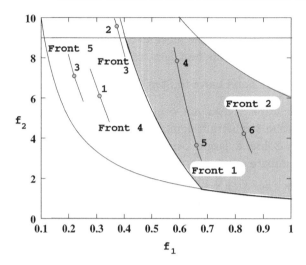

Figure 170   Non-constrain-dominated fronts.

because, solutions 1 and 3 have larger constraint violation values than solution 2. Continuing in this fashion, we obtain $P_4 = \{1\}$ and $P_5 = \{3\}$. Thus, the overall classification is $[(4,5),(6),(2),(1),(3)]$. All five fronts are shown in Figure 170. If the constraint was absent, the following would have been the non-dominated classification: $[(3,1,5),(2,4,6)]$. Now, since solutions 1, 2 and 3 are infeasible, the non-constrain-dominated classification is different and infeasible solutions have been pushed back into worse classes of non-constrain-domination.

The above definition of constrain-domination will allow a similar non-dominated classification in the feasible region, but will classify the infeasible solutions according to their constraint violation values. Usually, each infeasible solution will belong to a different non-constrain-dominated front in the order of their constraint violation values, except when more than one solutions have an identical constraint violation.

7.5.1   Constrained Tournament Selection Operator

By using the above definition of constrain-domination, we define a generic constrained tournament selection operator as follows.

Definition 7.2. *Given two solutions* $\mathbf{x}^{(i)}$ *and* $\mathbf{x}^{(j)}$*, choose solution* $\mathbf{x}^{(i)}$ *if any of the following conditions are true:*

1. *Solution* $\mathbf{x}^{(i)}$ *belongs to a better non-constrain-dominated set.*
2. *Solutions* $\mathbf{x}^{(i)}$ *and* $\mathbf{x}^{(j)}$ *belong to the same non-constrain-dominated set, but solution* $\mathbf{x}^{(i)}$ *resides in a less crowded region based on a niched-distance measure.*

The niched-distance measure refers to the measure of density of the solutions in the neighborhood of a solution. The niched-distance can be computed by using various metrics, as follows.

Niche count metric. Calculate the niche count nc_i and nc_j with solutions of the non-constrain-dominated set by using the sharing function method described earlier in Section 4.6. Since the niche count of a solution gives an idea of the number of crowded solutions and their relative distances from the given solution, a smaller value of the niche count means fewer solutions in the neighborhood. Thus, if this metric is used, the solution with the smaller niche count must be chosen. This metric requires a parameter σ_{share} and requires $O(N_p)$ computations, where N_p is the number of solutions in the non-constrain-dominated set. It is important to note that the relative distance measure for computing the niche count can be made in either the decision variable space or in the objective space.

Head count metric. Instead of computing the niche count, the total number of solutions (the head count) in the σ_{share}-neighborhood can be counted for each solution $\mathbf{x}^{(i)}$ and $\mathbf{x}^{(j)}$. Using this metric, choose the solution with the smaller head count. The complexity here is also $O(N_p)$. This metric may lead to ties, which can be avoided by using the niche count metric. This is because the niche count metric better quantifies the crowding of solutions. As for the niche count metric, the neighborhood checking can be performed either in the decision variable space or in the objective space.

Crowding distance metric. This metric is used in the NSGA-II. This metric estimates half of the perimeter of the maximum hypercube which can be allowed around a solution without including any other solution from the same obtained non-constrain-dominated front inside the hypercube. A large crowding distance for a solution implies that the solution is less crowded. The extreme solutions are assigned an infinite crowding distance. This measure requires $O(N_p \log N_p)$ computations, because of the involvement of the sorting of solutions in all M objectives. This crowding distance can be computed in the decision variable space as well.

Various other distance measures can also be used.

7.5.2 Hand Calculations

We now illustrate how the above constrained tournament selection operator acts on the six chosen solutions of the example problem Constr-Ex to create the mating pool.

Let us first choose solutions 1 and 2 to play the first tournament. Since solution 2 belongs to a better non-constrain-dominated front than solution 1 (see Figure 170 above), we choose solution 2. Next, we choose solutions 3 and 4 for the second tournament. Here, solution 4 is chosen, since it lies in the first front, whereas solution 3 lies in the fifth front. For the third tournament, we have solutions 5 and 6. Here, solution 5 wins. At the end of three tournaments, we have used all solutions exactly once and selected solutions 2, 4 and 5. Next, we shuffle the population members and perform three more tournaments to fill up six population slots. Let us say that after shuffling, the following sequence occurs: $(3, 2, 6, 1, 5, 4)$.

So, we compare solutions 3 and 2 next. Solution 2 wins. Next, among solutions 6 and 1, solution 6 wins. Finally, solutions 5 and 4 are compared. Since they belong to the same non-constrain-dominated set, we have to check their niched-distance values with all solutions in the set. By using the crowding distance metric we observe that solutions 4 and 5 are the only members of the first non-constrain-dominated front. Since these two are extreme solutions, both of them have an infinite crowding distance and thus both should survive. Let us say that we choose solution 4 at random.

Thus, after the constrained tournament selection, we obtain the following mating pool: $(2, 2, 4, 4, 5, 6)$. Solution 3 (the worst infeasible solution) gives away its place for another copy of solution 4, which lies in the best non-constrain-dominated front. Another infeasible solution (solution 1) is also eliminated.

7.5.3 Advantages and Disadvantages

In addition to the constraint violation computations, this strategy does not require any extra computational burden. The constrain-domination principle is generic and can be used with any other MOEAs. Since it forces an infeasible solution to be always dominated by a feasible solution, no other constraint handling strategy is needed.

The above constrain-domination definition is similar to that suggested elsewhere (Drechsler, 1998; Fonseca and Fleming, 1998a). However, these studies handle the constraint violations in a more similar way to that in the COMOGA approach (Surry et al., 1995) described later in Section 8.7.1. The only difference between the constrain-domination presented here and that in (Drechsler, 1998; Fonseca and Fleming, 1998a) is in the way domination is defined for the infeasible solutions. In the above definition, an infeasible solution having a larger overall constraint-violation is classified as a member of a larger non-constrain-domination level. On the other hand, in (Fonseca and Fleming, 1998a), infeasible solutions violating different constraints are classified as members of the same non-constrain-dominated front. Drechsler (1998) proposes a more detailed ordering approach. In these approaches one infeasible solution violating a constraint g_j marginally will be placed in the same non-constrain-dominated level with another solution violating a different constraint to a large extent but not violating g_j. This may cause an algorithm to wander in the infeasible search region for more generations before reaching the feasible region through some constraint boundaries. Moreover, since these approaches require domination checks to be performed with the constraint-violation values, they are supposedly computationally more expensive. However, a careful study is needed to investigate if the added complexity introduced by performing a non-domination check over the simple procedure described earlier in this section is beneficial to certain problems. Besides simply adding the constraint violations together, Binh and Korn (1997) also suggested a different constraint violation measure:

$$C(\mathbf{x}) = \left(\sum_{j=1}^{J} [c_j(\mathbf{x})]^p \right)^{\frac{1}{p}}, \qquad (7.11)$$

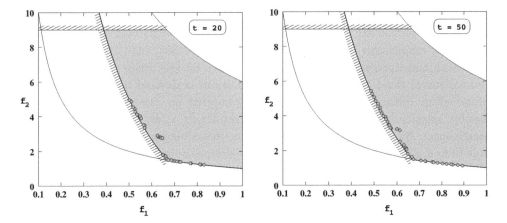

Figure 171   The NSGA-II population at   **Figure 172**   The NSGA-II population at
generation 20.   generation 50.

where $p > 0$ and $c_j(x) = |\min(0, g_j(x))|$ for all j. In the above approach, $p = 1$ is
used.

7.5.4   Simulation Results

We use the constraint tournament selection operator with the NSGA-II and employ the
crowded distance measure. As before, real-parameter GAs are used with a population
of size 40. Other NSGA-II parameters are the same as that used in the previous
chapter. Figures 171 to 174 show the best non-constrain-dominated solutions at
generations of 20, 50, 200 and 500. It is clear how solutions assemble closer to the
Pareto-optimal front and is distributed over the entire Pareto-optimal regions with an

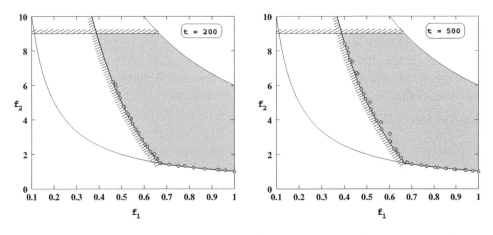

Figure 173   The NSGA-II population at   **Figure 174**   The NSGA-II population at
generation 200.   generation 500.

increasing number of generations. Although identical search operators to that used in the previous section are also used here, the constraint tournament selection operator adequately emphasizes correlated Pareto-optimal solutions, once they are discovered.

In Binh and Korn's constrained multi-objective ES (MOBES) approach, a careful comparison of the feasible and infeasible solutions is devised based on a diversity-preservation concept. These investigators preferred an infeasible solution over a feasible solution, if the infeasible solution resides in a less crowded region in the search space. By requiring a user-defined niche-infeasible parameter, the MOBES was able to find a widely distributed set of non-constrain-dominated solutions in a couple of test problems. Further studies are needed to justify the added complexities of the MOBES.

7.6   Ray–Tai–Seow's Method

Ray et al. (2001) suggested a more elaborate constraint handling technique, where the constraint violations of all constraints are not simply added together; instead, a non-domination check of the constraint violations is made. We describe this procedure here.

Three different non-dominated rankings of the population are first performed. The first ranking is performed by using M objective function values, and the resulting ranking is stored in an N-dimensional vector R_{obj}. The second ranking (R_{con}) is performed by using only the constraint violation values of all (J of them) constraints and no objective function information is used. Thus, the constraint violation of each constraint is used as a criterion and a non-domination classification of the population is performed with the constraint violation values. Notice that a feasible solution has zero constraint violation. Thus, all feasible solutions have a rank 1 in R_{con}. Such a ranking procedure will allow solutions violating only one independent constraint to lie in the same non-dominated front. The third ranking is performed by using a combined objective function and constraint violation values (a total of $(M + J)$ attributes). This produces the ranking R_{com}. Although objective function values and constraint violations are used together, interestingly there is no need of any penalty parameter. In the domination check, the criteria are individually compared, thereby eliminating the need of any penalty parameter. Once these rankings are complete, the following algorithm is used for handling the constraints.

Ray–Tai–Seow's Constraint-Handling Method

>   **Step 1** Using R_{com}, select all *feasible* solutions having a rank 1. This can be achieved by observing the feasibility of all rank 1 solutions from R_{con}. That is, $P' = \{q : R_{com}(q) = 1 \text{ and constraint violation } w_j(q) = 0, j = 1, 2, \ldots, J\}$. If $|P'| < N$, fill population P' by using Step 2.

>   **Step 2** Choose a solution A with R_{obj} by assigning more preference to low-ranked solutions. Choose its mate by first selecting two solutions B and C by using R_{con}. Then, use the following cases to choose either B or C:

Case 1 If B and C are both feasible, choose the one with better rank in R_{obj}. If both ranks are the same, a head-count metric is used to decide which of them resides in a less-crowded region in the population.

Case 2 If B and C are both infeasible, choose the one with a better rank in R_{con}. If both ranks are the same, then use a common-constraint-satisfaction metric (discussed later) to choose one of these. This metric measures the number of constraints that each of B or C satisfies along with A. The one satisfying the least number of common constraints is chosen.

Case 3 If B is feasible and C is not, choose B; otherwise, choose C. On the other hand, if B is infeasible and C is feasible, then choose C; otherwise, choose B.

Use parents to create offspring and put parents and offspring into the population.

In Step 2, solutions are chosen with different rankings R_{obj}, R_{con} and R_{com}. Here, we describe how a ranking is used to choose a solution from the population. Let us illustrate the procedure with R_{obj}. The maximum rank r^{max}, in the vector R_{obj} is first noted. Each rank is then mapped linearly between r^{max} and 1, so that the best non-dominated solutions get a rank r^{max} and the worst non-dominated solutions get a rank 1. Thereafter, a probability vector p_{obj} for selection is calculated in proportion to its mapped rank. In this way, the best non-dominated solutions are assigned the maximum probability of selection. This is similar to the roulette-wheel selection mechanism.

In Case 1, when the rank of both solutions B and C is the same, the head-count metric is used to resolve the tie. First, the average Euclidean distance $\bar{d}$ of the population (either in the decision variable space or in the objective space) is computed by averaging the Euclidean distance of each solution with all other solutions in the population, thereby requiring $N(N-1)/2$ distance computations. Now, for each of the solutions B and C, the number of solutions within a distance $\bar{d}$ from it is counted. This measures the head-count of the solution within a fixed neighborhood. The solution with a smaller head-count is chosen.

In Case 2, when the rank of both solutions B and C is the same, the following procedure is used to calculate the *common-constraint-satisfaction metric*. For each of solutions A, B and C, the set of constraints that are satisfied (not violated) are determined as S_A, S_B and S_C, respectively. Now the cardinality of each of the two intersecting set $S_A \cap S_B$ and $S_A \cap S_C$ is calculated. Let us say that they are $n_{AB} = |S_A \cap S_B|$ and $n_{AC} = |S_A \cap S_C|$. Now, if $n_{AB} < n_{AC}$, solution B is chosen; otherwise, solution C is chosen. This ensures that A and B together satisfy less common constraints than A and C do together. This, in general, allows two solutions from different regions of the search space to be mated. It also provides the needed diversity in the mated parents in the hope of creating diverse offspring. If both n_{AB} and n_{AC} are the same, one of either B or C is chosen at random.

Although a few other operators (such as a population shrinking and a mix-and-move crossover operator) are also suggested in the original study, we will leave out the details of these here and only discuss the constraint handling aspect of the algorithm. Investigators of this constrained handling approach have solved a number of engineering design problems, including a five-objective WATER problem (discussed later in Section 8.3). It remains to be investigated how the algorithm performs on more complex problems, particularly from the point of view of the computational burden associated with the method.

7.6.1 Hand Calculations

We now illustrate the working of this algorithm on six solutions on our constrained two-objective example problem Constr-Ex. The six population members are shown above in Table 25 and Figure 162.

First, we perform a non-dominated sorting by using the objective function values (columns 4 and 5 of Table 25). We obtain the following classification: $[(1,3,5),(2,4,6)]$. Thus, the rank vector is $R_{obj} = (1,2,1,2,1,2)$. This is because the first solution belongs to rank 1, the second solution is in rank 2, the third solution is in rank 1, and so on. Similarly, we obtain the non-dominated classification of the population with respect to constraint violations only (columns 6 and 7 of Table 25):

$$[(4,5,6),(1,2),(3)]$$

Thus, the rank vector for constraints is $R_{con} = (2,2,3,1,1,1)$. Finally, we obtain the non-dominated ranking of the combined objective function and constraints, totaling $(2+2)$, or four criteria. We obtain the following non-dominated classification having two fronts by using columns 4, 5, 6 and 7 of Table 25, as follows: $[(1,2,3,4,5),(6)]$. Thus, the corresponding ranking vector is $R_{com} = (1,1,1,1,1,2)$.

Now, we move to Step 1 of the algorithm. Of the five rank-1 solutions in R_{com}, we observe that solutions 4 and 5 are feasible. Thus, $P' = \{4,5\}$. These solutions act as the best solutions in the current population and are directly sent to the next population. Thus, these solutions act as elite solutions, which also participate in the genetic operations with other solutions in the population. Now, we move to Step 2 of the algorithm to complete the rest of the steps.

First, we choose a solution (we call it solution A) by using R_{obj}. Since there are only two different ranks in R_{obj} (or $r^{max} = 2$), we map each rank linearly between 2 and 1. Thus, we have the mapping $(2,1,2,1,2,1)$, obtained by swapping ranks 1 and 2 in R_{obj}, so that rank-1 solutions are assigned the value $r^{max} = 2$. We observe that the sum of all of these mapped values is $(2+1+2+1+2+1)$, or 9. Thus, the probability vector is $p_{obj} = (2/9,1/9,2/9,1/9,2/9,1/9)^{\mathsf{T}}$. With this probability vector, we choose a solution. A simple procedure is to first form the cumulative probability vector as $P_{obj} = (2/9,3/9,5/9,6/9,8/9,9/9)^{\mathsf{T}}$ and then choose a solution by using a random number between zero and one. Let us say the chosen random number is 0.432. Since, this lies between the first and second member of the cumulative probability vector, we choose solution 2 as A.

Similarly, we choose two solutions by using R_{con}. The probability vector is $p_{con} = (2/14,2/14,1/14,3/14,3/14,3/14)^{\mathsf{T}}$. Using random numbers 0.126 and 0.319, we obtain solution 1 as B and solution 3 as C. Now, we observe that both B and C are infeasible. Thus, we choose solution 1, because it has a smaller rank than solution 3 in R_{con}. Solution B is the winner.

Thus, solutions 1 and 2 mate with each other to create two offspring. The same procedure can be continued to fill up the rest of the population.

7.6.2   Computational Complexity

The non-dominated ranking procedure is governed by the ranking on the combined population, thereby requiring at most $O\left((M+J)N^2\right)$ comparisons. Each selection procedure of choosing a solution with any of the probability vectors requires $O(N)$ computations. The calculation of the average Euclidean distance $\bar{d}$ requires $O\left(N^2\right)$ calculations. Moreover, the head-count calculation procedure requires $O(N)$ comparisons. Thus, the overall complexity of the algorithm is $O\left((M+J)N^2\right)$.

7.6.3   Advantages

This algorithm handles infeasible solutions with more care than any other of the constraint-handling techniques we have discussed so far. When all three solutions A, B and C are infeasible (which may happen early on in a run), solutions B and C are chosen according to their constrained non-domination levels. This means that solutions violating different constraints are emphasized. In this way, diversity is maintained in the population.

7.6.4   Disadvantages

In a later generation, when all population members are feasible and belong to a suboptimal non-dominated front, the algorithm stagnates. This is because all of the population members will have a rank equal to 1 in R_{com}. Since all are also feasible, they fill up the population and no further processing is made. Most of the MOEAs described in this book handle these latter generations in a different way. Genetic operations are still allowed at this stage in the search for new solutions which will make the spread of the entire population better. In order to alleviate this problem, a simple change can be made. The selection of the best non-dominated feasible solutions using R_{com} constitutes a population P'. Thereafter, Step 2 of the algorithm can be used to create a new population P''. Now, populations P' and P'' can be combined together and the best N solutions can be chosen in a similar way to that in the NSGA-II, by first choosing the non-dominated solutions and then choosing the least crowded solutions from the last allowed non-dominated front.

There is another difficulty with this algorithm. During the crossover operation, three offspring are created. The first one is created by using a uniform crossover with an equal probability of choosing one variable value from each parent (solution A and its partner). The other two solutions are created by using a blend crossover, which uses a uniform probability distribution over a range that depends on a number of threshold parameter values. The difficulty arises in choosing parameter values related to each of these operators. Another difficulty arises because five solutions (three offspring and two parent) are accepted after each crossover operation. This process will cause the population to soon lose its diversity.

Three non-dominated ranking and head-count computations make the algorithm more computationally expensive than the other algorithms discussed so far.

7.6.5   Simulation Results

In order to investigate how well the above algorithm works, we apply it on the same two-objective, two-constraint optimization problem. Figure 175 shows the complete population after 50 generations. This figure shows that all 40 population members are feasible and non-dominated to each other. The solutions also have a good spread over the Pareto-optimal regions. However, as pointed out earlier, when all population members are non-dominated to each other, the algorithm gets stuck and cannot accept any new solutions. In our simulation run, this happened at the 31st generation and no change in the population members is observed after this stage.

7.7   Summary

In most practical search and optimization problems, constraints are evident. Often the constraints are many in numbers and are nonlinear. In this chapter, we have dealt with several multi-objective evolutionary algorithms which have been particularly suggested for handling constraints.

The first algorithm presented here is a usual penalty function approach, where

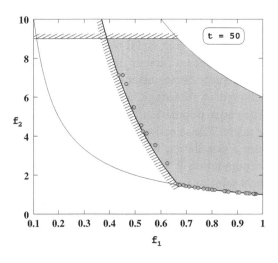

Figure 175   The population at the 50th generation with Ray-Tai-Seow's method and with their crossover operator. The population does not change thereafter.

the constraint violation in an infeasible solution is added to each objective function. Thereafter, the penalized objective function values are optimized. For relatively large penalty terms (compared to objective function values), this method practically compares infeasible solutions based on their constraint violations. Again, for the same reason, a feasible solution will practically dominate an infeasible solution. Both of these characteristics together allow the population members to become feasible from infeasible solutions and, thereafter, allow solutions to converge closer to the true Pareto-optimal solutions.

In Jiménez–Verdegay–Goméz-Skarmeta's constraint handling approach, feasible and infeasible solutions are carefully evaluated by ensuring that no infeasible solution gets a better fitness than any feasible solution. In a tournament played between two feasible or infeasible solutions, comparison is made based upon a niching strategy, thereby ensuring diversity among the obtained solutions.

In the constrained tournament approach discussed in this chapter, the definition of domination is modified. Before comparing two solutions for domination, they are checked for their feasibility. If one solution is feasible and the other is not, the feasible solution dominates the other. If two solutions are infeasible, the solution with the smaller normalized constraint violation dominates the other. On the other hand, if both solutions are feasible, the usual domination principle is applied. Although this definition (we called this the constrain-domination principle) can be used with any other MOEA, its usage has been demonstrated here with the NSGA-II.

In the Ray–Tai–Seow's constraint handling approach, three different non-dominated sorting procedures are used. In addition to a non-dominated sorting of the objective functions, a couple of non-dominated sortings using the constraint violation values and a combined set of objective function and constraint violation values are needed to construct the new population.

A recent study (Deb et al., 2001) has compared the last three constraint handling strategies on a number of test problems. For a brief description of their study, refer to Section 8.4.5 below. In all problems, the constrained tournament approach with the NSGA-II has been able to achieve a better convergence and maintain a better diversity than the other two approaches. However, more comparative studies are needed to establish the superiority of one algorithm over another. Further studies in this direction demand controllable constrained test problems, a number of which are suggested later in Section 8.3. It will also be interesting to investigate the effect of the constrain-domination principle on the performance of other commonly used MOEAs.

Exercise Problems

1. Identify the Pareto-optimal region of the following constrained optimization problem:

$$\text{Minimize} \quad f_1(\mathbf{x}) = x_1^2 + x_2^2,$$
$$\text{Maximize} \quad f_2(\mathbf{x}) = (x_1 - x_2)^2,$$
$$\text{subject to} \quad x_1 x_2 + 0.25 \geq 0,$$
$$-1 \leq x_1, x_2 \leq 1.$$

2. Calculate the fitness of the following points using the penalty function approach with $R = 10$ for the above problem.

$$\mathbf{x}^{(1)} = (0.5, 0.5)^T, \quad \mathbf{x}^{(2)} = (-0.5, 0.7)^T, \quad \mathbf{x}^{(3)} = (0.7, -0.8)^T,$$
$$\mathbf{x}^{(4)} = (-1.0, 0.0)^T, \quad \mathbf{x}^{(5)} = (0.5, -0.5)^T, \quad \mathbf{x}^{(6)} = (-0.8, 1.0)^T.$$

Sort the above points according to

(a) non-domination principle,
(b) constrain-domination principle.

3. Consider the problem:

$$\text{Minimize} \quad f_1(\mathbf{x}) = (x_1 - 2)^2 + x_2^2,$$
$$\text{Maximize} \quad f_2(\mathbf{x}) = 9x_1 - x_2^2,$$
$$\text{subject to} \quad x_1^2 + x_2^2 \leq 225,$$
$$x_1 - 3x_2 + 10 \leq 0,$$
$$-20 \leq x_1, x_2 \leq 20.$$

Normalize the constraints. Sort the following points in increasing level of constrain-non-domination:

$$\mathbf{x}^{(1)} = (0, 0)^T, \quad \mathbf{x}^{(2)} = (5, 10)^T, \quad \mathbf{x}^{(3)} = (-10, -15)^T,$$
$$\mathbf{x}^{(4)} = (-11, 0)^T, \quad \mathbf{x}^{(5)} = (10, 10)^T, \quad \mathbf{x}^{(6)} = (0, 15)^T.$$

Use Binh and Korn's method for calculating constraint violation and sort the points again.

4. Use Ray-Tai-Seow's constraint-handling method to choose two parents from the six population members given in the above problem. Use random numbers 0.739, 0.235, 0.125, 0.980, 0.681, 0.752, in this sequence.

5. In the problem 3, Jiménez–Verdegay–Goméz-Skarmeta's constraint-handling method is used. Use the following pairs of solutions and determine the winner:

(a) $\mathbf{x}^{(2)}$ and $\mathbf{x}^{(5)}$
(b) $\mathbf{x}^{(3)}$ and $\mathbf{x}^{(3)}$
(c) $\mathbf{x}^{(1)}$ and $\mathbf{x}^{(2)}$

Wherever needed, use the rest of the population members as the comparison set.

8

Salient Issues of Multi-Objective Evolutionary Algorithms

In this chapter, we will describe a number of issues related to design, development and application of multi-objective evolutionary algorithms. Although some of the issues are also pertinent to classical multi-objective optimization algorithms, they are particularly useful for population-based algorithms which intend to find multiple Pareto-optimal solutions in one single simulation run. These include the following:

1. Illustrative representation of non-dominated solutions.

2. Development of performance measures.

3. Test problem design for unconstrained and constrained multi-objective optimization.

4. Comparative studies of different MOEAs.

5. Decision variable versus objective space niching.

6. Preference of a particular region in the Pareto-optimal front.

7. Single-objective constraint handling using multi-objective EAs.

8. Scaling issues of MOEAs in more than two objectives.

9. Design of convergent MOEAs.

10. Controlled elitism in elitist MOEAs.

11. Design of MOEAs for scheduling problems.

Each of the above issues is important in the development of MOEAs. Although some attention has been given to some of the above, further comprehensive studies remain as imminent future research topics in the field of multi-objective evolutionary optimization. We will discuss these in the following sections.

8.1   Illustrative Representation of Non-Dominated Solutions

Multi-objective optimization deals with a multitude of information. In a single-objective optimization, although there may exist multiple decision variables, there is only one objective function. By plotting the objective function of the best solution at a current iteration, the performance of an algorithm can be demonstrated. However, in multi-objective optimization, there exists more than one objective and in most interesting cases they behave in a conflicting manner. Throughout this book, we have mostly considered two objectives and the performance of an algorithm has been shown by illustrating the obtained solutions on a two-dimensional objective space plot. When the number of objective functions is more than two, such an illustration becomes difficult. In this section, we show a number of different ways in which non-dominated solutions can be illustrated in such situations. A good discussion of some of these methods can also be found in the text by Miettinen (1999). Although it is customary to show the non-dominated solutions on an objective space plot, the decision variables of the obtained non-dominated solutions can also be shown by using the following illustration techniques.

8.1.1   Scatter-Plot Matrix Method

First Meisel (1973) and then Cleveland (1994) suggested plotting all $\binom{M}{2}$ pairs of plots among M objective functions. Figure 176 shows a typical example of such a plot with $M = 3$ objective functions. Different alternate solutions can be shown by different symbols or colors. Although $\binom{M}{2}$ or $M(M-1)/2$ plots are enough to show solutions in three pairs of objective spaces, usually all $M(M-1)$ plots are shown for better illustration. With $M = 3$ objectives, there are a total of 3×2 or 6 plots. The arrangement of the sub-plots is important. The diagonal sub-plots mark the axis for the corresponding off-diagonal sub-plots. For example, the sub-plot in position $(1,2)$ has its horizontal axis marked with f_2 and the vertical axis marked with f_1. If a user is not comfortable in viewing a plot with f_1 in the vertical axis, the sub-plot in position $(2,1)$ shows the same plot with f_1 marked in the horizontal axis. In this way, the axes of the sub-plots need not be labeled and the ranges of each axis can be shown only in the sub-plots. Thus, a plot in the (i, j) position of the matrix is identical to the plot in the (j, i) position, except that the plot is mirrored.

In the event of comparing the performance of two algorithms on an identical problem, the lower diagonal matrices can be plotted with non-dominated solutions obtained from one algorithm and the upper-diagonal matrices can be plotted with that of the other algorithm. In this way, the (i, j) position of the first algorithm can be compared with the (j, i) position of the other algorithm. For many objectives (more than five or so), plots where a clear front does not emerge may be omitted.

8.1.2   Value Path Method

This is a popular means of showing the obtained non-dominated solutions for a problem having more than two objectives. This method was suggested by Geoffrion

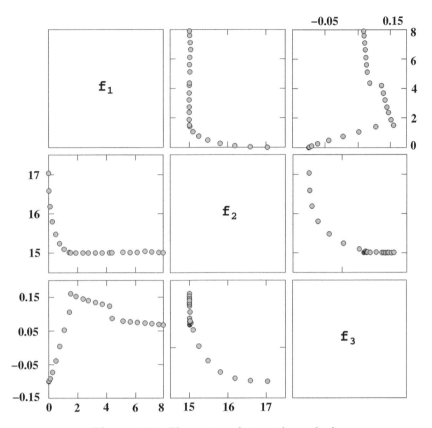

Figure 176   The scatter-plot matrix method.

et al. (1972). Figure 177 shows a typical value path plot representing the same set of non-dominated solutions as shown in the scatter-plot matrix method. The horizontal axis marks the identity of the objective function, and thus, must be ticked only at integers starting from 1. If there are M objectives, there would be M tick-marks on this axis. The vertical axis will mark the normalized objective function values. Two different types of information are plotted on the figure. The vertical shaded bar (can also be shown with range marks) for the objective function k represents the range covering the minimum and maximum values of the k-th objective function in the Pareto-optimal set, and not in the obtained non-dominated set. Although the figure is plotted with the normalized objective values, this is not mandatory. Each cross-line, connecting all three objective bars, as shown in Figure 177, corresponds to a solution from the obtained non-dominated set. Such lines simply join the values of different objective functions depicted by the solution. When all solutions from the obtained non-dominated set are plotted this way, the plot provides a number of types of information:

1. For each objective function, the extreme function values provide a qualitative assessment of the spread of the obtained solutions. An algorithm which spreads

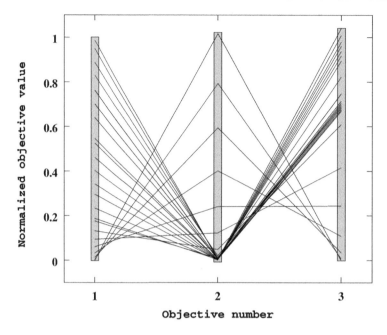

Figure 177 The value path method.

its solutions over the entire shaded bar is considered to be good in finding diverse solutions.

2. The extent to which the cross-lines 'zig-zag' shows the trade-off among the objective functions captured by the obtained non-dominated solutions. An algorithm having a large change of slope between two consecutive objective function bars is considered to be good in terms of finding good trade-off non-dominated solutions.

In order to show the diversity in each objective function, a histogram showing the frequency distribution can also be superimposed on this graph.

8.1.3 Bar Chart Method

Another useful way to represent different non-dominated solutions is to plot the solutions as a bar chart. First, the obtained non-dominated solutions are arranged in a particular order. Thereafter, for each objective function, the function value of each solution in the same order is plotted with a bar. Different colors or shades can be used to mark the bars corresponding to each solution. Since the objectives can take different ranges of values, it is customary to plot a bar chart diagram with the normalized objective values. In this way, if there are N obtained solutions, N different bars are plotted for each objective function. A gap can be provided between two consecutive sets of bars, marking clearly the start of bars for the next objective function. Figure 178 shows a typical bar chart plot. For ease of illustration, we only show three different non-dominated solutions. Since bars are plotted, the diversity in

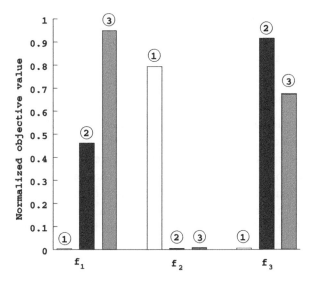

Figure 178   The bar chart method.

different solutions for each objective can be directly observed from the plot. However, if N is large, it becomes difficult to have an idea of the trade-offs among different objective functions captured in the obtained solutions.

8.1.4   Star Coordinate Method

Mañas (1982) suggested a star coordinate system to represent multiple non-dominated solutions (see Figure 179). For M objective functions, a circle is divided into M equal arcs. Each radial line connecting the end of an arc with the center of the circle represents the axis for each objective function. For each line, the center of the circle marks the minimum value of each objective and the circumference marks the maximum objective function value. Since the range of each objective function can be different, it is necessary to label both ends of each line. Normalization of objectives is not necessary here. Once the framework is ready, each solution can be marked on

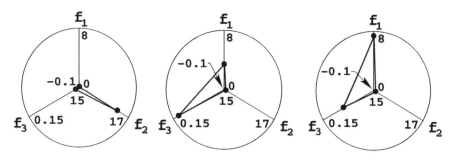

Figure 179   The star coordinate method.

one such a circle. The points marked on each line can be joined with straight lines to form a polygon. Thus, for N solutions, there will be N such circles. Visually, they will convey the convergence and diversity in obtained solutions.

8.1.5 Visual Method

This method is particularly suitable for design related problems. The obtained non-dominated solutions can be shown side-by-side with a tag of objective function values. In this way, a user can evaluate and compare different trade-off solutions. This will also allow the user to have a feel of how solutions change when a particular objective function value is changed. Figure 180 shows nine non-dominated solutions obtained for the shape design of a simply supported beam with a central-point loading (described below in Section 9.6). The figure shows solutions for two objectives: the weight of the beam and the maximum deflection of the beam. Starting from a low-weight beam, the obtained set of solutions shows how beams become more stiffened, incurring less deflection but with more weight (or cost) of the beam.

If possible, a pseudo-weight vector can also be given along with each solution. Since the true minimum and maximum objective function values of the Pareto-optimal set are usually not known, the pseudo-weights can be derived as follows:

$$w_i = \frac{f_i^{\max} - f_i}{f_i^{\max} - f_i^{\min}} \bigg/ \sum_{j=1}^{M} \frac{f_j^{\max} - f_j}{f_j^{\max} - f_j^{\min}}, \tag{8.1}$$

where $f_i^{\min}$ and $f_i^{\max}$ denote the minimum and maximum values of the i-th objective function among the obtained solutions (or Pareto-optimal solutions, if known). This weight vector for each solution provides a relative importance factor for each objective corresponding to the solution. The figure also shows the pseudo-weights calculated for all nine solutions.

Besides the obvious visual appearance, this method has another advantage. The decision variables and the corresponding objective function values (in terms of a pseudo-weight vector) are all shown in one set of plots. This provides a user with a plethora of information, which would hopefully make the decision-making easier.

8.2 Performance Metrics

When a new and innovative methodology is initially discovered for solving a search and optimization problem, a visual description is adequate to demonstrate the working of the proposed methodology. In such pioneering studies, it is important to establish, in the mind of the reader, a picture of the new suggested procedure. However, when the methodology becomes popular and a number of different implementations exist, it becomes necessary to compare these in terms of their performance on various test problems. This has been a common trend in the development of many successful solution methodologies, including multi-objective evolutionary algorithms.

Most earlier MOEAs demonstrated their working by showing the obtained non-

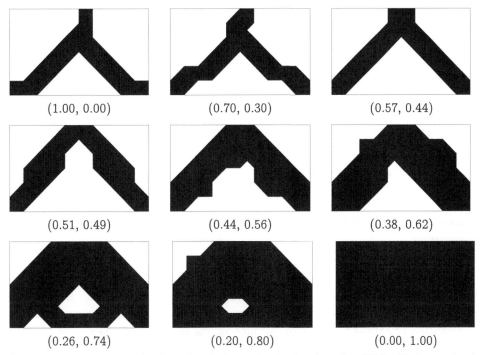

(1.00, 0.00)	(0.70, 0.30)	(0.57, 0.44)
(0.51, 0.49)	(0.44, 0.56)	(0.38, 0.62)
(0.26, 0.74)	(0.20, 0.80)	(0.00, 1.00)

Figure 180   Nine non-dominated solutions obtained using the NSGA-II for a simply supported beam design problem (see page 475). Starting from the top and moving towards the right, solutions are presented as increasing magnitude of weight (one of the objectives). This is a reprint of Figure 13 from Deb and Goel (2001b) (© Springer-Verlag Berlin Heidelberg 2001).

dominated solutions along with the true Pareto-optimal solutions in the objective space. In those studies, the emphasis has been given to demonstrate how closely the obtained solutions have converged to the true Pareto-optimal front. With the existence of many different MOEAs, it is necessary that their performances be quantified on a number of test problems. Before we discuss the performance metrics used in MOEA studies, we would like to highlight that similar to the importance of a choice of performance metric, there is a need to choose appropriate test problems for a comparative study. Test problems might be known for their nature of difficulties, the extent of difficulties and the exact location of the Pareto-optimal solutions (both in the decision variable and in the objective space). We shall discuss this important issue of test problem development for multi-objective optimization in Section 8.3 below.

It is amply mentioned in this book that there are two distinct goals in multi-objective optimization: (i) discover solutions as close to the Pareto-optimal solutions as possible, and (ii) find solutions as diverse as possible in the obtained non-dominated front. In some sense, these two goals are *orthogonal* to each other. The first goal requires a search *towards* the Pareto-optimal region, while the second goal requires a search *along* the Pareto-optimal front, as depicted in Figure 181. In this book, a

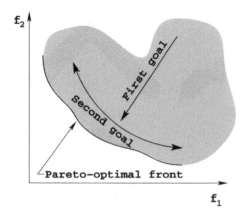

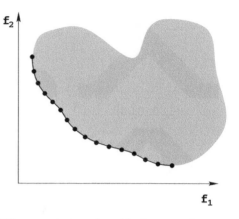

Figure 181 Two goals of multi-objective optimization.

Figure 182 An ideal set of non-dominated solutions.

diverse set of solutions is meant to represent a set of solutions covering the entire Pareto-optimal region uniformly. The measure of diversity can also be separated in two different measures of *extent* (meaning the spread of extreme solutions) and *distribution* (meaning the relative distance among solutions) (Zitzler et al., 2000).

An MOEA will be termed a good MOEA, if both goals are satisfied adequately. Thus, with a good MOEA, a user is expected to find solutions close to the true Pareto-optimal front, as well as solutions that span the entire Pareto-optimal region uniformly. In Figure 182, we show the performance of an ideal MOEA on a hypothetical problem. It is clear that all of the obtained non-dominated solutions lie on the Pareto-optimal front and they also maintain a uniform-like distribution over the entire Pareto-optimal region. However, because of the different types of difficulties associated with a problem or the inherent inefficiencies associated with the chosen algorithm, such a well-converged and well-distributed non-dominated set of solutions may not always be found by an MOEA in solving any arbitrary problem. Take two sets of non-dominated solutions obtained by using two algorithms on an identical problem, as depicted in Figures 183 and 184. With the first algorithm (Algorithm 1), the obtained solutions converge fairly well on the Pareto-optimal front, but clearly there is lack of diversity among them. This algorithm has failed to provide information about the intermediate Pareto-optimal region. On the other hand, the latter algorithm (Algorithm 2) has obtained a good diverse set of solutions, but unfortunately the solutions are not close to the true Pareto-optimal front. Although this latter set of solutions can provide a rough idea of different trade-off solutions, the exact Pareto-optimal solutions are not discovered. With such sets of obtained solutions, it is difficult to conclude which set is better in an absolute sense.

Since convergence to the Pareto-optimal front and the maintenance of a diverse set of solutions are two distinct and somewhat conflicting goals of multi-objective optimization, no single metric can decide the performance of an algorithm in an absolute sense. Algorithm 1 fairs well with respect to the first task of multi-objective

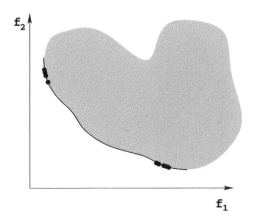 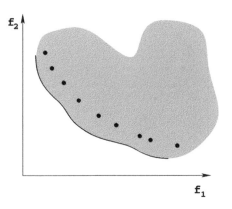

Figure 183 Convergence is good, but distribution is poor (by Algorithm 1).

Figure 184 Convergence is poor, but distribution is good (by Algorithm 2).

optimization, while Algorithm 2 fairs well with respect to the second task. We are again faced here with a two-objective scenario. If we define a metric responsible for finding the closeness of the obtained set of solutions to the Pareto-optimal front and another metric responsible for finding the spread of the solutions, the performance of the above described two algorithms are *non-dominated* to each other. There is a clear need of having at least two performance metrics for adequately evaluating both goals of multi-objective optimization.

Besides the two extreme cases of an ideal convergence with a bad diversity (Algorithm 1 in the previous example) and a bad convergence with an ideal diversity (Algorithm 2), there could be other scenarios. Figure 185 shows that the non-dominated set of solutions obtained by using Algorithm A dominates the non-dominated set of solutions obtained by using Algorithm B. In this case, Algorithm A has clearly performed better than Algorithm B. However, there could be a more confusing scenario (Figure 186) where a part of the solutions obtained by using Algorithm A dominates a part of the solutions obtained by using Algorithm B, and vice versa. This scenario introduces a third dimension of difficulty in designing a performance metric for multi-objective optimization. In this case, both algorithms have similar convergence and diversity properties. The outcome of the comparison will largely depend on the exact definition of the metrics used for these measures.

Based on these discussions, we realize that while comparing two or more algorithms, at least two performance metrics (one evaluating the progress towards the Pareto-optimal front and the other evaluating the spread of solutions) need to be used and the exact definitions of the performance metrics are important. In the following three subsections, we will categorize some of the performance metrics commonly used in the literature. The first type discusses metrics that can be used to measure the progress towards the Pareto-optimal front explicitly. The second type discusses metrics that can be used to measure the diversity among the obtained solutions explicitly. The

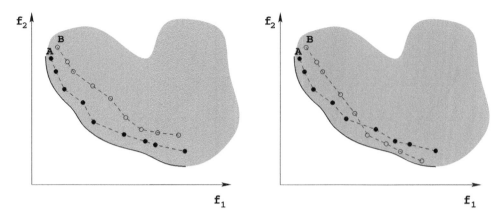

Figure 185 Algorithm A performs better than Algorithm B.

Figure 186 Algorithms A and B are difficult to compare.

third type uses two metrics which measure both goals of multi-objective optimization in an implicit manner.

8.2.1   Metrics Evaluating Closeness to the Pareto-Optimal Front

This metric explicitly computes a measure of the closeness of a set Q of N solutions from a known set of the Pareto-optimal set P*. In some test problems, the set P* may be known as a set of infinite solutions (for example, the case where an equation describing the Pareto-optimal relationship among the decision variables is known) or a set of finite solutions (only a few solutions are known or possible to compute). In order to find the proximity between two sets of different sizes, a number of metrics can be defined. The following metrics are already used for this purpose in different MOEA studies. They provide a good estimate of convergence if a large set for P* is chosen.

Error Ratio

This metric (ER) simply counts the number of solutions of Q which are not members of the Pareto-optimal set P* (Veldhuizen, 1999), or mathematically,

$$ER = \frac{\sum_{i=1}^{|Q|} e_i}{|Q|}, \tag{8.2}$$

where $e_i = 1$ if $i \notin$ P* and $e_i = 0$, otherwise. Figure 187 shows the Pareto-optimal set as filled circles and the obtained non-dominated set of solutions as open squares. In this case, the error ratio ER $= 3/5 = 0.6$, since there are three solutions which are not members of the Pareto-optimal set. Equation (8.2) reveals that a smaller value of ER means a better convergence to the Pareto-optimal front. The metric ER takes a value between zero and one. An ER $= 0$ means all solutions are members of the

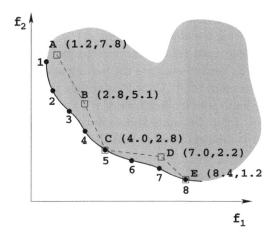

Figure 187   The set of non-dominated solutions Q are shown as open squares, while a set of the chosen Pareto-optimal set P* is shown as filled circles.

Pareto-optimal front P*, and an $ER = 1$ means no solution is a member of the P*.

It is worth mentioning here that although a member of Q is Pareto-optimal, if that solution does not exist in P*, it may be counted in equation (8.2) as a non-Pareto-optimal solution. Thus, it is essential that a large set for P* is used in the above equation. Another drawback of this metric is that if no member of Q is in the Pareto-optimal set, it does not distinguish the relative closeness of any set Q from P*. Because of this discreteness in the values of ER, this metric is not popularly used. The metric can be made more useful, by redefining e_i as follows. For each solution $i \in Q$, if the minimum Euclidean distance (in the objective space) between i and P* is larger than a threshold value δ, the parameter e_i is set to one. By using a suitable value for δ, this modified metric would then represent a measure of the proportion of the solutions close to the Pareto-optimal front.

Set Coverage Metric

A similar metric is suggested by Zitzler (1999). However, the metric can also be used to get an idea of the relative spread of solutions between two sets of solution vectors A and B. The set coverage metric $\mathcal{C}(A, B)$ calculates the proportion of solutions in B, which are weakly dominated by solutions of A:

$$\mathcal{C}(A, B) = \frac{|\{b \in B | \exists a \in A : a \preceq b\}|}{|B|}. \tag{8.3}$$

The metric value $\mathcal{C}(A, B) = 1$ means all members of B are weakly dominated by A. On the other hand, $\mathcal{C}(A, B) = 0$ means that no member of B is weakly dominated by A. Since the domination operator is not a symmetric operator (refer to Section 2.4.3), $\mathcal{C}(A, B)$ is not necessarily equal to $1 - \mathcal{C}(B, A)$. Thus, it is necessary to calculate both $\mathcal{C}(A, B)$ and $\mathcal{C}(B, A)$ to understand how many solutions of A are covered by B and vice versa. It is interesting to note that the cardinality of both vectors A and B need not be the same while using the above equation.

Although Zitzler (1999) used this metric for comparing the performance of two algorithms, it can also be used to evaluate the performance of an algorithm by using $A = P^*$ and $B = Q$. The metric $\mathcal{C}(P^*, Q)$ will determine the proportion of solutions in Q, which are weakly dominated by members of P^*. For the sets P^* and Q shown in Figure 187, $\mathcal{C}(P^*, Q) = 3/5 = 0.6$, since three solutions (A, B and D) are dominated by a member of P^*. It is needless to write that $\mathcal{C}(Q, P^*)$ is always zero.

Generational Distance

Instead of finding whether a solution of Q belongs to the set P^* or not, this metric finds an average distance of the solutions of Q from P^*, as follows (Veldhuizen, 1999):

$$GD = \frac{(\sum_{i=1}^{|Q|} d_i^p)^{1/p}}{|Q|}.$$ (8.4)

For $p = 2$, the parameter d_i is the Euclidean distance (in the objective space) between the solution $i \in Q$ and the *nearest* member of P^*:

$$d_i = \min_{k=1}^{|P^*|} \sqrt{\sum_{m=1}^{M} (f_m^{(i)} - f_m^{*\,(k)})^2},$$ (8.5)

where $f_m^{*\,(k)}$ is the m-th objective function value of the k-th member of P^*. For the obtained solutions shown in Figure 187, solution A is closest to the Pareto-optimal solution 1, solution B is closest to solution 3, solution C is closest to solution 5, solution D is closest to solution 7 and solution E is closest to 8. If the Pareto-optimal solutions have the following objective function values:

Solution	f_1	f_2	Solution	f_1	f_2
1	1.0	7.5	5	4.0	2.8
2	1.1	5.5	6	5.5	2.5
3	2.0	5.0	7	6.8	2.0
4	3.0	4.0	8	8.4	1.2

the corresponding Euclidean distances are as follows:

$$d_{A1} = \sqrt{(1.2 - 1.0)^2 + (7.8 - 7.5)^2} = 0.36,$$

$$d_{B3} = \sqrt{(2.8 - 2.0)^2 + (5.1 - 5.0)^2} = 0.81,$$

$$d_{C5} = \sqrt{(4.0 - 4.0)^2 + (2.8 - 2.8)^2} = 0.00,$$

$$d_{D7} = \sqrt{(7.0 - 6.8)^2 + (2.2 - 2.0)^2} = 0.28,$$

$$d_{E8} = \sqrt{(8.4 - 8.4)^2 + (1.2 - 1.2)^2} = 0.00.$$

Thus, the generational distance (GD) calculated using equation (8.4) with $p = 2$ is GD $= 0.19$. Intuitively, an algorithm having a small value of GD is better.

The difficulty with the above metric is that if there exists a Q for which there is a large fluctuation in the distance values, the metric may not reveal the true distance. In such an event, the calculation of the variance of the metric GD is necessary. Furthermore, if the objective function values are of differing magnitude, they should be normalized before calculating the distance measure. In order to make the distance calculations reliable, a large number of solutions in the P* set is recommended.

Because of its simplicity and average characteristics (with $p = 1$), other researchers have also suggested (Zitzler, 1999) and used this metric (Deb et al., 2000a). The latter investigators have used this metric in a recent comparative study. For each computation of this metric (they called it Υ), the standard deviations of Υ among multiple runs are also reported. If a small value of the standard deviation is observed, the calculated Υ can be accepted with confidence.

Maximum Pareto-Optimal Front Error

This metric (MFE) computes the worst distance d_i among all members of Q (Veldhuizen, 1999). For the example problem shown in Figure 187, the worst distance is caused by the solution B (referring to the calculations above). Thus, MFE $= 0.81$. This measure is a conservative measure of convergence and may provide incorrect information about the distribution of solutions. In this connection, a 'χ' percentile (where $\chi = 25$ or 50) of the distances d_i among all solutions of Q can be used as a metric. We will discuss more about such percentile measures later in this section.

8.2.2   Metrics Evaluating Diversity Among Non-Dominated Solutions

There also exists a number of metrics to find the diversity among obtained non-dominated solutions. In the following, we describe a few of them.

Spacing

Schott (1995) suggested a metric which is calculated with a relative distance measure between consecutive solutions in the obtained non-dominated set, as follows:

$$S = \sqrt{\frac{1}{|Q|} \sum_{i=1}^{|Q|} (d_i - \bar{d})^2}, \qquad (8.6)$$

where $d_i = \min_{k \in Q \wedge k \neq i} \sum_{m=1}^{M} |f_m^i - f_m^k|$ and $\bar{d}$ is the mean value of the above distance measure $\bar{d} = \sum_{i=1}^{|Q|} d_i / |Q|$. The distance measure is the minimum value of the sum of the absolute difference in objective function values between the i-th solution and any other solution in the obtained non-dominated set. Notice that this distance measure is different from the minimum Euclidean distance between two solutions.

The above metric measures the standard deviations of different d_i values. When the solutions are near uniformly spaced, the corresponding distance measure will be small. Thus, an algorithm finding a set of non-dominated solutions having a smaller spacing (S) is better.

For the example problem presented in Figure 187, we show the calculation procedure for d_A:

$$
\begin{aligned}
d_A &= \min\left((1.6+2.7),(2.8+5.0),(5.8+5.6),(7.2+6.6)\right) \\
&= 4.3.
\end{aligned}
$$

Similarly, $d_B = 3.5$, $d_C = 3.5$, $d_D = 2.4$ and $d_E = 2.4$. Figure 187 shows that the solutions A to E are almost uniformly spaced in the objective space. Thus, the standard deviation in the corresponding d_i values would be large. We observe that $\bar{d} = 3.22$ and the metric $S = 0.73$. For a set of non-dominated solutions which are randomly placed in the objective space, the standard deviation measure would be much small.

Conceptually, the above metric provides useful information about the spread of the obtained non-dominated solutions. However, if proper bookkeeping is not used, the implementational complexity is $O(|Q|^2)$. This is because for each solution i, all other solutions must be checked for finding the minimum distance d_i. Although by using the symmetry in distance measures, half the calculations can be avoided, the complexity is still quadratic to the number of obtained non-dominated solutions. Deb et al. (2000a) suggested calculating d_i between consecutive solutions in each objective function independently. The procedure is as follows.

First, sort the obtained non-dominated front in ascending order of magnitude in each objective function. Now, for each solution, sum the difference in objective function values between two nearest neighbors in each objective. For a detailed procedure, refer to Section 6.2 above. Since the sorting has a complexity $O(|Q|\log|Q|)$, this distance metric would be quicker to compute than the above distance metric. Since different objective functions are added together, normalizing the objectives before using equation (8.6) is essential. Moreover, the above metric does not take into account the extent of spread. As long as the spread is uniform within the range of obtained solutions, the metric S produces a small value. The following metric takes care of the extent of spread in the obtained solutions.

Spread

Deb et al. (2000a) suggested the following metric to alleviate the above difficulty:

$$
\Delta = \frac{\sum_{m=1}^{M} d_m^e + \sum_{i=1}^{|Q|} |d_i - \bar{d}|}{\sum_{m=1}^{M} d_m^e + |Q|\bar{d}}, \tag{8.7}
$$

where d_i can be any distance measure between neighboring solutions and $\bar{d}$ is the mean value of these distance measures. The Euclidean distance, the sum of the absolute differences in objective values or the crowding distance (defined earlier on page 248) can be used to calculate d_i. The parameter d_m^e is the distance between the extreme solutions of P^* and Q corresponding to m-th objective function.

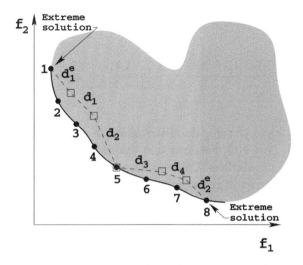

Figure 188   Distances from the extreme solutions.

For a two-objective problem, the corresponding d_1^e and d_2^e are shown in Figure 188 and d_i can be taken as the consecutive Euclidean distance between i-th and $(i+1)$-th solutions. Thus, the term $|Q|$ in the above equation may be replaced by the term $(|Q|-1)$. The metric takes a value zero for an ideal distribution, only when $d_j^e = 0$ and all d_i values are identical to their mean $\overline{d}$. The first condition means that in the obtained non-dominated set of solutions, the true extreme Pareto-optimal solutions exist. The second condition means that the distribution of intermediate solutions is uniform. Such a set is an ideal outcome of any multi-objective EA. Thus, for an ideal distribution of solutions, $\Delta = 0$. Consider another case, where the distribution of the obtained non-dominated solutions is uniform but they are clustered in one place. Such a distribution will make all $|d_i - \overline{d}|$ values zero, but will cause non-zero values for d_m^e. The corresponding Δ becomes $\sum_{m=1}^{M} d_m^e / (\sum_{m=1}^{M} d_m^e + (|Q|-1)\overline{d})$. This quantity lies within $[0, 1]$. Since the denominator measures the length of the piece-wise approximation of the Pareto-front, the Δ value increases with d_m^e. Thus, as the solutions get clustered more and more closer from the ideal distribution, the Δ value increases from zero towards one. For a non-uniform distribution of the non-dominated solutions, the second term in the numerator is not zero and, in turn, makes the Δ value more than that with a non-uniform distribution. Thus, for bad distributions, the Δ values can be more than one as well.

For the example problem in Figure 187, the extreme right Pareto-optimal solution (solution 8) is the same as the extreme non-dominated solution (solution E). Thus, $d_2^e = 0$ here. Since solution 1 (the extreme left Pareto-optimal solution) is not found, we calculate $d_1^e = 0.5$ (using Schott's difference distance measure). We calculate the d_i values accordingly:

$$d_1 = 4.3, \quad d_2 = 3.5,$$
$$d_3 = 3.5, \quad d_4 = 2.4.$$

The average of these values is $\overline{d} = 3.43$. We can now use these values to calculate the spread metric:

$$\Delta = \frac{0.5 + 0.0 + |4.3 - 3.43| + |3.5 - 3.43| + |3.5 - 3.43| + |2.4 - 3.43|}{0.5 + 0.0 + 4 \times 3.43},$$

$$= 0.18.$$

Since this value is close to zero, the distribution is not bad. If solution A were the same as solution 1, $\Delta = 0.15$, meaning that the distribution would have been better than the current set of solutions. Thus, an algorithm finding a smaller Δ value is able to find a better diverse set of non-dominated solutions.

Maximum Spread

Zitzler (1999) defined a metric measuring the length of the diagonal of a hyperbox formed by the extreme function values observed in the non-dominated set:

$$D = \sqrt{\sum_{m=1}^{M} \left(\max_{i=1}^{|Q|} f_m^i - \min_{i=1}^{|Q|} f_m^i \right)^2}. \tag{8.8}$$

For two-objective problems, this metric refers to the Euclidean distance between the two extreme solutions in the objective space, as shown in Figure 189.

In order to have a normalized version of the above metric, it can be modified as follows:

$$\overline{D} = \sqrt{\frac{1}{M} \sum_{m=1}^{M} \left(\frac{\max_{i=1}^{|Q|} f_m^i - \min_{i=1}^{|Q|} f_m^i}{F_m^{max} - F_m^{min}} \right)^2}. \tag{8.9}$$

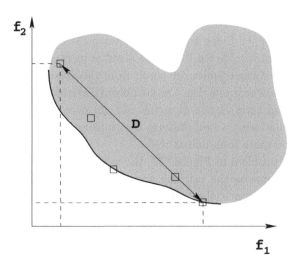

Figure 189   The maximum spread does not reveal true distribution of solutions.

Here, F_m^{max} and F_m^{min} are the maximum and minimum value of the m-th objective in the chosen set of Pareto-optimal solutions, P^*. In this way, if the above metric is one, a widely spread set of solutions is obtained. However, neither $\overline{D}$ nor D can evaluate the exact distribution of intermediate solutions.

Chi-Square-Like Deviation Measure

We proposed this metric for multi-modal function optimization and later used it to evaluate the distributing ability of multi-objective optimization algorithms (Deb, 1989). A brief description of this measure is also presented on page 158. In this metric, a neighborhood parameter ϵ is used to count the number of solutions, n_i, within each chosen Pareto-optimal solution (solution $i \in P^*$). The distance calculation can be made in either the objective space or in the decision variable space. The deviation between this counted set of numbers with an ideal set is measured in the chi-square sense:

$$\iota = \sqrt{\sum_{i=1}^{|P^*|+1} \left(\frac{n_i - \overline{n}_i}{\sigma_i}\right)^2}, \tag{8.10}$$

Since there is no preference to any particular Pareto-optimal solution, it is customary to choose a uniform distribution as an ideal distribution. This means that there should be $\overline{n}_i = |Q|/|P^*|$ solutions allocated in the niche of each chosen Pareto-optimal solution. The parameter $\sigma_i^2 = \overline{n}_i(1-\overline{n}_i/|Q|)$ is suggested for $i = 1, 2, \ldots, |P^*|$. However, the index $i = |P^*|+1$ represents all solutions which do not reside in the ϵ-neighborhood of any of the chosen Pareto-optimal solutions. For this index, the ideal number of solutions and its variance are calculated as follows:

$$\overline{n}_{|P^*|+1} = 0, \quad \sigma_{|P^*|+1}^2 = \sum_{i=1}^{|P^*|} \sigma_i^2 = |Q| \left(1 - \frac{1}{|P^*|}\right).$$

Let us illustrate the calculation procedure for the scenario depicted in Figure 190. There are five chosen Pareto-optimal solutions, marked as 1 to 5. The obtained non-dominated set of $|Q| = 10$ solutions are marked as squares. Thus, the expected number of solutions near each of the five Pareto-optimal solutions ($|P^*| = 5$) is $\overline{n} = 10/5$ or 2 and $\overline{n}_6 = 0$ (the expected number of solutions away from these five Pareto-optimal solutions). The corresponding standard deviations can be calculated by using these numbers: $\sigma_i^2 = 1.6$ for $i = 1$ to 5 and $\sigma_6^2 = 5 \times 1.6 = 8.0$. Now, we count the actual number of non-dominated solutions present near each Pareto-optimal solution (the neighborhood is shown by dashed lines). We observe from the figure that

$$n_1 = 1, \quad n_2 = 3, \quad n_3 = 1, \quad n_4 = 2, \quad n_5 = 1, \quad n_6 = 2.$$

With these values, the metric is:

$$\iota = \sqrt{\left(\frac{1-2}{\sqrt{1.6}}\right)^2 + \left(\frac{3-2}{\sqrt{1.6}}\right)^2 + \left(\frac{1-2}{\sqrt{1.6}}\right)^2 + \left(\frac{2-2}{\sqrt{1.6}}\right)^2 + \left(\frac{1-2}{\sqrt{1.6}}\right)^2 + \left(\frac{2-0}{\sqrt{8.0}}\right)^2},$$
$$= 1.73.$$

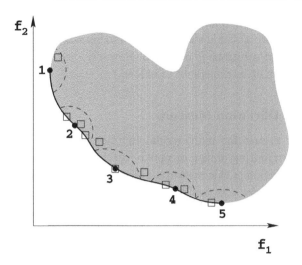

Figure 190   Non-dominated solutions in each niche of the five Pareto-optimal solutions.

Since the above metric is a deviation measure, an algorithm finding a smaller value of this metric is better able to distribute its solutions near the Pareto-optimal region. If the ideal distribution is found, this metric will have a value zero. If most solutions are away from the Pareto-optimal solutions, this metric cannot evaluate the spread of solutions adequately.

8.2.3   Metrics Evaluating Closeness and Diversity

There exist some metrics where both tasks have been evaluated in a combined sense. Such a metric can only provide a qualitative measure of convergence as well as diversity. Nevertheless, they can be used along with one of the above metrics to get a better overall evaluation.

Hypervolume

This metric calculates the volume (in the objective space) covered by members of Q (the region shown hatched in Figure 191) for problems where all objectives are to be minimized (Veldhuizen, 1999; Zitzler and Thiele, 1998b). Mathematically, for each solution $i \in Q$, a hypercube v_i is constructed with a reference point W and the solution i as the diagonal corners of the hypercube. The reference point can simply be found by constructing a vector of worst objective function values. Thereafter, a union of all hypercubes is found and its hypervolume (HV) is calculated:

$$\text{HV} = \text{volume}\left(\cup_{i=1}^{|Q|} v_i\right). \tag{8.11}$$

Figure 191 shows the chosen reference point W. The hypervolume is shown as a hatched region. Obviously, an algorithm with a large value of HV is desirable. For

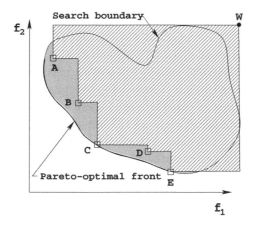

Figure 191   The hypervolume enclosed by the non-dominated solutions.

the example problem shown in Figure 187, the hypervolume HV is calculated with $W = (11.0, 10.0)^\mathsf{T}$ as:

$$
\begin{aligned}
\mathrm{HV} \; = \;\; & (11.0 - 8.4) \times (10.0 - 1.2) + (8.4 - 7.0) \times (10.0 - 2.2) \\
& + (7.0 - 4.0) \times (10.0 - 2.8) + (4.0 - 2.8) \times (10.0 - 5.1) \\
& + (2.8 - 1.2) \times (10.0 - 7.8) \\
= \;\; & 64.80.
\end{aligned}
$$

This metric is not free from arbitrary scaling of objectives. For example, if the first objective function takes values an order of magnitude more than that of the second objective, a unit improvement in f_1 would reduce HV much more than that a unit improvement in f_2. Thus, this metric will favor a set Q which has a better converged solution set for the least-scaled objective function. To eliminate this difficulty, the above metric can be evaluated by using normalized objective function values.

Another way to eliminate the bias to some extent and to be able to calculate a normalized value of this metric is to use the metric HVR which is the ratio of the HV of Q and of P*, as follows (Veldhuizen, 1999):

$$
\mathrm{HVR} = \frac{\mathrm{HV}(Q)}{\mathrm{HV}(P^*)}. \tag{8.12}
$$

For a problem where all objectives are to be minimized, the best (maximum) value of the HVR is one (when $Q = P^*$). For the Pareto-optimal solutions shown in Figure 187, $\mathrm{HV}(P^*) = 71.53$. Thus, $\mathrm{HVR} = 64.80/71.53 = 0.91$. Since this value is close to one, the obtained set is near the Pareto-optimal set. It is clear that values of both HV and HVR metrics depend on the chosen reference point W.

Attainment Surface Based Statistical Metric

Fonseca and Fleming (1996) suggested the concept of an attainment surface in the context of multi-objective optimization. In many studies, the obtained non-dominated solutions are usually shown by joining them with a curve. Although such a curve provides a better illustration of a front, there is no guarantee that any intermediate solution lying on the front is feasible, nor is there any guarantee that intermediate solutions are Pareto-optimal. Fonseca and Fleming argued that instead of joining the obtained non-dominated solutions by a curve, an envelope can be formed marking all those solutions in the search space which are sure to be dominated by the set of obtained non-dominated solutions. Figure 192 shows this envelope for a set of non-dominated solutions. The generated envelope is called an *attainment surface* and is identical to the surface used to calculate the hypervolume (discussed above). Like the hypervolume metric, an attainment surface also signifies a combination of both convergence and diversity of the obtained solutions.

The metric derived from the concept of attainment surface is a practical one. In practice, an MOEA will be run multiple times, each time starting the MOEA from a different initial population or parameter setting. Once all runs are over, the obtained non-dominated solutions can be used to find an attainment surface for each run. Figure 193 shows the non-dominated solutions for three different MOEA runs for the same problem with the same algorithm, but with different initial populations. With a stochastic search algorithm, such as an evolutionary algorithm, it is expected that variations in its performance over multiple runs may result. Thus, such a plot does not provide a clear idea of the true non-dominated front. When two or more algorithms are to be compared, the cluttering of the solutions near the Pareto-optimal front may not provide a clear idea about which algorithm performed better. Figure 194 shows the corresponding attainment surfaces, which can be used to define a metric for a reliable

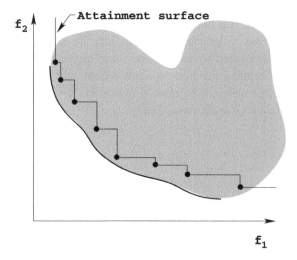

Figure 192   The attainment surface is created for a number of non-dominated solutions.

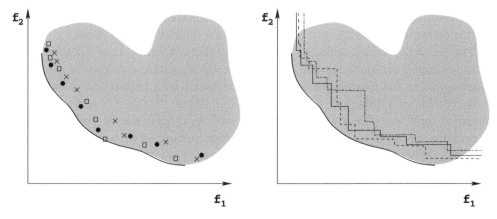

Figure 193 Non-dominated solutions obtained using three different runs of an MOEA.

Figure 194 Corresponding attainment surfaces provide a clear understanding of the obtained fronts.

comparison of two or more algorithms or for a clear understanding of the obtained non-dominated front.

First, a number of diagonal imaginary lines, running in the direction of the improvement in all objectives, are chosen. For each line, the intersecting points of all attainment surfaces for an algorithm are calculated. These points will lie on the chosen line and, thus will follow a frequency distribution. Using these points, a number of statistics, such as 25, 50 or 75% attainment surfaces, can be derived. Figure 195 shows an arbitrary cross-line AB and the corresponding intersecting points by the three attainment surfaces. The frequency distribution along the cross-line for a large

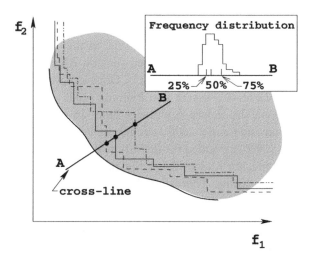

Figure 195 Intersection points on a typical cross-line. A frequency distribution or a histogram can be created from these points.

number of hypothetical attainment surfaces is also marked on the cross-line AB. The points corresponding to 25, 50 or 75% attainment surfaces are also shown.

The frequency distribution and different attainment surfaces for another set of non-dominated solutions obtained using a second algorithm can also be computed likewise. Once the frequency distributions are found on a chosen line, a statistical test (the Mann–Whitney U test (Knowles and Corne, 2000), or the Kolmogorov–Smirnov type test (Fonesca and Fleming, 1996), or others) can be performed with a confidence level β to conclude which algorithm performed better along that line. The same procedure can be repeated for other cross-lines (at different locations in the trade-off regions with different slopes). On every line, one of three decisions can be made with the chosen confidence level: (i) Algorithm A is better than Algorithm B, (ii) Algorithm B is better than Algorithm A, or (iii) no conclusion can be made with the chosen confidence level. With a total of L lines chosen, Knowles and Corne (2000) suggested a metric $[a, b]$ with a confidence limit β, where a is the percentage of times that Algorithm A was found better and b is the percentage of times that Algorithm B was found better. Thus, $100 - (a + b)$ gives the percentage of cases the results were statistically inconclusive. Thus, if two algorithms have performed equally well or equally bad, the percentage values a and b will be small, such as $[2.8, 3.2]$, meaning that, in 94% of the region found by two algorithms no algorithm is better than the other. However, if the metric returns values such as $[98.0, 1.8]$, it can be said that Algorithm A has performed better than Algorithm B. Noting the lines where each algorithm performed better or where the results were inconclusive, the outcome of this statistical metric can also be shown visually on the objective space. Figure 196 shows the regions (by continuous lines) where Algorithm A performed better and regions where the algorithm B performed better. The figure also shows the regions where inconclusive results were found (marked with dashed curves). The results on such a plot can be shown on the *grand* 50% attainment surface, which may be computed as the 50% attainment surface generated by using the combined set of points of two algorithms along any cross-line. In the absence of any preference to any objective function, it can be concluded that an Algorithm A has performed better than Algorithm B if $a > b$, and vice versa. Of course, the outcome will depends on the chosen confidence level. Fonesca and Fleming (1996) nicely showed how such a comparison depends on the chosen confidence limit. In most studies (Fonesca and Fleming, 1996; Knowles and Corne, 2000), confidence levels of 95 and 99% were used.

Knowles and Corne (2000) extended the definition of the above metric for comparing more than two algorithms. For K algorithms, the above procedure can be repeated for $\binom{K}{2}$ distinct pairs of algorithms. Thereafter, for each algorithm k the following two counts can be made:

1. The percentage (a_k) of the region where one can be statistically confident with the chosen confidence level that the algorithm K was not *beaten* by any other algorithm.
2. The percentage (b_k) of the region where one can be statistically confident with the chosen confidence level that the algorithm *beats* all other $(K - 1)$ algorithms.

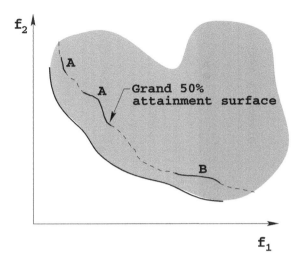

Figure 196   The regions where either Algorithm A or B performed statistically better are shown by continuous lines, while the regions where no conclusion can be made are marked with dashed lines.

It is interesting to note that for every algorithm, $a_k \geq b_k$, since the event described in item 2 is included in item 1 above. An algorithm with large values of a_k and b_k is considered to be good. Two or more algorithms having more or less the same values of a_k should be judged by the b_k value. In such an event, the algorithm having a large value of b_k is better.

Weighted Metric

A simple procedure to evaluate both goals would be to define a weighted metric of combining one of the convergence metrics and one of the diversity measuring metric together, as follows:

$$W = w_1 \, \text{GD} + w_2 \Delta, \tag{8.13}$$

with $w_1 + w_2 = 1$. Here, we have combined the generational distance (GD) metric for evaluating the converging ability and Δ to measure the diversity-preserving ability of an algorithm. We have seen in the previous two subsections that the GD takes a small value for a good converging algorithm and Δ takes a small value for a good diversity-preserving algorithm. Thus, an algorithm having an overall small value of W means that the algorithm is good in both aspects. The user can choose appropriate weights (w_1 and w_2) for combining the two metrics. However, if this metric is to be used, it is better that a normalized pair of metrics is employed.

Non-Dominated Evaluation Metric

Since both metrics evaluate two conflicting goals, it is ideal to pose the evaluation of MOEAs as a two-objective evaluation problem. If the metric values for one algorithm

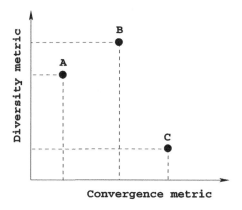

Figure 197 Algorithms A and C produce a non-dominated outcome.

dominate that of the other algorithm, then the former is undoubtedly better than the latter. Otherwise, no affirmative conclusion can be made about the two algorithms. Figure 197 shows the performance of three algorithms on a hypothetical problem. Clearly, Algorithm A dominates Algorithm B, but it cannot be said which is better between Algorithms A and C.

A number of other metrics and guidelines for comparing two non-dominated sets are discussed elsewhere (Hansen and Jaskiewicz, 1998).

8.3   Test Problem Design

In multi-objective evolutionary computation, researchers have used many different test problems with known sets of Pareto-optimal solutions. Veldhuizen (1999) in his doctoral thesis outlined many such problems. Here, we present a number of such test problems which are commonly used. Later, we argue that most of these test problems are not tunable and it is difficult to establish what feature of an algorithm has been tested by these problems. Based on these arguments, we shall present a systematic procedure of designing test problems for unconstrained and constrained multi-objective evolutionary optimization. Moreover, although many test problems were used in earlier studies, the exact locations of the Pareto-optimal solutions were not clearly shown. Here, we make an attempt to identify their exact location using the optimality conditions described in Section 2.5.

Although simple, the most studied single-variable test problem is Schaffer's two-objective problem (Schaffer, 1984):

$$\text{SCH1}: \begin{cases} \text{Minimize} & f_1(x) = x^2, \\ \text{Minimize} & f_2(x) = (x - 2)^2, \\ & -A \le x \le A. \end{cases} \qquad (8.14)$$

This problem has Pareto-optimal solutions $x^* \in [0, 2]$ and the Pareto-optimal set is a convex set:

$$f_2^* = (\sqrt{f_1^*} - 2)^2.$$

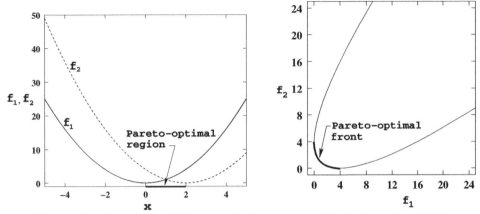

Figure 198 Decision variable and objective space in Schaffer's function SCH1.

in the range $0 \leq f_1^* \leq 4$. Figure 198 shows the objective space for this problem and the Pareto-optimal front. Different values of the bound-parameter A are used in different studies. Values as low as $A = 10$ to values as high as $A = 10^5$ have been used. As the value of A increases, the difficulty in approaching towards the Pareto-optimal front is enhanced.

Schaffer's second function, SCH2, is also used in many studies (Schaffer, 1984):

$$\text{SCH2:} \begin{cases} \text{Minimize} \quad f_1(x) = \begin{cases} -x & \text{if } x \leq 1, \\ x - 2 & \text{if } 1 < x \leq 3, \\ 4 - x & \text{if } 3 < x \leq 4, \\ x - 4 & \text{if } x > 4, \end{cases} \\ \text{Minimize} \quad f_2(x) = (x - 5)^2, \\ \quad -5 \leq x \leq 10. \end{cases} \quad (8.15)$$

The Pareto-optimal set consists of two discontinuous regions: $x^* \in \{[1, 2] \cup [4, 5]\}$. Figure 199 shows both functions and the objective space. The Pareto-optimal regions are shown by bold curves. Note that although the region BC produces a conflicting scenario with the two objectives (as f_1 increases, f_2 decreases, and vice versa), the scenario for DE is better than that for BC. Thus, the region BC does not belong to the Pareto-optimal region. The corresponding Pareto-optimal regions (AB and DE) are also shown in the objective space. The main difficulty an algorithm may face in solving this problem is that a stable subpopulation on each of the two disconnected Pareto-optimal regions may be difficult to maintain.

Fonseca and Fleming (1995) used a two-objective optimization problem having n variables:

$$\text{FON:} \begin{cases} \text{Minimize} \quad f_1(\mathbf{x}) = 1 - \exp\left(-\sum_{i=1}^{n}(x_i - \frac{1}{\sqrt{n}})^2\right), \\ \text{Minimize} \quad f_2(\mathbf{x}) = 1 - \exp\left(-\sum_{i=1}^{n}(x_i + \frac{1}{\sqrt{n}})^2\right), \\ \quad -4 \leq x_i \leq 4 \quad i = 1, 2, \ldots, n. \end{cases} \quad (8.16)$$

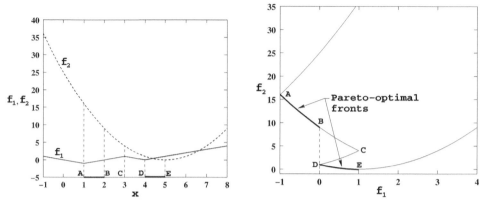

Figure 199 Decision variable and objective space in Schaffer's function SCH2.

The Pareto-optimal solution to this problem is $x_i^* \in [-1/\sqrt{n}, 1/\sqrt{n}]$ for $i = 1, 2, \ldots, n$. These solutions also satisfy the following relationship between the two function values:

$$f_2^* = 1 - \exp\left\{ -\left[2 - \sqrt{-\ln(1 - f_1^*)}\right]^2 \right\}$$

in the range $0 \le f_1^* \le 1 - \exp(-4)$. The interesting aspect is that the search space in the objective space and the Pareto-optimal function values do not depend on the dimensionality (the parameter n) of the problem. Figure 200 shows the objective space for $n = 10$. Solution A corresponds to $x_i^* = 1/\sqrt{n}$ for all $i = 1, 2, \ldots, n$, and solution B corresponds to $x_i^* = -1/\sqrt{n}$ for all $i = 1, 2, \ldots, n$. Another aspect of this problem is that the Pareto-optimal set is a nonconvex set. Thus, weighted sum approach will have difficulty in finding a diverse set of Pareto-optimal solutions.

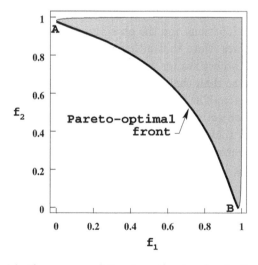

Figure 200 The objective space and Pareto-optimal region in the test problem FON.

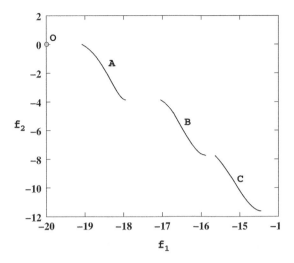

Figure 201   The objective space in the test problem KUR.

Kursawe (1990) used a two-objective optimization problem which is more complicated[1]:

$$
\text{KUR}: \begin{cases}
\text{Minimize} & f_1(x) = \sum_{i=1}^{2} \left[-10 \exp(-0.2\sqrt{x_i^2 + x_{i+1}^2}) \right], \\
\text{Minimize} & f_2(x) = \sum_{i=1}^{3} \left[|x_i|^{0.8} + 5 \sin(x_i^3) \right], \\
& -5 \leq x_i \leq 5, \quad i = 1, 2, 3.
\end{cases} \tag{8.17}
$$

The Pareto-optimal set is nonconvex as well as disconnected in this case. Figure 201 shows the objective space and the Pareto-optimal region for this problem. There are three distinct disconnected Pareto-optimal regions. The decision variable values corresponding to the true Pareto-optimal solutions are difficult to know here. Solution O is a Pareto-optimal solution having $x_i^* = 0$ for all $i = 1, 2, 3$. Some Pareto-optimal solutions (region A in Figure 201) correspond to $x_1^* = x_2^* = 0$, and some solutions (regions B and C) correspond to $x_1^* = x_3^*$. For these latter solutions, there are effectively two independent decision variables and we can use equation (2.9) earlier to arrive at the necessary condition for Pareto-optimality (assuming all x_i take negative values):

$$
2x_2^* \left[15(x_1^*)^2 \cos(x_1^*)^3 - 0.8(-x_1^*)^{-0.2} \right] = x_1^* \left[15(x_2^*)^2 \cos(x_2^*)^3 - 0.8(-x_2^*)^{-0.2} \right] \tag{8.18}
$$

A part of the region B is constituted for $x_2^* = 0$. Substituting $x_2^* = 0$ in the above equations, and using equation (2.9) with x_1 and x_3 variables, we obtain the necessary

---

[1] The original problem used the term $\sin^3(x_i)$, instead of $\sin(x_i^3)$, in $f_2(x)$. Moreover, the problem was defined for any number of decision variables and with no variable bounds (Kursawe, 1990). To simplify our discussion here, we follow the three-variable modified version used elsewhere (Veldhuizen, 1999).

condition for Pareto-optimality:

$$\exp(-0.2x_3^*)\left[15(x_1^*)^2\cos(x_1^*)^3 - 0.8(-x_1^*)^{-0.2}\right] =$$
$$\exp(-0.2x_1^*)\left[15(x_3^*)^2\cos(x_3^*)^3 - 0.8(-x_3^*)^{-0.2}\right]. \tag{8.19}$$

Figure 202 shows the decision variable values corresponding to Pareto-optimal solution O and Pareto-optimal regions A, B and C. The Pareto-optimal solutions for the overall three-dimensional decision variable space and for the three individual pair-wise decision variable spaces are shown. It is clear that they are constituted with disconnected set of solutions in the decision variable space. It is also interesting to note that the region B is constituted with a disconnected set of decision variables, although they form a continuous set of solutions in the objective space (Figure 201). Because of the discrete nature of the Pareto-optimal region, optimization algorithms may have difficulty in finding Pareto-optimal solutions in all regions. The difficulty in knowing the true Pareto-optimal front forced past studies to approximate the true Pareto-optimal front with a few experimentally found solutions (Veldhuizen, 1999).

Poloni et al. (2000) used the following two variable, two-objective problem which

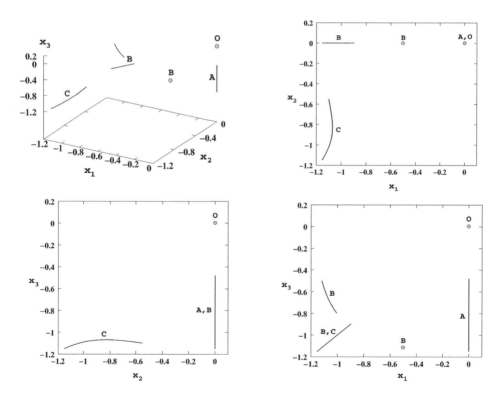

Figure 202 Pareto-optimal solutions in the decision variable space for the test problem KUR.

has been used by many researchers subsequently:

$$\text{POL}: \begin{cases} \text{Minimize} & f_1(x) = \left[1 + (A_1 - B_1)^2 + (A_2 - B_2)^2\right], \\ \text{Minimize} & f_2(x) = \left[(x_1 + 3)^2 + (x_2 + 1)^2\right], \\ \text{where} & A_1 = 0.5\sin 1 - 2\cos 1 + \sin 2 - 1.5\cos 2, \\ & A_2 = 1.5\sin 1 - \cos 1 + 2\sin 2 - 0.5\cos 2, \\ & B_1 = 0.5\sin x_1 - 2\cos x_1 + \sin x_2 - 1.5\cos x_2, \\ & B_2 = 1.5\sin x_1 - \cos x_1 + 2\sin x_2 - 0.5\cos x_2, \\ & -\pi \le (x_1, x_2) \le \pi. \end{cases} \qquad (8.20)$$

This function has a nonconvex and disconnected Pareto-optimal set, as shown in Figure 203. The true Pareto-optimal set of solutions is difficult to know in this problem. Figures 203 and 204 show that Pareto-optimal solutions are discontinuous in the objective as well as in the decision variable space. Like other problems having disconnected Pareto-optimal sets, this problem may also cause difficulty to many multi-objective optimization algorithms. However, it is interesting to note from Figure 204 that most parts (region A) of the Pareto-optimal region are constituted by the boundary solutions of the search space. If the lower bound on x_1 is relaxed, the convex Pareto-optimal front A gets wider and the Pareto-optimal front B vanishes. Thus, the existence of the Pareto-optimal region B is purely because of the setting up of the lower bound of x_1 to $-\pi$.

Viennet (1996) used two three-objective optimization problems. We present one of these in the following:

$$\text{VNT}: \begin{cases} \text{Minimize} & f_1(x) = 0.5(x_1^2 + x_2^2) + \sin(x_1^2 + x_2^2), \\ \text{Minimize} & f_2(x) = (3x_1 - 2x_2 + 4)^2/8 + (x_1 - x_2 + 1)^2/27 + 15, \\ \text{Minimize} & f_3(x) = \dfrac{1}{x_1^2 + x_2^2 + 1} - 1.1\exp\left[-(x_1^2 + x_2^2)\right], \\ & -3 \le (x_1, x_2) \le 3. \end{cases} \qquad (8.21)$$

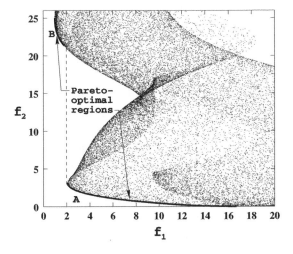

Figure 203   Pareto-optimal sets in the objective space for the test problem POL.

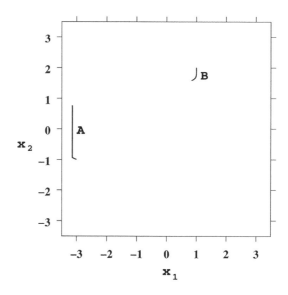

Figure 204 Pareto-optimal solutions in the decision variable space for the test problem POL.

This problem has two variables. The true location of the Pareto-optimal set can be found by realizing the parametric relationship between objectives 1 and 3:

$$f_1(r) = 0.5r + \sin r,$$
$$f_3(r) = \frac{1}{1+r} - 1.1\exp(-r).$$

where $r = x_1^2 + x_2^2$ and $0 \leq r \leq 18$. For each value of r, the minimum of the second objective function may be a candidate solution for the true Pareto-optimal set. The resulting Pareto-optimal solutions are shown in Figure 205. The decision variable region for the Pareto-optimal front is also shown in Figure 206. These plots show the exact location of the Pareto-optimal solutions. It is important to note that all studies in the past found inexact locations of the true Pareto-optimal region (Veldhuizen, 1999; Viennet, 1996). However, here we show their exact locations.

Although researchers have used a number of other test problems, the fundamental problem with all of these is that the difficulty caused by such problems cannot be controlled. In most problems, neither the dimensionality can be changed, nor the associated complexity (such as nonconvexity, the extent of discreteness of the Pareto-optimal region, etc.) can be changed in a simple manner. Moreover, some of the problems described above demonstrate how difficult it is to find the true nature of the Pareto-optimal front. It seems many researchers have developed such test problems without putting much thought into the outcome of the Pareto-optimal front. After the objective functions were developed, it was discovered that for some problems the resulting Pareto-optimal front is a union of a number of disconnected regions or a nonconvex region. These functions do not offer a way to make the problems more

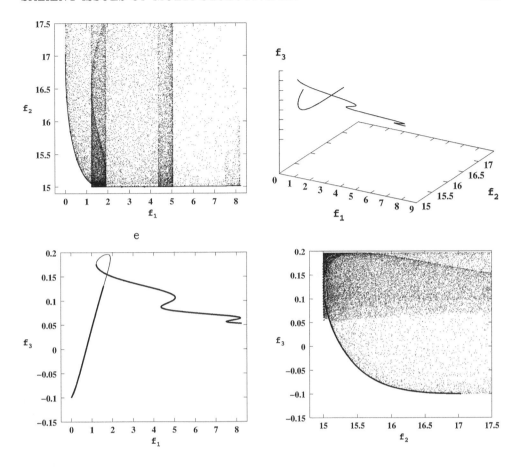

Figure 205 Pareto-optimal solutions (marked by bold curves) in the objective space for the test problem VNT.

difficult or more easy to solve.

The design of test problems is always important in evaluating the merit of any new algorithm. This is also true for multi-objective optimization algorithms. To really test the efficacy of any algorithm, it must be tried on *difficult* problems. This is because almost any algorithm will probably solve an easy problem, but the real advantage of one algorithm over another will show up in trying to solve difficult problems. For example, let us consider the scenario for a single-objective optimization algorithm. If we use a test problem which is easy (say a single-variable quadratic function), many algorithms, including a steepest-descent method (Deb, 1995), will do equally well in trying to find the minimum of the function. Does this mean that the steepest-descent method is as suitable as another non-gradient method, such as an evolutionary algorithm, in solving any problem? The answer to this question is a simple 'no'. It is well known that in solving multi-modal problems, the steepest-descent method is destined to proceed towards an optimum, in whose basin of attraction the

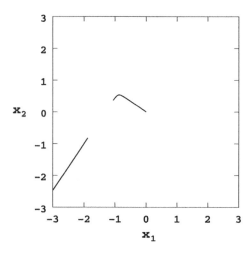

Figure 206   The decision variable space in the test problem VNT.

chosen initial solution lies. However, one can also argue that an algorithm 'A', which can solve difficult problems better than another algorithm 'B', need not solve easy problems better than 'B'. However, over a large sample of problems and algorithms, this argument is weak compared to the following counter-argument – an algorithm 'A', which can solve easy problems better than another algorithm 'B', need not solve difficult problems better than 'B'. Moreover, since most real-world problems are nonlinear, multi-modal and stochastic, we should prefer algorithms which are better in solving difficult problems.

Although the motivation for solving difficult problems is clear, the difficulty arises in defining which features make a problem difficult? The context of problem difficulty naturally depends on the nature of problems that the underlying algorithms are trying to solve. In the context of solving multi-objective optimization problems, we are interested in knowing the features that make a problem difficult for a multi-objective optimization algorithm. Fortunately, there are two tasks that a multi-objective optimization algorithm must do well:

1. Converge as close to the true Pareto-optimal region as possible.
2. Maintain as many widely spread non-dominated solutions as possible.

Keeping in mind these two tasks, we can now design a problem where each of the above tasks would be difficult to achieve. For example, by constructing a test problem with many local Pareto-optimal regions emerging on the way towards the global Pareto-optimal front, an optimization algorithm can be offered trouble in converging to the true Pareto-optimal front. Again, by constructing another test problem where the density of solutions is non-uniform along the Pareto-optimal region, an algorithm can be put into a test for finding a uniform spread of Pareto-optimal solutions. It is intuitive that a more difficult problem can be constructed by having both of the above features present in the test problem. In the following, we first suggest different difficulties that a multi-objective optimization algorithm may face and then construct test problems featuring these difficulties.

8.3.1 Difficulties in Converging to the Pareto-Optimal Front

A multi-objective optimization algorithm may face difficulty in converging to the true (or global) Pareto-optimal front because of various features that may be present in a problem (Deb, 1999c), including the following:

1. multi-modality;
2. deception;
3. isolated optimum;
4. collateral noise.

All of the above features are known to cause difficulty in single-objective EAs (Deb et al., 1993) and when present in a multi-objective problem may also cause difficulty to an MOEA. In tackling a multi-objective problem having multiple Pareto-optimal fronts, an MOEA, like many other search and optimization methods, may converge to a local Pareto-optimal front. Despite some criticism (Grefenstette, 1993), deception, if present in a problem, has been shown to cause GAs to get misled towards deceptive attractors (Deb et al., 1993; Goldberg et al., 1989). There may exist some problems where the optimum is surrounded by a fairly flat search space. Since there is no useful information that most of the search space can provide, no optimization algorithm will perform better than an exhaustive search method to find the optimum in these problems. Multi-objective optimization methods are also no exception to facing difficulty in solving such a problem. Collateral noise comes from the improper evaluation of low-order building blocks (partial solutions which may lead towards the true optimum) due to the excessive noise that may come from other part of the solution vector. These problems are usually 'rugged' with relatively large variation in the function landscapes. Multi-objective problems having such 'rugged' functions may also cause difficulties to MOEAs, if adequate population size (see Section 4.2.1 earlier) is not used.

8.3.2 Difficulties in Maintaining Diverse Pareto-Optimal Solutions

As it is important for an MOEA to find solutions near or on the true Pareto-optimal front, it is also necessary to find solutions as diverse as possible in the Pareto-optimal front. In most multi-objective EAs, a specific diversity-maintaining operator, such as a niching technique or a clustering technique is used to find diverse Pareto-optimal solutions. However, the following features might reasonably be thought to cause an MOEA difficulty in maintaining diverse Pareto-optimal solutions:

1. Convexity or nonconvexity in the Pareto-optimal front.
2. Discontinuity in the Pareto-optimal front.
3. Non-uniform distribution of solutions in the search space and in the Pareto-optimal front.

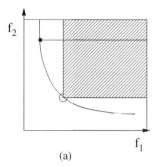
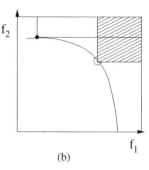

(a)   (b)

Figure 207   The fitness assignment proportional to the number of dominated solutions (the shaded area) favors intermediate solutions in a convex Pareto-optimal front (a), compared to that in a nonconvex Pareto-optimal front (b). Reproduced from Deb (1999c) (© 1999 by the Massachusetts Institute of Technology).

There exist multi-objective problems where the resulting Pareto-optimal front is nonconvex. Although it may not be readily apparent, in tackling such problems an MOEA's success in maintaining diverse Pareto-optimal solutions largely depends on the fitness assignment procedure. In some MOEAs, the fitness of a solution may be assigned proportional to the number of solutions it dominates. Figure 207 shows how such a fitness assignment favors intermediate solutions, in the case of problems with a convex Pareto-optimal front (Figure 207(a)). With respect to an individual champion solution (marked with small solid bullets on the figures), the proportion of dominated region covered by an intermediate solution is more in Figure 207(a) than in 207(b). Using such an MOEA which favors solutions having more dominated solutions, there is a natural tendency to find more intermediate solutions than solutions near individual champions, thereby causing an artificial bias towards some portion of the Pareto-optimal region.

In some multi-objective optimization problems, the Pareto-optimal front may not be continuous; instead it may be a collection of discretely spaced continuous sub-regions (as in test problems SCH2, POL and VNT). In such problems, although solutions within each sub-region may be found, competition among these solutions may lead to extinction of some of these regions.

It is also likely that the density of feasible solutions in the Pareto-optimal front is not uniform. Some regions in the front may be represented by a higher *density*[2] of solutions than other regions (see Figure 205 for such a non-uniformly dense search space). In such cases, there may be a natural tendency for MOEAs to find a biased distribution in the Pareto-optimal region. Using these concepts, a tunable two-objective test problem generator has been devised and presented elsewhere (Deb, 1999c).

---

[2] Density can be measured as the hypervolume of a sub-region in the parameter space representing a unit hypercube in the fitness space.

8.3.3   Tunable Two-Objective Optimization Problems

Let us consider the following n-variable two-objective problem:

$$\left.\begin{array}{ll} \text{Minimize} & f_1(\mathbf{x}) = f_1(x_1, x_2, \ldots, x_k), \\ \text{Minimize} & f_2(\mathbf{x}) = g(x_{k+1}, \ldots, x_n) \times h(f_1, g). \end{array}\right\} \tag{8.22}$$

The function f_1 is a function of k $(< n)$ variables $[\mathbf{x}_I = (x_1, \ldots, x_k)^T]$, while the function f_2 is a function of all n variables. The function g is a function of $(n-k)$ variables $[\mathbf{x}_{II} = (x_{k+1}, \ldots, x_n)^T]$ which do not appear in the function f_1. The function h is a function of the f_1 and g function values directly. We avoid complications by choosing the f_1 and g functions which take only positive values (or $f_1 > 0$ and $g > 0$) in the search space. By choosing appropriate functions for f_1, g and h, multi-objective problems having specific features can be created, as follows.

1. Convexity or discontinuity in the Pareto-optimal front can be affected by choosing an appropriate h function.
2. Convergence to the true Pareto-optimal front can be affected by using a difficult (multi-modal, deceptive, etc.) g function.
3. Diversity in the Pareto-optimal front can be affected by choosing an appropriate (nonlinear or multi-dimensional) f_1 function.

We describe each of the above constructions in the following subsections.

Convexity or Discontinuity in the Pareto-Optimal Front

By choosing an appropriate h function, multi-objective optimization problems with convex, nonconvex, or discontinuous Pareto-optimal fronts can be created. Specifically, if the following two properties of h are satisfied, the global Pareto-optimal set will correspond to the global minimum of the function g and to all values of the function f_1:

1. The function h is a monotonically non-decreasing function in g for a fixed value of f_1.
2. The function h is a monotonically decreasing function of f_1 for a fixed value of g.

The first condition ensures that the global Pareto-optimal front occurs for the global minimum value of the g function. The second condition ensures that there is a continuous 'conflicting' Pareto-optimal front. Although many different functions may exist, we present two such functions – one leading to a convex Pareto-optimal front and the other leading to a more generic problem having a control parameter which decides the convexity or the nonconvexity of the Pareto-optimal fronts.

Convex Pareto-Optimal Front

For the following function:

$$h(f_1, g) = \frac{1}{f_1}, \tag{8.23}$$

we only allow $f_1 > 0$. The resulting Pareto-optimal set is $(x_I^*, x_{II})^T = \{(x_I, x_{II})^T :$
$\nabla g(x_{II}) = 0\}$.

Convex and Nonconvex Pareto-Optimal Front

We choose the following function for h:

$$h(f_1, g) = \begin{cases} 1 - \left(\frac{f_1}{\beta g}\right)^\alpha, & \text{if } f_1 \leq \beta g, \\ 0, & \text{otherwise.} \end{cases} \tag{8.24}$$

With this function, we allow $f_1 \geq 0$, but $g > 0$. The global Pareto-optimal
set corresponds to the global minimum of the g function. The parameter β is a
normalization factor to adjust the range of values of the functions f_1 and g. In order to
have a significant Pareto-optimal region, β may be chosen as $\beta \geq f_{1,max}/g_{min}$, where
$f_{1,max}$ and g_{min} are the maximum value of the function f_1 and the minimum (or
global optimum) value of the function g, respectively. It is interesting to note from
equation (8.24) that when $\alpha > 1$, the resulting Pareto-optimal front is nonconvex.
In tackling these problems, the classical weighted sum approach cannot find any
intermediate Pareto-optimal solution by using any weight vector. The above function
can also be used to create multi-objective problems having a convex Pareto-optimal
set by setting $\alpha \leq 1$. Other interesting functions for h may also be chosen.

Test problems having local and global Pareto-optimal fronts of mixed type (some of
convex and some of nonconvex shapes) can also be created by making the parameter
α a function of g. These problems may cause difficulty to algorithms that work by
exploiting the shape of the Pareto-optimal front, simply because a search algorithm
needs to adopt to a different kind of front while moving from a local to a global
Pareto-optimal front. We illustrate one such problem, where the local Pareto-optimal
front is nonconvex, whereas the global Pareto-optimal front is convex. Consider the
following functions $(x_1, x_2 \in [0, 1])$ along with function h defined in equation (8.24):

$$g(x_2) = \begin{cases} 4 - 3\exp\left(-\frac{x_2 - 0.2}{0.02}\right)^2, & \text{if } 0 \leq x_2 \leq 0.4, \\ 4 - 2\exp\left(-\frac{x_2 - 0.7}{0.2}\right)^2, & \text{if } 0.4 < x_2 \leq 1, \end{cases} \tag{8.25}$$

$$f_1(x_1) = 4x_1, \tag{8.26}$$

$$\alpha = 0.25 + 3.75\frac{g(x_2) - g^{**}}{g^* - g^{**}}, \tag{8.27}$$

where g^* and g^{**} are the local and the global optimal function values of g, respectively.
Equation (8.27) is set to have a nonconvex local Pareto-optimal front at $\alpha = 4.0$
and a convex global Pareto-optimal front at $\alpha = 0.25$. The function h is given in
equation (8.24) with $\beta = 1$. A random set of 40 000 solutions $(x_1, x_2 \in [0.0, 1.0])$ is
generated and the corresponding solutions in the f_1–f_2 space are shown in Figure 208.
The figure clearly shows the nature of the convex global and nonconvex local Pareto-
optimal fronts. Notice that only a small portion of the search space leads to the

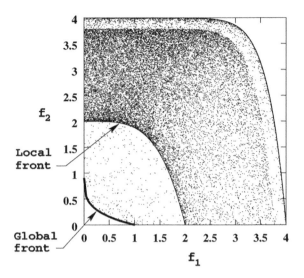

Figure 208   A two-objective function with a nonconvex local Pareto-optimal front and a convex global Pareto-optimal front. Reproduced from Deb (1999c) (© 1999 by the Massachusetts Institute of Technology).

global Pareto-optimal front. An apparent front at the top of the figure is due to the discontinuity in the $g(x_2)$ function at $x_2 = 0.4$.

By making α as a function of f_1, a Pareto-optimal front having partially convex and partially nonconvex regions can be created. We now illustrate a problem having a discontinuous Pareto-optimal front.

Discontinuous Pareto-Optimal Front

As mentioned earlier, we have to relax the condition for h being a monotonically decreasing function of f_1 in order to construct multi-objective problems with a discontinuous Pareto-optimal front. In the following, we show one such construction where the function h is a periodic function of f_1:

$$h(f_1, g) = 1 - \left(\frac{f_1}{g}\right)^\alpha - \frac{f_1}{g}\sin(2\pi q f_1). \tag{8.28}$$

The parameter q is the number of discontinuous regions in a unit interval of f_1. By choosing the following functions:

$$f_1(x_1) = x_1,$$
$$g(x_2) = 1 + 10x_2,$$

and allowing variables x_1 and x_2 to lie in the interval $[0,1]$, we have a two-objective optimization problem which has a discontinuous Pareto-optimal front. Since the h

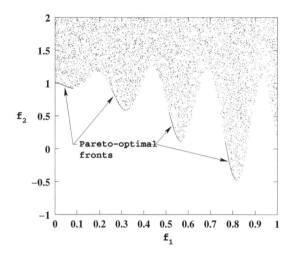

Figure 209   Random solutions are shown on an f_1–f_2 plot of a multi-objective problem having disconnected Pareto-optimal fronts.

function is periodic to x_1, certain portions of the search boundary are dominated. This introduces discontinuity in the resulting Pareto-optimal front. Figure 209 shows random solutions in the f_1–f_2 space and the resulting Pareto-optimal fronts. Here, we use $q = 4$ and $\alpha = 2$. In general, the discontinuity in the Pareto-optimal front may cause difficulty to MOEAs which do not have an efficient way of maintaining diversity among discontinuous regions. In particular, the function-space niching may face difficulties in these problems, because of the discontinuities in the Pareto-optimal front.

Hindrance to Reach the True Pareto-Optimal Front

It was discussed earlier that by choosing a difficult function for g alone, a difficult multi-objective optimization problem can be created. Test problems with standard multi-modal functions used in single-objective EA studies, such as Rastrigin's functions, NK landscapes, etc., can all be chosen for the g function.

Biased Search Space

With a simple monotonic g function, the search space can have an adverse density of solutions towards the Pareto-optimal region. Consider the following function for g:

$$g(x_{m+1},\ldots,x_n) = g_{\min} + (g_{\max} - g_{\min}) \left(\frac{\sum_{i=k+1}^{n} x_i - \sum_{i=k+1}^{n} x_i^{\min}}{\sum_{i=k+1}^{n} x_i^{\max} - \sum_{i=k+1}^{n} x_i^{\min}} \right)^{\gamma}, \quad (8.29)$$

where $g_{\min}$ and $g_{\max}$ are the minimum and maximum function values that the function g can take. The values $x_i^{\min}$ and $x_i^{\max}$ are the minimum and maximum values of the variable x_i. It is important to note that the Pareto-optimal region occurs when g takes the value g_{min}. The parameter γ controls the bias in the search space. If $\gamma < 1$, the density of solutions away from the Pareto-optimal front is large. We show this on a simple problem with $k = 1$, $n = 2$, and with the following functions: $f_1(x_1) = x_1$, and $h(f_1, g) = 1 - (f_1/g)^2$. We also use $g_{min} = 1$ and $g_{max} = 2$. Figures 210 and 211 show 50 000 random solutions, each with γ equal to 1.0 and 0.25, respectively. It is clear that for $\gamma = 0.25$, not even one solution is found on the Pareto-optimal front, whereas for $\gamma = 1.0$, many Pareto-optimal solutions exist in the set of 50 000 random solutions. Random search methods are likely to face difficulties in finding the Pareto-optimal front in the case with γ close to zero, mainly due to the low density of solutions towards the Pareto-optimal region.

Parameter Interactions

The difficulty in converging to the true Pareto-optimal front may also arise because of parameter interactions. It was discussed earlier that the Pareto-optimal set in the two-objective optimization problem described in equation (8.22) corresponds to all solutions of different f_1 values. Since the purpose in an MOEA is to find as many Pareto-optimal solutions as possible, and since in equation (8.22) the variables defining f_1 are different from the variables defining g, an MOEA may work in two stages. In one stage, all variables x_I may be found and in the other stage optimal x_{II} values may be obtained. This rather simple mode of working of an MOEA in two stages can face difficulty if the above variables are mapped to another set of variables. If $\mathcal{M}$ is a random orthonormal matrix of size $n \times n$, the true variables y can first be mapped to derive the variables x by using:

$$x = \mathcal{M}y. \tag{8.30}$$

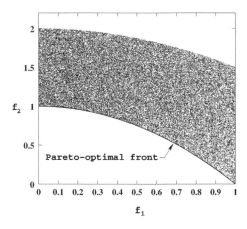

Figure 210   50 000 random solutions are shown for $\gamma = 1.0$. Reproduced from Deb (1999c) (© 1999 by the Massachusetts Institute of Technology).

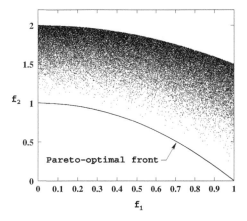

Figure 211   50 000 random solutions are shown for $\gamma = 0.25$. Reproduced from Deb (1999c) (© 1999 by the Massachusetts Institute of Technology).

Thereafter, the objective functions defined in equation (8.22) can be computed by using the variable vector x. Since the components of x can now be negative, care must be taken in defining the f_1 and g functions so as to satisfy the restrictions placed on them in previous subsections. A translation of these functions by adding a suitable large positive value may have to be used to force these functions to take non-negative values. Since an MOEA will be operating on the variable vector y and the function values depend on the interaction among variables of y, any change in one variable must be accompanied by related changes in other variables in order to remain on the Pareto-optimal front. This makes this mapped version of the problem difficult to solve.

Non-Uniformly Represented Pareto-Optimal Front

In previous test problems, we have used a linear, single-variable function for f_1. This helped us create a problem with a uniform distribution of solutions in f_1. Unless the underlying problem has discretely spaced Pareto-optimal regions, there is no difficulty for the Pareto-optimal solutions to get spread over the entire range of f_1 values. However, a bias for some portions of the range of values for f_1 may be created by choosing either of the following f_1 functions:

- the function f_1 is nonlinear;
- the function f_1 is a function of more than one variable.

It is clear that if a nonlinear f_1 function is chosen, the resulting Pareto-optimal region (or, for that matter, the entire search region) will have a bias towards some values of f_1. The non-uniformity in distribution of the Pareto-optimal region can also be created by simply choosing a multi-variable function (whether linear or nonlinear). Multi-objective optimization algorithms, which are not any good at maintaining diversity among solutions (or function values), will produce a biased Pareto-optimal front in such problems. Consider the single-variable, multi-modal function f_1:

$$f_1(x_1) = 1 - \exp(-4x_1)\sin^4(5\pi x_1), \quad 0 \le x_1 \le 1. \qquad (8.31)$$

The above function has five minima for different values of x_1, as shown in Figure 212. The right-hand figure shows the corresponding nonconvex Pareto-optimal front in a f_1-f_2 plot. We used $g = 1 + 10x_2$ and an h function defined in equation (8.24), having $\beta = 1$ and $\alpha = 4$. This produces a nonconvex Pareto-optimal front. The right-hand figure is generated on the Pareto-optimal front from 500 uniformly spaced solutions in x_1. The figure shows that the Pareto-optimal region is biased for solutions for which f_1 is near one.

Summary of Test Problems

The two-objective tunable test problems discussed above require three functions – f_1, g and h – which can be set to various complexity levels. In the following, we

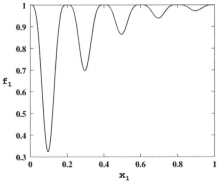
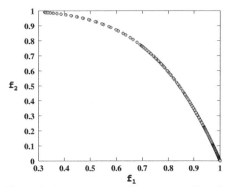

Figure 212   A multi-modal f₁ function (left) and the corresponding non-uniformly distributed nonconvex Pareto-optimal region (right). In the right plot, the Pareto-optimal solutions derived from 500 uniformly spaced x₁ solutions are shown. Reproduced from Deb (1999c) (© 1999 by the Massachusetts Institute of Technology).

summarize the properties of a two-objective optimization problem due to each of the above functions.

1. The function f_1 tests a multi-objective EA's ability to find diverse Pareto-optimal solutions. Thus, this function tests an algorithm's ability to handle difficulties *along* the Pareto-optimal front.
2. The function g tests a multi-objective EA's ability to converge to the true (or global) Pareto-optimal front. Thus, this function tests an algorithm's ability to handle difficulties *lateral* to the Pareto-optimal front.
3. The function h tests a multi-objective EA's ability to tackle multi-objective problems having convex, nonconvex or discontinuous Pareto-optimal fronts. Thus, this function tests an algorithm's ability to handle different *shapes* of the Pareto-optimal front.

The above construction makes an interesting connection between single-objective and multi-objective optimization. Since the convergence to the true Pareto-optimal front is directly related to minimization of the function g alone, all known problem difficulties that a single-objective optimization algorithm can possibly face (such as high-dimensionality, non-differentiability, ruggedness, noise, etc.) also become problem difficulties for a multi-objective optimization algorithm. Moreover, since the function g is responsible for convergence to the true Pareto-optimal front, the convergence properties of single-objective optimization algorithms now become valid for multi-objective optimization algorithms in solving problems constructed using the above procedure.

   In the light of the above discussion, a number of difficult functions are suggested in the original study (Deb, 1999c). Motivated by these suggestions, a recent study has constructed six difficult test problems, which we will describe next.

Zitzler–Deb–Thiele's (ZDT) Test Problems

Zitzler et al. (2000) framed six problems (ZDT1 to ZDT6) based on the above construction process. After this study was reported, these test problems have further been studied by other researchers. These problems have two objectives which are to be minimized:

$$\text{Minimize} \quad f_1(\mathbf{x}),$$
$$\text{Minimize} \quad f_2(\mathbf{x}) = g(\mathbf{x})h(f_1(\mathbf{x}), g(\mathbf{x})). \tag{8.32}$$

The six test problems vary in the way that the three functions $f_1(\mathbf{x})$, $g(\mathbf{x})$ and $h(\mathbf{x})$ are defined. In all problems except ZDT5, the Pareto-optimal front is formed with $g(\mathbf{x}) = 1$. Although these problems used f_1 as a single-variable function, the problem difficulty can be increased by using a multi-variate f_1 function or a mapped variable strategy (equation (8.30)).

ZDT1

This is a 30-variable ($n = 30$) problem having a convex Pareto-optimal set. The functions used are as follows:

$$\text{ZDT1:} \begin{cases} f_1(\mathbf{x}) &= x_1, \\ g(\mathbf{x}) &= 1 + \dfrac{9}{n-1}\sum_{i=2}^{n} x_i, \\ h(f_1, g) &= 1 - \sqrt{f_1/g}. \end{cases} \tag{8.33}$$

All variables lie in the range $[0, 1]$. The Pareto-optimal region corresponds to $0 \le x_1^* \le 1$ and $x_i^* = 0$ for $i = 2, 3, \ldots, 30$. Figure 213 shows a search region (which has a uniform density of solutions) and the Pareto-optimal region in the objective space. This is probably the easiest of all of the six problems, having a continuous Pareto-optimal front and a uniform distribution of solutions across the front. The

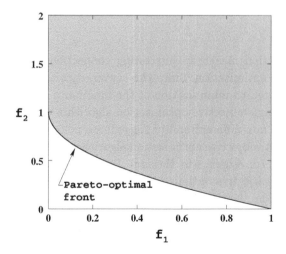

Figure 213 The search space near the Pareto-optimal region for ZDT1.

only difficulty an MOEA may face from this problem is in tackling a large number of decision variables.

ZDT2

This is also an $n = 30$ variable problem having a nonconvex Pareto-optimal set:

$$\text{ZDT2:} \begin{cases} f_1(x) & = & x_1, \\ g(x) & = & 1 + \dfrac{9}{n-1} \sum_{i=2}^{n} x_i, \\ h(f_1, g) & = & 1 - (f_1/g)^2. \end{cases} \tag{8.34}$$

All variables lie in the range $[0, 1]$. The Pareto-optimal region corresponds to $0 \le x_1^* \le 1$ and $x_i^* = 0$ for $i = 2, 3, \ldots, 30$. Figure 214 shows the search region (which has a uniform density of solutions) and the Pareto-optimal region on the objective function space. The only difficulty with this problem is that the Pareto-optimal region is nonconvex. Thus, weighted approaches will have difficulty in finding a good spread of solutions on the Pareto-optimal front.

ZDT3

This is an $n = 30$ variable problem having a number of disconnected Pareto-optimal fronts:

$$\text{ZDT3:} \begin{cases} f_1(x) & = & x_1, \\ g(x) & = & 1 + \dfrac{9}{n-1} \sum_{i=2}^{n} x_i, \\ h(f_1, g) & = & 1 - \sqrt{f_1/g} - (f_1/g) \sin(10\pi f_1). \end{cases} \tag{8.35}$$

All variables lie in the range $[0, 1]$. The Pareto-optimal region corresponds to $x_i^* = 0$

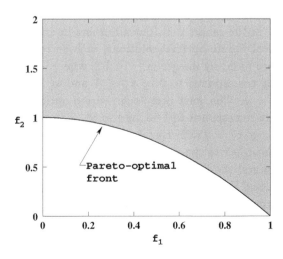

Figure 214   The search space near the Pareto-optimal region for ZDT2.

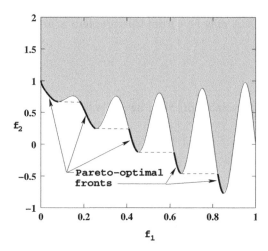

Figure 215   The search space near the Pareto-optimal region for ZDT3.

for $i = 2, 3, \ldots, 30$, and hence not all points satisfying $0 \le x_1^* \le 1$ lie on the Pareto-optimal front. Figure 215 shows the search region and the Pareto-optimal fronts in the objective space. The difficulty with this problem is that the Pareto-optimal region is discontinuous. The real test for MOEAs would be to find all discontinuous regions with a uniform spread of non-dominated solutions.

ZDT4

This is an $n = 10$ variable problem having a convex Pareto-optimal set:

$$\text{ZDT4:} \begin{cases} f_1(\mathbf{x}) &= x_1, \\ g(\mathbf{x}) &= 1 + 10(n-1) + \sum_{i=2}^{n}(x_i^2 - 10\cos(4\pi x_i)), \\ h(f_1, g) &= 1 - \sqrt{f_1/g}. \end{cases} \qquad (8.36)$$

The variable x_1 lies in the range $[0, 1]$, but all others lie in $[-5, 5]$. There exists 21^9 or about $8(10^{11})$ local Pareto-optimal solutions, each corresponding to $0 \le x_1^* \le 1$ and $x_i^* = 0.5m$, where m is any integer in $[-10, 10]$, where $i = 2, 3, \ldots, 10$. The global Pareto-optimal front corresponds to $0 \le x_1^* \le 1$ and $x_i^* = 0$ for $i = 2, 3, \ldots, 10$. This makes $g(\mathbf{x}^*) = 1$. The next-best local Pareto-optimal front corresponds to $g(\mathbf{x}) = 1.25$, the next corresponds to 1.50, and so on. The worst local Pareto-optimal front corresponds to $g(\mathbf{x}) = 25.00$, thereby making a total of 100 distinct Pareto-optimal fronts in the objective space, of which only one is global. Figure 216 shows a partial search region and the Pareto-optimal front in the objective space. The sheer number of multiple local Pareto-optimal fronts produces a large number of hurdles for an MOEA to converge to the global Pareto-optimal front.

ZDT5

This is a Boolean function defined over bit-strings. There are a total of 11 discrete variables. The first variable x_1 is presented by a 30-bit substring and the rest 10

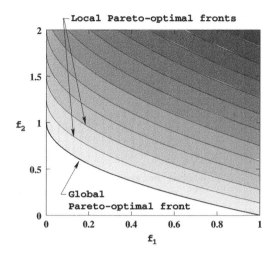

Figure 216   The partial search space near the Pareto-optimal region for ZDT4.

variables $(x_2 - x_{11})$ are represented by 5 bits each. The function $u(x_i)$ denotes the number of '1s' in the substring used to represent the variable x_i.

$$\text{ZDT5:} \begin{cases} f_1(x) & = & 1 + u(x_1) \\ g(x) & = & \sum_{i=2}^{11} v(u(x_i)) \\ v(u(x_i)) & = & \begin{cases} 2 + u(x_i) & \text{if } u(x_i) < 5, \\ 1 & \text{if } u(x_i) = 5, \end{cases} \\ h(f_1, g) & = & 1/f_1(x). \end{cases}$$

The global Pareto-optimal front corresponds to the minimum value of g, or $g(x^*) = 10$. This happens when variables $x_2 - x_{10}$ are represented by a 5-bit substring of all '1s', or when $u(x_i) = 5$ for $i = 2, 3 \ldots, 10$. The diversity among the Pareto-optimal solutions is maintained by the presence of different substrings representing x_1. Since $u(x_1)$ can take 31 different integer values $(0-30)$, the first objective f_1 takes any integer value in $[1, 31]$. Figure 217 shows the discrete global Pareto-optimal set. The best local Pareto-optimal front corresponds to the next-best value of g, or $g(x^*) = 11$. It is important to note that this next-best solution corresponds to any one of the variables x_2 to x_{10} having a string maximally different from that in the global Pareto-optimal set. Here, any of the variables $x_2 - x_{10}$ is a substring with all '0s'. The $v(u(x_i))$ function suggests that most of the search space leads towards this local optimal solution, whereas the globally optimal substring is an isolated maximum. In the context of single-objective optimization, these problems have been shown to deceive a simple GA. Hence, they are called *deceptive* problems. Multi-objective GAs are also likely to face difficulty from deception and may easily converge to the best local Pareto-optimal solutions, instead of the global Pareto-optimal solutions. In the above problem, there are a total of 1023 local Pareto-optimal sets. These fronts lie within the two dashed curves shown in the figure. It becomes difficult for an optimization algorithm to reach the global Pareto-optimal set without getting attracted to one of these local Pareto-optimal fronts.

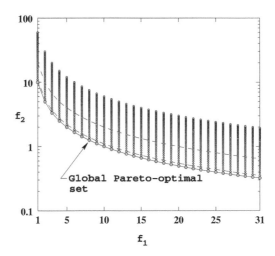

Figure 217 The search space and the Pareto-optimal region for ZDT5.

ZDT6

This is a 10-variable problem having a nonconvex Pareto-optimal set. Moreover, the density of solutions across the Pareto-optimal region is non-uniform and the density towards the Pareto-optimal front is also thin:

$$\text{ZDT6:} \begin{cases} f_1(\mathbf{x}) &= 1 - \exp(-4x_1)\sin^6(6\pi x_1), \\ g(\mathbf{x}) &= 1 + 9\left[(\sum_{i=2}^{10} x_i)/9\right]^{0.25}, \\ h(f_1, g) &= 1 - (f_1/g)^2. \end{cases} \tag{8.37}$$

All variables lie in the range $[0, 1]$. The Pareto-optimal region corresponds to $0 \le x_1^* \le 1$ and $x_i^* = 0$ for $i = 2, 3, \dots, 10$. Figure 218 shows the Pareto-optimal region on the objective function space. A set of 100 uniformly distributed x_1^* solutions (in $[0, 1]$) on the Pareto-optimal front are also shown in the figure by circles. The adverse density of solutions across the Pareto-optimal front, coupled with the nonconvex nature of the front, may cause difficulties for many multi-objective optimization algorithms to converge to the true Pareto-optimal front.

8.3.4   Test Problems with More Than Two Objectives

Most research on multi-objective evolutionary algorithms have been restricted to two-objective problems. There have been two reasons for this, as follows:

1. Both tasks of converging near the Pareto-optimal front and of finding a widespread of solutions can be tested adequately with as few as two objectives and
2. Graphical ways for visualizing trade-off solutions in problems with more than two objectives become difficult to achieve.

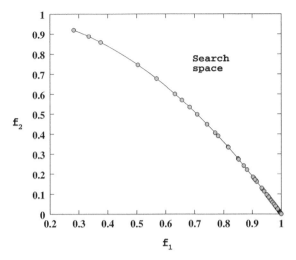

Figure 218   The Pareto-optimal region for ZDT6.

Although two-objective test problems can test both of the desired aspects of an MOEA, there should be some test problems which also test if an algorithm can be applied to more than two objectives efficiently. For a large number of objectives, the Pareto-optimal set becomes multi-dimensional in the objective space. Algorithms then require better strategies for finding a set of widespread solutions. It would be an interesting research project to investigate if the currently known MOEAs are able to find a set of widespread non-dominated solutions closer to the true Pareto-optimal front within a reasonable computational effort.

All of the basic difficulties highlighted in the previous subsection for two-objective problems may also be present in a problem having a larger number of objectives. In the following, we construct a test problem generator having M objective functions, similar in construction to the two-objective test problem generator described earlier:

$$\left. \begin{aligned} &\text{Minimize} \quad f_1(\mathbf{x}_1), \\ &\text{Minimize} \quad f_2(\mathbf{x}_2), \\ &\qquad \vdots \quad \vdots \\ &\text{Minimize} \quad f_{M-1}(\mathbf{x}_{M-1}), \\ &\text{Minimize} \quad f_M(\mathbf{x}) = g(\mathbf{x}_M) h\left(f_1(\mathbf{x}_1), f_2(\mathbf{x}_2), \dots, f_{M-1}(\mathbf{x}_{M-1}), g(\mathbf{x}_M)\right), \\ &\text{subject to} \quad \mathbf{x}_i \in \mathbb{R}^{|\mathbf{x}_i|}, \quad \text{for } i = 1, 2, \dots, M. \end{aligned} \right\} \quad (8.38)$$

Here, the decision variable vector $\mathbf{x}$ is partitioned into M non-overlapping blocks as follows:

$$\mathbf{x} \equiv (\mathbf{x}_1, \mathbf{x}_2, \dots, \mathbf{x}_{M-1}, \mathbf{x}_M)^{\mathsf{T}}.$$

Each vector $\mathbf{x}_i$ can be of different size. The objective functions $f_1 - f_{M-1}$ can be chosen in a similar way as the function f_1 chosen in the previous subsection. The function g can be similar to the function g described earlier and has the effect of producing difficulty in progressing towards the true Pareto-optimal front. However, the function h is now different and must include all objective function values f_1 – f_{M-1} and g. However, the structure of the function h may be similar to the function h described earlier and has the effect of causing difficulty along the Pareto-optimal front. For example, the following h function will produce a continuous Pareto-optimal region:

$$h(f_1, f_2, \ldots, f_{M-1}, g) = 1 - \left(\frac{\sum_{i=1}^{M-1} f_i}{\beta g} \right)^{\alpha}.$$
(8.39)

The Pareto-optimal front will be convex for $\alpha < 1$. The parameter β is a normalization parameter. In order to create a problem with a discontinuous set of Pareto-optimal fronts, a periodic h function as used in the previous subsection can also be developed here. Since the first $(M - 1)$ objectives are functions of the non-overlapping set of decision variables, the Pareto-optimal solutions correspond to the values of x_M^* for which g is minimum and for all permissible values of the first $(M - 1)$ set of variables, which satisfy $f_M = g(x_M^*)h(f_1, \ldots, f_{M-1}, g(x_M^*))$ (the Pareto-optimal surface) and are non-dominated by each other.

In order to construct more difficult test problems, the decision variable vector x can be mapped into a different variable vector y, as suggested above in equation (8.30).

8.3.5   Test Problems for Constrained Optimization

In addition to the above difficulties, the presence of 'hard' constraints in a multi-objective problem may cause further hurdles. Constraints may cause hindrance for MOEAs to converge to the true Pareto-optimal region and may also cause difficulty in maintaining a diverse set of Pareto-optimal solutions. It is intuitive that the success of an MOEA in tackling both of these hindrances will largely depend on the constraint-handling technique used.

However, first we will present a number of test problems commonly used in the literature and then discuss a systematic procedure for developing difficult test problems. Veldhuizen (1999) have cited a number of constrained test problems used by several researchers. It is evident from their survey that most constrained test problems used only two to three variables and constraints are not sufficiently nonlinear. In the following, we present three such test problems and discuss why they may not provide adequate difficulty to an MOEA.

Test Problem BNH

Binh and Korn (1997) used the following two-variable constrained problem:

$$
\begin{aligned}
\text{Minimize} \quad & f_1(x) = 4x_1^2 + 4x_2^2, \\
\text{Minimize} \quad & f_2(x) = (x_1 - 5)^2 + (x_2 - 5)^2, \\
\text{subject to} \quad & C_1(x) \equiv (x_1 - 5)^2 + x_2^2 \leq 25, \\
& C_2(x) \equiv (x_1 - 8)^2 + (x_2 + 3)^2 \geq 7.7, \\
& 0 \leq x_1 \leq 5, \\
& 0 \leq x_2 \leq 3.
\end{aligned}
$$
(8.40)

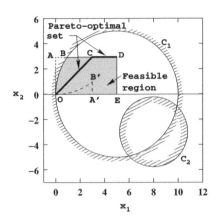

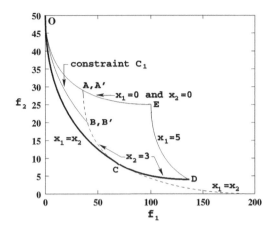

Figure 219   Decision variable space for the test problem BNH.

Figure 220   Objective space for the test problem BNH.

Figure 219 shows the feasible decision variable space (the shaded region enclosed by OBDEO). Both constraints (C_1, C_2) are also shown on the figure. It is interesting to note that the second constraint C_2 is redundant and does not make any part of the bounded space infeasible. On the other hand, constraint C_1 makes the region OABO infeasible. Figure 220 shows the corresponding objective space. Since constraint C_2 does not affect the feasible region, we do not show this constraint in the objective space. The constraint boundary of C_1 is also shown in this figure. There exists an equivalent region OA'B'O in the decision variable space which maps to the same region OABO in the objective space. Thus, although some portion of the objective space is eliminated because of the constraint C_1, no solution of the unconstrained bounded search space gets eliminated in the objective space. Thus, the presence of the first constraint is also not critical in this problem. The only difficulty that this constraint creates is that it reduces the density of solutions in the region OABO in the objective space. The shape and continuity of the Pareto-optimal set is unchanged by the inclusion of both constraints.

The Pareto-optimal solutions are constituted by solutions $x_1^* = x_2^* \in [0,3]$ (region OC) and $x_1^* \in [3,5]$, $x_2^* = 3$ (region CD). These solutions are marked by using bold continuous curves. The addition of both constraints in the problem does not make any solution in the unconstrained Pareto-optimal front infeasible. Thus, constraints may not introduce any additional difficulty in solving this problem.

Test Problem OSY

Osyczka and Kundu (1995) used the following six-variable test problem:

$$\text{Minimize} \quad f_1(\mathbf{x}) = -\left[25(x_1-2)^2 + (x_2-2)^2 + (x_3-1)^2 + (x_4-4)^2 + (x_5-1)^2\right],$$

$$\text{Minimize} \quad f_2(\mathbf{x}) = x_1^2 + x_2^2 + x_3^2 + x_4^2 + x_5^2 + x_6^2,$$

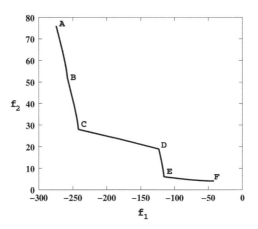

Figure 221   The constrained Pareto-optimal front for the problem OSY. This is a reprint
of Figure 4 from Deb et al. (2001) (© Springer-Verlag Berlin Heidelberg 2001).

$$
\begin{aligned}
\text{subject to}\quad & C_1(\mathbf{x}) \equiv x_1 + x_2 - 2 \geq 0, \\
& C_2(\mathbf{x}) \equiv 6 - x_1 - x_2 \geq 0, \\
& C_3(\mathbf{x}) \equiv 2 - x_2 + x_1 \geq 0, \\
& C_4(\mathbf{x}) \equiv 2 - x_1 + 3x_2 \geq 0, \\
& C_5(\mathbf{x}) \equiv 4 - (x_3 - 3)^2 - x_4 \geq 0, \\
& C_6(\mathbf{x}) \equiv (x_5 - 3)^2 + x_6 - 4 \geq 0, \\
& 0 \leq x_1, x_2, x_6 \leq 10, \quad 1 \leq x_3, x_5 \leq 5, \quad 0 \leq x_4 \leq 6.
\end{aligned}
\tag{8.41}
$$

There are six constraints, four of which are linear. Since this is a six-variable problem,
it is difficult to show the feasible decision variable space. However, a careful analysis
of the constraints and the objective function reveals the constrained Pareto-optimal
front, as shown in Figure 221. The Pareto-optimal region is a concatenation of five
regions. Every region lies on some of the constraints. However, for the entire Pareto-
optimal region, $x_4^* = x_6^* = 0$. Table 28 shows the other variable values in each of
the five regions and the constraints that are active in each region. Since the entire
Pareto-optimal region demands an MOEA to maintain its subpopulations at different
intersections of constraint boundaries, this may be a difficult problem to solve.

Test Problem SRN

Srinivas and Deb (1994) borrowed the following function from Chankong and Haimes
(1983):

$$
\begin{aligned}
\text{Minimize}\quad & f_1(\mathbf{x}) = 2 + (x_1 - 2)^2 + (x_2 - 1)^2, \\
\text{Minimize}\quad & f_2(\mathbf{x}) = 9x_1 - (x_2 - 1)^2, \\
\text{subject to}\quad & C_1(\mathbf{x}) \equiv x_1^2 + x_2^2 \leq 225, \\
& C_2(\mathbf{x}) \equiv x_1 - 3x_2 + 10 \leq 0, \\
& -20 \leq x_1 \leq 20, \\
& -20 \leq x_2 \leq 20.
\end{aligned}
\tag{8.42}
$$

Table 28   Pareto-optimal solutions for the problem OSY. This is a reprint of Table 1 from Deb et al. (2001) (© Springer-Verlag Berlin Heidelberg 2001).

| Region | Optimal values | | | | Active |
	x_1^*	x_2^*	x_3^*	x_5^*	constraints
AB	5	1	$(1, \dots, 5)$	5	2,4,6
BC	5	1	$(1, \dots, 5)$	1	2,4,6
CD	$(4.056, \dots, 5)$	$(x_1^* - 2)/3$	1	1	4,5,6
DE	0	2	$(1, \dots, 3.732)$	1	1,3,6
EF	$(0, \dots, 1)$	$2 - x_1^*$	1	1	1,5,6

Figure 222 shows the feasible decision variable space and the corresponding Pareto-optimal set. By calculating the derivatives of both objectives and using equation (2.8), we obtain:

$$\begin{vmatrix} 2(x_1^* - 2) & 9 \\ 2(x_2^* - 1) & -2(x_2^* - 1) \end{vmatrix} = 0.$$

The above equation is satisfied for two cases: (i) $x_1^* = -2.5$ and (ii) $x_2^* = 1$. In the feasible region, only the first case prevails. Thus, the Pareto-optimal solutions correspond to $x_1^* = -2.5$ and $x_2^* \in [-14.79, 2.50]$. The feasible objective space, along with the Pareto-optimal solutions, are shown in Figure 223. The only difficulty the constraints introduce in this problem is that they eliminate some parts of the

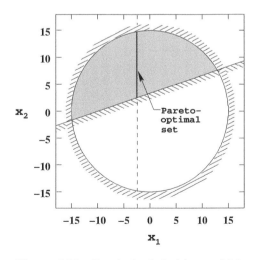
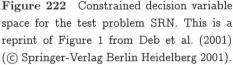

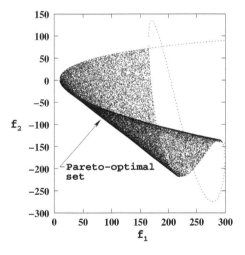

Figure 222   Constrained decision variable space for the test problem SRN. This is a reprint of Figure 1 from Deb et al. (2001) (© Springer-Verlag Berlin Heidelberg 2001).

Figure 223   Feasible objective space for the test problem SRN. This is a reprint of Figure 2 from Deb et al. (2001) (© Springer-Verlag Berlin Heidelberg 2001).

unconstrained Pareto-optimal set (shown by dashed lines).

Test Problem TNK

Tanaka (1995) suggested the following two-variable problem:

$$
\begin{aligned}
\text{Minimize} \quad & f_1(\mathbf{x}) = x_1, \\
\text{Minimize} \quad & f_2(\mathbf{x}) = x_2, \\
\text{subject to} \quad & C_1(\mathbf{x}) \equiv x_1^2 + x_2^2 - 1 - 0.1 \cos\left(16 \arctan \tfrac{x_1}{x_2}\right) \geq 0, \\
& C_2(\mathbf{x}) \equiv (x_1 - 0.5)^2 + (x_2 - 0.5)^2 \leq 0.5, \\
& 0 \leq x_1 \leq \pi, \\
& 0 \leq x_2 \leq \pi.
\end{aligned}
\tag{8.43}
$$

The feasible decision variable space is shown in Figure 224. Since $f_1 = x_1$ and $f_2 = x_2$, the feasible objective space is also the same as the feasible decision variable space. The unconstrained decision variable space consists of all solutions in the square $0 \leq (x_1, x_2) \leq \pi$. Thus, the only unconstrained Pareto-optimal solution is $x_1^* = x_2^* = 0$. However, the inclusion of the first constraint makes this solution infeasible. The constrained Pareto-optimal solutions lie on the boundary of the first constraint. Since the constraint function is periodic and the second constraint function must also be satisfied, not all solutions on the boundary of the first constraint are Pareto-optimal. The disconnected Pareto-optimal set is shown in Figure 224. Since the Pareto-optimal solutions lie on a nonlinear constraint surface, an optimization algorithm may have difficulty in finding a good spread of solutions across all of the discontinuous Pareto-optimal sets.

The above test problems are not tunable for introducing varying degrees of

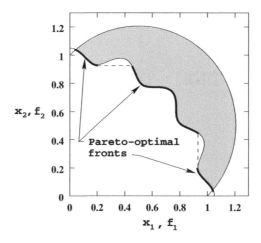

Figure 224 The feasible decision variable and objective spaces for the TNK problem. This is a reprint of Figure 3 from Deb et al. (2001) (© Springer-Verlag Berlin Heidelberg 2001).

complexity in constrained optimization. As for the tunable unconstrained test problems, we also suggest here a number of test problems where the complexity of the constrained search space can be controlled. The proposed problems are designed to cause two different kinds of tunable difficulties in a multi-objective optimization algorithm, as follows:

- difficulty near the Pareto-optimal front;
- difficulty in the entire search space.

We will discuss both of these in the following subsections.

Difficulty in the Vicinity of the Pareto-Optimal Front

In these test problems, constraints do not make any major portion of the search region infeasible, except near the Pareto-optimal front. Constraints are designed in a way so that some portion of the unconstrained Pareto-optimal region is now infeasible. In this way, the overall Pareto-optimal front will constitute some part of the unconstrained Pareto-optimal region and some part of the constraint boundaries. In Chapter 7, we have used one such problem to illustrate the working principle of a number of constrained handling MOEAs. In the following, we present a generic test problem with J constraints:

$$
\text{CTP1:} \quad
\begin{cases}
\text{Minimize} & f_1(x_I), \\
\text{Minimize} & f_2(x) = g(x_{II}) \exp(-f_1(x_I)/g(x_{II})), \\
\text{subject to} & C_j(x) \equiv f_2(x) - a_j \exp[-b_j f_1(x_I)] \geq 0, \quad j = 1, 2, \ldots, J.
\end{cases}
$$
$$(8.44)$$

Here, $x = (x_I, x_{II})^T$ and the function $f_1(x_I)$ and $g(x_{II})$ can be any multi-variable functions. There are J inequality constraints and the parameters (a_j, b_j) must be chosen in a way so that at least some portion of the unconstrained Pareto-optimal region is infeasible. We now describe a procedure to calculate the (a_j, b_j) parameters for J constraints.

Procedure for Calculating a_j and b_j

    Step 1 Set $j = 0$, $a_j = b_j = 1$; also set $\Delta = 1/(J+1)$ and $\alpha = \Delta$.

    Step 2 Calculate $\beta = a_j \exp(-b_j \alpha)$ and

$$
a_{j+1} = (a_j + \beta)/2, \quad b_{j+1} = -\frac{1}{\alpha} \ln(\beta/a_{j+1}).
$$

    Increment $\alpha = \alpha + \Delta$ and $j = j + 1$.

    Step 3 If $j < J$, go to Step 2. Otherwise, the process is complete.

For two constraints ($J = 2$), the above procedure finds the following parameter values:

$$a_1 = 0.858, \quad b_1 = 0.541, \quad a_2 = 0.728, \quad b_2 = 0.295.$$

Figure 225 shows the unconstrained Pareto-optimal region (with a dashed line), and the two constraints. With the presence of both constraints, the figure demonstrates

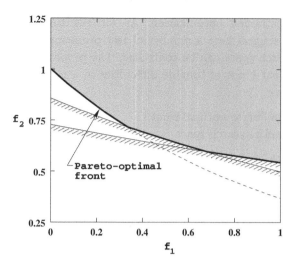

Figure 225 Constrained test problem CTP1 with two constraints. This is a reprint of Figure 5 from Deb et al. (2001) (© Springer-Verlag Berlin Heidelberg 2001).

that only about one-third of the original unconstrained Pareto-optimal region is now feasible. The other two-thirds part of the constrained Pareto-optimal region comes from the two constraints.

The reason why this problem may cause difficulty to an MOEA is as follows. Since a part of the constraint boundary of each constraint now constitutes the Pareto-optimal region, a spread in the Pareto-optimal solutions requires the decision variables (x) to satisfy the inequality constraints with the equality sign. Each constraint is an implicit nonlinear function of the decision variables. Thus, it may be difficult to discover and maintain a number of solutions on a nonlinear constraint boundary. The presence of more numbers of such constraints will demand the algorithm to discover and maintain many such correlations among the decision variables. The complexity of the test problem can be further increased by using a multi-modal function g.

Furthermore, besides finding and maintaining correlated decision variables to fall on several constraint boundaries, there could be other difficulties near the Pareto-optimal front. The constraint functions can be such that the unconstrained Pareto-optimal region is now infeasible and the resulting Pareto-optimal set is a collection of a number of discrete regions. Let us first present such a function mathematically and then describe the difficulties:

$$
\begin{aligned}
\text{CTP2–} \\
\text{CTP8:}
\end{aligned}
\left\{
\begin{aligned}
&\text{Minimize} && f_1(\mathbf{x}) = x_1, \\
&\text{Minimize} && f_2(\mathbf{x}) = g(\mathbf{x})\left(1 - \tfrac{f_1(\mathbf{x})}{g(\mathbf{x})}\right), \\
&\text{subject to} && C(\mathbf{x}) \equiv \cos(\theta)[f_2(\mathbf{x}) - e] - \sin(\theta)f_1(\mathbf{x}) \geq \\
& && \quad a\left|\sin\left\{b\pi\left[\sin(\theta)(f_2(\mathbf{x}) - e) + \cos(\theta)f_1(\mathbf{x})\right]^c\right\}\right|^d.
\end{aligned}
\right.
\tag{8.45}
$$

The decision variable x_1 is restricted in $[0, 1]$ and the bounds of other variables depend on the chosen $g(\mathbf{x})$ function. It is important to note that the problem can be made harder by choosing a multi-variate f_1 function. The constraint $C(\mathbf{x})$ has six parameters (θ, a, b, c, d and e). In fact, the above problem can be used as a constrained test problem generator by tuning these six parameters. We use the above problem to construct different test problems.

First, we use the following parameter values:

$$\theta = -0.2\pi, \quad a = 0.2, \quad b = 10, \quad c = 1, \quad d = 6, \quad e = 1.$$

The resulting feasible objective space is shown in Figure 226. It is clear from this figure that the unconstrained Pareto-optimal region (shown by dashes) is now infeasible. The periodic nature of the constraint boundary makes the Pareto-optimal region discontinuous, having a number of disconnected continuous regions. The task of an optimization algorithm would be to find as many such disconnected regions as possible. The number of such regions can be controlled by increasing the value of the parameter b. It is also clear that with the increase in number of disconnected regions, an algorithm will have difficulty in finding representative solutions in all disconnected regions.

The above problem can be made more difficult by using a small value of d, so that in each disconnected region there exists only one Pareto-optimal solution. Figure 227 shows the feasible objective space for $d = 0.5$ and $a = 0.1$ (while other parameters are the same as that in the previous test problem). Although most of the search space is feasible, near the Pareto-optimal region the feasible search regions are disconnected, with finally each sub-region leading to a singular feasible Pareto-optimal solution. An algorithm will face difficulty in finding all discrete Pareto-optimal solutions because of the changing nature from a continuous to a discontinuous feasible search space near the Pareto-optimal region.

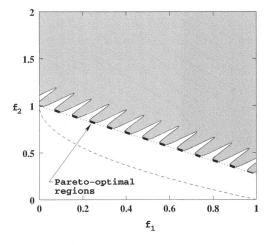

Figure 226 The constrained test problem CTP2. This is a reprint of Figure 6 from Deb et al. (2001) (© Springer-Verlag Berlin Heidelberg 2001).

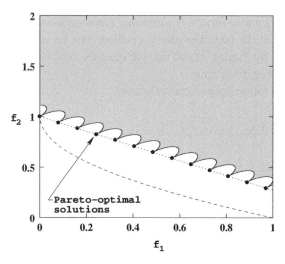

Figure 227 The constrained test problem CTP3. This is a reprint of Figure 7 from Deb et al. (2001) (© Springer-Verlag Berlin Heidelberg 2001).

The problem can be made more difficult by increasing the value of the parameter a, which has an effect of making the transition from the continuous to the discontinuous feasible region far away from the Pareto-optimal region. Since an algorithm now has to travel a long narrow feasible *tunnel* in search of the lone Pareto-optimal solution at the end of the tunnel, this problem will be much more difficult to solve compared to the previous problem. Figure 228 shows one such problem with $a = 0.75$ with the rest of the parameters being the same as those in the previous test problem.

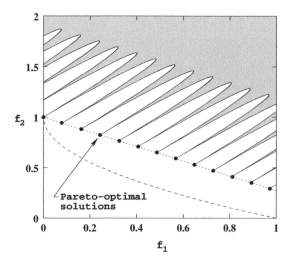

Figure 228 The constrained test problem CTP4. This is a reprint of Figure 8 from Deb et al. (2001) (© Springer-Verlag Berlin Heidelberg 2001).

In all of the three above problems, the disconnected regions are equally distributed in the objective space. The discrete Pareto-optimal solutions can be scattered non-uniformly by using $c \neq 1$. Figure 229 shows the feasible objective space for a problem with $c = 2$ and with the other parameters the same as those in Figure 227. Since $c > 1$, more Pareto-optimal solutions lie towards the right (higher values of f_1). If, however, $c < 1$ is used, more Pareto-optimal solutions will lie towards the left. For more Pareto-optimal solutions towards the right, the problem can be made more difficult by using a large value of c. The difficulty will arise in finding all of the many closely packed discrete Pareto-optimal solutions.

It is important to mention here that although the above test problems will cause difficulty in the vicinity of the Pareto-optimal region, an algorithm has to maintain an adequate diversity well before it comes close to the Pareto-optimal region. If an algorithm approaches the Pareto-optimal region without much diversity, it may be too late to create diversity among the population members, as the feasible search region in the vicinity of the Pareto-optimal region is discontinuous.

Difficulty in the Entire Search Space

The above test problems cause difficulty to an algorithm in the vicinity of the Pareto-optimal region. Difficulties may also come from the infeasible search region in the entire search space. Fortunately, the same constrained test problem generator can also be used for this purpose.

Figure 230 shows the feasible objective search space for the following parameter values:

$$\theta = 0.1\pi, \quad a = 40, \quad b = 0.5, \quad c = 1, \quad d = 2, \quad e = -2.$$

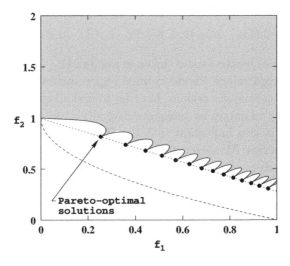

Figure 229   The constrained test problem CTP5. This is a reprint of Figure 9 from Deb et al. (2001) (© Springer-Verlag Berlin Heidelberg 2001).

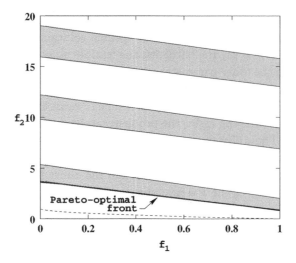

Figure 230   The constrained test problem CTP6. This is a reprint of Figure 10 from Deb et al. (2001) (© Springer-Verlag Berlin Heidelberg 2001).

The objective space has infeasible bands of differing widths towards the Pareto-optimal region. Since an algorithm has to overcome a number of such infeasible bands before coming to the island containing the Pareto-optimal front, an MOEA may face difficulty in solving this problem. The unconstrained Pareto-optimal region is now not feasible. The entire constrained Pareto-optimal front lies on a part of the constraint boundary. The difficulty can be increased by widening the infeasible regions (by using a small value of d).

The infeasibility in the objective search space may also exist in some part of the Pareto-optimal front. Using the following parameter values:

$$\theta = -0.05\pi, \quad a = 40, \quad b = 5, \quad c = 1, \quad d = 6, \quad e = 0.$$

we obtain the feasible objective space shown in Figure 231. This problem makes some portions of the unconstrained Pareto-optimal region infeasible, thereby providing a disconnected set of continuous regions. In order to find all such disconnected regions, an algorithm has to maintain an adequate diversity right from the beginning of a simulation run. It is also important that the Pareto-optimal solutions must lie on the constraint boundary. Moreover, the algorithm also has to maintain its solutions feasible as it proceeds towards the Pareto-optimal region.

With the constraints mentioned earlier, a combination of more than one effect can be achieved together in a problem. For example, with two constraints $C_1(\mathbf{x})$ and $C_2(\mathbf{x})$ having the following parameter values:

$$C_1: \quad \theta = 0.1\pi, \quad a = 40, \quad b = 0.5, \quad c = 1, \quad d = 2, \quad e = -2.$$
$$C_2: \quad \theta = -0.05\pi, \quad a = 40, \quad b = 2.0, \quad c = 1, \quad d = 6, \quad e = 0.$$

we obtain the feasible search space shown in Figure 232. Both constraints produce

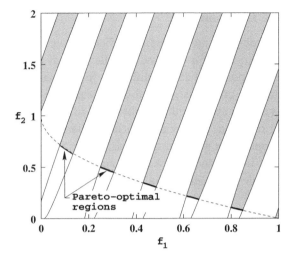

Figure 231 The constrained test problem CTP7. This is a reprint of Figure 11 from Deb et al. (2001) (© Springer-Verlag Berlin Heidelberg 2001).

disconnected islands of feasible objective space. It will become difficult for any optimization algorithm to find the correct feasible islands and converge to the Pareto-optimal solutions. As before, the difficulty can be controlled by the above six parameters. Although some of the above test problems can be questioned for their practical significance, it is believed here that an MOEA solving these difficult problems will also be able to solve any easier problem. However, the attractive aspect of these test problems is that the level of difficulty can be controlled to a desired level.

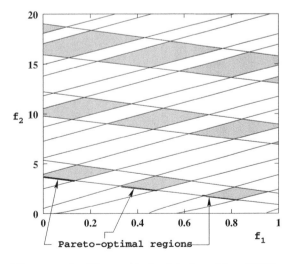

Figure 232 The constrained test problem CTP8.

Difficulty with the Function $g(\mathbf{x})$

In all of the above problems, an additional difficulty can be introduced by using a nonlinear and difficult function for $g(\mathbf{x})$, very similar to that used in the unconstrained test problems. The function $g(\mathbf{x})$ causes difficulty in progressing towards the Pareto-optimal front. In the test problems given above in equation (8.45), if a linear $g(\mathbf{x})$ is used, the decision variable space will resemble that of the objective space. However, for a nonlinear $g(\mathbf{x})$ function, the decision variable space can be very different. Since evolutionary search operators work on the decision variable space, the shape and the extent of infeasibility in the decision variable space is important. We illustrate the feasible decision variable space for the following two $g(\mathbf{x})$ functions:

$$
\begin{aligned}
g_1(\mathbf{x}) &= 1 + x_2, \\
g_2(\mathbf{x}) &= 11 + x_2^2 - 10\cos(2\pi x_2).
\end{aligned}
$$

In both cases, we vary $x_2 \in [0, 1]$. The Pareto-optimal solutions will correspond to $x_2^* = 0$. We use the test problem CTP7. Figures 233 and 234 show the feasible decision variable spaces for both cases. It is clear that with the function $g_2(\mathbf{x})$, the feasible decision variable space is more complicated. Since an MOEA search is performed in the decision variable space, it is expected that in solving the latter case, an MOEA may face more difficulty than the former.

More Than Two Objectives

By using the above concept, test problems having more than two objectives can also be developed. We shall modify equation (8.38) as follows. Using a M-dimensional

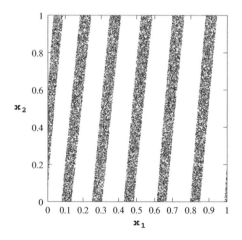

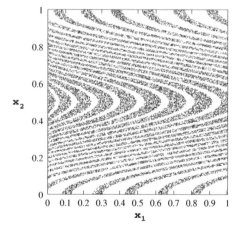

Figure 233 Random feasible solutions in the decision variable space for CTP7 with $g_1(\mathbf{x})$.

Figure 234 Random feasible solutions in the decision variable space for CTP7 with $g_2(\mathbf{x})$.

transformation (rotational R and translational e), we compute:

$$\mathbf{f}' = R^{-1}(\mathbf{f} - \mathbf{e}).$$

The matrix R will involve $(M-1)$ rotation angles. Thereafter, the following constraint set can be used:

$$C(\mathbf{x}) \equiv f_M'(\mathbf{x}) - \sum_{j=1}^{M-1} a_j \left| \sin\left(b_j \pi f'(\mathbf{x})_j^{c_j} \right) \right|^{d_j} \geq 0. \qquad (8.46)$$

Here, a_j, b_j, c_j, d_j and θ_j are all parameters that must be set to get a desired effect. As before, a combination of more than one such constraints can also be used.

8.4   Comparison of Multi-Objective Evolutionary Algorithms

With the availability of many multi-objective evolutionary algorithms, it is natural to ask which of them (if any) performs better when compared to other algorithms on various test problems. To settle on an answer to the above question, one has to be careful. Since a settlement based on analytical means is not easy, one has to carefully choose test problems for such a comparative study. In particular, we argue that the test problems must be hard enough to really distinguish the differences in the performances of the various algorithms. The previous section discussed in detail some such test problems. In the following, we will discuss a few significant studies where comparisons of different MOEAs have been made.

8.4.1   Zitzler, Deb and Thiele's Study

Zitzler (1999) and Zitzler et al. (2000) compared the following eight multi-objective EAs on the six test problems ZDT1 to ZDT6 described earlier.

RAND This is a random search method, where the non-dominated solutions from a randomly chosen set of solutions are reported. The size of the random population is kept the same as the total number of function evaluations used with the rest of the algorithms described below.

SOGA This is a single-objective genetic algorithm, which is used to solve a weighted sum of objectives. SOGA is applied many times, every time finding the optimum solution for a different weighted sum of objectives. For each weighted objective, SOGA is run for 250 generations.

VEGA This is the vector evaluated genetic algorithm suggested by Schaffer (1984). This algorithm was described earlier in Section 5.4.

HLGA This is the weighted genetic algorithm (Hajela and Lin, 1992), which was described earlier in Section 5.6.

MOGA This is Fonseca and Fleming's multi-objective genetic algorithm (Fonseca and Fleming, 1993), which was described earlier in Section 5.8.

NSGA This is the non-dominated sorting GA of Srinivas and Deb (1994). This algorithm was also described earlier in Section 5.9.

NPGA This is Horn, Nafploitis and Goldberg's niched Pareto genetic algorithm (Horn et al., 1994), described earlier in Section 5.10.

SPEA This is Zitzler and Thiele's strength Pareto evolutionary algorithm (Zitzler and Thiele, 1998b), described earlier in Section 6.4. It is interesting to note that this is the only *elitist* algorithm used in this comparative study.

For all algorithms, the following parameter values were used:

Population size	100,
Crossover rate	0.8,
Mutation rate	0.01,
Maximum number of generations	250.

Wherever a sharing strategy is used (in the NSGA, MOGA and NPGA), a niche radius $\sigma_{share} = 0.4886$ is used. For the ZDT5 problem defined over binary strings, a genotypic niche parameter of $\sigma_{share} = 34$ is used. In all problems except ZDT5, 30 bits are used to represent each decision variable. For the NPGA, a domination pressure of $t_{dom} = 10$ is used. For the SPEA, an external population of size 20 and the regular EA population of size 80 are used.

In all studies, five independent runs were made for each algorithm to solve each problem. The final populations from all five simulations are combined to create a set of 500 solutions. Thereafter, a non-domination sorting is performed to declare the best non-dominated set of solutions. In the case of the SOGA only, 100 different weight combinations were considered, each of which is run with a population size of 100 and for 250 generations. Thus, compared to the other seven algorithms, the SOGA uses 100 times more function evaluations in solving each problem.

The summary of the simulation results is nicely captured in a set of box plots, as shown in Figure 235. In each box, the upper and lower rows are the upper and lower quartiles of a pair-wise set coverage metric, described in equation (8.3) earlier. This metric measures the proportion of members in the set B which is dominated by members of set A. Thus, $C(A, B) = 1$ means all members of B are weakly dominated by A. On the other hand, $C(A, B) = 0$ means that no member of B is weakly dominated by A. Since domination is not a symmetric operator, $C(A, B)$ is not necessarily equal to $1 - C(B, A)$. The bold lines in the figure show the medians of $C(A, B)$. Dashed appendages show the shape of the distribution of the metric, while dots show the outliers. The following conclusions were made:

- All multi-objective EAs do better than the random search method (RAND).
- The NSGA outperforms all other non-elitist algorithms (all others except the SPEA) on all problems.
- The nonconvexity in the Pareto-optimal front of ZDT2 caused difficulty to the HLGA, VEGA and SOGA. Since all of these three algorithms used the weighted

Figure 235 Box plots showing the $\mathcal{C}(A, B)$ metric. Algorithms A and B refer to two algorithms in a row and column, respectively. Reproduced from Zitzler et al. (2000) (© 2000 by the Massachusetts Institute of Technology).

sum of the objectives, it was expected that they would face difficulty in solving a problem with a nonconvex Pareto-optimal front.

- Discreteness in the Pareto-optimal front for ZDT3 caused difficulty to most algorithms in finding a diverse set of solutions.
- Multi-modality (ZDT4) and deception (ZDT5) caused ample difficulty for all algorithms to converge to the true Pareto-optimal front. It was clear that a different parameter setting (perhaps with a large population size) was needed to find the true Pareto-optimal front in these problems.
- A search space with non-uniform density (in ZDT6) caused difficulty to all algorithms except the SPEA in maintaining a diverse set of solutions.
- A hierarchy of algorithms emerged in terms of their performances in finding a set of solutions close to the true Pareto-optimal front. The hierarchy in the descending

order of merit observed was as follows:

1. SPEA;
2. NSGA;
3. VEGA;
4. NPGA, HLGA;
5. MOGA;
6. RAND.

The lone elitist algorithm, SPEA, clearly outperformed any other algorithm. It was difficult to judge the superiority between the NPGA and HLGA. On some problems, the NPGA performed better than the HLGA, while on other problems the HLGA did better than the NPGA. On most problems, the SPEA performed better than the SOGA.

• Elite-preservation is an important matter for converging to the Pareto-optimal front.

Based on the above outcome, elitism was introduced with the NSGA, VEGA, HLGA, MOGA and NPGA. In general, the performances of all of the MOEAs with elitism improved considerably. Interestingly, both the SPEA and the elitist NSGA (different from the NSGA-II) performed equally well.

8.4.2   Veldhuizen's Study

Veldhuizen (1999) compared four algorithms (NSGA, NPGA, MOGA and MOMGA) on a set of seven test problems, including SCH1, FON, KUR, POL, VNT, and a two-variable version of ZDT3. The population size used in the first four algorithms was 50 and in the MOMGA was 100. A single-point crossover with a crossover probability of one and a bit-wise mutation with a mutation probability of 1/24 were used. All variables were coded in equal number of bits so that the overall string length becomes 24. For the NPGA, $t_{dom} = 5$ and for the NSGA, the σ_{share} values were calculated by using Fonseca and Fleming's update rule (see equation (5.18) above). He used a number of performance metrics covering the convergence and spread issues. Based on the generational distance and spacing metrics, his results showed the superiority of the MOMGA to the NSGA, NPGA and MOGA on the above test problems. The NSGA performed the worst among all four algorithms. Between the MOGA and NPGA, the former performed better. Although these results are contradictory to the conclusions reported by Zitzler–Deb–Thiele (1998b), different problems were tried in the respective studies. Moreover, the parameter settings used in the two studies were also not identical. For example, the overall size of the search space considered in the Veldhuizen's study (2^{24}) is much smaller than that used in Zitzler–Deb–Thiele's study (2^{300} to 2^{900}).

8.4.3   Knowles and Corne's Study

On a set of six test problems, including SCH1, SCH2 and KUR, Knowles and Corne (2000) compared a total of 13 algorithms. Besides the PAES, these investigators

considered the NSGA and NPGA with and without an elitist strategy. Some
algorithms were different implementations of the $(\mu + \lambda)$-PAES (with different values
of μ and λ). A population of size 100 (or archive size of 100) was used. For the NPGA,
a $t_{dom} \in [4, 10]$ was used; however, for the PAES a tournament size of 2 was used.
Crossover and mutation probabilities of 0.9 and 1/(string-length) were chosen. For
the NSGA and NPGA, the niche size parameter σ_{share} was calculated by performing
several experiments. Using the statistical metric based on the attainment surface
method, the conclusion was that the elitist NSGA with an archiving strategy is best
among all 13 algorithms, followed by the $(1 + 1)$-PAES. In the elitist NSGA, the five
best shared fitness solutions are directly copied into the next generation, thereby not
allowing the best five solutions to be lost in the genetic processing. For introducing
an archiving, an archive of size 100 is maintained with the non-dominated solutions,
as in the PAES, and the archive at the end of the final generation is presented as the
obtained non-dominated set. It was also observed that the computational complexity
for the elitist NSGA is more than that of the $(1 + 1)$-PAES. The use of a multiple-
membered PAES was found to be disadvantageous. Moreover, the computational
overhead was also more. The reason for the elitist NSGA's better performance was due
to its less-noisy non-dominated sorting and less-noisy fitness assignment procedures.
With the introduction of the NSGA-II, which uses a faster non-dominated sorting
procedure (compared to NSGA) and an elitist strategy, better solutions with a less
computational complexity are also possible. In the following, we will discuss a recent
comparative study involving the NSGA-II.

8.4.4   Deb, Agrawal, Pratap and Meyarivan's Study

From the above studies, one matter has become absolutely clear. Elitism plays a major
role in the performance of an MOEA. Motivated by these studies, newly suggested
algorithms are all developed with an explicit elite-preserving strategy. As discussed
earlier in Section 6.2, the NSGA-II is one such implementation. In the studies of Deb et
al. (2000a, 2000b), three different elitist MOEAs are compared: NSGA-II, SPEA and
PAES. A set of nine test problems (SCH1, FON, POL, KUR, ZDT1, ZDT2, ZDT3,
ZDT4 and ZDT6) are considered. For unconstrained multi-objective optimization,
some of these are, by far, the most difficult problems suggested in the literature. All
algorithms are run for a maximum of 250 generations with a population size of 100 (for
the SPEA, an EA population of size 80 and an external population of size 20 are used).
A crossover probability of 0.9 and a mutation probability of 1/(string-length) (for
binary-coded EAs) or $1/n$ (for real-parameter EAs) are used. For the real-parameter
NSGA-II, the SBX and real-parameter mutation operators are used with distribution
indices (discussed on page 113) $\eta_c = 20$ and $\eta_m = 20$. For binary-coded EAs, 30 bits
are used to code each variable.

   Table 29 shows the mean and variance values of a convergence metric (generational
distance (GD), redefined as Υ in the original study) obtained by using the four
algorithms NSGA-II (real-parameter), NSGA-II (binary-coded), SPEA and PAES.

Table 29　Mean $\overline{\Upsilon}$ and variance σ_Υ^2 values of the convergence metric Υ.

Algorithm		SCH	FON	POL	KUR	
NSGA-II	$\overline{\Upsilon}$	0.003391	0.001931	0.015553	0.028964	
Real-coded	σ_Υ^2	0	0	0.000001	0.000018	
NSGA-II	$\overline{\Upsilon}$	0.002833	0.002571	0.017029	0.028951	
Binary-coded	σ_Υ^2	0.000001	0	0.000003	0.000016	
SPEA	$\overline{\Upsilon}$	0.003465	0.010611	0.054531	0.049077	
	σ_Υ^2	0	0.000005	0.000179	0.000081	
PAES	$\overline{\Upsilon}$	0.001313	0.151263	0.030864	0.057323	
	σ_Υ^2	0.000003	0.000905	0.000431	0.011989	
Algorithm		ZDT1	ZDT2	ZDT3	ZDT4	ZDT6
NSGA-II	$\overline{\Upsilon}$	0.033482	0.072391	0.114500	0.513053	0.296564
Real-coded	σ_Υ^2	0.004750	0.031689	0.007940	0.118460	0.013135
NSGA-II	$\overline{\Upsilon}$	0.000894	0.000824	0.043411	3.227636	7.806798
Binary-coded	σ_Υ^2	0	0	0.000042	7.307630	0.001667
SPEA	$\overline{\Upsilon}$	0.001249	0.003043	0.044212	9.513615	0.020166
	σ_Υ^2	0	0.000020	0.000019	11.321067	0.000923
PAES	$\overline{\Upsilon}$	0.082085	0.126276	0.023872	0.854816	0.085469
	σ_Υ^2	0.008679	0.036877	0.00001	0.527238	0.006664

In all problems, a set of $|P^*| = 500$ uniformly spaced Pareto-optimal solutions are chosen. The generational distance of each of the $|Q|$ obtained non-dominated solutions from this set of 500 Pareto-optimal solutions is computed. This table shows that the NSGA-II (real-parameter or binary-coded) is able to converge better in all problems except for ZDT3 and ZDT6, where the PAES was found to have the better convergence. In all cases with the NSGA-II, the variance of Υ in 10 runs is also small, except for ZDT4, where the NSGA-II (binary coded) is the best. The fixed archive strategy of the PAES allows better convergence to be achieved in two out of the nine problems.

Table 30 shows the mean and variance values of the diversity metric (spread metric defined on page 328) Δ obtained using all three algorithms. The NSGA-II (real- or binary-coded) performs the best in all nine test problems in terms of Δ. For problems with disconnected Pareto-optimal regions, the Δ calculation is performed in each continuous region and averaged. The table shows that the worst performance is observed with the PAES. For illustration, we show only one of the ten runs of the PAES with an arbitrary run of the NSGA-II (real-parameter) on problem SCH1 in Figure 236. On most problems, the real-parameter NSGA-II is able to find a better spread of solutions than any other algorithm, including the binary-coded NSGA-II.

In order to demonstrate the working of these algorithms, we also show typical simulation results of the PAES, SPEA and NSGA-II on the test problems KUR, ZDT2, ZDT4 and ZDT6. The problem KUR has three disconnected regions in the

Table 30   Mean $\overline{\Delta}$ and variance $\sigma_{\overline{\Delta}}^2$ values of the diversity metric Δ.

Algorithm		SCH	FON	POL	KUR	
NSGA-II	$\overline{\Delta}$	0.477899	0.378065	0.452150	0.411477	
Real-coded	$\sigma_{\overline{\Delta}}^2$	0.003471	0.000639	0.002868	0.000992	
NSGA-II	$\overline{\Delta}$	0.449265	0.395131	0.503721	0.442195	
Binary-coded	$\sigma_{\overline{\Delta}}^2$	0.002062	0.001314	0.004656	0.001498	
SPEA	$\overline{\Delta}$	0.818346	0.804113	0.954327	0.880424	
	$\sigma_{\overline{\Delta}}^2$	0.004497	0.002961	0.013170	0.009066	
PAES	$\overline{\Delta}$	1.063288	1.162528	1.020007	1.079838	
	$\sigma_{\overline{\Delta}}^2$	0.002868	0.008945	0	0.013772	
Algorithm		ZDT1	ZDT2	ZDT3	ZDT4	ZDT6
NSGA-II	$\overline{\Delta}$	0.390307	0.430776	0.738540	0.702612	0.668025
Real-coded	$\sigma_{\overline{\Delta}}^2$	0.001876	0.004721	0.019706	0.064648	0.009923
NSGA-II	$\overline{\Delta}$	0.463292	0.435112	0.575606	0.479475	0.644477
Binary-coded	$\sigma_{\overline{\Delta}}^2$	0.041622	0.024607	0.005078	0.009841	0.035042
SPEA	$\overline{\Delta}$	0.730155	0.678127	0.665726	0.732097	0.900793
	$\sigma_{\overline{\Delta}}^2$	0.009066	0.004483	0.000666	0.011284	0.004124
PAES	$\overline{\Delta}$	1.229794	1.165942	0.789920	0.870458	1.153052
	$\sigma_{\overline{\Delta}}^2$	0.004839	0.007682	0.001653	0.101399	0.003916

Pareto-optimal front. Figure 237 shows all non-dominated solutions obtained after 250 generations with the NSGA-II (real-parameter). This figure demonstrates the abilities of the NSGA-II in converging to the true front and in finding diverse set of solutions in this front. Figure 238 shows the obtained non-dominated solutions with the SPEA, which is the next-best algorithm for this problem (see Tables 29 and 30). Although the convergence is adequate, the spread in the obtained non-dominated front is not as good as that obtained with the NSGA-II.

Next, we show the non-dominated solutions for the problem ZDT2 in Figures 239 and 240. This problem has a nonconvex Pareto-optimal front. The performances of the binary-coded NSGA-II and SPEA on this function are presented in these figures. Although the convergence is not a difficulty here, both the real and binary-coded NSGA-IIs are better able to spread solutions in the entire Pareto-optimal region than the SPEA (the next-best algorithm observed for this problem).

The problem ZDT4 has 21^9 or $7.94(10^{11})$ different local Pareto-optimal fronts in the search space, of which only one corresponds to the global Pareto-optimal front. The Euclidean distance in the decision space between solutions of two consecutive local Pareto-optimal sets is 0.25. Figure 241 shows that both of the real-parameter NSGA-II and PAES get stuck at different local Pareto-optimal sets, although the convergence and ability to find a diverse set of solutions are definitely better with the NSGA-

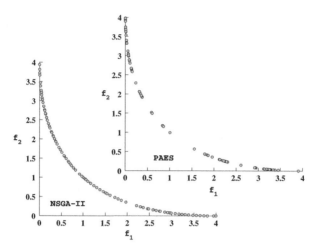

Figure 236 The NSGA-II finds a better spread of solutions than the PAES on SCH1.

II. Binary-coded GAs have difficulties in converging near the global Pareto-optimal front, a matter which was also observed in previous single-objective studies (Deb and Agrawal, 1999b). On a similar 10-variable Rastrigin's function (which is used as $g(\mathbf{x})$ in ZDT4), that study clearly showed that a population of size of about at least 500 is needed for single-objective binary-coded GAs (with a tournament selection, a single-point crossover and a bit-wise mutation) to find the global optimum solution in more than 50% of the simulation runs. Since we have used a population of size 100, it is not expected that a multi-objective GA would find the global Pareto-optimal solution in this case.

Finally, Figure 242 shows that the SPEA finds a better converged set of non-

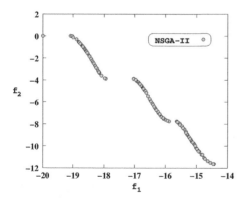

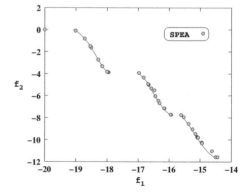

Figure 237 Non-dominated solutions with the NSGA-II (real-parameter) on problem KUR.

Figure 238 Non-dominated solutions with the SPEA on problem KUR.

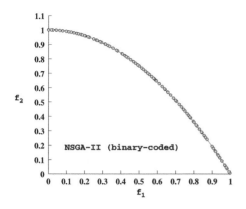

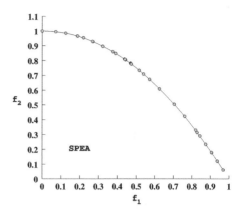

Figure 239 Non-dominated solutions with NSGA-II (binary-coded) on ZDT2.

Figure 240 Non-dominated solutions with SPEA on ZDT2.

dominated solutions in problem ZDT6 compared to any other algorithm. However, the distribution in solutions is better with the real-parameter NSGA-II. Furthermore, by using different parameter settings (running NSGA-II for longer generations and by using different η_m values), investigators have been able to improve the performance of the NSGA-II in all problems (Deb et al., 2000a).

In our original study, we have also compared three algorithms on a rotated EC4 problem (Deb et al., 2000a), in which the decision variables are rotated before calculating the objective function values (discussed above on page 353). The rotation of the decision variables causes all of the latter to vary in a particular fashion in order to remain on the Pareto-optimal front, a matter which provides additional difficulty (the

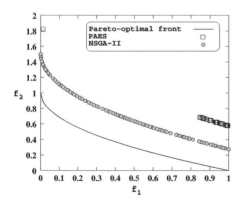

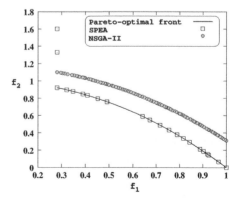

Figure 241 The real-parameter NSGA-II finds a better convergence and spread of solutions than the PAES on problem ZDT4.

Figure 242 The real-parameter NSGA-II finds a better distribution of solutions than the SPEA on problem ZDT6, although the SPEA has a better convergence.

so-called linkage problem) to any optimization algorithm. As expected, no algorithm performed well on this problem.

8.4.5 Constrained Optimization Studies

Unlike in the case of unconstrained multi-objective optimization, there do not exist many systematic studies on constrained multi-objective optimization. One of the reasons for this is probably the unavailability of suitable test problems for constrained multi-objective optimization. With the test problems suggested above in Section 8.3.5 of this book, a consideration of more constrained optimization algorithms and further comparative studies may be of potential interest to researchers.

In the following, we will describe one such recent study (Deb et al., 2000a), where the test problems, Constr-Ex, SRN and TNK, and an engineering design problem, WATER (Ray et al., 2001) (as described below) were studied:

$$
\begin{aligned}
\text{Minimize} \quad & f_1(\mathbf{x}) = 106\,780.37(x_2 + x_3) + 61\,704.67, \\
\text{Minimize} \quad & f_2(\mathbf{x}) = 3000x_1, \\
\text{Minimize} \quad & f_3(\mathbf{x}) = (305\,700)2289x_2/(0.06 \times 2289)^{0.65}, \\
\text{Minimize} \quad & f_4(\mathbf{x}) = (250)2289 \exp(-39.75x_2 + 9.9x_3 + 2.74), \\
\text{Minimize} \quad & f_5(\mathbf{x}) = 25[1.39/(x_1x_2) + 4940x_3 - 80], \\
\text{Subjected to} \quad & g_1(\mathbf{x}) \equiv 0.001\,39/(x_1x_2) + 4.94x_3 - 0.08 \leq 1, \\
& g_2(\mathbf{x}) \equiv 0.000\,306/(x_1x_2) + 1.082x_3 - 0.0986 \leq 1, \\
& g_3(\mathbf{x}) \equiv 12.307/(x_1x_2) + 49\,408.24x_3 \\
& \qquad\quad + 4051.02 \leq 50\,000, \\
& g_4(\mathbf{x}) \equiv 2.098/(x_1x_2) + 8046.33x_3 \\
& \qquad\quad - 696.71 \leq 16\,000, \\
& g_5(\mathbf{x}) \equiv 2.138/(x_1x_2) + 7883.39x_3 \\
& \qquad\quad - 705.04 \leq 10\,000, \\
& g_6(\mathbf{x}) \equiv 0.417(x_1x_2) + 1721.26x_3 - 136.54 \leq 2000, \\
& g_7(\mathbf{x}) \equiv 0.164/(x_1x_2) + 631.13x_3 - 54.48 \leq 550, \\
& 0.01 \leq x_1 \leq 0.45, \\
& 0.01 \leq x_2 \leq 0.10, \\
& 0.01 \leq x_3 \leq 0.10.
\end{aligned}
\tag{8.47}
$$

In the Constr-Ex problem, a part of the unconstrained Pareto-optimal region is made infeasible by a constraint. Thus, the resulting constrained Pareto-optimal region is a concatenation of the first constraint boundary and some part of the unconstrained Pareto-optimal region. In SRN, the constrained Pareto-optimal set is only a sub-set of the unconstrained Pareto-optimal set. The third problem TNK, has a discontinuous Pareto-optimal region, entirely falling on the first constraint boundary. The fourth problem, WATER, is a five-objective and seven-constraint problem, which Ray et al. (Ray et al., 2001) have attempted to solve. With five objectives, it is difficult to discuss the effect of constraints on the unconstrained Pareto-optimal region, but below we will show all $\binom{5}{2}$ or 10 pair-wise plots presenting the obtained non-dominated solutions.

In all of the problems, a population size of 100, and distribution indices for the real-parameter crossover and mutation operators of $\eta_c = 20$ and $\eta_m = 100$, respectively, were used. The constrain-domination concept and the constrained handling technique of Ray–Tai–Seow (RTS) (Ray et al., 2000, 2001) are implemented in two NSGA-II implementations. These NSGA-IIs are run for a maximum of 500 generations. Crossover and mutation probabilities were the same as before. Figure 243 shows the obtained set of 100 non-dominated solutions after 500 generations. This figure shows that the NSGA-II with the constrain-domination principle is able to maintain solutions in the entire constrained Pareto-optimal region. It is important to note that in order to maintain a spread of solutions on the constraint boundary, solutions must be modified in a particular manner dictated by the constraint function. This makes the task difficult for a search operator. Figure 244 shows the obtained solutions using the RTS procedure after 500 generations of NSGA-II. It is clear that this NSGA-II performs poorer than the constrain-domination-based NSGA-II (see Figure 243) in terms of converging to the true Pareto-optimal front, and also in terms of maintaining a diverse set of non-dominated solutions.

Simulation results on the SRN problem using the two constraint handling approaches are shown in Figures 245 and 246. Figures 247 and 248 show the feasible objective spaces and the obtained non-dominated solutions with the constrain-domination-based NSGA-II and the RTS-based NSGA-II on the problem TNK. In both of the above problems (SRN and TNK), the superiority of the constrain-domination principle in both aspects of multi-objective optimization is clear.

Ray et al. (2001) normalized the objective functions of the problem WATER in the following manner:

$$f_1/8(10^4), \quad f_2/1500, \quad f_3/3(10^6), \quad f_4/6(10^6), \quad f_5/8000.$$

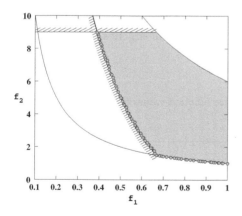

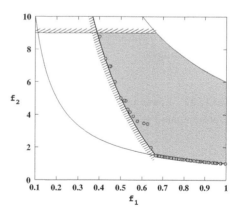

Figure 243 Obtained non-dominated solutions with the constrain-domination-based NSGA-II on the constrained problem Constr-Ex.

Figure 244 Obtained non-dominated solutions with the RTS-based NSGA-II on the constrained problem Constr-Ex.

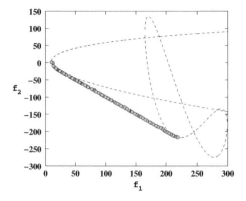

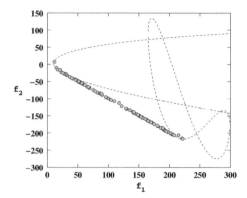

Figure 245   Obtained non-dominated solutions with the constrain-domination-based
NSGA-II on the constrained problem SRN.

Figure 246   Obtained non-dominated solutions with the RTS-based NSGA-II on the
constrained problem SRN.

Since there are five objective functions in WATER, we used the scatter-plot matrix
method of comparing the solutions, as shown in Figure 249. Table 31 shows the
comparison of the constrain-domination-based and RTS-based NSGA-IIs in terms of
the spread of obtained non-dominated solutions. In most multi-objective problems,
constrain-domination-based NSGA-II has found a better spread of solutions than RTS-
based NSGA-II. The former results are shown in the upper diagonal portion of the
figure, while the latter results are shown in the lower diagonal portion. The axes of
any plot can be obtained by looking at the corresponding diagonal boxes and their
ranges. We observe that the constrain-domination-based NSGA-II plots have better-
formed patterns than the RTS plots, meaning that better convergence is achieved by
the former approach. For example, the interactions in figures f_1–f_3, f_1–f_4, and f_3–f_4
are very clear from the constrain-domination results. Although similar patterns exist
in the results obtained by using the RTS technique, the convergence to the true fronts
is not adequate.

Table 31   Lower and upper bounds of the objective function values observed in the obtained
non-dominated solutions.

Approach	f_1	f_2	f_3
Constrain-domination	0.798–0.920	0.027–0.900	0.095–0.951
RTS	0.810–0.956	0.046–0.834	0.067–0.934

Approach	f_4	f_5
Constrain-domination	0.031–1.110	0.001–3.124
RTS	0.036–1.561	0.211–3.116

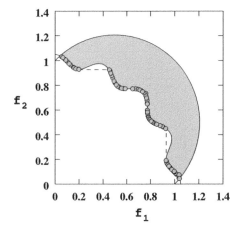

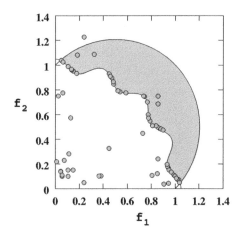

Figure 247 Obtained non-dominated solutions with the constrain-domination-based NSGA-II on the constrained problem TNK.

Figure 248 Obtained non-dominated solutions with the RTS-based NSGA-II on the constrained problem TNK.

8.5   Objective Versus Decision-Space Niching

The issue of objective versus decision-space niching is important in the context of obtaining diversity among non-dominated solutions. It is clear that when objective-space niching is performed, diversity in objective function values is expected, whereas when parameter-space niching is performed, diversity in the decision variable values is expected. We illustrate the difference by comparing the performance of the NSGA on a test problem with two different niching strategies. The problem is given below:

$$\left.\begin{array}{ll} \text{Minimize} & f_1(\mathbf{x}) = 1 - \exp(-4x_1)\sin^4(5\pi x_1), \\[2mm] \text{Minimize} & f_2(\mathbf{x}) = g(x_2)\left[1 - \left(\dfrac{f_1(\mathbf{x})}{g(x_2)}\right)^4\right], \\[2mm] \text{where} & g(x_2) = 1 + 9x_2^2, \\[1mm] & 0 \le x_1 \le 1, \quad -1 \le x_2 \le 1. \end{array}\right\} \qquad (8.48)$$

The NSGA with a reasonable parameter setting (population size of 100, 15-bit coding for each variable, σ_{share} of 0.2236, crossover probability of one, and no mutation) is run for 500 generations. Typical runs for both niching methods are shown in Figures 250 and 251. Although it seems that both methods are able to maintain diversity in the function space (with a better distribution in the f_1–f_2 space with function-space niching), the inset for the first plot (Figure 250) shows that the NSGA with parameter-space niching has found diverse values of the decision variable, whereas the NSGA with function-space niching (Figure 251) converges to only about 50% of the entire decision variable space. Since the first minimum and its basin of attraction span the complete space for the function f_1, the function-space niching does not have the motivation to find other important solutions. Thus, in problems like this, function-space niching

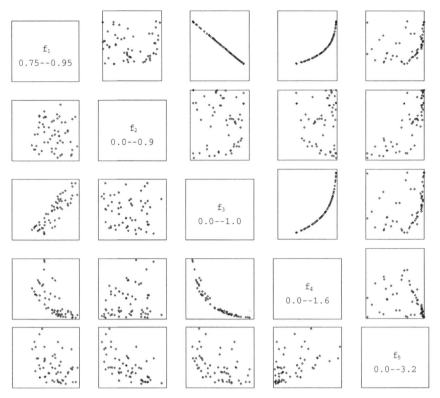

Figure 249 Upper diagonal plots for the constrain-domination-based NSGA-II and lower diagonal plots are for RTS-based NSGA-II. Compare the (i,j) plot (RTS procedure with $i > j$) with the (j,i) plot (constrain-domination procedure). The labels and ranges used for each axis are shown in the diagonal boxes.

may hide information about other important Pareto-optimal solutions in the search space.

It is important to understand that the choice between parameter-space or function-space niching entirely depends on what is desired for a set of Pareto-optimal solutions in the underlying problem. In some problems it may be important to have solutions with a trade-off in objective function values, without much regard for how similar or diverse the actual solutions (**x** vectors or strings) are. In such cases, function-space niching will, in general, provide solutions with a better trade-off in objective function values. Since there is no induced pressure for the solutions to differ from each other in the decision variable space, the Pareto-optimal solutions may not be very different from each other, unless the underlying objective functions demand them to be so. On the other hand, in some problems the emphasis could be on finding a set of solutions which are diverse enough in the decision-variable space, as well as maintaining a trade-off among the objective functions. In such cases, parameter-space niching would be better. This is because, in some sense, categorizing a population by using non-

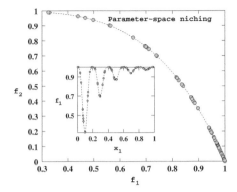

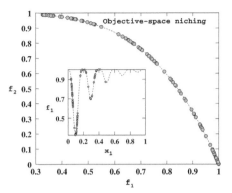

Figure 250 Parameter-space niching finds diverse solutions in both of the objective and decision-variable spaces. Reproduced from Deb (1999c) (© 1999 by the Massachusetts Institute of Technology).

Figure 251 Function-space niching finds diverse solutions in the objective space, but not in the decision-variable space. Reproduced from Deb (1999c) (© 1999 by the Massachusetts Institute of Technology).

domination helps to preserve some diversity among the objective functions and an explicit parameter-space niching helps to maintain diversity in the solution vector.

The fitness-space niching has an advantage over parameter-space niching. Since diversity is maintained in the objective space, MOEAs which employ fitness-space niching can be extended naturally to any other EA, such as to a GP or an EP. A GP, for example, uses parse trees as a solution. Finding a similarity measure between two parse trees for implementing niching is more computationally complex than carrying out the same between their objective vectors (Langdon, 1995). Whatever the way solutions are represented, since every EA solution would be evaluated as a real vector of objectives, the MOEAs with fitness-space niching described in this book can be used directly with other EAs.

8.6 Searching for Preferred Solutions

The past few years of research on multi-objective evolutionary optimization algorithms have amply demonstrated that they are capable of finding multiple and diverse Pareto-optimal (or near Pareto-optimal) solutions in a single simulation run. It is then natural to ask: 'How does one choose a particular solution from the obtained set of non-dominated solutions?'. In the following, we first review a few techniques often followed in the context of multiple criterion decision-making and then suggest a number of techniques which have been used in MOEA research.

We categorize the techniques into two types: post-optimal techniques and optimization-level techniques. In the former approach, the solutions obtained after the optimization technique are analyzed to choose a preferred solution. On the other hand, in the second approach the optimization technique is directed towards a preferred

Pareto-optimal region.

8.6.1 Post-Optimal Techniques

Once a set of non-dominated solutions is obtained, usually some higher-level decision-making considerations (often societal or political) are used to choose a solution. The following methods are often used in classical multi-objective optimization.

Compromise Programming Approach

The method of compromise programming, sometimes known as 'the method of global criteria', picks a solution which is minimally located from a given reference point (Yu, 1973; Zeleny, 1973). The user has to fix a distance metric d() and a reference point $\mathbf{z}$ for this purpose. A couple of commonly used metrics are presented below:

$$l_p\text{-metric:} \qquad d(\mathbf{f}, \mathbf{z}) = \left(\sum_{m=1}^{M} |f_m(\mathbf{x}) - z_m|^p \right)^{1/p}, \qquad (8.49)$$

$$\text{Tchebycheff metric:} \qquad d(\mathbf{f}, \mathbf{z}) = \max_{m=1}^{M} \frac{|f_m(\mathbf{x}) - z_m|}{\max_{\mathbf{x} \in \mathcal{S}} f_m(\mathbf{x}) - z_m}. \qquad (8.50)$$

Here, $\mathcal{S}$ is the entire search space. The reference point can be chosen as the *ideal* point (see Section 2.4.1 above), which is usually comprised of the individual best objective function values $\mathbf{z} = (f_1^*, f_2^*, \ldots, f_M^*)^\mathsf{T}$. Since this solution is usually a 'non-existent solution', the user is interested in choosing a feasible solution, which is closest to this reference solution. Figure 252 shows the reference point and the chosen solution in a hypothetical set of obtained Pareto-optimal solutions (marked by filled circles) for a two-objective minimization problem. The l_2-metric becomes the Euclidean distance metric.

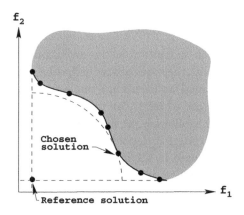

Figure 252 The reference point and the chosen optimal solution.

Marginal Rate of Substitution Approach

The marginal rate of substitution is the amount of improvement in one objective function which can be obtained by sacrificing an unit decrement in any other objective function (Miettinen, 1999). The solution having the maximum marginal rate of substitution is the one chosen by this method. Since pair-wise comparisons have to be made with all M objectives for each Pareto-optimal solution, this method may be computationally expensive. Figure 253 shows the preferred 'knee' point, where the marginal rate of substitution is maximum among a set of obtained Pareto-optimal solutions for both objectives.

Pseudo-Weight Vector Approach

In this approach, a pseudo-weight vector is calculated for each obtained solution. Although there may exist a number of strategies to calculate a pseudo-weight vector, we propose a simple procedure. Here, we will assume minimization problems only. However, a little effort is needed to modify the procedure for maximization problems.

From the obtained set of solutions, the minimum f_i^{min} and maximum f_i^{max} values of each objective function i are noted. Thereafter, the following equation is used to compute the weight w_i for the i-th objective function:

$$w_i = \frac{(f_i^{max} - f_i(\mathbf{x})) / (f_i^{max} - f_i^{min})}{\sum_{m=1}^{M} (f_m^{max} - f_m(\mathbf{x})) / (f_m^{max} - f_m^{min})} \tag{8.51}$$

This equation calculates the relative distance of the solution from the worst (maximum) value in each objective function. Thus, for the best solution for the i-th objective, the weight w_i is a maximum. The numerator in the right side of the above equation ensures that the sum of all weight components for a solution is equal to one. Figure 254 shows a set of non-dominated solutions and their pseudo-weights for both objectives.

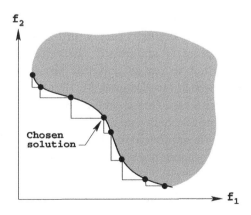

Figure 253   The chosen solution having the maximum marginal rate of substitution.

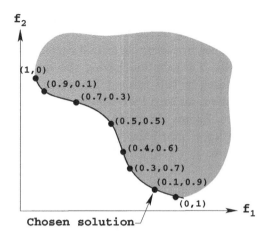

Figure 254 The chosen optimal solution using the pseudo-weight vector approach.

We call the weight vector calculated by equation (8.51) a *pseudo* weight vector for a particular purpose. The weight calculated by the above equation for any Pareto-optimal solution need not necessarily be identical to the true Pareto-optimal solution corresponding to the same weight vector. Either because of the presence of constraints or because of nonconvexity in the Pareto-optimal region, the extreme solutions in the true Pareto-optimal set may not be the solutions found by using extreme weight vectors (such as $w_i = 1$ and all other $w_j = 0$). However, for problems having a convex Pareto-optimal region, the weight computed using the above equation gives a relative idea of the location of the solution in the Pareto-optimal region. The interesting part is that although there is no formal definition of a weight vector for a solution situated in the nonconvex region of the Pareto-optimal region, equation (8.51) allows us a way to compute a relative trade-off value between objectives for all obtained non-dominated solutions.

Once the weight vectors are calculated, a simple strategy would be to choose the solution closer to a user-preferred weight vector. In Figure 254, if a 90% weightage for f_2 and a 10% weightage for f_1 are desired, the corresponding choice is marked on the figure. It is important to realize that this procedure is different from the classical weighted sum approach (see Chapter 3). Here, a solution is chosen from a set of obtained solutions, with each corresponding to a different weight vector. In the classical weighted sum approach, only one solution optimizing the weighted average of the objectives is found. We argue that in the presence of multiple non-dominated solutions, it is convenient to choose one solution with a preferred weight vector.

8.6.2   Optimization-Level Techniques

In this section, we describe a number of methods which can be used during the optimization phase to find a preferred set of solutions in the Pareto-optimal region.

Finding a biased distribution, based on a user's preference for objectives, has two advantages:

1. The search effort can be reduced.
2. Better precision in the non-dominated solutions can be obtained.

Since the search is directed to a particular region in the search space, it is intuitive that a search effort would be reduced. Secondly, since a smaller region is searched, the density of the obtained solutions is expected to be high, thereby increasing the precision in such solutions.

Utility Functions

Multiple objective functions can be used to construct a utility function $U(\mathbf{f})$ (Keeney and Raiffa, 1976). The meaning of a utility function is that any two solutions having the same utility function value have the same preference to a user. In this way, multiple objectives are reduced to a single objective of maximizing the utility function. It is obvious that the construction of a utility function is problem-dependent and is highly subjective to the user. However, if a utility function can be constructed, a solution maximizing the utility function can be obtained.

Biased Sharing Approach

Here, we propose a sharing approach which uses a biased distance metric. In calculating the distance metric (discussed in Section 5.9 above) in the fitness-space sharing, the following normalized distance metric is suggested:

$$d(i,j) = \left[\sum_{k=1}^{M} \frac{(f_k^{(i)} - f_k^{(j)})^2}{(f_k^{max} - f_k^{min})^2} \right]^{\frac{1}{2}}. \tag{8.52}$$

This distance metric is nothing but the normalized Euclidean distance between two objective vectors. In the proposed biased sharing approach, an unequal weightage is given to each objective in computing the Euclidean distance. For example, if w_k ($\in (0,1)$) is the weight assigned to the k-th objective function, then the normalized w_k' is calculated for a convex problem as follows:

$$w_k' = \frac{(1 - w_k)}{max_{m=1}^{M}(1 - w_m)}, \tag{8.53}$$

and the modified distance metric is computed as follows:

$$d(i,j) = \left[\sum_{k=1}^{M} w_k' \frac{(f_k^{(i)} - f_k^{(j)})^2}{(f_k^{max} - f_k^{min})^2} \right]^{\frac{1}{2}}. \tag{8.54}$$

The fitness-based sharing can then be used with this distance metric. It is important to realize that when a sharing is performed with the second objective alone (with a large w_2), more solutions near the optimum value of f_1 would be obtained. This is why a transformation of weights, as used in equation (8.53), is necessary. Although this new distance metric requires a new estimate for σ, equation (4.64) (see earlier) can also be used. With this approach, the highest-priority objective function always gets a weight equal to one, whereas all others get a weight between zero and one. It is interesting to note that if equal weights are assigned to each objective, equation (8.54) reduces to equation (8.52).

For convex Pareto-optimal regions, a higher weight for an objective function will produce more dense solutions near the individual optimum. Figure 255 explains this fact. For a two-objective optimization problem, if an extreme case of $w_1 = 0$ and $w_2 = 1$ is used, the corresponding $\mathbf{w}'$ is as follows: $w'_1 = 1$ and $w'_2 = 0$. Since the effect of f_2 is absent in calculating the distance metric, equal number of solutions are expected to be created in each equal partition of the f_1 search space. Thus, the density of the Pareto-optimal solutions in the partition closer to the individual best f_1^* will be less. For the nonconvex Pareto-optimal region, w'_k can be calculated as $w'_k = w_k / \max_{i=1}^{M} w_k$.

In the following, we apply a modified real-parameter NSGA with the above biased sharing approach to the test problem SCH1. Other applications can be found elsewhere (Deb, in press). We use a population of size 100, crossover probability of 0.9, mutation probability of $1/n$, and distribution indices for the simulated binary crossover and the polynomial mutation operators of $\eta_c = 30$ and $\eta_m = 500$, respectively. The NSGAs are run up until 500 generations. In SCH1, the Pareto-optimal solutions lie in $x \in (0, 2)$. We divide this region into 10 equal divisions and count the number of Pareto-optimal solutions found in each division under different weight vectors. Figure 256 shows the number of such individuals versus x (the number of solutions in $(0,0.2)$ are shown at $x = 0.2$ in the figure). An average of 20 runs is plotted. This figure clearly shows that when weights $w_1 = w_2 = 0.5$ are used, a more or less uniform distribution is observed.

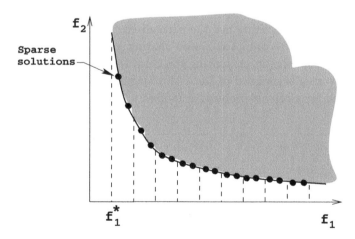

Figure 255 The density of Pareto-optimal solutions near the individual champion solution of f_1 is less with $w_1 = 0$ and $w_2 = 1$.

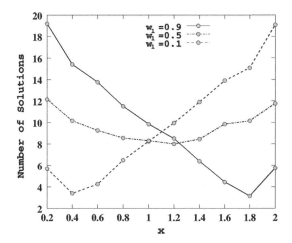

Figure 256 The number of solutions in each partition of the Pareto-optimal region under different weight vectors for the problem SCH1.

However, when $w_1 = 0.1$ and $w_2 = 0.9$ are used, more solutions are found closer to the $x = 2$ solution (the individual champion to the single-objective optimization problem: minimize $f_2 = (x-2)^2$). However, when $w_1 = 0.9$ and $w_2 = 0.1$ are used, an opposing trend emerges. This experiment shows how by changing the weight vector, the density of solutions along the Pareto-optimal front can be changed. In order to investigate how the Pareto-optimal solutions are distributed, we have also plotted the solutions obtained for one run in Figures 257 and 258 for $w_1 = 0.9$ and $w_1 = 0.1$, respectively. Figure 257, with $w_1 = 0.9$, shows that more solutions are near f_1^*, while Figure 258, with $w_1 = 0.1$, shows that more solutions are near f_2^*.

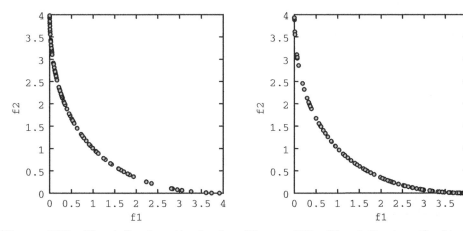

Figure 257 Biased Pareto-optimal solutions with $w_1 = 0.9$ on SCH1.

Figure 258 Biased Pareto-optimal solutions with $w_1 = 0.1$ on SCH1.

It is interesting to note that the proposed approach does not find a single compromised solution; instead, it finds a biased distribution of solutions. If a user wants to have a bias towards a particular objective, this biased sharing approach produces more solutions towards the preferred region in the search space. Finding more dense solutions in the region of preference is a much better approach than predefining a weighted sum of objective and then finding only one optimum solution. With a biased population, there exist many solutions in the desired region. One solution can then be chosen by using the previously described methods. The above biased niching concept can also be implemented with other niching procedures, such as with the crowded distance operator used in the NSGA-II.

Although the above technique is able to find a biased distribution of solutions towards a particular objective, the procedure is not appropriate for finding solutions focused in any arbitrary part of the Pareto-optimal region. For example, if one is interested in finding solutions only in the middle portion of the Pareto-optimal region, the above procedure would not be able to find a biased distribution. The following approach uses the generalized convex cone approach (Miettinen, 1999), often practiced in the classical multi-objective optimization.

Guided Domination Approach

In this approach (Branke et al., 2000), a weighted function of the objectives is defined as follows:

$$\Omega_i(\mathbf{f}(\mathbf{x})) = f_i(\mathbf{x}) + \sum_{j=1, j \neq i}^{M} a_{ij} f_j(\mathbf{x}), \quad i = 1, 2, \ldots, M. \tag{8.55}$$

where a_{ij} is the amount of gain in the j-th objective function for a loss of one unit in the i-th objective function. The above set of equations require fixing the matrix **a**, which has a one in its diagonal elements. Now, we define a different domination concept for minimization problems as follows.

Definition 8.1. *A solution* $\mathbf{x}^{(1)}$ *dominates another solution* $\mathbf{x}^{(2)}$, *if* $\Omega_i(\mathbf{f}(\mathbf{x}^{(1)})) \leq \Omega_i(\mathbf{f}(\mathbf{x}^{(2)}))$ *for all* $i = 1, 2, \ldots, M$ *and the strict inequality is satisfied at least for one objective.*

Let us illustrate the concept for two ($M = 2$) objective functions. The two weighted functions are as follows:

$$\Omega_1(f_1, f_2) = f_1 + a_{12} f_2, \tag{8.56}$$
$$\Omega_2(f_1, f_2) = a_{21} f_1 + f_2. \tag{8.57}$$

Figure 259(b) shows the contour lines corresponding to the above two linear functions passing through a solution A in the objective space. All solutions in the hatched region are dominated by A according to the above definition of domination. It is interesting to note that when using the usual definition of domination (Figure 259(a)), the region marked by a horizontal and a vertical line will be dominated by A. Thus, it is clear

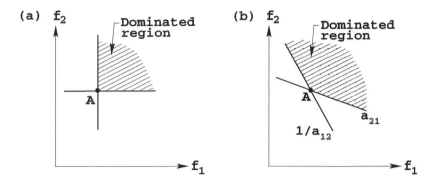

Figure 259 The regions dominated by solution A: (a) the dominated region using the usual definition; (b) the dominated region using definition 8.1.

from these figures that the modified definition of domination allows a larger region to become dominated by any solution than the usual definition. It is also interesting to realize that since a larger region is now dominated, the complete Pareto-optimal front (as per the original domination definition) may not be non-dominated according to this new definition of domination. For the same value of the matrix $\mathbf{a}$ in the two-objective function illustration, the resulting non-dominated front is depicted in Figure 260. This figure shows that regions near the individual champions are now not non-dominated. It is clear from this figure that some portion (shown by a thin continuous curve) of the Pareto-optimal region is dominated by a member in the middle portion of the Pareto-optimal region (shown by a bold curve). Thus, an MOEA is expected to find only the middle portion of the Pareto-optimal region, thereby biasing the search towards a particular region of the Pareto-optimal front. Thus, by choosing appropriate values for

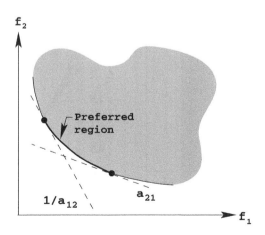

Figure 260 The non-dominated portion of the Pareto-optimal region.

the elements of the matrix a, a part of the Pareto-optimal region can be emphasized.

We would like to highlight that this approach can also be viewed in a different manner. Instead of using a modified domination principle, this procedure can be viewed as an multi-objective optimization approach with the original domination principle acting on a linearly transformed set of objective functions. A little thought will reveal that the above definition of domination on the objective vector f is the same as the original domination definition on the transformed vector Ω. Thus, the inability of weight-based approaches to handle problems with nonconvex Pareto-optimal region still holds for this modified approach.

In order to demonstrate the working of the above procedure, we apply an NSGA with the modified domination principle. All NSGA parameters are chosen the same as that used in the biased sharing approach. For the SCH1 problem, we first use the following values: $a_{12} = a_{21} = 0.75$. The corresponding extreme contour lines are shown in Figure 261. It is clear that the Pareto-optimal region enclosed between these two extreme contour lines dominates any solution in the other regions of the original Pareto-optimal front. The NSGA simulation shows how the simple change in the domination principle allows us to find a biased distribution of the Pareto-optimal solutions.

Next, we change the parameters as follows: $a_{12} = 0$ and $a_{21} = 0.75$. This allows a region closer to the minimum f_1 solution to be found. Figure 262 shows that an NSGA with identical parameter settings has found the corresponding portion of the Pareto-optimal region.

Finally, we apply an NSGA with the guided domination principle on the welded beam design problem (discussed earlier on page 128). Here, two objectives (weight

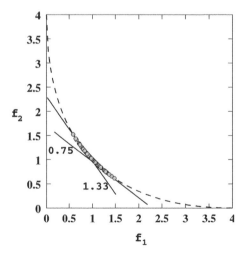

Figure 261 An intermediate portion of the Pareto-optimal region for the problem SCH1.

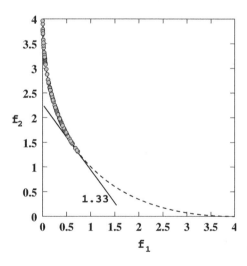

Figure 262 A region near the minimum f_1 solution for the problem SCH1.

and the deflection) are minimized. The parameter settings and the problem details are the same as before. By using $a_{12} = 0.50$ and $a_{21} = 0.25$, we obtain the portion of the Pareto-optimal region shown in Figure 263. A simulation of the NSGA with the usual domination principle is also shown. This figure clearly shows how the guided domination approach finds the intermediate portion of the non-dominated front. It is important to realize that for any weight-based method, the objective functions must be normalized. In the welded beam problem, we have multiplied the deflection function by 10^3 before using the guided domination operation. When such normalization is used, it becomes somewhat difficult to choose an appropriate weight a_{ij}. Nevertheless, this is a useful approach for finding a biased distribution of solutions in the Pareto-optimal region.

Weighted Domination Approach

Somewhat different from the above approach, Parmee et al. (2000) suggested a weighted dominance principle. These investigators defined an index function $I_i(x^{(1)}, x^{(2)})$ between two solutions $x^{(1)}$ and $x^{(2)}$ for the i-th objective function as follows:

$$I_i(x^{(1)}, x^{(2)}) = \begin{cases} 1, & f_i(x^{(1)}) \le f_i(x^{(2)}); \\ 0, & \text{otherwise.} \end{cases} \tag{8.58}$$

Now, if solution $x^{(1)}$ dominates solution $x^{(2)}$, then each I_i function would take a value equal to one. In other words, we can write the usual condition for domination in a different way:

$$\sum_{i=1}^{M} I_i(x^{(1)}, x^{(2)}) = M, \tag{8.59}$$

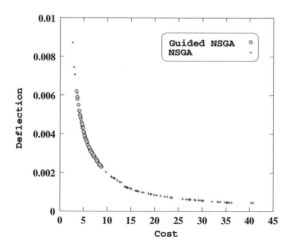

Figure 263 An intermediate region of the Pareto-optimal front for the welded beam design problem.

with the condition that $f_i(x^{(1)}) < f_i(x^{(2)})$ for at least one objective i. In the original definition of domination, all objectives are given equal importance. Rewriting the above equation and extending the relationship to inequality, one may write:

$$\sum_{i=1}^{M} \frac{1}{M} I_i(x^{(1)}, x^{(2)}) \geq 1. \tag{8.60}$$

Generalizing the above concept for a weight vector $\mathbf{w}$ (such that $\sum_{i=1}^{M} w_i = 1$) which indicates a preference relationship among different objectives, we can write the above inequality for the $\mathbf{w}$-dominating condition as follows:

$$\sum_{i=1}^{M} w_i I_i(x^{(1)}, x^{(2)}) \geq 1. \tag{8.61}$$

If all objectives are of equal importance, each $w_i = 1/M$ and we have the inequality shown in equation (8.60). However, for any other generic weight vector, we have the above condition for a solution $x^{(1)}$ to be $\mathbf{w}$-dominating solution $x^{(2)}$. Generalizing further, these investigators suggested the condition for a $(\mathbf{w}, \tau)$-dominance (with $\tau \leq 1$) between two solutions, as follows:

$$\sum_{i=1}^{M} w_i I_i(x^{(1)}, x^{(2)}) \geq \tau. \tag{8.62}$$

Based on these conditions for $\mathbf{w}$-dominance and $(\mathbf{w}, \tau)$-dominance, a corresponding non-dominated and a Pareto-optimal set can also be identified. However, if the above weak dominance conditions are used, the obtained set would be a strict non-dominated one. However, if any one of the index functions is restricted for the strict inequality condition, a weak non-dominated front will be found.

Although the above two definitions for dominance introduce flexibilities, they are associated with additional parameters which a user has to supply. The investigators suggested a preference-based procedure for selecting these parameters (the weight vector $\mathbf{w}$ and the τ parameter).

The above definitions for dominance are certainly interesting and may be used to introduce bias in the Pareto-optimal region. The index function I_i takes Boolean values of one or zero. For a better implementation, a real-valued index function can also be constructed based on the actual difference in objective function values of the two solutions.

8.7 Exploiting Multi-Objective Evolutionary Optimization

The previous sections have demonstrated how a number of conflicting objectives can be given varying importance and how the population approach of evolutionary algorithms can be used to find many optimal solutions corresponding to the resulting multi-objective optimization problem. The multi-objective optimization concept can be exploited to solve other search and optimization problems including single-objective constraint handling and goal programming problems. In this section, we will discuss these two applications where evolutionary MOEA can be directly applied for this purpose.

However, a number of other possibilities also exist. In a recent study (Bleuler et al., 2001), the bloating of genetic programs often encountered in genetic programming (GP) applications is controlled by converting the problem into a two-objective optimization problem of optimizing the underlying objective function and minimizing the *size* of a genetic program. Since the minimization of program size is also an important objective, the GP attempts to find the optimal program without making the program unnecessarily large. Another study (Knowles et al., 2001) shows that by carefully decomposing the original single objective function into multiple functionally different objectives and treating the problem as a multi-objective optimization problem makes the problem easier to solve than the usual single-objective optimization procedure.

8.7.1 Constrained Single-Objective Optimization

First, we will discuss different ways that multi-objective optimization techniques have been used as an alternate strategy for single-objective constrained optimization. In the latter, there exist a single objective function and a number of constraints (inequality or equality):

$$
\left.
\begin{aligned}
\text{Minimize} \quad & f(\mathbf{x}), \\
\text{subject to} \quad & g_j(\mathbf{x}) \geq 0, && j = 1, 2, \ldots, J, \\
& h_k(\mathbf{x}) = 0, && k = 1, 2, \ldots, K, \\
& x_i^{(L)} \leq x_i \leq x_i^{(U)}, && i = 1, 2, \ldots, n.
\end{aligned}
\right\}
\tag{8.63}
$$

Without loss of generality, we assume that the objective function $f(\mathbf{x})$ is minimized. For a maximization problem, the duality principle can be used to convert the problem into a minimization problem. In most difficult and real-world problems, the constraints $g_j(\mathbf{x})$ and $h_k(\mathbf{x})$ are nonlinear and make most of the search space infeasible. This causes difficulty even in finding a single feasible solution. Ideally, the problem is best posed as a multi-objective optimization problem of minimizing the objective function and minimizing all constraint violations. Thus, if there are $C = J + K$ constraints, there are a total of $M = (C+1)$ objectives in the corresponding multi-objective constrained optimization problem:

$$
\left.
\begin{aligned}
\text{Minimize} \quad & f(\mathbf{x}), \\
\text{Minimize} \quad & \langle g_j(\mathbf{x}) \rangle^2, && j = 1, 2, \ldots, J, \\
\text{Minimize} \quad & [h_k(\mathbf{x})]^2, && k = 1, 2, \ldots, K, \\
& x_i^{(L)} \leq x_i \leq x_i^{(U)}, && i = 1, 2, \ldots, n.
\end{aligned}
\right\}
\tag{8.64}
$$

Here, the bracket operator $\langle \alpha \rangle$ returns α if α is negative; otherwise, it returns zero. In this way, the constraint violations ($g_j(\mathbf{x}) < 0$) return a non-zero value. For all solutions, satisfying an inequality or equality constraint returns a value equal to zero.

Since each constraint violation is minimized in the above multi-objective formulation, the resulting Pareto-optimal region contains the constrained minimum solution, for which all constraint violations are zero. Figure 264 illustrates the concept with the following example problem:

$$\left.\begin{array}{ll}\text{Minimize} & f(x_1, x_2) = 1 + \sqrt{x_1^2 + x_2^2}, \\ \text{subject to} & g(x_1, x_2) \equiv 1 - (x_1 - 1.5)^2 - (x_2 - 1.5)^2 \geq 0.\end{array}\right\} \qquad (8.65)$$

A little calculation will show that the constrained minimum is $x_1^* = x_2^* = 1.5 - 1/\sqrt{2} = 0.793$. The corresponding function value is $f^* = 2.121$. This minimum solution is shown in the figure as solution A. The horizontal axis marks the constraint violation. Thus, all solutions with zero constraint violation are feasible. These solutions are also shown in the figure, along with the entire objective space of the resulting two-objective optimization problem. The corresponding Pareto-optimal solution set for the two-objective optimization problem is marked. It is clear that the constraint minimum solution A belongs to one end of the Pareto-optimal solution set.

Although the above multi-objective formulation is an ideal way to treat constrained optimization problems, the classical single-objective optimization literature converts the above problem into a weighted sum of objectives. This method is largely known as the *penalty function* approach (refer to Section 4.2.3), where the original objective function $f(x)$ and all constraint violations are added together with a weight vector consisting of penalty parameters, as follows:

$$P(\mathbf{x}, \mathbf{R}, \mathbf{r}) = f(\mathbf{x}) + \sum_{j=1}^{J} R_j \langle g_j(\mathbf{x}) \rangle^2 + \sum_{k=1}^{K} r_k [h_k(\mathbf{x})]^2. \qquad (8.66)$$

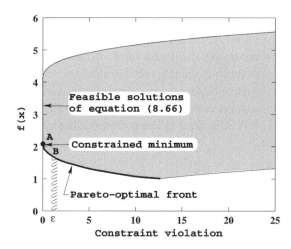

Figure 264 The constrained minimum, feasible solutions of the original single-objective optimization problem, and the Pareto-optimal set of the multi-objective problem.

In the parlance of multi-objective optimization, the above function is a weighted sum of $(J + K + 1)$ objectives with a weight vector $\mathbf{w} = (1, \mathbf{R}, \mathbf{r})^\mathsf{T}$. The classical optimization literature witnessed an enormous difficulty in fixing a weight vector for successful working of the above method. The components of the weight vector determine a fixed path from anywhere in the search space towards the constrained minimum. Sometimes, instead of converging to the true constrained minimum, the path terminates to a local and artificial minimum. Thus, for sufficiently nonlinear problems, not all weight vectors will allow a smooth convergence towards the true constrained minimum. Often, the user has to experiment with various weight vectors to solve the constrained optimization problem. This difficulty is similar in principle to the difficulty faced with the weighted approaches in solving multi-objective optimization problems. We have discussed earlier in Chapter 3 that an arbitrary weight vector may not lead to a specific Pareto-optimal solution for nonconvex and even for convex search space problems. With the above connection between the penalty-function approach for constraint handling for single-objective optimization and the weighted approach for multi-objective optimization, we can explain why users of the penalty function based approach face difficulty in finding the true constrained minimum.

With the availability of Pareto-based multi-objective optimization methods, there exist a number of advantages of using these in solving constrained optimization problems, as follows:

1. The constraint handling problem can be solved in a natural way. No artificial formulation of a penalized objective function $P(\mathbf{x}, \mathbf{R}, \mathbf{r})$ is needed.
2. There is no need of any penalty parameters $(\mathbf{R}, \mathbf{r})$. This would make the approach non-subjective of the user.
3. In many real-world problems, a constraint is never 'hard', that is, a solution with a permissible constraint violation can still be considered if there is a substantial gain in the objective function. Since the Pareto-optimal solution set corresponding to the resulting multi-objective problem (equation (8.64)) would contain trade-off solutions, including solutions that violate each constraints marginally or by a large extent, the obtained non-dominated set will contain solutions corresponding to the problem with 'soft' constraints. Figure 264 above shows that if $g_j(\mathbf{x}) \leq \sqrt{\epsilon}$ (or $|h_k(\mathbf{x})| \leq \sqrt{\epsilon}$) were allowed, all solutions within the region AB would be of interest to a user. Since the objective function values monotonically improve towards B, the solution B would be preferred. What is important is that since many solutions can be found on the Pareto-optimal front by using an MOEA, such post-optimality decision-making is possible to carry out in the presence of multiple optimal solutions.

When multiple constraints exist, a similar phenomenon happens in more than two dimensions. Since most of the MOEAs discussed earlier can also be used to solve problems having more than two objectives with a complexity (at worst) as linear to the number of objectives, such a computation is practical. However, for a very large number of constraints, it is not necessary that all constraints need to be considered as a separate objective function. The so-called hard constraints can be used as constraints, while soft constraints can be treated as additional objectives. Ideally, constraints which are *active* (Deb, 1995) at the constrained minimum may be considered as objectives, in order to find solutions closer to the constrained optimum solution.

Biased MOEA for Constrained Single-Objective Optimization

It is clear from the above discussion that in order to solve the constrained optimization problem by using a multi-objective optimization problem formulation, we are interested in solutions which must be biased towards one end of the Pareto-optimal set, namely towards the region for which all constraint violations are small. Thus, we need to use an algorithm which introduces a predefined bias in finding Pareto-optimal solutions. In Section 8.6 above, we have outlined a number of approaches to enable MOEAs to achieve this very task. One such technique can be used to find a biased population of solutions towards the desired end of the Pareto-optimal region. In this way, more solutions at the desired end will be found when compared to the rest of the true Pareto-optimal set.

Another approach would be to use an MOEA which has a natural bias in finding Pareto-optimal solutions towards the individual champion solutions. We have seen earlier in Section 5.4 that the VEGA is one such algorithm for doing the task. This algorithm has a tendency to find solutions near the minimum solutions of each objective. Although some unconstrained minimum solutions (which are often infeasible by a large margin) may also be found by using this procedure, our interest would be on that part of the population which makes all constraint violations close to zero. Alternatively, since we know which way to introduce bias, the VEGA can be modified with a biasing technique. For example, instead of using an equal proportion of the population for evaluating with each objective function, a smaller proportion can be evaluated with the original objective function $f(\mathbf{x})$. In this way, more emphasis will be given for minimizing the constraint violations. A biased sharing or a guided domination approach can also be applied in this context.

In the following, we present two different implementations, where similar principles are used quite satisfactorily.

Surry, Radcliffe and Boyd's Constrained MOGA

In this implementation (Surry et al., 1995), in addition to calculating the objective function value, each solution in the population is also checked for constraint violation. If there is a constraint violation, its amount is noted; otherwise, a value zero is used as the constraint violation. Thereafter, the population is classified according to a non-dominated ranking by using the amount of constraint violations only. In a C-dimensional space of constraint violations ($C = J + K$), each solution i is assigned a rank r_i based on how many solutions dominates it. It is interesting to note that although there are C constraint violations, the Pareto-ranking produces a univariate metric defining the level of overall constraint violation. Thus, it is enough to construct a two-objective optimization problem with the original objective function f_i and the ranking r_i as the two objectives. Figure 265 shows solutions in the constraint violation space (CV_1 and CV_2) and their corresponding Pareto-rankings. Solutions A to E are infeasible, while solution F is feasible. Solution C is dominated by two solutions (solution B and F) in the constraint violation space.

Once the Pareto-ranking is completed, a binary tournament selection is used to create the mating pool in a special way. For the tournament selection, a user-defined parameter p_{cost} is used. This parameter denotes the proportion of the population to be selected when using the original objective function $f(\mathbf{x})$. The tournament selection operator is implemented as follows. For two solutions chosen for a tournament, one of the two objectives (the original objective function f_i) is chosen with a probability p_{cost} as the fitness. The Pareto-ranking r_i is chosen with a probability $(1 - p_{cost})$. Thus the constrained MOGA (COMOGA) works with the right-side plot of the two objectives shown in Figure 265. If the objective function is the fitness, solution E is the best choice in the population shown. On the other hand, if constraint violation is important, solutions F is the best choice. For a tie in any tournament played with the chosen objective, the other objective value is used to break the tie. After the mating pool is created, the crossover and mutation operators are applied on the complete population.

The effect of choosing a small value of p_{cost} is to give more importance to constraint satisfaction and less importance to minimization of the objective function. This may be a good strategy to follow in the beginning of a simulation run, where the emphasis is to create more and more feasible solutions. The effect of choosing a large value of p_{cost} is the opposite. More solutions are compared based on the original objective function value $f(\mathbf{x})$ and hence there is an emphasis on finding the minimum value of the objective function. This property is desired towards the latter generations of a

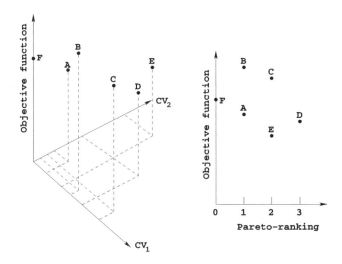

Figure 265   The Pareto-ranking of a few sample solutions.

simulation run, when most solutions are feasible and the only task remaining is to find the constrained minimum. Since the parameter p_{cost} needs to be dynamically varied during a simulation run, the investigators have also proposed a controlled approach. They defined a parameter τ to indicate the desired proportion of feasible solutions in the population (a value of 0.1 was suggested). The COMOGA begins with an initial value of $p_{cost} = 0.5$. Thereafter, in each iteration, it is updated as follows. If the actual proportion of feasible solutions is less than τ, the parameter p_{cost} is reduced to encourage the generation of more feasible solutions. They suggested a simple reduction rule:

$$p_{cost} \leftarrow (1 - \epsilon)p_{cost}.$$

On the other hand, if the actual proportion of feasible solutions is more than the target τ, p_{cost} is increased by using the following rule:

$$p_{cost} \leftarrow 1 - (1 - \epsilon)(1 - p_{cost}).$$

Thus, this method is essentially an extension of the VEGA in that an equal proportion of the subpopulation is not used for each objective function. Instead, depending on the proportion of the feasible solutions present in the population, the subpopulation size for each of the two objectives (original objective function and the Pareto-ranking) is varied.

On a pipeline optimization problem, investigators have concluded that the COMOGA worked similarly to the best known penalty function approach for handling constraints in terms of computational complexity and reliability in finding the best solution. However, compared to the experimentations needed to find a good set of penalty parameters, COMOGA requires fewer of these with its parameters, such as τ and ϵ, and the update rules for choosing p_{cost}.

Although this method allows more flexibility in the way that infeasible solutions can become feasible and feasible solutions can approach the constrained minimum than the classical penalty function approach, this method is not entirely flexible. The Pareto-ranking method adopted here steers the infeasible solutions towards the feasible region and there may exist some constrained optimization problems where this approach may fail, because in such problems non-dominated solutions based on constraint violations may steer the search in the wrong direction. Coello (2000) suggested a somewhat more flexible strategy, which we will discuss next.

Coello's Approach

Instead of using the Pareto-ranking as the only objective of handling all constraints, Coello (2000) suggested a method where each constraint violation is explicitly used as an objective. The population is divided into $(C + 1)$ subpopulations of equal size. As in the VEGA, each subpopulation deals with one of the objectives (either the original objective function or a constraint violation). For ease of illustration, let us say the subpopulations are numbered as $0, 1, 2, \ldots, C$. The first subpopulation numbered 0 is dealt with by the objective function $f(x)$. Thereafter, the i-th subpopulation is dealt with by the i-th constraint.

For minimization problems, the first subpopulation dealing with $f(x)$ is evaluated a fitness solely by using the objective function value. No constraint violation is checked for these solutions. However, the evaluation procedure for other subpopulations is a little different. For a solution x in a subpopulation dealing with the j-th constraint $g_j(x)$ and violating a total of $v(x)$ constraints, a fitness is assigned hierarchically as follows:

$$\text{If} \qquad g_j(x) < 0, \quad F(x) = -g_j(x).$$
$$\text{Otherwise, if} \quad v(x) > 0 \qquad F(x) = v(x).$$
$$\text{Otherwise} \qquad\qquad\qquad F(x) = f(x).$$

First, the solution is checked for violation of the i-th constraint. If it is violated, the fitness is assigned as the negative of the constraint violation, so that when minimized the violation of the constraint assigned for its own subpopulation is minimized. If the solution does not violate the i-th constraint, it is checked for violation of all other constraints. Here, only the number of violated constraints $v(x)$ is counted. The one with a smaller number of violated constraints is assigned a smaller fitness. If the solution is feasible, the fitness is assigned based on its objective function value and the solution is moved to the first subpopulation.

Since each subpopulation except the first one emphasizes solutions that do not violate a particular constraint and encourages solutions with a minimum number of constraint violations, the overall effect of the search is to move towards the feasible region. By evaluating the first subpopulation members (whether feasible or infeasible) in terms of $f(x)$ alone will mostly cause a preference to solutions close to the unconstrained minimum. These solutions may not be directly of interest to us, but the presence of them in a population may be useful for maintaining diversity in the population. Since in most problems the number of constraints are usually large, the proportion of such solutions may not be overwhelming. Although it seems to be more flexible than the COMOGA approach, this method is also not entirely free from an artificial bias in the search process. The above procedure of hierarchical fitness assignment is artificial and may prohibit a convergence to the correct minimum in certain problems.

Nevertheless, Coello (2000) has found new and improved solutions to a number of single-objective constrained engineering design problems by using this approach. In most cases, the solutions are better than or equivalent to the best solutions reported in the literature. Since no non-dominated sorting and niching strategies are used, the approach would be computationally fast.

However, in order to develop an algorithm which is free from any artificial bias, we suggest using a state-of-the-art Pareto-based MOEA technique, but employ a biasing technique suggested in an earlier section so that a biased *distribution* of solutions emerges near the constrained minimum solution, instead of guiding the search in an artificial manner. Since the nature of the problem demands finding a biased set of solutions near the constrained minimum, such a biasing would not be artificial to the problem. It is important to realize that it is better to design an algorithm for finding a biased distribution of solutions, rather than using a biased fitness assignment procedure to lead the search towards any particular region. Once a biased distribution is found, a preferred solutions can always be chosen.

8.7.2 Goal Programming Using Multi-Objective Optimization

In Section 3.6 above, a brief description of different approaches to goal programming was presented. Recall that the task in goal programming is different to that in an optimization problem. With specified targets for each objective, the purpose in goal programming is to find a solution that achieves all of the specified targets, if possible. If not, the purpose is to find a solution (or a set of solutions) which violates each target minimally. In this present section, we suggest one procedure for converting a multiple goal programming problem into a multi-objective optimization problem and show some simulation results using an MOEA. The results clearly demonstrate the usefulness of MOEAs in goal programming, particularly in terms of finding multiple, widespread, trade-off solutions without the need of any user-supplied weight vector. In fact, the proposed approach simultaneously finds solutions to the same goal programming problem formed for different weight factors, thereby making this procedure both practical and different from classical approaches.

Each goal is converted into an objective function of minimizing the difference between the goal and its target. The conversion procedure depends on the type of goals used. We present these in the following table (Deb, 2001).

Type	Goal	Objective function
$\leq$	$f_j(\mathbf{x}) \leq t_j$	Minimize $\langle f_j(\mathbf{x}) - t_j \rangle$
$\geq$	$f_j(\mathbf{x}) \geq t_j$	Minimize $\langle t_j - f_j(\mathbf{x}) \rangle$
$=$	$f_j(\mathbf{x}) = t_j$	Minimize $|f_j(\mathbf{x}) - t_j|$
Range	$f_j(\mathbf{x}) \in [t_j^l, t_j^u]$	Minimize $\max(\langle t_j^l - f_j(\mathbf{x}) \rangle, \langle f_j(\mathbf{x}) - t_j^u \rangle)$

Here, the bracket operator $\langle \ \rangle$ returns the value of the operand if the operand is positive; otherwise, it returns zero. The operator $| \ |$ returns the absolute value of the operand. In this way, a goal programming problem is formulated as a multi-objective problem. Although other similar methods have been suggested in classical goal programming texts (Romero, 1991; Steuer, 1986), the advantage with the above formulation is that (i) there is no need of any additional constraint for each goal, and (ii) since GAs do not require objective functions to be differentiable, the above objective function can be used.

Although somewhat obvious, we shall show that the nonlinear programming (NLP) problem of solving the weighted goal programming for a fixed set of weight factors is exactly the same as solving the above reformulated problem. We shall only consider the 'less-than-equal-to' type goal; however, the same conclusion can be made for other types of goal as well. Consider a goal programming problem having one goal of finding solutions in the feasible space S for which the criterion is $f(\mathbf{x}) \leq t$. We use equation (3.20) (see earlier) to construct the corresponding NLP problem:

$$\left.\begin{array}{ll} \text{Minimize} & p \\ \text{subject to} & f(\mathbf{x}) - p \leq t, \\ & p \geq 0, \\ & \mathbf{x} \in \mathcal{S}. \end{array}\right\} \tag{8.67}$$

We can rewrite both constraints involving p as $p \geq \max[0, (f(\mathbf{x}) - t)]$. When the difference $(f(\mathbf{x}) - t)$ is negative, the above problem has the solution $p = 0$ and when the difference $f(\mathbf{x}) - t$ is positive, the above problem has the solution $p = f(\mathbf{x}) - t$. This is exactly achieved by simply solving the problem: Minimize $\langle f(\mathbf{x}) - t \rangle$.

Since we now have a way to convert a goal programming problem into an equivalent multi-objective optimization problem, we can use an MOEA to solve the resulting goal programming problem. In certain cases, a unique solution to a goal programming problem may exist, no matter what weight factors are chosen. In such cases, the equivalent multi-objective optimization problem is similar to a problem without conflicting objectives and the resulting Pareto-optimal set contains only one solution. However, in most cases, goal programming problems are sensitive to the chosen weight factors, and the resulting solution to the problem largely depends on the specific weight factors used. The advantage of using the multi-objective reformulation is that each Pareto-optimal solution corresponding to the multi-objective problem becomes the solution of the original goal programming problem for a specific set of weight factors. Thus, by using MOEAs, we can get multiple solutions to the goal programming problem simultaneously.

After multiple solutions are found, designers can then use higher-level decision-making approaches or compromise programming (Romero, 1991) to choose one particular solution. Each solution $\mathbf{x}$ can be analyzed to estimate the relative importance of each criterion function as follows:

$$w_j = \frac{|f_j(\mathbf{x}) - t_j|/|t_j|}{\sum_{j=1}^{M} |f_j(\mathbf{x}) - t_j|/|t_j|}. \tag{8.68}$$

For a 'range' type goal, the target t_j can be substituted by either t_j^l or t_j^u depending on which is closer to $f(\mathbf{x})$.

Moreover, the proposed approach also does not pose any other difficulties which the weighted goal programming method may have. Since solutions are compared criterion-wise, there is no danger of comparing 'apples with oranges'; nor is there any difficulty of scaling in criterion function values. Furthermore, this approach does not pose any difficulty in solving nonconvex goal programming problems.

Simulation Results

We show the working of the proposed approach on a number of problems.

Test Problem P1

We first consider the example problem given earlier in equation (3.21). The goal programming problem is converted into a two-objective optimization problem P1 as follows:

$$\left.\begin{array}{ll} \text{Minimize} & \langle f_1(x_1, x_2) - 2\rangle, \\ \text{Minimize} & \langle f_2(x_1, x_2) - 2\rangle, \\ \text{subject to} & S \equiv (0.1 \leq x_1 \leq 1, \quad 0 \leq x_2 \leq 10). \end{array}\right\} \tag{8.69}$$

Here, the criterion functions are $f_1 = 10x_1$ and $f_2 = (10 + (x_2 - 5)^2)/(10x_1)$. We use a population of size 50 and run the NSGA for 50 generations. A $\sigma_{\text{share}} = 0.158$ (equation (4.64) earlier with $n = 2$ and $q = 10$) is used. As discussed above, the feasible decision space lies above the hyperbola. All 50 solutions in the initial population and all non-dominated solutions at the final population are shown in Figure 266, which is plotted using criterion function values f_1 and f_2. All final solutions have $x_2 = 5$ and $0.2 \leq x_1 \leq 0.5$. This figure also marks the region (with dashed lines) of true solutions of this goal programming problem with different weight factors. The figure shows that the NSGA in a single run has been able to find different solutions in the desired range. Although other regions (for $f_1 < 2$ and $f_1 > 5$) on the hyperbola are Pareto-optimal solutions of the two-objective optimization problem of minimizing f_1 and f_2, the reformulation of the objective functions allows the NSGA to find only the required region, which are also solutions of the goal programming problem. Table 32 shows five different solutions obtained by the NSGA. Relative weight factors for each solution are also computed by using equation (8.68) (see above). If the first criterion is of more importance, solutions in first or second row can be chosen, whereas if the second criterion is the important one, solutions in the fourth or fifth rows can be chosen. The solution in the third row shows the situation where both criteria are of more or

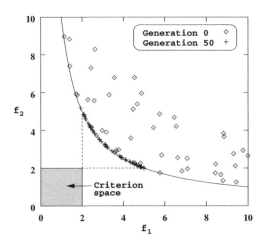

Figure 266   The NSGA solutions are shown on an f_1–f_2 plot for problem P1 (Deb, 2001). Reproduced by permission of Operational Research Society Ltd.

Table 32   Five solutions to the goal programming problem P1 are shown (Deb, 2001).
Reproduced by permission of Operational Research Society Ltd.

x_1	x_2	$f_1(\mathbf{x})$	$f_2(\mathbf{x})$	w_1	w_2
0.2029	5.0228	2.0289	4.9290	0.0098	0.9902
0.2626	5.0298	2.6260	3.8083	0.2572	0.7428
0.3145	5.0343	3.1448	3.1802	0.4923	0.5076
0.3690	5.0375	3.6896	2.7107	0.7027	0.2972
0.4969	5.0702	4.9688	2.0135	0.9955	0.0045

less equal importance. The advantage of using the proposed technique is that all such
(and many more as shown in Figure 266) solutions can be found simultaneously in
one single run.

Test Problem P2

We alter the above problem to create a different goal programming problem P2:

$$\left.\begin{array}{ll} \text{goal} & (f_1 = x_1 \in [0.25, 0.75]), \\ \text{goal} & \left[f_2 = (1 - \sqrt{x_1(1 - x_1)})(1 + 10x_2^2) \le 0.4\right], \\ \text{subject to} & 0 \le x_1 \le 1, \quad 0 \le x_2 \le 1. \end{array}\right\} \qquad (8.70)$$

The feasible decision space and the criterion space are shown in Figure 267. There
exists only one solution ($x_1 = x_2 = 5$) to this problem, no matter what non-zero weight

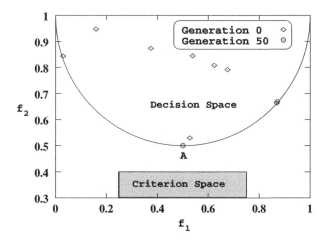

Figure 267   NSGA solutions are shown on an f_1–f_2 plot for problem P2 (Deb, 2001).
Reproduced by permission of Operational Research Society Ltd.

factors are chosen. This is because this solution (marked as 'A' on the figure) makes the shortest deviation from the criterion space. An NSGA with identical parameter settings and an identical initial population to that used in test problem P1 are used. It is observed that after 50 generations, all 50 population members converge at the most satisficing solution marked by 'A'. This test problem shows that although a multi-objective optimization technique is used, the use of a reformulated objective function allows us to find the sole optimal solution of the goal programming problem.

Test Problem P3

Next, we choose a problem P3 similar to that used by Ignizio (1976):

$$\left.\begin{array}{rl} \text{goal} & (f_1 = x_1 x_2 \geq 16), \\ \text{goal} & (f_2 = (x_1 - 3)^2 + x_2^2 \leq 9), \\ \text{subject to} & 6x_1 + 7x_2 \leq 42. \end{array}\right\} \qquad (8.71)$$

This investigator considered the constraint as the third goal, and emphasized that the first-level priority in the problem is to find solutions which will satisfy the constraint. However, here we argue that such a first-priority goal can be taken care of by using it as a *hard* constraint so that any solution violating the constraint will receive a large penalty. Since the constraint is explicitly taken care of, the next-level priority is to find solution(s) which will minimize the deviation in the two goals presented in equation (8.71).

Figures 268 and 269 show the problem in the solution and in the function space, respectively. The feasible search space is shown by plotting about 25 000 points in Figure 269. This figure shows that there exists no feasible solution which satisfies both goals. In order to solve this problem, we use a real-parameter NSGA, but the variables are coded directly. A simulated binary crossover (SBX) with $\eta_c = 30$ and a polynomial mutation operator with $\eta_m = 100$ are used (Deb and Agrawal, 1995). A population size of 100 is employed and the NSGA is run for 50 generations. Other parameters identical to those used in the previous test problem are used. The only solution obtained by the NSGA is as follows:

$$x_1 = 3.568, \quad x_2 = 2.939, \quad f_1 = 10.486, \quad f_2 = 8.961.$$

This solution is marked on both figures with a circle. Such a solution is feasible and lies on the constraint boundary. It does not violate the second goal; however, it does violate the first goal by an amount of $(16 - 10.486)$ or 5.514. Figure 269 shows that it violates the first goal $(f_1 \geq 16)$ minimally (keeping the minimum distance from the feasible search space).

An Engineering Design

Finally, we apply the technique to a goal programming problem constructed from the welded beam design problem discussed earlier on page 128. It is intuitive that

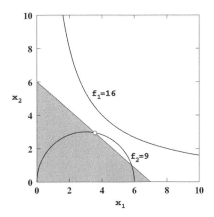

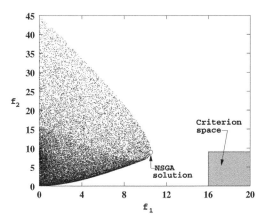

Figure 268 Criterion and decision spaces are shown for the test problem P3 (Deb, 2001). Reproduced by permission of Operational Research Society Ltd.

Figure 269 The NSGA solution is shown on a f_1–f_2 plot for the test problem P3 (Deb, 2001). Reproduced by permission of Operational Research Society Ltd.

an optimal design for cost will cause all four design variables to take small values. When the beam dimensions are small, it is likely that the deflection at the end of the beam is going to be large. Thus, the design solutions for minimum cost $(f_1(\mathbf{x}))$ and minimum end deflection $(f_2(\mathbf{x}))$ conflict with each other. In the following, we present a goal programming problem from the cost and deflection considerations:

$$\left.\begin{array}{rl} \text{goal} & [f_1(\mathbf{x}) = 1.104\,71h^2\ell + 0.048\,11tb(14.0 + \ell) \leq 5], \\ \text{goal} & (f_2(\mathbf{x}) = \dfrac{2.1952}{t^3b} \leq 0.001), \\ \text{subject to} & g_1(\mathbf{x}) \equiv 13\,600 - \tau(\mathbf{x}) \geq 0, \\ & g_2(\mathbf{x}) \equiv 30\,000 - \sigma(\mathbf{x}) \geq 0, \\ & g_3(\mathbf{x}) \equiv b - h \geq 0, \\ & g_4(\mathbf{x}) \equiv P_c(\mathbf{x}) - 6000 \geq 0, \\ & 0.125 \leq h, b \leq 5.0 \text{ and } 0.1 \leq \ell, t \leq 10.0. \end{array}\right\} \quad (8.72)$$

All of the terms have been explained earlier (see page 128). Here, we would like to have a design for which the cost is smaller than 5 units and the deflection is smaller than 0.001 inch. If there exists any such solution, then that solution is the desired solution. However, if such a solution does not exist, we are interested in finding a solution which will minimize the deviation in cost and deflection from 5 and 0.001 in, respectively.

Here, constraints are handled by using the bracket-operator penalty function (Deb, 1995). Penalty parameters of 100 and 0.1 are used for the first and second criterion functions, respectively. A violation of any of the above four constraints will make the design unacceptable. Thus, in terms of the discussion presented in Ignizio (1976), satisfaction of these constraints is the first priority.

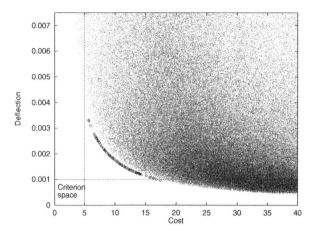

Figure 270 NSGA solutions (each marked with a 'diamond') are shown on the objective function space for the welded beam problem (Deb, 2001). Reproduced by permission of Operational Research Society Ltd.

In order to investigate the search space, we plot many random feasible solutions in the f_1–f_2 space in Figure 270. The corresponding criterion space (marking the region with cost ≤ 5 and deflection ≤ 0.001) is also shown in this figure. The figure shows that there exists no feasible solution in the criterion space, meaning therefore that the solution to the above goal programming problem will have to violate at least one goal. A real-parameter NSGA with 100 population members and an SBX operator with $\eta_c = 30$ and a polynomial mutation operator with $\eta_m = 100$ are employed. We also use a σ_{share} value of 0.281 (see equation (4.64) earlier, with $P = 4$ and $q = 10$). Figure 270 shows the solutions (each marked with a 'diamond') obtained after 500 generations. The existence of multiple solutions is accounted for by the fact that no knowledge of weight factor for each goal is assumed here, and that each solution can be accounted for by different combinations of weight factors for cost and deflection quantities.

We construct another goal programming problem by changing the targets to $t_1 = 2.0$ and $t_2 = 0.05$. Since there exists no solution with a cost smaller than 2 units, and the deflection of 0.05 in is also large enough, the resulting most satisficing solution should represent the minimum-cost solution. Figure 271 shows that the NSGA with identical parameter settings converges to one solution:

$$h = 0.222, \quad \ell = 7.024, \quad t = 8.295, \quad b = 0.244.$$

This solution has a cost of 2.431 units and deflection of 0.0157 in. Figure 271 shows that such a solution is very close to the minimum-cost solution.

8.8   Scaling Issues

In multi-objective optimization, the problem difficulty varies rather interestingly with the number of objectives. In the previous chapters, we have cited examples mostly

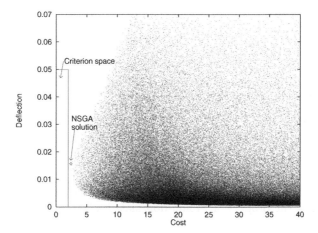

Figure 271 The NSGA solution is shown on the objective function space for the modified welded beam problem (Deb, 2001). Reproduced by permission of Operational Research Society Ltd.

with two objectives, primarily because of the ease in which two-dimensional Pareto-optimal fronts can be visually demonstrated. When the number of objectives increases, the dimensionality of the objective space also increases. With M objectives, the Pareto-optimal front can be at most an M-dimensional surface. Figure 272 shows the Pareto-optimal surface in a typical three-objective optimization problem. Any pair of solutions on such a surface are non-dominated to each other. In the problem shown in the figure, all objectives are to be minimized. Thus, the feasible search space lies above this surface. The task of an MOEA here is to reach this surface from the interior of the search space and distribute solutions as uniformly as possible over

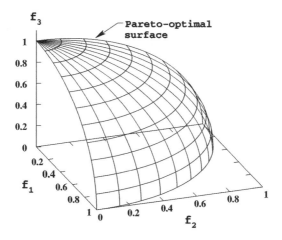

Figure 272 A typical Pareto-optimal surface for a three-objective minimization problem.

the surface. Although the above example shows a three-dimensional Pareto-optimal surface, conflicting objectives can also degenerately produce a three-dimensional Pareto-optimal curve.

8.8.1   Non-Dominated Solutions in a Population

With an increase in the number of objective functions, the dimensionality of the Pareto-optimal set increases. It is also intuitive that with an increase in number of objective functions, the number of non-dominated solutions in the initial random population will also increase. This has a serious implication in choosing an appropriate population size. However, before we discuss the effect of M on the population sizing, we investigate how the proportion of non-dominated solutions in a random population increases with M.

In a multi-objective optimization problem having M objective functions, we are interested in counting the number of non-dominated solutions $|\mathcal{F}_1|$ in a randomly created population of size N. One of the ways to do this is to calculate the probability $P(K)$ of having a population with exactly K non-dominated solutions, where K can be varied from one to N. Thereafter, the expected value of K can be found by using the obtained probability distribution. In order to find $P(K)$, we have the scenario depicted in Figure 273. The following two conditions must be fulfilled to calculate this probability:

1. All pairs of solutions ($\binom{K}{2}$ of them) in the cluster A (non-dominated set) must not dominate each other.
2. Every solution in cluster B (dominated set) must be dominated by at least one solution from cluster A.

The probability calculation for the second case is difficult to carry out, because the dominance check for any two pairs of solutions may not be independent. For example, let us consider the dominance checks with solution 1 from cluster A and solutions a and b from cluster B. Let us also assume that solution a dominates solution b. Now, while checking the dominance between solutions 1 and a, we would find that

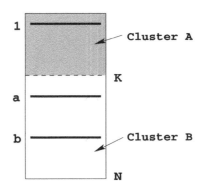

Figure 273   The procedure for counting non-dominated solutions.

solution 1 dominates solution a. Since solution a inherently dominates solution b, the transitivity property for dominance assures us that solution 1 also dominates solution b. This happens with a probability of one. Thus, it would be erroneous to consider that the probability of solution 1 dominating solution a and probability of solution 1 dominating solution b are independent. Since there exist many such *chains*, the exact probability computation becomes difficult to achieve.

However, we can attempt to investigate how the proportion of non-dominated solutions increase with M by randomly creating a population of solutions and explicitly counting the non-dominated solutions in a computer simulation. We can experiment with different values of N and M. In order to get a good estimate of the mean proportion of non-dominated solutions, we use one million random populations ($f_i \in [0, 1]$) for each combination of N and M. Figure 274 shows how the proportion of non-dominated solutions ($|\mathcal{F}_1|/N$) varies with M for a number of fixed population sizes. We show the results with N = 50, 100 and 200. It is clear from this figure that as the number of objective functions increases, most solutions in the population belong to the non-dominated front. The growth in the number of non-dominated solutions is similar to a logistic growth pattern. In all cases, the standard deviation from each combination is calculated and is found to be reasonably small. It is interesting to note that for larger population sizes, the growth is delayed. This aspect is important in the context of choosing an appropriate population size, a matter which we discuss in the next subsection.

In order to investigate the effect of the population size N, we show the variation of the proportion of non-dominated solutions with population size for values of M = 2, 5 and 10 in Figure 275. We observe that as the population size increases, the proportion of non-dominated solutions decreases. The standard deviation in each case is small.

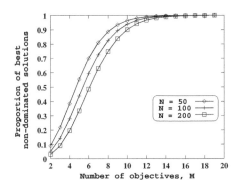

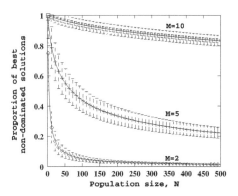

Figure 274 The proportion of the best non-dominated solutions is shown varying with the number of objective functions.

Figure 275 The proportion of the best non-dominated solutions is shown varying with the population size.

8.8.2   Population Sizing

We have observed that as the number of objective functions increases, more and more solutions tend to lie in the first non-dominated front. In particular, when the initial population is randomly created, this may cause a difficulty to most multi-objective evolutionary algorithms. Recall that most MOEAs described in the previous chapters emphasize all solutions of the first non-dominated front equally by assigning a similar fitness. In fact, the NSGA assigns exactly the same fitness to all solutions of the first non-dominated front before the niching operator is applied. If all population members lie in the first non-dominated front, most MOEAs also assign the same (or similar) fitness to all solutions. When this happens, there is no selection advantage to any of these solutions. The recombination and mutation operators must need to create solutions in a better front in order for the search to proceed towards the Pareto-optimal region. In the absence of any selection pressure for better solutions, the task of recombination of mutation operators to find better solutions may be difficult in general. In implementing an appropriate elitism, these algorithms will also fail, simply because most of the population members belong to the best non-dominated front and there may not be any population slot left to include any new solution. There are apparently two solutions to this problem:

- use a large population size;
- use a modified algorithm.

We have seen in Figure 275 above that for a particular M, the proportion of non-dominated solutions decreases with population size. Thus, if we require a population with a user-specified maximum proportion of non-dominated solutions (say p_1), then Figure 275 can be used to estimate what would be a reasonable population size. In fact, we have simulated the proportions of best non-dominated solutions for different M values and plotted the data obtained in Figure 276. For example, let us say that we require at most 30% population members ($p_1 = 0.3$) in the best non-dominated front in the initial random population. This figure shows with arrows the corresponding minimum population sizes for different values of M. It is clear that the required population size increases exponentially with the number of objectives.

Another approach would be to use a modified technique for assigning fitness. Instead of assigning fitness based on the non-dominated rank of a solution, some other criterion could be used. The amount of spread in objective space may be used to assign fitness. For example, the NSGA can be used with objective space sharing, instead of parameter space sharing. In this way, solutions that are closely packed in one part of the non-dominated front will not be favored when compared to those that lie in less dense regions on the non-dominated front. Nevertheless, more careful studies must be performed to investigate whether the currently known evolutionary algorithms can scale well to solve MOOPs with large M values.

Besides an appropriate proportion of non-dominated solutions, an adequate population size for a random initial population should also depend on other factors, such as the signal-to-noise ratio and the degree of nonlinearity associated with the

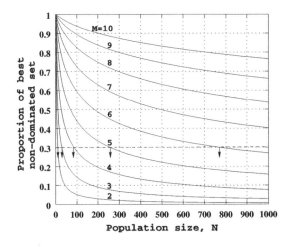

Figure 276   Chart for finding the minimum population size.

problem (Goldberg et al., 1992; Harik et al., 1999). Some of these quantities may be difficult to compute in an arbitrary MOOP, but an initial guess of an adequate population size obtained from Figure 276 may be a good starting point for an iterative population sizing task. As often used in the case of single-objective optimization problems, an EA simulation needs to be started with a random population. If information about better regions of the decision space is known, an EA population can be initialized there, instead of initializing randomly in the entire decision space. In such an event, the above figures are of not much use, as they were obtained by assuming a random initial population. Nevertheless, the figures do show how the dimensionality in objective functions causes many solutions to belong to the same non-dominated front.

As in single-objective EAs, dynamic population sizing strategies have also been suggested in MOEAs. In one implementation (Tan et al., 2001), an MOEA population is adaptively sized based on the difference between the distribution of solutions in the best non-dominated front and a user-defined distribution. In such strategies, the reduction of an existing population can be achieved using a niching strategy. However, a technique for adding new yet good solutions in a reasonable manner remains as a challenging task to such MOEAs.

8.9   Convergence Issues

Algorithm development and the application of multi-objective evolutionary algorithms date back to the 1980s. Many new and improved algorithms, with a better understanding of their working behaviors, are fairly recent phenomena. Thus, it is not surprising that there does not exist a multitude of mathematical convergence theories related to multi-objective evolutionary optimization. Part of the reason is also due to the fact that mathematical convergence theories related to single-objective evolutionary algorithms have only recently begun to emerge (Rudolph, 1994; Vose, 1999). However, with the overwhelming and increasing interest in the area of multi-objective optimization, many such theories should appear on the horizon before too long.

Finding such a theory for multi-objective optimization is one step harder than in the case of single-objective optimization. This is because, in the ideal approach to multi-objective optimization there are two tasks: convergence to the Pareto-optimal front and maintenance of a diverse Pareto-optimal set. Thus, in addition to finding a proof for convergence to the Pareto-optimal front, it is also necessary to have a proof of diversity among solutions along the Pareto-optimal front. Since achievement of one does not automatically ensure achievement of the other, both proofs are necessary for multi-objective evolutionary optimization.

8.9.1 Convergent MOEAs

Most of the credit in attempting to outline convergence theories related to MOEAs goes to G. Rudolph (Rudolph, 1998a, 2001; Rudolph and Agapie, 2000). In his first study (Rudolph, 1998a), he extended the theory of convergence for the single-objective canonical evolutionary algorithm to multi-objective optimization. His proofs rely on the positiveness of the *variation kernel*, which relates to the search power of the chosen genetic operators. Simply stated, if the transition probability from a set of parents to any solution in the search space is non-zero positive, then the transition process of the variation operators (crossover and mutation operators) is said to have a positive variation kernel. For variation operators having a positive variation kernel and with elitism, Rudolph (2001) has shown that an MOEA can be designed in order to have a property to converge to the true Pareto-optimal front in a finite number of function evaluations in finite search space problems. For continuous search space problems, designing such convergent MOEAs is difficult (Rudolph, 1998b). Although this has been an important step in the theoretical studies of MOEAs, what is missing from the proof is a component which implies that in addition to the convergence to the Pareto-optimal front, there should be an adequate diversity present among the obtained solutions.

In their recent study, Rudolph and Agapie (2000) extended the above outline of a convergence proof for elitist MOEAs. First, they proposed a base-line elitist MOEA where the elite population size is infinite.

Rudolph and Agapie's Base Algorithm VV

> **Step 1** A population $P(0)$ is drawn at random and $t = 0$ is set. An elite set $E(0)$ is formed with non-dominated solutions of $P(0)$.

> **Step 2** Generate a new population $P(t+1) = \text{generate}(E(t))$ by usual selection, recombination, and mutation operators.

> **Step 3** Combine $P(t+1)$ and $E(t)$. Set the new elitist set $E(t+1)$ with the

non-dominated solutions from the combined set $P(t+1) \cup E(t)$.

Step 4 If stopping criterion is not satisfied, set $t = t + 1$ and go to Step 2. Otherwise, declare $E(t+1)$ is the set of obtained non-dominated solutions.

With a homogeneous finite Markov chain having a positive transition matrix, these investigators have shown that all Pareto-optimal solutions will be members of the set $E(t)$ in a finite time with probability one. If $E(t)$ contains any non-Pareto-optimal solutions, it will be eventually replaced by a new Pareto-optimal solution in Step 3. When $E(t)$ contains Pareto-optimal solutions only, no dominated solution can enter the set $E(t)$. The algorithm does not check for the spread of non-dominated solutions stored in $E(t)$. When run for a long time, the above algorithm will eventually find each and every Pareto-optimal solution, but to achieve this the required size of $E(t)$ may be impractical. Realizing this fact, the investigators have modified the above algorithm for a finite-sized archive set $E(t)$.

Rudolph and Agapie's Base Algorithm AR1

Step 1 A population $P(0)$ is drawn at random and $t = 0$ is set. An elite set $E(0)$ is formed with non-dominated solutions of $P(0)$.

Step 2 Generate a new population $P(t+1) = \text{generate}(E(t))$ by usual selection, recombination, and mutation operators.

Step 3 Store non-dominated solutions of $P(t+1)$ in $P^*(t)$. Initialize a new set $Q(t) = \emptyset$.

Step 4 For each element $y \in P^*(t)$, perform the following steps:

    Step 4a Collect all solutions from $E(t)$ which are dominated by y:
    $D_y = \{e \in E(t) : y \prec e\}$.

    Step 4b If $D_y \neq \emptyset$, delete those solutions from $E(t)$ and include y in $E(t)$.

    Step 4c If y is non-dominated with all members of $E(t)$, then update
    $Q(t) = Q(t) \cup \{y\}$.

Step 5 Calculate $k = \min(N - |E(t)|, |Q(t)|)$, the minimum of unoccupied positions in the elite set or the number of new solutions that are non-dominated with members of $E(t)$.

Step 6 Update $E(t+1) = E(t) \cup \text{draw}(k, Q(t))$ with newly found good solutions.

Step 7 If stopping criterion is not satisfied, set $t = t + 1$ and go to Step 2. Otherwise, declare $E(t+1)$ as the obtained set of non-dominated solutions.

The function $\text{draw}(k, Q(t))$ returns a set of, at most, k *distinct* solutions from the set $Q(t)$. Thus, essentially the non-dominated solutions from each iteration are checked with the external population (archive $E(t)$). If any new solution *strongly* dominates a member of $E(t)$, this dominated solution is eliminated from $E(t)$. Furthermore, a copy of the new solution is included in the archive $E(t)$. If any new solution is non-dominated with all members of $E(t)$, then it is moved to the special set $Q(t)$. Later, depending on the available population slots, some members from $Q(t)$ are copied to the archive $E(t)$. Here, the set $E(t)$ is an external population and does not participate in any genetic operations.

These investigators have also shown that once a Pareto-optimal solution moves into $E(t)$, either by replacing a dominated solution from $E(t)$ or by filling the remaining slots of $E(t)$ from $Q(t)$, that particular solution cannot be deleted. Since a fixed size N is maintained for the set $E(t)$, once N Pareto-optimal solutions enter $E(t)$, no other Pareto-optimal solutions will be accepted.

It is important to note that by copying distinct elements of $Q(t)$ in $E(t)$ in Step 6 does not guarantee finding a distinct set of solution in $E(t)$. There may well be duplicate solutions entering $E(t)$ in Step 4b. Even if there is no duplicate solution in the final $E(t)$, there is no guarantee for their good spread across the Pareto-optimal front. Thus, with a positive transition matrix for the genetic operators and with the use of the above elitist strategy, convergence to the Pareto-optimal set is guaranteed in a finite time with probability one. However, there are two difficulties, as follows:

1. There is no guarantee of an underlying spread among the members of the set $E(t)$.
2. The proof does not imply a particular time complexity; it only proves that solutions in the optimal set will only be found in a finite time.

The investigators also suggested two other algorithms which introduce elitism and make use of the elite solutions in genetic operations. These algorithms also do not guarantee maintaining a spread of solutions in the obtained set of non-dominated solutions. Nevertheless, the proof of convergence to the Pareto-optimal front in these algorithms itself is an important achievement in its own right.

8.9.2   An MOEA with Spread

Motivated by the above studies, we suggest an algorithm which attempts to converge to the true Pareto-optimal front and simultaneously attempts to maintain the best spread of solutions. First, we outline the algorithm.

An Elitist Steady-State MOEA

Step 1 A population $P(0)$ is drawn at random and $t = 0$ is set.

Step 2 Create a single solution $y = \text{generate}(P(t))$ by usual selection, recombination, and mutation operators.

Step 3 Collect all solutions from $P(t)$ which are dominated by y:
$$D_y = \{p \in P(t) : y \preceq p\}.$$

Step 4 If $D_y \neq \emptyset$, delete one of the members of D_y at random and include y in $P(t)$ and go to Step 5. Otherwise, if $D_y = \emptyset$ and y is non-dominated in $P(t)$, then decide to replace a solution from $P(t)$ with y by using a crowding routine: $P(t+1) = \text{Crowding}(y, P(t))$

Step 5 If a stopping criterion is not satisfied, set $t = t + 1$ and go to Step 2. Otherwise, declare $P(t+1)$ as the set of obtained non-dominated solutions.

The crowding routine in Step 4 decides inclusion of y in P(t) based on whether by doing so the distribution of solutions gets better or not. Before we check this, we find the non-dominated front of P(t). Let us call this front $F_1(t)$ of size $\eta = |F_1(t)|$. A diversity measure $\mathcal{D}(P)$ of a population P is used to decide whether to accept y or not. If the inclusion of y (in place of an existing solution) in P(t) improves this measure, the solution y will be included in P(t) by replacing a solution from $F_1(t)$. There could be several strategies used for finding the replacing member. We suggest choosing that member which if replaced by y produces the maximum improvement in the diversity measure. If replacement of any existing solution by y does not improve the diversity measure, P(t) is unchanged.

Crowding(y, P(t))

    Step C1 Find the non-dominated set $F_1(t)$ of p(t). Set $\eta = |F_1(t)|$. Include y in $F_1(t)$.

    Step C2 Choose each solution $p \in F_1(t)$ except the extreme solutions. Calculate $\mathcal{D}_p(F_1)$ as the diversity measure resulting by temporarily excluding p from $F_1(t)$.

    Step C3 Find $p^* = \{p : \mathcal{D}_p(F_1) \trianglerighteq \mathcal{D}_i(F_1)$ for all i$\}$, for which the $\mathcal{D}$-metric has the worst value.

    Step C4 Delete p^* from $F_1(t)$.

The diversity metric $\mathcal{D}$ can be calculated in either of the two spaces: decision variable space or objective space. Depending on this choice, the spread in solutions will be achieved.

The above algorithm accepts a new solution y only when the following two situations arise:

- solution y dominates at least one member of P(t);
- solution y is non-dominated to all members of P(t) and solution y improves the diversity metric.

Thus, with the second condition it is likely that a Pareto-optimal solution residing in P(t) may get replaced by a non-Pareto-optimal yet non-dominated solution y in trying to make the diversity measure better. However, when the parent population P(t) contains a set of Pareto-optimal solutions which are maximally diverse, no new solution can update P(t), thereby achieving convergence as well as maximum diversity among obtained Pareto-optimal solutions. Thus, such a vector of solutions is a stable *attractor* to the above algorithm. However, the search procedure to reach to such a solution vector may not be straightforward. The use of an archive to remember the previously found good solutions or a dynamically updated definition of dominance may be some ways to improve the performance. The attractive aspect is that if any member p of the parent population P(t) is non-Pareto-optimal, there exists a finite probability to find a Pareto-optimal solution y (for a search operator having a positive variation kernel) dominating p. In such a scenario, the first condition above will allow acceptance of solution y, thereby allowing the above algorithm to progress towards the Pareto-optimal front.

It is important to note that the spread of solutions obtained by this algorithm depends on the chosen diversity metric. Any diversity metric, described earlier in Section 8.2, can be used for this purpose. Here, we illustrate a diversity metric for two-objective optimization problems. First, we include y in $F_1(t)$, thereby increasing the size of $F_1(t)$ to $\eta + 1$. Thereafter, except the two end solutions, we exclude one, say the j-th solution of the remaining $(\eta - 1)$ solutions, and calculate the diversity metric $\mathcal{D}_j(F_1)$:

$$\mathcal{D}_j = \sum_{i=1}^{\eta-1} \frac{|d_i - \overline{d}|}{\eta - 1}, \tag{8.73}$$

where $\overline{d} = \sum_{i=1}^{\eta-1} d_i/(\eta - 1)$. In the above equation, we calculate the consecutive distances d_i of all η solutions (excluding the j-th solution). The above procedure is continued for every solution in $F_1(t)$ except the end solutions. The solution with the worst diversity measure (the largest $\mathcal{D}_j$ value) is eliminated from $F_1(t)$. It is clear that the smaller the above diversity measure $\mathcal{D}_j$, then the better is the distribution. For the above diversity measure $\mathcal{D}_j$, the algorithm will eventually lead to a distribution for which the distance between all consecutive pairs of solutions will be identical (or, when $\mathcal{D}_j = 0$), thereby ensuring a uniform distribution.

Although the above discussion does not sketch a proof of convergence to a maximally diverse set of Pareto-optimal solutions, it argues that such a set of solutions is a stable attractor of the above algorithm. More studies in this direction are needed to design MOEAs having a proof of convergence to the Pareto-optimal set as well as a proof of maximal diversity among them. More interesting studies should include a complexity analysis of such MOEAs.

Simulation Results

In order to demonstrate the working of the above algorithm, we attempt to solve the following two-objective problem:

$$\left. \begin{array}{ll} \text{Minimize} & f_1(x_1, x_2) = x_1, \\ \text{Minimize} & f_2(x_1, x_2) = 1 - x_1 + x_2^2, \\ \text{subject to} & 0 \leq x_1 \leq 1, \quad -1 \leq x_2 \leq 1. \end{array} \right\} \tag{8.74}$$

The problem has a Pareto-optimal front with $x_2^* = 0$ and $0 \leq x_1^* \leq 1$. The functional relationship for the Pareto-optimal solutions is a straight line: $f_2^* = 1 - f_1^*$.

We use a real-parameter implementation with population of size 10, binary tournament selection, the SBX operator with $\eta_c = 10$ and the polynomial mutation operator with $\eta_m = 5$. Crossover and mutation probabilities are 0.9 and 0.5, respectively. The SBX operator is used to create two offspring, of which one is selected at random and mutated. The algorithm is run for 80 000 function evaluations (8000 generations) to record the progress of the algorithm. Figure 277 shows the diversity

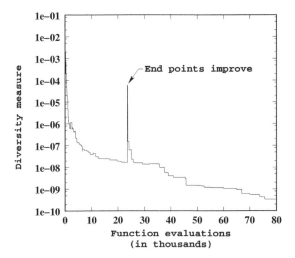

Figure 277 Increasingly better distribution is achieved with the proposed algorithm.

measure $\mathcal{D}$ with function evaluations. It is clear that the above algorithm finds an increasingly better distribution with generation number. At the end of 80 000 function evaluations, the diversity measure is very close to zero, meaning that a uniform-like distribution is achieved. The population members at the end of 80 000 function evaluations are shown in Figure 278. This figure does indeed show a uniform-like distribution. The sudden worsening in the diversity metric in Figure 277 occurs due to the discovery of improved extreme solutions.

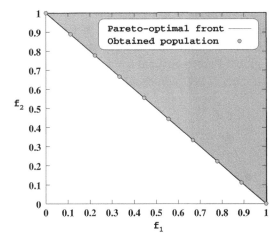

Figure 278 A uniform-like distribution is achieved with the proposed algorithm.

Computational Complexity

The above algorithm requires one sorting in one of the two objective functions requiring $O(N \log N)$ computations for each new solution. In order to create N new solutions in a generational sense, $O(N^2 \log N)$ computations are necessary. However, if a special bookkeeping strategy is used for tracking the unaffected distance measures, the computational complexity can be reduced.

Distances in More Than Two Objectives

Finding the neighbors and then calculating the consecutive Euclidean distances among non-dominated solutions for problems having more than two objectives is computationally expensive. For these problems, the obtained non-dominated solutions can be used to construct a higher-dimensional surface by employing the so-called triangularization method. As shown in Figure 279, several distance measures can be associated with such a triangularized surface. The average distance of all edges (shown by bold lines) including a solution i can be used as the distance d_i. Alternatively, the hypervolume (shown hatched in the figure) enclosed by all triangular elements where the current solution is a node can be used as d_i. The crowding distance metric used in the NSGA-II can also be used in problems with more than two objectives. Thereafter, a diversity measure similar to that given above in equation (8.73) can also be employed.

8.10   Controlling Elitism

The elitist MOEAs described in Chapter 6 raise an important issue relating to EA research, namely the concept of exploitation versus exploration (Goldberg, 1989). Let us imagine that at a generation, we have a population R_t where most of the members lie on the non-dominated front of rank one and this front is not close to the true Pareto-optimal front. This will happen in the case of multi-modal multi-objective problems, where a population can get attracted to a local Pareto-optimal front away from the global Pareto-optimal front. Since a large number of population members belong to the current best non-dominated front, the elite-preserving operator

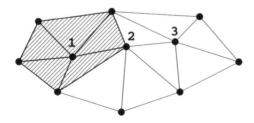

Figure 279   Triangularization of a non-dominated surface approximated by a set of obtained non-dominated solutions (marked by filled circles).

will result in not accepting many dominated solutions. For example, in NSGA-II (Deb et al., 2000b), elite solutions are emphasized on two occasions, i.e. once in the usual tournament selection operation and again during the elite-preserving operation in the `until` loop (see page 246). The former operation involves the crowded tournament selection operator, which emphasizes the elite solutions (the current-best non-dominated solutions). In the latter case, solutions are selected starting from the current-best non-dominated solutions until all population slots are filled. This dual emphasis of the elite solutions will cause a rapid deletion of solutions belonging to the non-elitist fronts. Although the crowding tournament operator will ensure diversity along the current non-dominated front, lateral diversity will be lost. In many problems, when this happens the search slows down, simply because there may be a lack of diversity in the particular decision variables left to push the search towards better regions of optimality. Thus, in order to ensure better convergence, a search algorithm may need diversity in both aspects – along the Pareto-optimal front and lateral to the Pareto-optimal front, as shown in Figure 280.

A recent study (Parks and Miller, 1998) of a specific problem has suggested that the variability present in non-dominated solutions may allow a strong elitist selection pressure to be used without prematurely converging to a sub-optimal front. Apparently this makes sense, but in general it may be beneficial to maintain both kinds of population diversities in a multi-objective EA. In the test problems discussed above in Section 8.3, the lateral diversity is ensured by the function $g()$ through $(n-1)$ decision variables. Solutions converging to any local Pareto-optimal front will cause all of these decision variables to take an identical value. A strong elitism will reduce the lateral variability in these solutions and may eventually prevent the algorithm moving towards the true Pareto-optimal front. Thus, it is important to explicitly preserve both kinds of diversities in an MOEA in order to handle different kinds of multi-objective optimization problems.

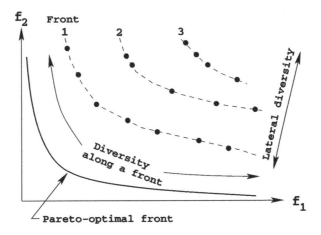

Figure 280   The controlled elitism procedure. This is a reprint of Figure 2 from Deb and Goel (2001a) (© Springer-Verlag Berlin Heidelberg 2001).

Although most MOEAs use an explicit diversity preserving mechanism among the non-dominated solutions, there is not much emphasis on maintaining diversity laterally. We view this problem as a problem of balancing exploration and exploitation issues in an MOEA. In the above discussion for NSGA-II, it is clear that in certain complex problems, this algorithm, in the absence of a lateral diversity-preserving operator such as mutation, causes too much exploitation of the currently-best non-dominated solutions. In order to counteract this excessive selection pressure, an adequate exploration by means of the search operators must be used. Achieving a proper balance of these two issues is not possible with the uncontrolled elitism used in NSGA-II. In a recent study (Deb et al., 2000a), it was observed that in the test problem ZDT4 with Rastrigin's multi-modal function as the g functional, NSGA-II could not converge to the global Pareto-optimal front. However, when a mutation operator with a larger mutation strength is used, the algorithm succeeds in achieving convergence. Increasing variability through mutation enhances the exploration power of an MOEA and a balance between enhanced exploitation and the modified exploration can be maintained.

Although many researchers have adopted an increased exploration requirement by using a large mutation strength in the context of single-objective EAs, the extent of needed exploration is always problem-dependent. In the following subsection, instead of concentrating on changing the search operators, we suggest a controlled elitism mechanism for NSGA-II which will control the extent of exploitation rather than controlling the extent of exploration. However, we highlight that a similar controlled elitism mechanism is needed and can also be introduced into other elitist MOEAs (Laumanns et al., 2001).

8.10.1 Controlled Elitism in NSGA-II

In the proposed controlled NSGA-II, we restrict the number of individuals in the current best non-dominated front adaptively. We attempt to maintain a predefined distribution of number of individuals in each front. Specifically, we use a geometric distribution for this purpose:

$$N_i = rN_{i-1}, \tag{8.75}$$

where N_i is the maximum number of allowed individuals in the i-th front and r (< 1) is the reduction rate. This is in agreement with another independent observation (Kumar and Rockett, in press). Although the parameter r is user-defined, the procedure is adaptive, as follows.

First, the population $R_t = P_t \cup Q_t$ is sorted for non-domination. Let us say that the number of non-dominated fronts in the combined population (of size 2N) is K. Thus, according to the geometric distribution, the maximum number of individual allowed in the i-th front ($i = 1, 2, \ldots, K$) in the new population of size N is:

$$N_i = N\frac{1-r}{1-r^K}r^{i-1}. \tag{8.76}$$

Since $r < 1$, the maximum allowable number of individuals in the first front is the highest. Thereafter, each front is allowed to have an exponentially reducing number of solutions. The distribution considered above is an assumption; however, other distributions, such as an arithmetic distribution or a harmonic distribution, may also be tried. Nevertheless, the main concept of the proposed approach is to forcibly allow solutions from different non-dominated fronts to co-exist in the population.

Although equation (8.76) denotes the maximum allowable number of individuals N_i in each front i in a population, there may not exist exactly N_i individuals in such fronts. We resolve this problem by starting a procedure from the first front. First, the number of individuals in the first front is counted. Let us say that there are N_1^t individuals. If $N_1^t > N_1$ (that is, there are more solutions than allowed), we only choose N_1 solutions by using the crowded tournament selection. In this way, exactly N_1 solutions that are residing in a less crowded region are selected. On the other hand, if $N_1^t \leq N_1$ (that is, there are less or equal number of solutions in the population than allowed), we choose all N_1^t solutions and count the number of remaining slots $\rho_1 = N_1 - N_1^t$. The maximum allowed number of individuals in the second front is now increased to $N_2 + \rho_1$. Thereafter, the actual number of solutions N_2^t present in the second front is counted and is compared with N_2 as above. This procedure is continued until N individuals are selected. Figure 281 shows that a population of size $2N$ (having four non-dominated fronts with the top-most subpopulation representing front one and so on) is reduced to a new population P_{t+1} of size N by using the above procedure. In the transition shown on the right, all four fronts have representative solutions in P_{t+1}. Besides this controlled elite-preserving procedure, the rest of the procedure is kept the same as that used in NSGA-II. The left figure also shows the new population P_{t+1} (having only two fronts) which would have been obtained by using the usual NSGA-II procedure. It is clear that the new population obtained under the

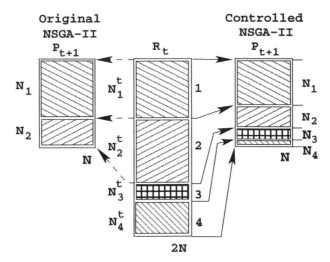

Figure 281 The controlled elite-preserving procedure in NSGA-II. This is a reprint of Figure 3 from Deb and Goel (2001a) (© Springer-Verlag Berlin Heidelberg 2001).

controlled NSGA-II procedure will, in general, be more diverse than that obtained by using the usual NSGA-II approach. Since the population is halved, it is likely that in each front there would be more solutions than allowed. However, there could be some situations where after all 2N solutions are processed as above, there are still some slots left in the new population to be filled. This may happen particularly when r is large. In such cases, we make another pass with those individuals left out from the first front, continue to other fronts, and then start including them until we fill up the remaining slots.

Discussion

As mentioned earlier, keeping individuals from many non-dominated fronts in the population helps the recombination operator to create diverse solutions. NSGA-II and many other successful MOEAs thrive at maintaining diversity among solutions of individual non-dominated fronts. The controlled elitism procedure suggested above will help maintain diversity in the solutions across such fronts. In solving difficult multi-objective optimization problems, this additional feature may be helpful in progressing towards the true Pareto-optimal front.

It is intuitive that the parameter r is important in maintaining the correct balance between the exploitation and exploration issues discussed above. This parameter sets up the extent of exploration allowed in an MOEA. If r is small, the extent of exploration is large, and vice versa. In general, the optimal value of r will depend on the problem and it will be difficult to determine it theoretically. In the following section, we present simulation results on a number of difficult problems to investigate if there exists any value of r where NSGA-II performs well.

A Test Problem

In order to investigate the effect of controlled elitism alone, we do not use the mutation operator. In addition, for controlled NSGA-II runs, we do not use the selection operator in make_new_pop() to create the offspring population. We use a population size of 100, a crossover probability of 0.95, and a distribution index for the SBX operator (Deb and Agrawal, 1995) of 20. The algorithms are run until 200 generations are completed. Instead of using the usual definition of domination, we have used the strong dominance condition in these studies. The problem is a biased test function:

$$\left.\begin{aligned}
\text{Minimize} \quad & f_1(x) = x_1, \\
\text{Minimize} \quad & f_2(x) = g(x)\left(1 - \sqrt{x_1/g(x)}\right), \\
\text{where} \quad & g(x) = 1 + \left(\sum_{i=2}^{10} x_i\right)^{0.25}, \\
& x_1 \in [0,1], \quad x_i \in [-5,5], \quad i = 2, \dots, 10.
\end{aligned}\right\} \quad (8.77)$$

In all simulations, 25 independent runs from different initial populations are taken. Not all simulations with the NSGA-II and the controlled NSGA-II have converged to the true Pareto-optimal front with the above parameter settings. However, one advantage of working with the above construction of a multi-objective test problem is that the function g() indicates the convergence of the obtained front near the true Pareto-optimal front. For an ideal convergence to the latter, the g() value would be one. A simulation with a smaller g() is better. Figure 282 shows that NSGA-II could not converge close to the true Pareto-optimal front. The average of the best value of g() in a population is calculated and is also plotted in the figure. It is clear that all controlled elitism runs (with different r values except r = 0.9) have converged better than runs with the original NSGA-II. The average g() value is closer to one than that in NSGA-II. Among the different r values, r = 0.65 performed the best. For smaller r values, an adequate number of fronts are not allowed to survive, thereby not maintaining enough diversity to proceed to near the true Pareto-optimal front. For r close to 1, there is not enough selection pressure allowed for the non-dominated solutions, thereby slowing down the progress. Figure 282 clearly shows that there is a trade-off reduction rate (around r = 0.65, in this problem), which makes a good balance of the two aspects.

In order to investigate the composition of a population with controlled elitism, we count the number of fronts and the number of best non-dominated solutions in 25 different runs of the NSGA-II and the controlled NSGA-II with r = 0.65. The average values of these numbers are plotted in Figures 283 and 284, respectively. Figure 283 shows that in the case of the original NSGA-II, the number of fronts grows for a few generations, but then eventually drops to one. On the other hand, the controlled NSGA-II with r = 0.65 steadily finds and maintains solutions in more and more fronts with increasing numbers of generations. This keeps an adequate diversity in the population to enable progress towards the true Pareto-optimal front.

Figure 284 shows that all 100 population members reside in the non-dominated front

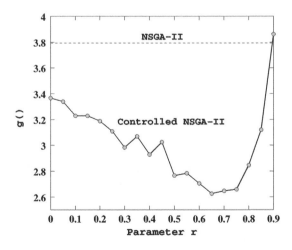

Figure 282 Convergence observed for the usual and controlled NSGA-II. This is a reprint of Figure 8 from Deb and Goel (2001a) (© Springer-Verlag Berlin Heidelberg 2001).

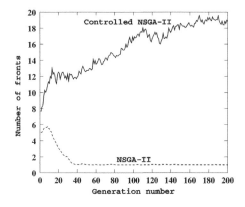

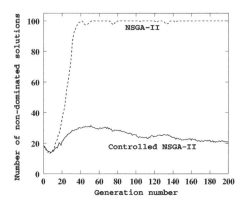

Figure 283   The average number of fronts present in a NSGA-II population, with and without controlled elitism.

Figure 284   Average number of best non-dominated solutions present in a NSGA-II population, with and without controlled elitism.

with the NSGA-II after only a few generations, thereby loosing the lateral diversity. However, only about 20–25 solutions are maintained in the best non-dominated set in the case of the controlled NSGA-II with $r = 0.65$. The rest of the population members belong to other fronts, thereby maintaining a good lateral diversity.

There may exist other ways to control the lateral diversity in an MOEA. Alternative distributions, such as an arithmetic distribution, can be tried instead of a geometric distribution. A similar concept can be used with other MOEAs. In archived MOEAs, such as the PAES and its successors, the number of individuals allotted in the non-dominated archives compared to dominated archives can be explicitly controlled. Although more studies are needed, the experimental results shown above clearly demonstrate the need for introducing lateral diversity in an MOEA. As diversity among non-dominated solutions is required to be maintained for obtaining diverse Pareto-optimal solutions, lateral diversity is also needed to be maintained for achieving a better convergence to the true Pareto-optimal front.

8.11   Multi-Objective Scheduling Algorithms

Unlike the major interest shown in applying evolutionary algorithms to single-objective scheduling problems (Davis, 1991; Gen and Cheng, 1997; Reeves, 1993a; Starkweather et al., 1991), the application of multi-objective evolutionary optimization in multi-objective scheduling problems has so far received a lukewarm response. With the demonstration of efficient multi-objective function optimization algorithms, this is probably the time that researchers interested in scheduling problems (such as the traveling salesperson problem, job-shop scheduling, flow-shop scheduling and other combinatorial optimization problems) might pay attention to the possibilities of extending the ideas of MOEAs to multi-objective scheduling problems.

However, there do exist a number of studies where MOEAs are applied to multi-objective flow-shop, job-shop and open-shop scheduling problems. We will discuss them briefly in this section. In addition, there exist other studies which have also been used in multi-objective scheduling and planning problems using multi-objective EAs and related strategies such as ant colony search algorithms (Iredi et al., 2001; Krause and Nissen, 1995; Shaw et al., 1999; Tamaki et al., 1995, 1999; Zhou and Gen, 1997).

8.11.1 Random-Weight Based Genetic Local Search

As early as the mid 1990s, Murata and Ishibuchi (1995), applied the random-weight GA (RWGA), described earlier in Section 5.7, to a two-objective flow-shop scheduling problem. Before we discuss this application, let us briefly outline the objectives in a flow-shop scheduling problem.

In such a scheduling problem, a total of n jobs must be finished by using m machines. However, each job has exactly m operations, each of which must be processed in a different machine. Thus, each job has to pass through each machine in a particular order. Moreover, the order of machines needed to complete a job is the same for all of the jobs. For each job, the time required to complete the task in each machine is predefined. Figure 285 shows a typical schedule for a five-job and three-machine problem. Note that all jobs follow the same order of machines: M1 to M2 and then to M3. Each job i has an overall completion time, called the *flow time* F_i. This figure shows the flow time for the first and fifth jobs. There is also an overall completion time for all jobs, assuming that the first job in the first machine started at time zero. This overall completion time of all jobs is known as the *make-span*. The figure also marks

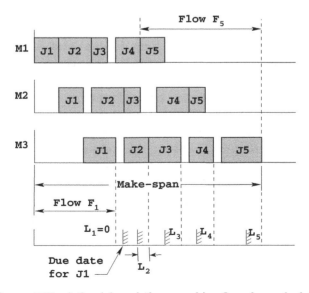

Figure 285 A five-job and three-machine flow-shop schedule.

the make-span for the illustrated schedule. Furthermore, each job has a due date of completion. If the actual completion time is more than the due date, the *tardiness* for that job is defined as the extra time taken to complete the job. The tardiness for a job i is shown as L_i in the figure. It is interesting to note that if a job is completed before the due date, the tardiness is zero. The flow time of a job reflects the time taken to complete the job. The smaller the flow time for a job, then the better is the schedule for that job. However, since jobs can have different flow times, it is better to minimize the mean flow time $\overline{F}$, which is defined as the average of the flow times for all of the jobs:

$$\overline{F} = \frac{1}{n} \sum_{i=1}^{n} F_i. \tag{8.78}$$

Some careful thought will reveal that the absolute minimum of the mean flow time occurs when all operations for a job are performed without any time delay. In this way, each job requires a minimum possible flow time, thereby minimizing the mean flow time. Since the operation time of a job on a machine is different for different jobs, the overall schedule for achieving the minimum mean flow time would stagger the operations of each job so that the individual mean flow time is a minimum. This process will lead to a large make-span. However, minimizing the make-span is also an important concern in a flow-shop scheduling. Thus, both optimization problems of minimizing the mean flow time and minimizing the make-span have conflicting optimal solutions (Bagchi, 1999).

The overall tardiness of a schedule can be measured by calculating the mean tardiness as follows:

$$\overline{L} = \frac{1}{n} \sum_{i=1}^{n} L_i. \tag{8.79}$$

In a good schedule, it is desired to have a minimum mean tardiness, which ensures a minimum delay of completion of all jobs from the due dates. Although minimization of the mean tardiness and minimization of the make-span are somewhat related for reasonable due dates, the minimization of mean tardiness and the minimization of mean flow time will produce conflicting optimal solutions due to a similar reason as that given above. In general, all three objectives of minimizing make-span, minimizing mean flow time, and minimizing tardiness produce conflicting optimal solutions. In most single-objective scheduling problems, the make-span is minimized, keeping the mean flow time and mean tardiness restricted to certain values.

Murata and Ishibuchi (1995) formulated a two-objective scheduling problem for minimizing the make-span and the mean tardiness. These investigators used a two-point order crossover operator and a shift-change mutation operator for creating valid feasible offspring from feasible parents. In the two-point order crossover operator, two random genes are chosen at random. An offspring schedule is created by copying the genes outside the region bounded by the chosen genes from one parent and arranging the inside genes as they are ordered in the second parent. Figure 286 illustrates this crossover operator on two parent strings representing two schedules. Jobs 1, 2, 7,

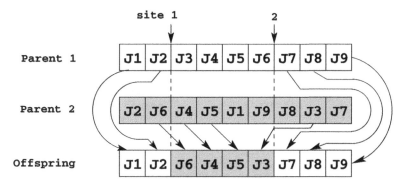

Figure 286   Two-point order crossover operator.

8 and 9 are copied from the first parent and jobs 3 to 6 are copied in the same order as they appear in the second parent. In the shift-change mutation operator, two genes are chosen at random and are placed side by side, thereby making the linkage between these two genes better. Figure 287 illustrates this mutation operator. Jobs 3 and 6 come together in the mutated schedule. In comparison with the VEGA, these investigators reported better converged solutions with their RWGA.

Later, Ishibuchi and Murata (1998a) introduced a local search technique in their RWGA in the search for finding better converged solutions. The algorithm is identical to that mentioned earlier in Section 5.7, except that each solution x created by using the crossover and mutation operators is sent to a local search method to find an improved local solution. A fixed number of solutions are created in the neighborhood of x by exchanging places between two jobs. To determine the improvement, the weighted objective function value (with the same random weight vector used by its parents in the selection operation) is used. The local search operator is simple. If a neighboring solution improves the weighted objective, it is accepted; otherwise, another solution is tried. When a pre-specified number of solutions are tried, the best solution becomes a new population member.

As before, an external population with non-dominated solutions is maintained and updated in each iteration with the non-dominated solutions of the new population. In addition, elitism is maintained by copying a fixed number of solutions (chosen at random) from the elite population to create an offspring population. Figure 288 shows

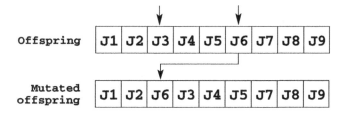

Figure 287   Shift-change mutation operator.

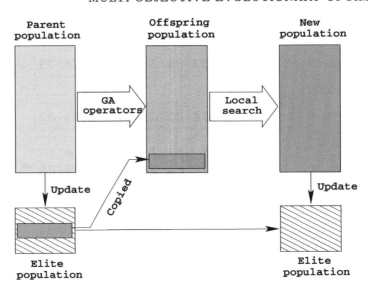

Figure 288   The hybrid RWGA procedure.

a schematic of one generation of the hybrid RWGA with the local search method.
This figure shows that by genetic operations, the offspring population is partially
filled. Some solutions from the elite population are directly copied into the offspring
population. Thereafter, each offspring population member is modified by using the
local search technique. For elite solutions introduced into the offspring population, a
random weight vector is used to calculate the objective function needed during the
local search method.

On a 10-job and five-machine flow-shop problem of minimizing the make-span and
minimizing the maximum tardiness (instead of minimizing the mean tardiness), these
investigators have used the following parameter settings: GA population of size 20,
elite population of size 3, and a maximum number of function evaluations of 10 000. In
comparison with the VEGA (shown by squares in Figure 289) and a constant weight
GA (CWGA) (shown by squares in Figure 290), their hybrid RWGA (shown by circles)
is able to find more solutions in the final non-dominated set. In both figures, the trade-
off between make-span and maximum tardiness is evident. The VEGA is unable to find
intermediate solutions. The CWGA is used with an elite set proposed in the hybrid
RWGA, thereby collecting multiple and diverse non-dominated solutions along the
way. Furthermore, these investigators have solved a 20-job and 10-machine flow-shop
scheduling problem having three objectives and similar observations have been made.

8.11.2   Multi-Objective Genetic Local Search

Jaskiewicz (1998) suggested a multi-objective genetic local search (MOGLS) approach,
where each solution, created either in the initial population or in subsequent genetic
operations is modified with a local search technique. In his approach, a scalarizing

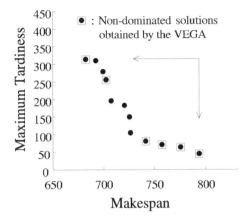

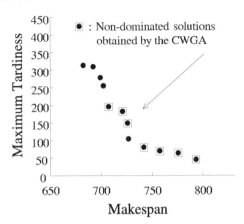

Figure 289 Non-dominated solutions obtained with the hybrid RWGA and VEGA (Ishibuchi and Murata, 1998a). Reprinted with permission (© 1998 IEEE).

Figure 290 Non-dominated solutions obtained with the hybrid RWGA and a constant weight GA ($w_1 = w_2 = 0.5$) (Ishibuchi and Murata, 1998a). Reprinted with permission (© 1998 IEEE).

function is chosen at random. The best N solutions from the population are chosen according to the chosen scalarizing function. Two solutions from these N solutions are chosen at random and a new offspring is created by genetic operations. A local search procedure is initiated from the offspring until a better offspring is created. If the new solution makes the distribution better, it is accepted in the main GA population. An external elite set is then updated with this new solution. By considering different cost values for each arc, Jaskiewicz (1998) formed many three- and four-objective 50 and 100-city traveling salesperson problems (TSPs). He used the standard ordering-GA and its operators and showed that the MOGLS is able to find more trade-off solutions in the final generation than a repetitive single-objective genetic local search (GLS) in these permutation problems.

8.11.3 NSGA and Elitist NSGA (ENGA)

Bagchi (1999) extended Srinivas and Deb's NSGA (Srinivas and Deb, 1994) (see Section 5.9 above) to solve flow-shop, job-shop and open-shop scheduling problems. In all cases, this investigator used three objective functions: (i) make-span, (ii) mean flow time, and (iii) mean tardiness. In an effort to find more solutions in the obtained non-dominated set, Bagchi (1999) introduced an elitist version of the NSGA; he called this an elitist NSGA (or simply ENGA). In this approach, the offspring population obtained after genetic operations is combined with the parent population. Thereafter, the combined population is sorted in different non-dominated levels. The best 50% solutions from this combined sorted population is chosen as the new population. The rest of the procedures of assigning fitness and implementing the sharing approach are identical to that used in the original NSGA.

Simulation results on 21 test problems, including a 49-job and 15-machine flow-shop scheduling problem, suggested that although both the NSGA and ENGA (with identical computational effort) found solutions with a *comparable degree of convergence*, the ENGA found more numbers of non-dominated solutions than the NSGA. This result is not surprising because an elitist algorithm is supposed to preserve and propagate more non-dominated solutions than the non-elitist version. The presence of more non-dominated solutions in the parent population helps create better offspring. Moreover, once a set of non-dominated solutions are found, they can only be eliminated if a better set of non-dominated solutions is discovered. This observation supports similar observations reported in other studies (Leung et al., 1998; Zitzler, 1999) and suggests the use of elitist algorithms even in the case of multi-objective scheduling problems.

This particular study also reported other simulation results using the NSGA and ENGA in different job-shop and open-shop scheduling problems. In a job-shop scheduling problem, the unidirectional flow of jobs is relaxed. However, each job has a particular order of operations. Before a job is completely finished, some operations of another job can be started. Thus, job-shop scheduling allows more flexibility in designing a schedule than flow-shop scheduling. An open-shop scheduling is even more flexible in that the order of operations of any job is also not fixed. Any operation can be performed at any point of time. On a number of problems, including a 20-job and 10-machine problem (minimizing make-span, mean flow time, and mean tardiness), the study has shown that the elitist NSGA finds more non-dominated solutions in the obtained non-dominated front than the non-elitist NSGA.

8.12 Summary

In this chapter, we have dealt with a number of salient issues regarding multi-objective evolutionary algorithms. Although there exist many other issues, the topics discussed here are of immense importance.

Most multi-objective studies performed so far involve only two objectives. While dealing with more than two objectives, it becomes essential to illustrate the trade-off solutions in a meaningful manner. We have reviewed a number of different ways in which the non-dominated solutions can be represented. Of these, the scatter-plot and value path methods are the most commonly used.

With the success of MOEAs in different problem domains, many new techniques have been suggested. This demands a proper method of assessing the performance of a newly suggested algorithm. Since an MOEA is supposed to perform the two tasks of (i) converging close to the true Pareto-optimal front and (ii) maintaining a diverse set of non-dominated solutions, an algorithm must be assessed with respect to both of these tasks. Ironically, it is difficult to have one performance metric to evaluate both of the above issues adequately. In this chapter, we have reviewed a number of performance measures which have been suggested in the literature. While developing new performance metrics, their scalability to problems having more than two objectives must be kept in mind.

For evaluating a new algorithm, there is also a need to test it with problems possessing known complexities of the search space and with a known Pareto-optimal set. We have presented a number of test problems which are suggested in the literature and for the first time have determined the exact location of Pareto-optimal solutions for many of these. Knowledge of the exact locations of the Pareto-optimal solutions is helpful in investigating the search abilities of an algorithm. Test problems assess the ability of a multi-objective optimization algorithm to successfully perform different tasks. For the first time, we have also suggested a number of test problems for constrained multi-objective optimization with known set of constrained Pareto-optimal solutions. Such test problems allow a multi-objective optimization algorithm to be evaluated for its ability to reach the feasible search space and then to converge to the correct constrained Pareto-optimal solutions.

With the development of a number of MOEAs over the past few years, there have been some studies in comparing them systematically. The next section in this chapter has outlined these comparisons and presented their outcome. Although each study has considered different MOEAs with different performance measures, a common conclusion has emerged from all of this work. Those MOEAs which properly implemented elite-preservation, emphasized non-dominated solutions, and maintained diversity among non-dominated solutions, all performed well. In several studies, it was clear that elite-preservation is an important operation in converging as well as sustaining a good diverse set of non-dominated solutions.

An important aspect of maintaining diversity among non-dominated solutions is the space in which the diversity is required. In some problems, diversity among solutions is more important in the decision variable space than in the objective space. It has been shown that on such occasions, the diversity preserving operator (such as sharing, clustering, etc.) must treat the proximity of the solutions in the decision variable space. On the other hand, if the diversity in the objective space is more important, the proximity must be measured in the objective space. It is important to keep in mind that the proximity in one space may not mean that a proximity in the other space would be automatically obtained. This has been found to be particularly true in certain nonlinear and complex problems. Most MOEAs suggested in the previous two chapters have attempted to maintain diversity in the objective space, although they can be easily modified to find better distributed solutions in the decision variable space by simply calculating the proximity measure (distance between two solutions) in the decision variable space.

Throughout this book, we have been arguing the need for finding a diverse set of non-dominated solutions. However, in a particular problem, the user may not be interested in the complete Pareto-optimal set; instead, the user may be interested in a certain region of the Pareto-optimal set. Such a bias can arise if not all objectives are of equal importance to the user. MOEAs can be used to find a preferred distribution of solutions in the Pareto-optimal region with such information. We have argued that finding a preferred distribution in the region of interest is more practical and less subjective than finding *one* biased solution in the region of interest. In this chapter, we have discussed a number of techniques for achieving this task. In these MOEAs, the user needs to specify a relative preference structure of the objectives. Algorithms use this information to find a biased set of solutions in the Pareto-optimal front. If such knowledge is available, there are two advantages: (i) the search effort is reduced, and (ii) better precision in solutions in the desired region is achieved.

The concept of multi-objective optimization can also be used to solve other kinds of optimization problems in an efficient way. For example, a constrained single-objective optimization problem can be considered as being a multi-objective optimization problem of optimizing the objective function and minimizing all constraint violations. Most single-objective constraint handling techniques, including the penalty function approach, introduce artificial means of penalizing infeasible solutions. The search path to the feasible region largely depends on the way that the constraints are handled. Treating constraint violations as objectives does not restrict the search path in any way, thereby making the overall approach flexible and generic. In addition, the principle of finding multiple optimal solutions can also be extended to other similar problems, such as goal programming. Because of the lack of an optimization algorithm which can find multiple optimal solutions simultaneously, goal programming approaches traditionally use relative weights of objectives and resort to finding one solution corresponding to one weight vector at a time. Here, we have suggested a simple procedure to find multiple trade-off solutions in a goal programming problem by converting the problem into a multi-objective optimization problem. Simulation results demonstrate the efficacy of the proposed approach.

As mentioned earlier, most studies in multi-objective evolutionary algorithms have considered only two objectives. In this chapter, we have discussed some issues regarding the scalability of MOEAs in problems having more than two objectives. An interesting aspect is that as the number of objectives increase, a larger proportion of a randomly chosen population becomes non-dominated. When this happens, introduction of elitism becomes tricky. This is because a large number of population members are candidate elite solutions, therefore not allowing many new solutions to be accepted in any generation. Moreover, we have suggested a way to choose an adequate population size such that a reasonable proportion of the population members belongs to the dominated fronts for initially introducing variability in the population.

In MOEAs, convergence to the Pareto-optimal front and simultaneous maintenance of a good distribution are both important. Although there exist a number of MOEAs with theoretical convergence properties to the true Pareto-optimal front, they do not guarantee maintaining any spread of solutions. Moreover, such studies also do not suggest any upper bound of the computational complexities needed to converge to the Pareto-optimal front. More studies to develop MOEAs with properties of convergence as well as spread of solutions remain as an imminent challenge to the researchers of MOEAs.

Most elitist MOEA implementations do not control the extent of elitism adequately.

For example, both Rudolph's MOEA and the NSGA-II do not allow choosing any dominated solution before all of the non-dominated solutions are chosen. When most population members belong to the best non-dominated front (either towards later generations or in solving problems having a large number of objectives), the lateral diversity needed to proceed towards the true Pareto-optimal front may have been lost in certain problems. By allowing a certain proportion of other dominated fronts to co-exist in the population, the lateral diversity can be restored. In this chapter, we have proposed a strategy where elitism can be controlled by a single parameter. By forcefully allowing solutions from dominated fronts to exist in the population, the effective selective advantage to the best non-dominated solutions is controlled. The SPEA controls elitism by choosing the size of the elite population to be small. In this way, there is always a fixed ratio of elite to population members (1:4 ratio was suggested by the original investigators) is maintained. The PAES controls elitism by limiting the maximum number of solutions allowed in any grid of the archive. Such controlled elitism techniques will be more useful in complex problems, particularly in those problems having a large number of objectives.

Finally, we have discussed a number of multi-objective scheduling techniques using evolutionary algorithms. Although to date there do not exist many studies in this direction, the successful simulations' demonstration of finding well-distributed trade-off solutions in a number of applications suggest further immediate studies on such scheduling MOEAs.

Exercise Problems

1. The following non-dominated solutions are found by an MOEA in minimizing f_1 and maximizing f_2:

$$\mathbf{f}^{(1)} = (1,1)^\mathsf{T}, \quad \mathbf{f}^{(2)} = (4,5)^\mathsf{T}, \qquad \mathbf{f}^{(3)} = (6,7)^\mathsf{T},$$
$$\mathbf{f}^{(4)} = (2,2)^\mathsf{T}, \quad \mathbf{f}^{(5)} = (2.5,3.5)^\mathsf{T}.$$

Find a pseudo-weight vector for each of these solutions.

2. Consider the following problem:

$$\begin{aligned}
\text{Minimize} \quad & f_1(\mathbf{x}) = (x_1 - 2)^2 + x_2^2, \\
\text{Minimize} \quad & f_2(\mathbf{x}) = (x_1 - x_2)^2, \\
& -5 \le x_1, x_2 \le 5.
\end{aligned}$$

Find the Pareto-optimal front. Choose 11 Pareto-optimal points as P^* uniformly distributed in f_1. Consider the following points found by an MOEA:

$$\mathbf{x}^{(1)} = (1.0, 0.8)^\mathsf{T}, \quad \mathbf{x}^{(2)} = (1.5, 0.6)^\mathsf{T}, \quad \mathbf{x}^{(3)} = (1.2, 0.8)^\mathsf{T},$$
$$\mathbf{x}^{(4)} = (1.7, 0.5)^\mathsf{T}, \quad \mathbf{x}^{(5)} = (2, 0)^\mathsf{T}, \qquad \mathbf{x}^{(6)} = (1.9, 0.2)^\mathsf{T}.$$

Calculate the following performance metrics for the above set of points:

(a) Error ratio,

(b) Set coverage metric,
(c) Spread,
(d) Hypervolume.

3. Consider the problem having the Pareto-optimal front defined as follows:

$$f_1^* + f_2^* + f_3^* = 1.$$

Use a uniformly distributed set of points in a step of 0.2 in f_1 and f_2 as P^*. For the following points

$$f^{(1)} = (1, 0, 0)^T, \qquad f^{(2)} = (0.5, 0.6, 0.3)^T, \quad f^{(3)} = (0.1, 0.7, 0.4)^T,$$
$$f^{(4)} = (0.2, 0.5, 0.3)^T, \quad f^{(5)} = (0.0, 0.9, 0.2)^T, \quad f^{(6)} = (0.1, 0.0, 0.9)^T.$$

calculate the following performance metrics:

(a) Hypervolume
(b) Spread
(c) Generalized distance

4. The following pairs of points are found in 10 runs of an MOEA for minimizing both objectives:

	Run 1	Run 2	Run 3	Run 4
$f^{(1)}$	$(2.0, 4.0)^T$	$(3.0, 3.0)^T$	$(2.5, 4.5)^T$	$(3.5, 4.0)^T$
$f^{(2)}$	$(4.0, 2.0)^T$	$(5.0, 1.0)^T$	$(4.0, 1.8)^T$	$(5.2, 1.5)^T$
	Run 5	Run 6	Run 7	Run 8
$f^{(1)}$	$(4.2, 1.2)^T$	$(1.5, 5.0)^T$	$(3.7, 2.0)^T$	$(5.5, 1.0)^T$
$f^{(2)}$	$(3.6, 3.8)^T$	$(4.0, 1.0)^T$	$(4.5, 1.5)^T$	$(2.0, 4.2)^T$
	Run 9	Run 10		
$f^{(1)}$	$(5.1, 1.1)^T$	$(4.9, 1.9)^T$		
$f^{(2)}$	$(2.0, 4.8)^T$	$(2.0, 4.1)^T$		

Calculate the 50% attainment point along the $f_1 = f_2$ cross-line.

5. In order to create a three-objective test problem for minimization, we would like to have the Pareto-optimal surface as a three-dimensional plane: $\sum_{i=1}^{3} f_i^* = 1$. Considering a uniform distribution of solutions lateral to the Pareto-optimal surface, construct the test problem for the following cases:

(a) Assume a uniform density of solutions along the Pareto-optimal front.
(b) Assume an increasing density of Pareto-optimal solutions towards increasing f_3.

6. The following non-dominated solutions are found by an MOEA in solving a two-objective minimization problem:

$$f^{(1)} = (2.0, 4.0)^T, \quad f^{(2)} = (5.0, 2.0)^T, \quad f^{(3)} = (1.0, 5.0)^T,$$
$$f^{(4)} = (4.0, 3.5)^T, \quad f^{(5)} = (7.0, 1.0)^T.$$

Choose a solution from the above set based on

(a) compromised programming approach using $p = 2$,
(b) marginal rate of substitution approach,
(c) pseudo-weight vector approach with target weight vector $(0.3, 0.7)^T$.

7. In a two-objective optimization problem of maximizing both f_1 and f_2, the Pareto-optimal solutions lie on the circle:

$$f_1^{*2} + f_2^{*2} = 1.$$

(a) It is desired to find Pareto-optimal solutions in the range $0.2 \leq f_1^* \leq 0.6$. How would the objective functions be weighed in the guided domination approach.
(b) If the Pareto-optimal solutions are desired to be found in $0 \leq f_2^* \leq 0.2$, how would the objectives be weighed?

8. In a problem of minimizing f_1 and maximizing f_2, the Pareto-optimal solutions are supposed to satisfy

$$f_2^* = f_1^{*2}.$$

If $a_{12} = a_{21} = 0.25$ are used in the guided domination approach, which Pareto-optimal solutions are expected to be found?

9. Consider the following single-variable problem:

$$\text{Minimize} \quad f(x) = (x - 5)^2,$$
$$\text{subject to} \quad x \leq 2.$$

Convert the problem into a two-objective problem with a second objective of minimizing the constraint violation.

(a) Find the Pareto-optimal front and identify the constrained minimum solution of the original problem.
(b) In order to allow flexibility in optimization, the constraint can be relaxed as $x \leq 2.5$ to solve the above problem. If the guided domination approach is used to find trade-off solutions up to the optimal solution corresponding to $x \leq 2.5$ constraint, how would the two objectives be weighed?
(c) For the two-objective problem, identify the penalty parameter which will correspond to the solution $x = 2.5$.

10. Consider the following problem:

$$\text{Minimize} \quad f_1(x) = (x_1 - 2)^2 + (x_2 - 5)^2,$$
$$\text{subject to} \quad x_1 + x_2 = 4.$$

Convert the problem into a two-objective problem.

(a) If the classical weighted-sum method is to be used with a weight vector $(0.5, 0.5)^T$, which solution would be found?
(b) If the classical weighted-sum method is to be used with a weight vector $(0.75, 0.25)^T$, which solution would be found?

11. In the above problem, the user expects to find a solution for which the objective function is less than 1 unit and the constraint violation is less than 1 unit. What is(are) the resulting compromised solution(s)?

    Hint: Convert the problem into a goal programming problem.

12. Using the goal programming approach find the compromised point(s) on the ellipse

$$x^2 + 4y^2 = 1,$$

which are not less than 0.75 units and not more than 1.5 units away from the origin and within a distance 0.75 units from the point $(2,0)^T$. Construct the goal programming problem and solve the problem graphically.

    If the first goal is changed to find points not more than 0.25 units away from the origin, what are the compromised solutions?

13. For a random population of size 200, estimate the proportion of individuals in the first three non-dominated fronts using the given chart. A random population is assumed.

    (a) Assume six objectives
    (b) Assume eight objectives.

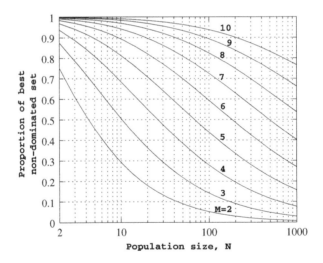

14. Assuming all solutions in the best non-dominated front receive two copies and the next-best non-dominated solutions receive one copy under a selection operator, what minimum population size would be appropriate to maintain all solutions of the first two non-dominated fronts in the mating pool for

    (a) three objectives,
    (b) five objectives.

    Use the chart given in the previous problem.

15. Use the following populations $P(t+1)$ and $E(t)$ to create the new archive $E(t+1)$ using Rudolph's convergent MOEA for minimizing two objectives:

Soln.	P(t + 1)	
	f_1	f_2
1	2.0	5.0
2	1.0	3.0
3	6.0	4.0
4	3.0	2.0
5	5.0	0.8
6	4.0	0.9

Soln.	E(t)	
	f_1	f_2
a	6.0	0.5
b	2.0	2.0
c	3.0	1.0
d	5.0	0.5
e	1.0	4.0

16. Consider the parent and offspring populations for a two-objective minimization problem:

	P_t									
Soln.	1	2	3	4	5	6	7	8	9	10
f_1	5.0	2.0	0.5	2.0	5.0	1.8	1.0	1.5	3.5	4.0
f_2	0.0	4.0	5.0	2.0	3.0	3.0	4.0	4.0	4.5	1.0

	Q_t									
Soln.	a	b	c	d	e	f	g	h	i	j
f_1	2.5	3.0	4.5	3.0	4.0	4.0	1.5	6.0	5.5	2.5
f_2	1.0	6.0	3.5	3.0	2.0	4.0	5.0	2.0	1.5	2.5

Using the NSGA-II procedure, construct the new parent population.
Compare the above population with the new parent population P_{t+1} constructed using the controlled elitism procedure with

(a) $r = 0.3$,
(b) $r = 0.8$.

9

Applications of Multi-Objective Evolutionary Algorithms

In the previous chapters, multi-objective evolutionary algorithms are described and applied to test problems. The advantage of investigating the performance of MOEAs on test problems as opposed to application problems is that in test problems the problem complexity, including the location of entire Pareto-optimal set, is precisely known. Thus, while evaluating an MOEA, it becomes useful and informative to apply these algorithms on test problems. On the other hand, testing an MOEA's performance on application problems is also necessary to demonstrate the use of the algorithm in practice. In this chapter, we will present five applications of a number of elitist and non-elitist MOEAs.

First, a number of mechanical component design problems are solved. Although a few decision variables are used in each problem, the application of MOEAs shows how multiple trade-off solutions can be found involving two conflicting design objectives: cost and the reliability of designs. Although reliability of a design is not directly used in the formulations, the minimization of maximum deflection anywhere in the component or maximization of the stiffness of a component is equivalent to maximizing the reliability of a design. In all of the case studies, solutions having each objective varying over a wide range of values are found.

Secondly, an MOEA is applied to a number of truss-structure design problems with a novel representation scheme, which allows both topology and member sizing optimization problems to be handled simultaneously. In these problems, minimization of the weight of the truss and the minimization of the maximum deflection of the truss are two conflicting objectives. Interestingly, the entire non-dominated set of solutions consists of a number of regions with differing topologies. Each contiguous set of solutions have identical topologies but vary in member sizing only. Such properties among trade-off solutions are not intuitive and can emerge from the obtained non-dominated solutions.

Thirdly, a microwave absorber is designed for two conflicting objectives – minimization of reflectivity in a frequency band and minimization of the overall thickness of the absorber. On a couple of different frequency-band absorber design problems, three MOEAs are compared for their convergence and their ability to find diverse non-dominated sets of solutions. In all cases, solutions having a widely varying range of both objectives are found.

Fourthly, MOEAs are applied to find interplanetary trajectories for three objectives – delivered payload to the destination, time of flight and heliocentric revolutions involved in the trajectories. On an Earth–Mars rendezvous mission, a number of different non-dominated trajectories are found by an MOEA. Conflicting characteristics for the first two objectives have been observed in the obtained solutions. Interesting spiral trajectories, of fairly long lengths, are also observed.

Finally, a hybrid optimization technique with an MOEA, assisted by a local search technique, is suggested for real-world problem solving. In most application problems in this and previous chapters, a large number of non-dominated solutions are found by an MOEA. However, in practice, one is interested in only a handful of Pareto-optimal solutions. The hybrid technique proposed a clustering approach to group a number of non-dominated solutions together and finally suggested a desired number of well-distributed trade-off solutions. The proposed technique is demonstrated by applying it to a number of engineering shape design problems. With no knowledge of any information about the optimal shapes in these problems, the proposed hybrid MOEA can find different trade-off shapes, some of which are intuitive to an experienced designer.

9.1   An Overview of Different Applications

Interested readers may refer to other application case studies for different real-world applications of MOEAs. In Table 33, we outline a number of such studies. However, we do not claim this list to be complete.

Table 33   Some real-world application studies of MOEAs.

Researcher(s) (Year)	Application area
S. S. Rao (1993)	Structural optimization
A. D. Belegundu et al. (1994)	Laminated ceramic composites
T. J. Stanley and T. Mudge (1995)	Microprocessor chip design
C. S. Chang et al. (1995)	Traction for DC railway system
F. Jiménez and J. L. Verdegay (1995)	Solid transportation
C. A. Coello et al. (1995)	Counterweight balancing of a robot arm
A. J. Chipperfield and P. J. Fleming (1996)	Gas turbine engine controller design
T. Arslan et al. (1996)	VLSI circuit design
S. Y. Hahn (1996)	Permanent magnet motor design
D. S. Weile et al. (1996)	Broad-band microwave absorber design

D. S. Todd and P. Sen (1997)	Containership loading design
D. H. Loughlin and S. Ranjithan (1997)	Air pollution management
D. Lee (1997)	Marine vehicle design
E. Zitzler and L. Thiele (1998a)	Synthesis of digital hardware-software multiprocessor system
G. T. Parks and I. Miller (1998)	Pressurized water reactor reload design
K. Fujita et al. (1998)	Automotive engine design
S. Obayashi et al. (1998)	Aircraft wing plan-form shape design
K. Mitra et al. (1998)	Dynamic optimization of an industrial nylon 6 semi-batch reactor
H. A. Guvenir and E. Erel (1998)	Inventory classification
T. Bagchi (1999)	Multi-criterion flow-shop scheduling
D. Cvetkovic and I. Parmee (1998)	Airframe design and conceptual design
R. Kumar and P. I. Rockett (1998)	Hierarchical learning of pattern spaces
C. M. Fonseca and P. J. Fleming (1998b)	Gas turbine engine design
S. Mardle et al. (1998)	Fishery modeling
B. Paechter et al. (1998)	Class timetabling of a university
A. G. Cunha et al. (1999)	Extruder screw design
S. Garg and S. K. Gupta (1999)	Free radical bulk polymerization reactor
N. Eklund and M. Embrechts (1999)	Color-efficiency trade-off in filtered light
C. Poloni et al. (2000)	Aerodynamic shape design
E. Schlemmer et al. (2000)	Hydroelectric generator design
P. Di Barba et al. (2000)	Electrostatic micro-motor design
L. Costa and P. Oliveira (2000)	Laminated composite plate design
A. J. Blumel et al. (2000)	Autopilot controller design
K. B. Matthews et al. (2000)	Land use planning
H. Meunier et al. (2000)	Radio network optimization
X. Li et al. (2000)	Medical image reconstruction
F. B. Zhou et al. (2000)	Continuous casting process
A. Petrovski and J. McCall (2001)	Cancer chemotherapy
M. Lahanas et al. (2001)	Dose optimization in brachytherapy
W. El Moudani et al. (2001)	Airlines crew rostering
M. Erickson et al. (2001)	Groundwater quality management
M. Thompson (2001)	Analog filter tuning

N. Laumanns et al. (2001)	Road train design
D. Sasaki et al. (2001)	Supersonic wing design
Ishibuchi et al. (2001)	Linguistic rule extraction
I. F. Sbalzarini et al. (2001)	Microchannel flow optimization
H. E. Aguirre et al. (2001)	Halftone image generation

9.2 Mechanical Component Design

In this section, we discuss four mechanical component design problems which were originally solved as single-objective optimization problems. However, ideally these problems are better posed as two-objective optimization problems. Because of the lack of an adequate multi-objective optimization method for finding multiple Pareto-optimal solutions, they were not solved as single-objective optimization problems. Through these representative problems, we will demonstrate how easily an MOEA can be used to find a diverse set of non-dominated solutions.

In all of these problems (originally appeared in Deb et al. (2000d)), we have kept the GA parameters the same: population size of 100, crossover probability of 1.0, and mutation probability of $1/n$ (where n is the number of variables). An SBX operator (discussed earlier on page 113) with a distribution index of $\eta_c = 10$, and a real-parameter mutation operator (discussed above on page 124) with a distribution index of $\eta_m = 500$, are employed. In all simulations, we run the NSGA-II for a maximum of 100 generations. All constraints are normalized and the sum of all constraint violations is added to all objective functions (discussed in Section 7.3).

9.2.1   Two-Bar Truss Design

This problem was originally studied using the ϵ-constraint method (Palli et al., 1999). The truss (see Figure 291) has to carry a certain load without elastic failure. Thus, in addition to the objective of designing the truss for minimum volume, there are additional objectives of minimizing the stresses in each of the two members AC and BC. We construct the following two-objective optimization problem for three variables: y (vertical distance between B and C in m), and x_1 and x_2 (cross-sectional areas of

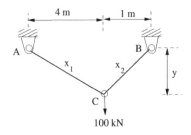

Figure 291   A two-bar truss. This is a reprint of Figure 1 from Deb et al. (2000d) (© Springer-Verlag Berlin Heidelberg 2000).

AC and BC, respectively, in m^2):

$$\left.\begin{array}{ll} \text{Minimize} & f_1(x,y) = x_1\sqrt{16+y^2} + x_2\sqrt{1+y^2}, \\ \text{Minimize} & f_2(x,y) = \max(\sigma_{AC}, \sigma_{BC}), \\ \text{subject to} & \max(\sigma_{AC}, \sigma_{BC}) \leq 1(10^5), \\ & 1 \leq y \leq 3, \quad \text{and} \quad x \geq 0. \end{array}\right\}$$ (9.1)

The stresses are calculated as follows:

$$\sigma_{AC} = \frac{20\sqrt{16+y^2}}{yx_1}, \qquad \sigma_{BC} = \frac{80\sqrt{1+y^2}}{yx_2}.$$

The original study reported only five solutions with the following two extreme values: (0.004 445 m^3, 89 983 kPa) and (0.004 833 m^3, 83 268 kPa). In order to restrict solutions with stress in the above range, we have included an additional constraint of maximum stress being smaller than 1(10^5). A penalty parameter of R = 10^3 is used to handle this constraint. We also add the following variable bounds: $0 \leq x_i \leq 0.01$ m^2 for i = 1 and 2. Figure 292 shows the optimized front found using NSGA-II. The solutions are spread between the following two extreme values: (0.004 07 m^3, 99 755 kPa) and (0.053 04 m^3, 8439 kPa), which indicates the power of the NSGA-II in finding a wider spread of solutions when compared to the ϵ-constraint method. The latter method could not find a wide variety of solutions in terms of the second objective. If minimization of stress is important, the NSGA-II finds a solution with a stress as low as 8439 kPa, whereas the ϵ-constraint method has found a solution with a minimum stress of 83 268 kPa, an order of magnitude higher than that found in the NSGA-II. What is also important is that all of these solutions have been found in just one simulation run of the NSGA-II.

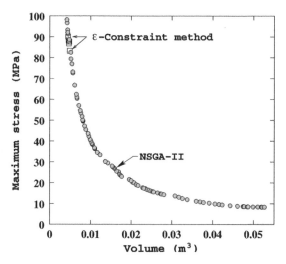

Figure 292 Optimized solutions obtained using the NSGA-II for the two-bar truss problem. Five solutions found using the ϵ-constraint method are also shown for comparison. This is a reprint of Figure 2 from Deb et al. (2000d) (© Springer-Verlag Berlin Heidelberg 2000).

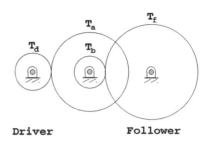

Figure 293 A compound gear train.

9.2.2 *Gear Train Design*

A compound gear train is to be designed to achieve a specific gear ratio between the driver and driven shafts (see Figure 293). The objective of the gear train design is to find the number of teeth in each of the four gears so as to minimize (i) the error between the obtained gear ratio and a required gear ratio of 1/6.931 (Kannan and Kramer, 1994) and (ii) the maximum size of any of the four gears. Since the number of teeth must be integers, all four variables are thus strictly integers. By denoting the variable vector $x = (x_1, x_2, x_3, x_4) = (T_d, T_b, T_a, T_f)$, we write the two-objective optimization problem as follows:

$$
\left.
\begin{aligned}
\text{Minimize} \quad & f_1(x) = \left(\frac{1}{6.931} - \frac{x_1 x_2}{x_3 x_4} \right)^2, \\
\text{Minimize} \quad & f_2(x) = \max(x_1, x_2, x_3, x_4), \\
\text{subject to} \quad & 12 \le x_1, x_2, x_3, x_4 \le 60, \\
& \text{all } x_i\text{s are integers.}
\end{aligned}
\right\} \tag{9.2}
$$

Here, a module of 1 cm is assumed. A discrete version of the SBX operator (Deb and Goyal, 1998) is used to make sure that only integer offspring are created from the two integer parents. Since only integer values are allowed, distribution indices of $\eta_c = 2$ and $\eta_m = 10$ are used for the SBX and polynomial mutation operators, respectively.

Figure 294 shows the obtained non-dominated solutions. The solutions obtained by the single-objective GAs (GeneAS-I and GeneAS-II) (Deb and Goyal, 1998), by the augmented Lagrangian (AL), and by the branch-and-bound (BB) methods (Kannan and Kramer, 1994) for the error minimization are also shown. This figure shows that although an NSGA-II could not find the best single-objective solutions (GeneAS-I and II), the solution E is close to them and is better than all other single-objective optimization algorithms in terms of the second objective.

The figure also shows that widely spread solutions are obtained by the NSGA-II. The solutions marked as 'E' and 'D' (extreme obtained solutions) are shown in the following table.

Solution	x_1	x_2	x_3	x_4	Error	Max. diameter (cm)
E	12	12	27	37	$1.83(10^{-8})$	37
D	12	12	13	13	$5.01(10^{-1})$	13

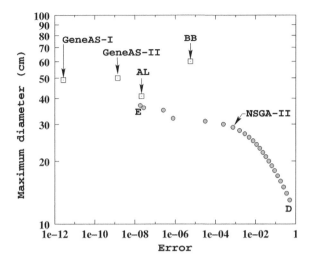

Figure 294   Optimized solutions obtained using the NSGA-II for the gear train design problem. The previously found minimum-error solutions with the GeneAS-I and GeneAS-II, the AL and BB methods are also shown. This is a reprint of Figure 3 from Deb et al. (2000d) (© Springer-Verlag Berlin Heidelberg 2000).

9.2.3   Spring Design

A helical compression spring needs to be designed for minimum volume and for minimum stress. Three variables are identified: the number of spring coils N, the wire diameter d, and the mean coil diameter D. Of these variables, N is an integer variable, d is a discrete variable having 42 non-equispaced values (as given in Kannan and Kramer (1994)), and D is a real-parameter variable. Denoting the variable vector $x = (x_1, x_2, x_3) = (N, d, D)$, we write the two-objective optimization problem as follows:

$$\text{Minimize } f_1(\mathbf{x}) = 0.25\pi^2 x_2^2 x_3 (x_1 + 2),$$

$$\text{Minimize } f_2(\mathbf{x}) = \frac{8KP_{\max}x_3}{\pi x_2^3},$$

$$\text{subject to } g_1(\mathbf{x}) \equiv \ell_{\max} - \frac{P_{\max}}{k} - 1.05(x_1 + 2)x_2 \geq 0,$$

$$g_2(\mathbf{x}) \equiv x_2 - d_{\min} \geq 0,$$

$$g_3(\mathbf{x}) \equiv D_{\max} - (x_2 + x_3) \geq 0, \qquad (9.3)$$

$$g_4(\mathbf{x}) \equiv C - 3 \geq 0,$$

$$g_5(\mathbf{x}) \equiv \delta_{pm} - \delta_p \geq 0,$$

$$g_6(\mathbf{x}) \equiv \frac{P_{\max} - P}{k} - \delta_w \geq 0,$$

$$g_7(\mathbf{x}) \equiv S - \frac{8KP_{\max}x_3}{\pi x_2^3} \geq 0,$$

$$g_8(\mathbf{x}) \equiv V_{max} - 0.25\pi^2 x_2^2 x_3 (x_1 + 2) \geq 0,$$

x_1 is integer, x_2 is discrete, x_3 is continuous.

The parameters used above are as follows:

$$K = \frac{4C - 1}{4C - 4} + \frac{0.615 x_2}{x_3}, \qquad P = 300 \text{ lb}, \qquad D_{max} = 3 \text{ in},$$

$$k = \frac{G x_2^4}{8 x_1 x_3^3}, \qquad\qquad P_{max} = 1000 \text{ lb}, \qquad \delta_w = 1.25 \text{ in},$$

$$\delta_p = \frac{P}{k}, \qquad\qquad\qquad \ell_{max} = 14 \text{ in}, \qquad \delta_{pm} = 6 \text{ in},$$

$$S = 189 \text{ ksi}, \qquad\qquad d_{min} = 0.2 \text{ in}, \qquad C = D/d.$$

We add the last two constraints to restrict the stress to be within an allowable strength and the volume to be within a pre-specified volume of $V_{max} = 30$ in^3. The discrete version of the SBX is used to handle the first two variables and the continuous version of SBX is used to handle the third variable. A penalty parameter of $R = 10^3$ is used for each normalized constraint.

Figure 295 shows the non-dominated front obtained using the NSGA-II with the penalty function approach, and the best solutions obtained by using two single-objective optimization algorithms – the GeneAS (Deb and Goyal, 1998) and the branch-and-bound (BB) method (Kannan and Kramer, 1994). The NSGA-II is able to find solutions close to these single-objective (volume) optima and, most importantly, is able to maintain a wide spread of different solutions. The extreme NSGA-II solutions are presented in the following table.

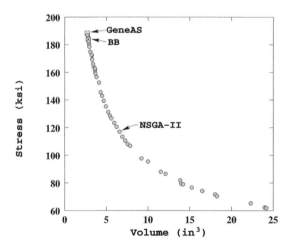

Figure 295 Optimized solutions obtained using the NSGA-II are shown by circles for the spring design problem. The previously found minimum-volume solutions with the GeneAS and BB methods are also shown. This is a reprint of Figure 4 from Deb et al. (2000d) (© Springer-Verlag Berlin Heidelberg 2000).

Solution	x_1	x_2	x_3	f_1	f_2
Min. volume	5	0.307	1.619	2.690	1 87 053
Min. stress	24	0.500	1.865	24.189	61 949

9.3 Truss-Structure Design

Optimal design of truss-structures has always been an active area of research in the field of search and optimization. Various techniques based on classical and evolutionary optimization techniques have been developed to find optimal truss-structures for a single objective (Kirsch, 1989; Ringertz, 1985; Topping, 1983; Vanderplaat and Moses, 1972). Most of these techniques can be classified into three main categories: (i) sizing, (ii) configuration, and (iii) topology optimization. In the sizing optimization of trusses, the cross-sectional areas of the members are considered as design variables and the coordinates of the nodes and connectivity among the various members are considered to be fixed (Goldberg and Samtani, 1986). A three-objective formulation (weight, deflection and stress) of a number of sizing optimization problems is solved elsewhere by using the MOSES optimization tool (Coello and Christiansen, 1999). In the configuration optimization of trusses, the change in nodal coordinates are also kept as design variables (Imai and Schmit, 1981). In the topology optimization, the connectivity between the members in a truss is to be determined (Kirsch, 1989; Ringertz, 1985). Classical optimization methods have not been used adequately in topology optimization, simply because they lack efficient ways to represent the connectivity of the members. Although the above three optimization problems are discussed separately, the most efficient way to design truss-structures optimally is to consider all three optimization problems simultaneously. In most attempts, multi-level optimization methods have been used (Rajan, 1995; Ringertz, 1985). In such a method, when topology optimization is performed, member areas and the truss configuration are assumed to be fixed. Once an optimized topology is found, the member areas and/or configuration of the obtained topology are optimized. In the context of multi-objective optimization, there also exists at least one study (Ruy et al., 2000), where a level-wise search strategy is used. In this approach, first, the cross-sectional area of each member and the coordinate of the nodes are assumed to be fixed and only the topology (or connectivity) of the truss members is varied. This first-level optimization will produce a number of different topologies, each corresponding to a trade-off between conflicting objectives. Thereafter, in the second-level optimization, each of the obtained topologies are considered one at a time and member sizing and node coordinates are varied to find a non-dominated set of solutions. When all such simulation runs are over, all solutions are grouped together and the best non-dominated solution set is found. First of all, it is obvious that such a multi-level optimization technique may not always provide the globally best solution, since the optimization problems considered in both levels may not be linearly separable. Secondly, in the context of multi-objective optimization, there may exist many non-dominated topologies for a reasonably large truss and performing the second-level optimization from each of these solutions would be computationally expensive.

In this study, we propose a representation scheme which naturally allows all three optimizations to be performed simultaneously.

9.3.1   A Combined Optimization Approach

In a truss-structure design, certain nodes carry a load or are used as supports, while other nodes are used for the purpose of better distributing the loads. The former nodes (say n' of them) must be present in a feasible truss and are called *basic* nodes. The remaining nodes (say n'' of them) are optional and called *non-basic* nodes. The objective in a truss-structure design (with all three design considerations – sizing, configuration and topology) is to find which optional nodes are necessary in a truss, what are the coordinates of these optional nodes, and which members must be present, so that certain objectives (usually the weight of the truss and the maximum deflection in the truss) are minimized by satisfying certain constraints (often the stresses in the members and the displacements of nodes).

The proposed approach assumes a *ground* structure, which is a complete truss with all possible member connections among all nodes (basic or non-basic) in the structure. A truss is represented by specifying a cross-sectional area for each member in the ground structure. Thus, a solution represented in the GA population is a vector of m real numbers within a specified range $(-A, A)$. The presence or absence of a member in the ground structure is determined by comparing the cross-sectional area of the member with a user-defined small critical cross-sectional area, ϵ. If an area is smaller than ϵ, that member is assumed to be absent in the realized truss. This is how trusses with differing topologies can be obtained with a fixed-length representation of the truss member areas. Figure 296 proposes the coding procedure to be used. This figure shows how a five-element vector of cross-sectional areas can represent a two-member truss. Since elements 1 and 3 have positive cross-sectional areas and values larger than a small value ϵ, there exist only two members. This representation scheme has another advantage. Since member areas are directly used, the values higher than ϵ specify the actual member cross-sectional area. Taking account of the above discussion, we now present the formulation of the truss-structure optimization problem as a nonlinear

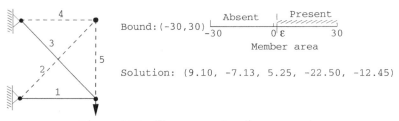

Figure 296   The proposed coding approach.

programming (NLP) problem:

$$
\left.
\begin{aligned}
&\text{Minimize} \quad f_1(\mathbf{A}, \xi) = \sum_{j=1}^{m} \rho_j \ell_j A_j, \\
&\text{Minimize} \quad f_2(\mathbf{A}, \xi) = \max_{k=1}^{n} \delta_k(\mathbf{A}, \xi), \\
&\text{subject to} \quad G1 \equiv \text{Truss is acceptable to the user,} \\
&\phantom{\text{subject to} \quad} G2 \equiv \text{Truss is kinematically stable,} \\
&\phantom{\text{subject to} \quad} G3 \equiv S_j - \sigma_j(\mathbf{A}, \xi) \geq 0, \qquad j = 1, 2, \ldots, m, \\
&\phantom{\text{subject to} \quad} G4 \equiv \delta_k^{\max} - \delta_k(\mathbf{A}, \xi) \geq 0, \qquad k = 1, 2, \ldots, n, \\
&\phantom{\text{subject to} \quad} A_i^{\min} \leq A_i \leq A_i^{\max}, \qquad i = 1, 2, \ldots, m, \\
&\phantom{\text{subject to} \quad} \xi_i^{\min} \leq \xi_i \leq \xi_i^{\max}, \qquad i = 1, 2, \ldots, n''.
\end{aligned}
\right\} \quad (9.4)
$$

In the above NLP problem, the design variables are the cross-sectional areas of members present in a truss (denoted as $\mathbf{A}$) and the coordinates of all n'' non-basic nodes (denoted as ξ). The first objective function denotes the overall weight of the truss, while the second objective function denotes the maximum deflection at any node in the truss. The first constraint makes sure that all basic nodes are present in a truss. The second constraint ensures that the connectivity represented by a GA solution represents a structure and not a mechanism. The third set of constraints forces the member stress values not to exceed the allowable strength of the material. The fourth set of constraints restricts the deflection in any node within an allowable limit. In this study, we have not used G4. It is important to highlight that we have used a finite element method to find the stress developed in any member and the deflection of any node for every GA solution. The parameters S_j and $\delta_k^{\max}$ are the allowable strength of the j-th member and the allowable deflection of the k-th node, respectively. In order to achieve a significant effect of all of the constraints, we normalize all of these. All constraints are handled by using the constrain-domination principle. In the simulations, we use the real-parameter NSGA-II, with the SBX operator and the polynomial mutation operator.

Three-Bar Truss

Figure 297 shows the three-bar truss with three loading cases. The allowable strength in the slanted members is 5 kPa and in the vertical member is 20 kPa. It is interesting to note that the loading is not symmetric in this problem. The cross-sectional areas are initialized in $[-20, 20]$ in^2. We use a population of size 50, a crossover probability of 0.9 and a mutation probability of 0.1. For the SBX and mutation operators, we use $\eta_c = 10$ and $\eta_m = 50$, respectively. The NSGA-II is run for 150 generations. The population of 50 solutions is plotted in Figure 298. This figure shows that the NSGA-II is able to find multiple non-dominated solutions in one single simulation run. The structure having minimum weight (solution A) of 7.16 lb has a deflection of 0.035 in. The structure having maximum weight (solution F) of 76.57 lb has a deflection of 0.0022 in. The corresponding topologies are also shown in this figure. The entire non-dominated set of solutions reveals much salient information about this truss design problem. First, out of all possible topologies, the three topologies shown

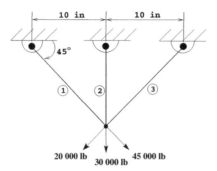

Figure 297   The three-bar truss.

in the figure are found to be Pareto-optimal. Secondly, each Pareto-optimal topology has its niche in the obtained front. Neighboring solutions under each topology vary in their cross-sectional sizes. Table 34 lists the six extreme solutions marked in the figure. It is clear that for the minimum weight topology (solutions within A and B), a truss with only the left-most and middle members is optimal. The variation in the solutions occurs because of variation in the cross-sectional sizes. The same is true for all three topologies. By investigating these six solutions (A to F), the original study (Deb et al., 2000c) argued that NSGA-II has found the truly extreme solutions possible for each topology in the Pareto-optimal front. For example, solution F is the minimum deflection solution, because all members have reached their maximum allowable cross-sectional areas (Table 34). Ruy et al. (2000) attempted to solve the

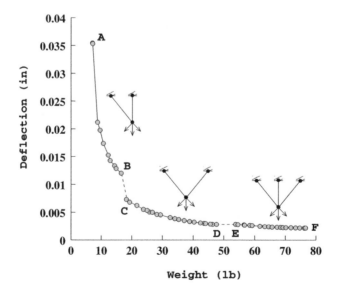

Figure 298   Obtained non-dominated solutions using the NSGA-II.

Table 34   Six important solutions marking the beginning and end of different topologies.

| Solution | Cross-sectional area (in²) | | | Weight (lb) | Deflection (in) |
	Left	Middle	Right		
A	4.046	1.442	—	7.163	0.035
B	8.132	4.981	—	16.481	0.012
C	8.105	—	4.759	18.192	0.007
D	20.000	—	13.667	47.612	0.003
E	14.975	7.096	19.246	53.824	0.003
F	20.000	20.000	20.000	76.568	0.002

same problem using a multi-stage multi-objective GA. The range of solutions obtained in their study for the weight objective is approximately [13, 38] lb and for the deflection objective is approximately [0.0015, 0.011] in. The plot drawn with the solutions found using NSGA-II clearly shows that a much better spread of solutions is found. It is important to highlight that their study on the same problem could not find the intermediate (in the range from C to D) non-dominated solutions.

Ten-Bar Truss

Next, we consider the well-studied ten-bar truss design problem for two objectives – weight of the truss and deflection at node 5 (see Figure 299). The loading is also shown in the structure. All possible members are shown on the figure. The allowable strength in all members is assumed to be 25 kPa. Here, we use a population of size 100 and run NSGA-II for 150 generations. The SBX and mutation parameters are the same as before. The decision variables are initialized in [−35, 35] in². Figure 300 shows the non-dominated solutions obtained by NSGA-II. It is interesting to note that of

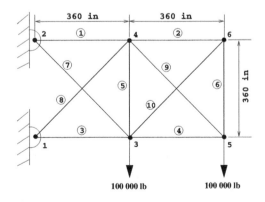

Figure 299   The ten-bar truss.

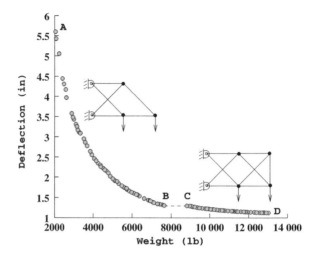

Figure 300   Obtained non-dominated solutions for the 10-bar truss.

the many possible topologies, NSGA-II has found only two different topologies in the obtained non-dominated front. The structure, having a minimum weight of 2046.186 lb, has a deflection of 5.599 in. Once again, in comparison to Ruy et al. (2000), the NSGA-II has been able to find a wide range of solutions. By investigating the extreme solutions marked in the figure by using various theories of the mechanics of solids, the original study (Deb et al., 2000c) argued that the obtained non-dominated front is close to the true Pareto-optimal front.

It is clear from both simulations that in one single simulation run, the NSGA-II is capable of finding a number of trade-off solutions. What is interesting is that the obtained non-dominated front has distinct regions with solutions of different topologies. Within such a region, solutions vary by having different cross-sectional areas of members, but while still keeping the same topology. Each region has a limit in the value of the objective functions. To improve a solution away from these limits, a different topology is desired. Knowledge of these trade-off solutions in a problem is highly informative in real-world design activities.

9.4   Microwave Absorber Design

A microwave absorber is made of a number of absorptive coatings stacked on a perfect electric conductor (PEC) (Weile et al., 1996), as shown schematically in Figure 301. This device is usually designed to suppress reflection over a pre-specified wide band of frequencies, $\nu \in B$. However, reflection over other frequencies are allowed. When optimized for such a scenario (for example, suppression in the range 2 to 8 GHz), the optimum design solution tends to be thick, thereby making the absorber expensive. On the other hand, if a thin absorber is designed, the absorption quality is sacrificed. Thus,

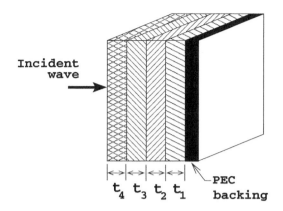

Figure 301   A schematic of a microwave absorber.

the design of a microwave absorber for minimum thickness (or minimum cost) and maximum absorption quality (or minimum reflectivity) resorts to conflicting objectives and hence becomes a multi-objective optimization problem.

Weile et al. (1996) used the NSGA, NPGA and a crowding MOEA to find multiple non-dominated solutions in a number of microwave absorber design problem. These investigators used a fixed number N_l of absorbing layers. The material of each layer is chosen from a set of 16 possible materials, so that four bits can be used to code the material for each layer. The thicknesses of each layer are also used as decision variables. From a table of material properties, the permittivity and permeability of each material is chosen and a total reflection coefficient $R(\nu)$ is calculated for an incident wave of frequency ν. The incident wave can be oriented in any direction. The maximum $R(\nu)$ for any incident wave over a set of discrete frequency values $\nu \in B$ is found and a term $R = 20 \log_{10}[\max(R(\nu), \nu \in B)]$ is minimized. This becomes the first objective function. The second objective function is the sum of all layer thicknesses, or $t = \sum_{i=1}^{N_l} t_i$.

In addition to the NSGA and NPGA, investigators introduced a crowded tournament Pareto GA (CTPGA), where diversity is maintained by using DeJong's crowding method (DeJong, 1975), along with a tournament selection method with non-dominated ranking. Crossover and mutation operators are used as usual. In the first experiment, non-dominated solutions are obtained for a frequency absorption in the range 0.2 to 2 GHz. A population of size 8000, a crossover probability of 0.9, and a mutation probability of 0.005 are used. Binary-coded multi-objective GAs are run for 150 generations. All absorbers have $N_l = 5$ layers. The thickness of each layer is coded in 7 bits so that the overall string length is $5 \times (4 + 7)$ or 55. For the CTPGA, a generation gap of 0.6 and a crowding factor of 40 are used (DeJong, 1975). For the NPGA, $t_{dom} = 100$ is used. For both the NSGA and NPGA, a σ_{share} value of 0.025 is employed. Sharing is performed in the decision variable space. However, instead of using the complete population to compute the niche count, only 2.5% of the population (200 randomly chosen solutions from the population) is used. Figure 302 shows the

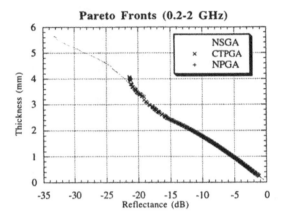

Figure 302 Obtained non-dominated solutions are shown using three algorithms. The NSGA (solutions shown by dots) finds the best spread of solutions (Weile et al., 1996). Reprinted with permission (© 1996 IEEE).

obtained non-dominated front using all three approaches. It is clear that the NSGA is able to find a better distribution of solutions among the three approaches. Although algorithms were run for 150 generations, the NSGA is able to find a good distribution of solutions only after 10 generations. However, the investigators have also observed that the NSGA is the most computationally expensive among the three algorithms.

Next, we reproduce the obtained non-dominated solutions for the 2–8 GHz band. As shown in Figure 303, the NSGA also outperformed the NPGA and CTPGA in this problem. For the two solutions (Designs 1 and 2, as shown in the figure), the objective function values are as follows:

Design	Max. reflectance (dB)	Total thickness (mm)
1	-23.16	4.134
2	-13.63	1.394

The frequency responses of these two designs are shown in Figure 304. This figure clearly shows the suppression of frequencies in the range 2–8 GHz. Design 1 is better for absorption, while design 2 is better for cost (or for the thickness of the absorber).

9.5   Low-Thrust Spacecraft Trajectory Optimization

The 'faster, cheaper and better' mission of the space program has placed emphasis on better designs of spacecraft trajectories for achieving shorter flight times, smaller launch vehicles and simpler flight systems. All of these objectives have an overall goal of minimizing the cost of interplanetary missions. Due to their high propellant efficiency, low-thrust propulsion systems have been recently paid renewed interest.

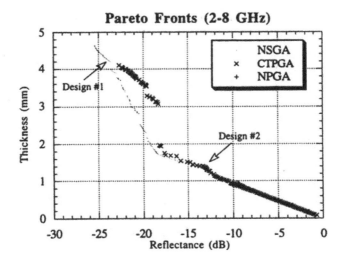

Figure 303   Obtained non-dominated solutions are shown using three algorithms for the design of a 2–8 GHz microwave absorber. The NSGA solutions (shown by dots) span the objective space better than the CTPGA and NPGA solutions (Weile et al., 1996). Reprinted with permission (© 1996 IEEE).

The Jet Propulsion Laboratory (JPL), in Pasadana, CA, has developed a calculus-of-variations (COV) technique to formulate the problem of finding the optimal spacecraft trajectory for maximizing the mass of the spacecraft arriving at the destination. The objective function and constraints on maximum possible engine thrust and rendezvous are derived from the fundamental equations of motion for a spacecraft subject to a single gravitational source (Coverstone-Carroll et al., 2000; Rauwolf and Coverstone-Carroll, 1996). At the core of the COV approach, a Solar Electric Propulsion Trajectory Optimization (SEPTOP) software program is used. Other parameters, such as the origin, destination, trajectory type and spacecraft engine characteristics, are all specified by the user. A Lagrange multiplier concept is used to handle the constraints. The optimization and the solution of the equations of motion are achieved by using numerical methods, thereby requiring a starting solution. The convergence of such a procedure often depends on the starting solution, particularly when the trajectory must be determined for a complex scenario.

A particular difficulty with the COV approach is its inability to perform multi-objective optimization. As mentioned above, the determination of the true optimal trajectory cannot be done with just one objective of maximizing the mass of the payload delivered at the destination. Other objectives, such as minimizing the time of travel, which is conflicting to the above objective, are also equally important. For a rendezvous task, the spacecraft has to align itself with the planet concerned before landing. Specifically, at the destination, the velocity vector of the spacecraft has to match with that of the destination planet. For short trajectories, a spacecraft has to move out radially fast to reach the orbit of the destination planet and then has to

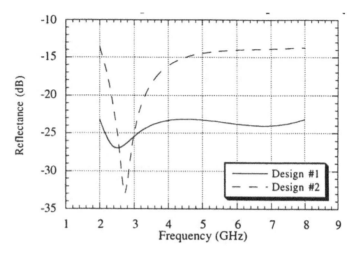

Figure 304   Design 1 is better able to suppress frequency in the range 2–8 GHz than design 2, but design 2 has a smaller thickness than design 1 (Weile et al., 1996). Reprinted with permission (© 1996 IEEE).

spend a considerable amount of energy in aligning itself for a smooth rendezvous with the latter. This process actually burns more fuel, thereby causing a lesser amount of delivered payload than that in a longer trajectory. In the latter, the spacecraft gets enough time to align itself with the destination planet, a process in which the spacecraft can coast for a substantial amount of time. Such a process leads to the delivery of a large payload at the destination. Thus, it is interesting to know which size of trajectory will deliver which size of mass. Since the COV technique cannot consider multiple objectives easily, researchers have used other objectives as constraints.

Coverstone-Carroll et al. (2000) proposed a GA-based multi-objective optimization technique using an NSGA to find multiple trade-off solutions to this problem. To evaluate a solution (trajectory), the SEPTOP software is called for, and the delivered payload mass and the total time of flight are calculated. In order to reduce the computational complexity, the SEPTOP program is run for a fixed number of generations. At the end of a SEPTOP run, the extent of violations of all of the COV constraints are determined. This overall violation is appended to each objective function as a constraint. A large penalty coefficient is used to put a large emphasis on achieving solutions with proper satisfaction of the COV equations and constraints. In addition, so as to achieve optimal solutions quickly, these authors have used information about the problem and included a third objective of maximizing heliocentric revolutions – the number of revolutions around the sun in the entire trajectory. Since longer flights can be achieved in various ways, the third objective provides a meaningful way (observed in past studies) to achieve the same.

Thus, the multi-objective optimization problem had eight decision variables controlling the trajectory, three objective functions, i.e. (i) maximize the delivered payload at destination, (ii) maximize the negative of the time of flight, and (iii) maximize the total number of heliocentric revolutions in the trajectory, and three constraints, i.e. (i) limiting the SEPTOP convergence error, (ii) limiting the minimum heliocentric revolutions, and (iii) limiting the maximum heliocentric revolutions in the trajectory.

On the Earth–Mars rendezvous mission, the study found interesting trade-off solutions. Using a population of size 150, the NSGA with $\sigma_{share} = 0.033$ was run for 30 generations on a Sun Ultra 10 Workstation with a 333 MHz ULTRA Sparc IIi processor. The obtained non-dominated solutions are shown in Figure 305 for two of the three objectives. It is clear that there exist short-time flights with smaller delivered payloads (solution marked as 44) and long-time flights with larger delivered payloads (solution marked as 36). To the surprise of the original investigators, two different types of trajectories emerged. The representative solutions of the first set of trajectories are shown in Figure 306. Solution 44 can deliver a mass of 685.28 kg and requires about 1.12 years. On other hand, solution 72 can deliver almost 862 kg with a travel time of about 3 years. In these figures, each continuous part of a trajectory represents a thrusting arc and each dashed part of a trajectory represents a coasting arc. It is interesting to note that only a small improvement in delivered mass occurs in the solutions between 73 and 72. To move to a somewhat improved delivered mass, a different strategy for the trajectory must be found. Near solution 72, an additional burn is added, causing the trajectories to have better delivered masses. Solution 36 can

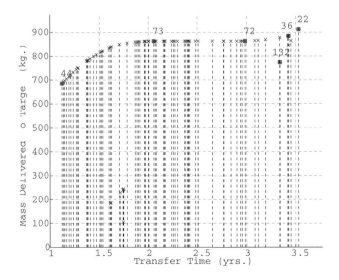

Figure 305 Obtained non-dominated solutions. Reprinted from Computer Methods in Applied Mechanics and Engineering, Volume 186, V. Coverstone-Carroll, J. W. Hartmann and W. J. Mason, 'Optimal multi-objective low-thrust spacecraft trajectories', pages 387–402, Copyright 2000, with permission from Elsevier Science.

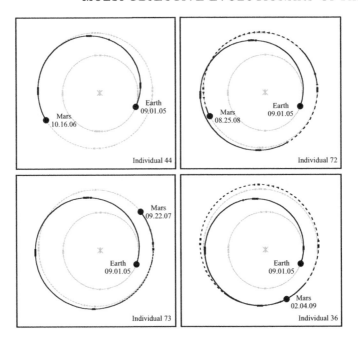

Figure 306   The first set of trajectories. Reprinted from Computer Methods in Applied Mechanics and Engineering, Volume 186, V. Coverstone-Carroll, J. W. Hartmann and W. J. Mason, 'Optimal multi-objective low-thrust spacecraft trajectories', pages 387–402, Copyright 2000, with permission from Elsevier Science.

deliver a mass of 884.10 kg. The discovery of a completely different type of solutions when the trade-off between objectives has been fully exploited in one type of solutions is very similar to that observed in the truss design problem discussed earlier.

The second set of trade-off solutions occurs with solutions 22 to 132, shown in Figure 307. These solutions start with an inward journey to perform a number of heliocentric revolutions around the Sun, primarily to make the spacecraft's orbit inclinations to match more closely to that of Mars, before spiraling out to the latter planet. Starting with a delivered payload of 773.33 kg (solution 132), the set contains a solution (22) with a delivered payload of 911.78 kg. Because of the number of heliocentric revolutions, the flight time is long in these trajectories. For more detailed discussions of this Earth–Mars rendezvous and other interplanetary trajectories, readers may refer to the original study (Coverstone-Carroll et al., 2000).

9.6   A Hybrid MOEA for Engineering Shape Design

It is clear from the previous chapters that the use of an MOEA is an efficient procedure for finding a wide-spread and a well-converged set of solutions in a multi-objective optimization problem. In this section, we would like to take MOEAs a step closer to practice by considering the following:

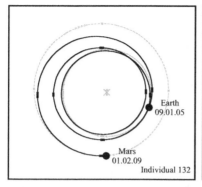 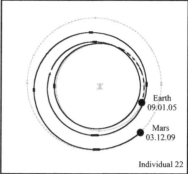

Figure 307   The second set of trajectories. Reprinted from Computer Methods in Applied Mechanics and Engineering, Volume 186, V. Coverstone-Carroll, J. W. Hartmann and W. J. Mason, 'Optimal multi-objective low-thrust spacecraft trajectories', pages 387–402, Copyright 2000, with permission from Elsevier Science.

- ensuring convergence closer to the true Pareto-optimal front;
- reducing the size of the obtained non-dominated set.

We illustrate the proposed hybrid strategy by addressing both of the above issues in the following subsections.

9.6.1   Better Convergence

In a real-world problem, one does not usually have any knowledge of the Pareto-optimal front. Although MOEAs have demonstrated good convergence properties in test problems, we enhance the probability of its true convergence by using a hybrid approach. A local search strategy is suggested from each obtained solution of the MOEA in order to find a better solution. Since a local search strategy requires a single-objective function, a weighted objective, a Tchebycheff metric, or any other weighted metric which will convert multiple objectives into a single objective, can be used. In this study, we use a weighted objective:

$$F(\mathbf{x}) = \sum_{j=1}^{M} \bar{w}_j^{\mathbf{x}} f_j(\mathbf{x}), \tag{9.5}$$

where weights are calculated from the obtained set of solutions in a special way. First, the minimum f_j^{min} and maximum f_j^{max} values of each objective function f_j are noted. Thereafter, for any solution $\mathbf{x}$ in the obtained set, the weight for each objective function is calculated as follows:

$$\bar{w}_j^{\mathbf{x}} = \frac{(f_j^{max} - f_j(\mathbf{x}))/(f_j^{max} - f_j^{min})}{\sum_{k=1}^{M} (f_k^{max} - f_k(\mathbf{x}))/(f_k^{max} - f_k^{min})}. \tag{9.6}$$

In the above calculation, minimization of the objective functions is assumed. When a solution $\mathbf{x}$ is close to the individual minimum of the function f_j, the numerator becomes one, thus causing a large value of the weight for this function. For an objective which has to be maximized, the term $(f_j^{max} - f_j(x))$ needs to be replaced with $(f_j(x) - f_j^{min})$. The division of the numerator with the denominator ensures that the calculated weights are normalized, or $\sum_{j=1}^{M} \bar{w}_j^x = 1$.

In order to distinguish the above calculated weight from the usual user-specified weight needed in weighted sum multi-objective optimization procedures, we refer to these calculated weights $\bar{w}$ as pseudo-weights. Once the pseudo-weights are calculated from the obtained non-dominated set of solutions, the local search procedure is simple. Begin the search from each solution x independently with the purpose of optimizing $F(x)$, as given in equation (9.5). Figure 308 illustrates this procedure. Since the pseudo-weight vector $\bar{w}^x$ dictates roughly the importance of different objective functions at solution x, optimizing $F(x)$ will produce a Pareto-optimal or a near Pareto-optimal solution in convex problems. However, for nonconvex Pareto-optimal regions, a different metric, such as the Tchebycheff metric (see Section 3.3 earlier), can be used. Nevertheless, the overall idea is that once an MOEA finds a set of solutions close to the true Pareto-optimal region, we use a local search technique for each of these solutions with a differing emphasis on objective functions in the hope of better convergence to the true Pareto-optimal front. Since independent local search methods are tried for each solution obtained using an MOEA, all optimized solutions obtained by the local search method need not be non-dominated to each other. Thus, we find the non-dominated set of solutions from the obtained set of solutions before proceeding further.

The complete procedure of the proposed hybrid strategy is shown in Figure 309. Starting from the MOEA results, we first apply a local search technique, followed by a non-domination check. After non-dominated solutions are found, a clustering

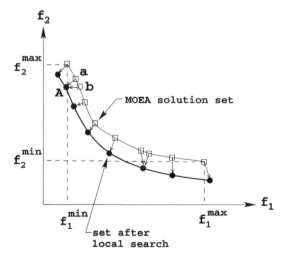

Figure 308 The local search technique may find better solutions. This is a reprint of Figure 1 from Deb and Goel (2001b) (© Springer-Verlag Berlin Heidelberg 2001).

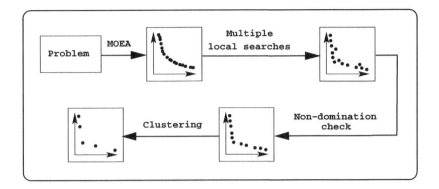

Figure 309 The proposed hybrid procedure of using a local search technique, a non-domination check, and a clustering technique. This is a reprint of Figure 2 from Deb and Goel (2001b) (© Springer-Verlag Berlin Heidelberg 2001).

technique is used to reduce the size of the optimal set, as discussed in the next subsection.

9.6.2 Reducing the Size of the Non-Dominated Set

In an ideal scenario, a user is interested in finding a good spread of non-dominated solutions close to the true Pareto-optimal front. From a practical standpoint, the user would be interested in a handful of solutions (in most cases, 5 to 10 solutions are probably enough). Interestingly, most MOEA studies use a population of size 100 or more, thereby finding about 100 different non-dominated solutions. The interesting question to ask is: 'Why are MOEAs used to find many more solutions than desired?'

The answer is fundamental to the working of an EA. The population size required in an EA depends on a number of factors related to the number of decision variables, the complexity of the problem, etc. (Goldberg et al., 1992; Harik et al., 1999). An MOEA population cannot be sized according to the desired number of non-dominated solutions in a problem. Since most real-world problems have a large number of decision variables, a large population size is usually necessary for a successful EA run. The irony is that when an MOEA works well with such a population size N, eventually it finds N different non-dominated solutions, particularly if the niching mechanism used in the MOEA works well. Thus, we need to devise a separate procedure for identifying a handful of solutions from the large obtained set of non-dominated solutions. One approach would be to use a clustering technique similar to that used by Zitzler (1999) for reducing the size of the obtained non-dominated set of solutions to only a handful. In this technique, each of the N solutions is assumed to belong to a separate cluster. Thereafter, the distance d_c between all pairs of clusters is calculated by first finding the centroid of each cluster and then calculating the Euclidean distance between the centroids. Two clusters having a minimum distance between them are merged together

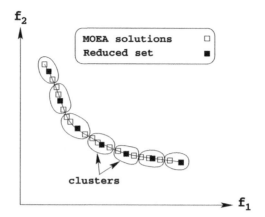

Figure 310 The clustering method for reducing the set of non-dominated solutions. This is a reprint of Figure 3 from Deb and Goel (2001b) (© Springer-Verlag Berlin Heidelberg 2001).

into a bigger cluster. This procedure is continued until the desired number of clusters are identified. Finally, with the remaining clusters, the solution closest to the centroid of the cluster is retained and all other solutions from each cluster are deleted. This is how the clusters can be merged and the cardinality of the solution set can be reduced.[1] Figure 310 shows the MOEA solution set as open boxes and the reduced set as filled boxes. Care should be taken to include the extreme solutions of the extreme clusters. In the following subsection, we illustrate the working of the proposed hybrid strategy in the optimal shape design of engineering components.

9.6.3 Optimal Shape Design

Designing the shape of engineering components is not a new activity. The use of optimization in engineering shape design has also received a lot of attention. However, in optimal shape design problems, the most popular approach has been to predefine a parametric mathematical function for the boundary describing the shape and use an optimization technique to find the optimal values of the parameters describing the mathematical function. Although this procedure requires prior knowledge of the shape, the classical optimization methods facilitated the optimization of parameters describing the mathematical shape functions.

   With the advent of evolutionary algorithms as alternate optimization methods, there exists a number of applications to optimal shape design problems, where shapes are evolved by deciding the presence or absence of a number of small elements (Hamda and Schoenauer, 2000; Jakiela et al., 2000; Sandgren et al., 1990). A predefined area (or

---

[1] It may be a better idea to consider the extreme solutions residing in a separate cluster. This way, extreme solutions are always retained.

volume) is divided into a number of small regular elements. The task of an evolutionary algorithm is to find which elements should be kept and which should be sacrificed so that the resulting shape is optimal with respect to an objective function. This procedure has a number of advantages.

1. The use of a numerical finite element method (or boundary element method) is a usual method for analyzing an engineering component. Since these methods require a component to be divided into small elements, this approach reduces one computational step in the finite element procedure.
2. Since no a priori knowledge about the shape is required, this method is not subjective to the user.
3. By simply using three-dimensional elements, this approach can be extended to three-dimensional shape design problems.
4. The number and the shape of holes in a component can evolve naturally without explicitly fixing them by the user.

Most studies of this approach, including those involving evolutionary algorithms, have concentrated on optimizing a single objective. In this present study, we apply this evolutionary procedure for multiple conflicting objectives.

Representation

Here, we consider two-dimensional shape design problems only. However, the procedure can be easily extended to three-dimensional shape design problems as well. We begin with a rectangular plate, describing the maximum overall region where the shapes will be confined. Thereafter, we divide the rectangular plate into a finite number of small elements (see Figure 311). Here, we consider square elements, although any other shape including triangular or rectangular elements can also be considered. Since the presence or absence of every element is a decision variable, we use a binary coding for describing a shape. For the shape shown in Figure 312, the

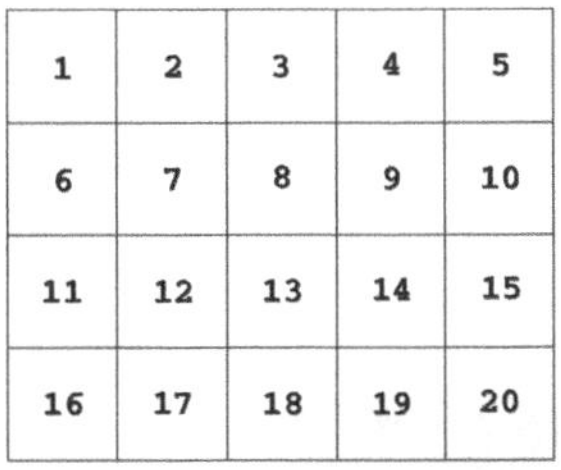

1	2	3	4	5
6	7	8	9	10
11	12	13	14	15
16	17	18	19	20

 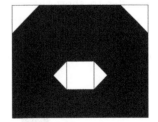

Figure 311 A rectangular plate is divided into small elements. This is a reprint of Figure 4 from Deb and Goel (2001b) (© Springer-Verlag Berlin Heidelberg 2001).

Figure 312 The skeleton of the shape. This is a reprint of Figure 5 from Deb and Goel (2001b) (© Springer-Verlag Berlin Heidelberg 2001).

Figure 313 The final smoothened shape. This is a reprint of Figure 6 from Deb and Goel (2001b) (© Springer-Verlag Berlin Heidelberg 2001).

corresponding binary coding is as follows:

01110 11111 10001 11111

The presence is denoted by a '1' and the absence is shown by a '0'. A left-to-right coding procedure, as shown in Figure 311, is adopted here. In order to smoothen the 'stair case'-like shape denoted by the basic skeleton representation, we add triangular elements (shown lightly shaded) for different cases in Figure 314. The skeleton shape shown in Figure 312 represents the true shape shown in Figure 313.

Evaluation

When the shape is smoothened, it is further divided into smaller elements. Here, we divide all squares and other shaped elements into triangular elements. A component is evaluated by finding the maximum stress and deflection developed at any point in the component due to the application of the given loads. Since no connectivity check is made while creating a new string or while creating the initial random population, a string may represent a number of disconnected regions in the rectangle. In such a case, we proceed with the biggest cluster of connected elements (where two elements are defined to be connected if they have at least one common corner). The string is repaired by assigning a '0' to all elements which are not a part of the biggest cluster.

In all applications here, two conflicting objectives are chosen: the weight of the resulting component and the maximum deflection anywhere in it. These two objectives are conflicting because a minimum weight design is usually not stiff and produces a large deflection, whereas a minimum deflection design has densely packed elements, thereby causing a large weight of the resulting component. The maximum stress and deflection values are restricted to lie within specified limits of the design by using them as constraints.

Simulation Results

In order to show the efficacy of the proposed hybrid multi-objective optimization procedure in solving optimal shape design problems, we use a number of mechanical components. In all cases, we use the NSGA-II as the multi-objective optimization algorithm. Since binary-coded strings are used to represent a shape, we use a bit-

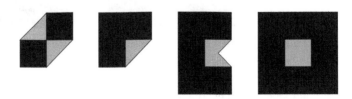

Figure 314 Different cases of smoothening through triangular elements. This is a reprint of Figure 7 from Deb and Goel (2001b) (© Springer-Verlag Berlin Heidelberg 2001).

wise hill-climbing strategy as the local search operator. The procedure is simple. Starting from the left of the string, every bit is flipped to check if it improves the design. If it does, the flipped bit is retained; otherwise, the bit is unchanged. This procedure is continued until no bit-flipping over the entire string length has resulted in an improvement.

Since the shapes are represented in a two-dimensional grid, we introduce a new crossover operator which respects the rows or columns of two parents. Whether to swap rows or columns is decided with a probability of 0.5. Each row or column is swapped with a probability of 0.95/d, where d is the number of rows or columns, as the case may be. In this way, on an average, about one row or one column gets swapped between the parents. A bit-wise mutation with a probability of 1/(string-length) is used. The NSGA-II is run for 150 generations. It is important to highlight that the NSGA-II does not require any extra parameter setting. In all problems, a population of size 30 and the following material properties are used:

$$\begin{array}{ll} \text{Plate thickness: 10 mm,} & \text{Yield strength: 150 MPa,} \\ \text{Young's modulus: 200 GPa,} & \text{Poisson's ratio: 0.25.} \end{array}$$

In all figures shown below, the 'weight' represents the area of the plate (since the thickness and density of material are assumed uniform throughout the plate) and the 'scaled deflection' refers to the maximum deflection multiplied by a suitable constant number so as to make it of the same order as the weight.

Cantilever Plate Design

First, we consider a cantilever plate design problem, where an end load $P = 10$ kN is applied, as shown in Figure 315. The rectangular plate of size 60×100 mm^2 is divided into 60 small square elements. Thus, 60 bits are used to construct a binary string representing a shape.

Figure 316 shows the four steps of the proposed hybrid approach in designing the cantilever plate. The first plot shows the non-dominated solutions obtained by using the NSGA-II. Since the population size is 30, the NSGA-II is able to find 30 different non-dominated solutions. Thereafter, the local search method is applied for each non-

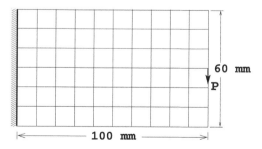

Figure 315 The loading and support of the cantilever plate. This is a reprint of Figure 8 from Deb and Goel (2001b) (© Springer-Verlag Berlin Heidelberg 2001).

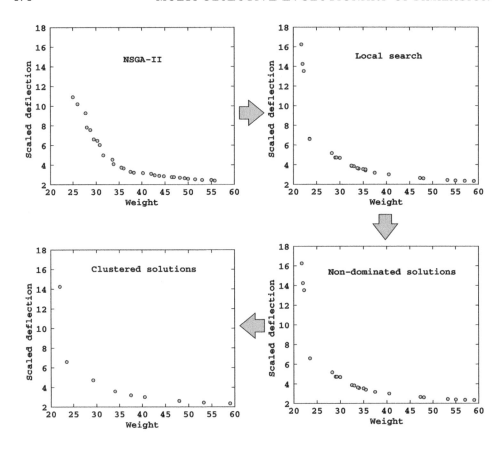

Figure 316   The hybrid procedure finds nine trade-off solutions for the cantilever plate design problem. This is a reprint of Figure 9 from Deb and Goel (2001b) (© Springer-Verlag Berlin Heidelberg 2001).

dominated solution, and a new and improved set of solutions are obtained (the second plot). The third plot is the result of the non-domination check on the solutions shown in the second plot. Three dominated solutions are eliminated by this process. The final plot is obtained after the clustering operation with a choice of nine solutions. This plot shows how nine well-distributed set of solutions are found from the third plot of 27 solutions. If fewer than nine solutions are desired, the clustering mechanism can be set accordingly.

In order to visualize the obtained set of nine solutions having a wide range of trade-offs in the weight and scaled deflection values, we show the shapes in Figure 317. It is clear that starting from a low-weight solution (with large deflection), how large-weight (with small deflection) shapes are found by the hybrid method. It is interesting to note that the minimum weight solution eliminated one complete row (the bottom-most row) in order to reduce the overall weight. The second solution (the element (1,2) in the 3×3 matrix) plot corresponds to the second-best weight solution. It is well known that

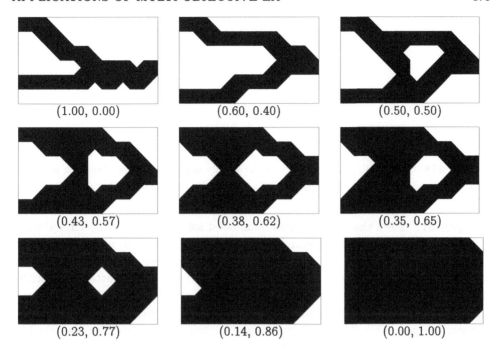

(1.00, 0.00)	(0.60, 0.40)	(0.50, 0.50)
(0.43, 0.57)	(0.38, 0.62)	(0.35, 0.65)
(0.23, 0.77)	(0.14, 0.86)	(0.00, 1.00)

Figure 317   Nine trade-off shapes for the cantilever plate design. This is a reprint of Figure 10 from Deb and Goel (2001b) (© Springer-Verlag Berlin Heidelberg 2001).

for an end load cantilever plate, a parabolic shape is optimal. Both shapes (elements (1,1) and (1,2)) refer to a parabolic shape. As the importance of deflection increases, the shapes tend to include more and more elements, thereby making the plate rigid enough to have a lesser deflection. For the third solution, the development of a vertical stiffener is interesting. This is a compromise between the minimum weight solution and a minimum deflection solution. By adding a stiffener, the weight of the structure does not increase much, whereas the stiffness of the plate increases considerably (hence the deflection reduces). Finally, the complete plate (with the right top and bottom ends chopped off) is found to be the smallest deflection solution.

We would like to reiterate here that the above nine solutions are not results of multiple runs of a multi-objective optimization algorithm. All nine solutions (and if needed, more can also be obtained) with interesting trade-offs between weight and deflection are obtained by using one simulation run of the hybrid method.

Simply Supported Plate Design

Next, we consider a simply supported plate design, starting from a rectangular plate with identical dimensions to that in the previous design. The plate is supported at two nodes, as shown in Figure 318, and a vertical load $P = 10$ kN acts upon the top-middle node of the plate.

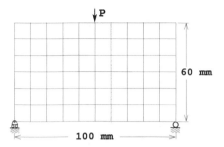

Figure 318   The loading and support of the simply supported plate. This is a reprint of Figure 11 from Deb and Goel (2001b) (© Springer-Verlag Berlin Heidelberg 2001).

Figure 319 shows the obtained non-dominated solutions (shown by circles) using the NSGA-II. After the local search method is completed, the obtained solutions (shown by squares) have a wider distribution. The number of solutions have been reduced from 30 to 22 by the non-domination check. Finally, the clustering algorithm finds nine widely separated solutions (shown by diamonds) from 22 non-dominated solutions. The shapes of these nine solutions are shown in Figure 180 in the previous chapter (on page 320). The minimum weight solution tends to use one row (the top-most row) less, but since the load is acting on the top of the plate, one element is added so as to transfer the load to the plate. The third solution (shown in the (1,3)-th position in the matrix) is interesting. A careful look at Figure 319 reveals that this solution is a knee solution. From this solution, a small advantage in weight can

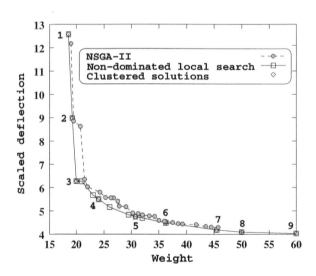

Figure 319   The hybrid procedure finds nine trade-off solutions for the simply supported plate design problem. This is a reprint of Figure 12 from Deb and Goel (2001b) (© Springer-Verlag Berlin Heidelberg 2001).

only be achieved with a large sacrifice in the deflection. Similarly, in order to achieve a small advantage in deflection, a large sacrifice in weight is needed. The shapes in positions (1,2) and (2,1) can be compared with respect to the shape in position (1,3). The shape in position (3,1), or solution 7, is also interesting. In order to achieve a further reduction in deflection, stiffening of the two slanted arms is needed. Finally, the absolute minimum-deflection shape is the complete rectangle with the maximum possible weight.

Starting with the minimum weight design having two slim slanted legs, down to thickening the legs to make them stiff, followed by joining the legs with a stiffener, and finally finding the complete rectangular plate having the minimum deflection, are all well-engineered trade-off solutions. In the absence of any such explicit engineering knowledge, it is interesting how the hybrid procedure with the NSGA-II is able to find the whole family of different trade-off solutions.

9.6.4 Hybrid MOEAs

The above procedure of using an MOEA with a local search technique is not the only hybrid MOEA suggested in the literature. We have already discussed earlier in Section 8.11 a technique where a local search method is used with an RWGA. Other implementations of local search methods are also suggested in MOEA studies (Jaskiewicz, 1998; Menzcer et al., 2000; Vicini and Quagliarella, 1999).

The main purpose of using a local search technique in an EA is to ensure proper convergence to the true optimum. Because of their stochastic operators, EAs can quickly steer their searches to near the true optimum. However, converging to the exact optimum may be best achieved if a local search technique is used for the solution obtained by an EA (Goldberg, 1989). Although there are several different ways to implement a local search strategy with an EA, the following two procedures are mainly used:

1. Use a local search method during EA generations.
2. Use a local search method at the end of an EA run.

In the first case, a parent population is operated by the usual genetic operations and then followed by a local search method to create the new population (Jaskiewicz, 1998; Vicini and Quagliarella, 1999). In the second case, a local search method is applied only to the solutions obtained after a complete run of an EA (Deb and Goel, 2001b). The hybrid MOEA suggested in this section is an example of the second procedure.

Both implementations have their advantages and disadvantages. In the first implementation, local searches originating from a below-average solution in the population may not lead to a better region in the search space, thereby wasting the effort spent in the local search procedures. Since for every new solution in every generation a local search procedure is applied, this implementation tends to be more computationally expensive. On the other hand, since only local optimum solutions are used in each generation, many EA generations may not be required. In the second implementation, local searches are applied only to a handful of solutions obtained after an EA run. Since the obtained EA solutions are supposed to be closer to the true optimum, many iterations of the local search method may not be necessary to reach to the true optimum. However, the number of generations needed in the EA run to find a solution close to the true optimum may be considerable.

The overall computational complexity of both implementations depend on two important factors: (i) the number of function evaluations spent in each application of the local search procedure, and (ii) the number of function evaluations spent in an EA run. A proper balance between these two parameters must be made in order to take advantage of both local search and EA search principles (Goldberg, 1999). With a maximum of h_{ls} function evaluations allocated to each local search application, the above two hybrid EA implementations with a population of size N have overall function evaluations as given on the left and right sides of the following equation:

$$N h_{ls} t_1 = N t_2 + N h_{ls}, \tag{9.7}$$

where t_1 and t_2 are the numbers of EA generations allowed in the first and second hybrid procedures, respectively. In the case of multi-objective optimization, the second procedure will find approximately N non-dominated solutions at the end of an MOEA run. Hence, $N h_{ls}$ function evaluations are considered in the right-side term of the above equation. For an identical computational complexity, it is clear from the above equation that:

$$\frac{t_1}{t_2} = \frac{1}{h_{ls}} + \frac{1}{t_2}. \tag{9.8}$$

Since both h_{ls} and t_2 must be greater than 2, the right side is less than one and $t_1 < t_2$, meaning that the EA generations allowed in the first procedure are smaller than those in the second procedure, thereby allowing more importance to EAs in the second procedure.

In the context of multi-objective optimization, the hybrid method suggested in this section has another advantage. Once a set of widely spread non-dominated solutions is found, they are used to derive a set of pseudo-weights based on their locations in the front. The use of such pseudo-weights in a local search method will take each solution closer to the corresponding weight much sooner than the use of random weights suggested in the hybrid RWGA method.

9.7 Summary

In this chapter, we have presented a number of applications of MOEAs in solving different real-world multi-objective optimization problems, including several mechanical component designs, truss-structure designs, a microwave absorber design, and a satellite trajectory design. In all of the problems, the power of MOEAs in finding a well-distributed and well-converged set of non-dominated solutions has been clearly demonstrated.

For real-world problem-solving, the use of a local search technique is often suggested with single-objective EAs. The proper use of a local search technique may make an

EA fast and reliable in converging close to the true optimum. In this chapter, we have extended this idea to multi-objective optimization and suggested a hybrid MOEA technique, where a local search method is used to update each non-dominated solution obtained after an MOEA simulation run. Since a local search method works with only one objective function, a composite objective function (using a weighted sum or a weighted metric method) is suggested for the local search procedure. Instead of using a random weight vector as in other studies (Ishibuchi and Murata, 1998a), the proposed hybrid approach uses a pseudo-weight vector calculated from the location of each solution in the entire obtained set of non-dominated solutions. We have argued that this approach would be computationally quicker than the random weight approach in converging near to the true Pareto-optimal front. The efficacy of this hybrid approach has been demonstrated on a number of engineering shape design problems.

This chapter has clearly demonstrated that the MOEAs discussed in this book are ready and waiting to be tested on more complex real-world multi-objective optimization problems.

10

Epilogue

We began Chapter 1 by classifying the application of multi-objective optimization algorithms into two fundamentally different approaches. In the ideal approach, no special importance is given to any particular objective and a set of trade-off or Pareto-optimal solutions are desired to be found. After a set of Pareto-optimal (or near Pareto-optimal) solutions is found, some higher-level information regarding the problem is used for choosing one solution from the obtained set of solutions. On the other hand, in the preference-based approach, instead of finding a set of Pareto-optimal solutions, the focus is to find one of the Pareto-optimal solutions based on a user-specified relative importance vector for the objectives. From a practical standpoint, we have argued that finding a reliable importance vector is difficult in the absence of any knowledge of the Pareto-optimal solutions. However, the presence of a number of Pareto-optimal (or near Pareto-optimal) solutions helps a user to choose a particular solution. In this respect, the preference-based approaches are more subjective to the user than the ideal approach and are highly sensitive to the chosen importance vector.

Most classical multi-objective optimization algorithms follow the preference-based approach, whereas evolutionary multi-objective optimization algorithms follow the ideal approach. Since the second task of choosing a particular solution from an obtained set of non-dominated solutions requires further problem information, most evolutionary algorithms described in this book have dealt with finding a set of widespread Pareto-optimal or near-Pareto-optimal solutions.

In Chapter 2, we defined a dominance relationship between any two solutions in the context of multiple objectives and then presented a number of procedures for finding non-dominated solutions with varying degrees of computational complexity. Of these, the procedure suggested by Kung et al. (1975) is computationally the best. Since the task of identifying the non-dominated set of solutions in a population is repeatedly used in most MOEAs, Kung et al.'s approach is recommended. Although these methods are suggested for identifying the best non-dominated set of solutions in a population, some MOEAs require a population to be completely sorted according to different levels of non-domination. Since classical multi-objective optimization studies do not require this task, there exist no studies related to non-dominated sorting. In MOEA studies, more efforts must be made in finding efficient ways to perform non-dominated sorting. In the light of the ideal approach to multi-objective optimization, it is clearly argued that there are two tasks that a multi-objective optimization algorithm must do well, as follows:

1. converge as close to the true Pareto-optimal solutions as possible;
2. maintain as diverse a population as possible.

The first task ensures that the obtained solutions are near optimal, while the second task ensures that a wide range of trade-off solutions is obtained.

The ideal approach of choosing a compromised solution from multiple Pareto-optimal solutions can be better appreciated if the preference-based approach of choosing a compromised importance vector leading to a specific Pareto-optimal solution is understood. In this regard, we have presented a number of preference-based optimization techniques which are commonly used in the classical multi-objective optimization studies. Although these methods have been in use for many years, the main difficulty arises in choosing a relative importance factor for each objective. Most of the preference-based methods discussed in Chapter 3 have a proof of convergence to a Pareto-optimal solution for certain classes of problems. If a relative importance vector is reliably known in a problem, it is advisable to use one of these methods. However, for a more exploratory search and investigation of a problem for the first time, the choice of a relative importance vector is not easy and is often subjective. This book has discussed algorithms for handling such problems which can be reliably solved with less subjectivity from the user.

Since MOEAs use an evolutionary algorithm as the baseline optimization procedure, we have devoted a significant portion of this book to describing various evolutionary algorithms (EAs) used for solving single-objective optimization problems. Although not all EAs are discussed, nor all EAs are treated with equal rigor, Chapter 4 has outlined the working principles of several EAs, their parameter interactions, and their broad-range applicability to different types of optimization problems. Specifically, EAs that have been used to handle multi-modal optimization problems are discussed in detail. This has been done due to the similarities in their working principles with those of multi-objective optimization. In both problems, finding and maintaining multiple optimal solutions from one generation to another is one of the primary tasks. Most MOEAs have borrowed the principles used in solving multi-modal optimization problems. Specifically, different niching techniques and clustering methods have been discussed.

Non-elitist MOEAs, which do not use any elite-preserving operator, have been discussed in Chapter 5. The descriptions of the algorithms in this chapter demonstrates how a single-objective EA can be modified to achieve both tasks needed in multi-objective optimization. In most MOEAs of this chapter, convergence towards the Pareto-optimal front is achieved by assigning a fitness based on the non-domination ranking of solutions. Diversity among solutions is achieved by using an explicit niching or crowding operation. The way that different MOEAs vary is the way that these issues are implemented. Although most methods have used the non-domination concept, more research is needed in using other concepts, such as the predator–prey model also discussed in the chapter. Horn (1997) has provided a good review of other techniques that have been used in multi-objective evolutionary optimization.

Elitist MOEAs incorporate an elite-preserving operator, where previously found good solutions are used in genetic operations in the current generation. This helps the search to progress faster towards the Pareto-optimal solutions and helps to achieve a monotonically non-deteriorating performance of an MOEA. Elitism is preserved in various ways. Combining the parent population with the offspring population and then accepting the best non-dominated solutions of the combined population is a common approach to elitist MOEAs. Some MOEAs also use and maintain an external elite population which gets updated with the new populations at every generation. In addition to the two tasks needed in a multi-objective optimization, Chapter 6 has demonstrated different ways that elitism can be incorporated. Simulation results on a simple test problem have shown that MOEAs with elitism have a better converging ability than those without elitism.

For constraint handling, a number of different constrained MOEAs have been discussed in Chapter 7. Starting from the simple penalty function approach, MOEAs have included more sophisticated constraint-based domination techniques. In this regard, a definition of constrain-domination between two solutions has been suggested for handling constrained optimization problems. Although an implementation of this definition with an elitist MOEA – NSGA-II – has shown impressive results, this approach can also be used with other MOEAs. The important matter in constrained optimization is that infeasible solutions should not be simply discarded. A strategy must be used to prefer one infeasible solution to the other.

In Chapter 8, we have discussed a number of important issues which must be addressed well in the context of multi-objective evolutionary optimization. This chapter has raised these issues and discussed the current state-of-the-art research into each issue. However, more research is necessary to understand each of these better.

Because of the two-pronged task needed in multi-objective optimization, there is also a need to have two different performance measures while evaluating an algorithm. One measure should assess the ability of an MOEA to converge to the Pareto-optimal front, while the other measure should assess the ability of an MOEA to maintain a diverse set of trade-off solutions. It seems difficult to have one measure for achieving both of these tasks, but only further research will reveal if this is a possibility.

With the development of different MOEAs, there is a need for them to be compared. For a proper evaluation of algorithms, there is a need to have test problems which adequately test an algorithm's ability to carry out various tasks. Since an MOEA is supposed to do two tasks well, test problems can be developed to investigate an MOEA's ability to overcome the different kinds of hurdles that may exist in a problem when trying to achieve each of the two tasks. In Chapter 8, we have discussed a particular test problem generator to develop a number of such test problems, some of which have already been used by many researchers. More research could be done in this direction. Although a similar test problem generator has also been suggested for constrained optimization, many others are certainly possible.

A number of studies comparing different MOEAs have also been outlined in Chapter 8. Some conclusive results have appeared from these comparisons. However, detailed studies involving constrained and unconstrained optimization, computational time comparison, and parameter sensitivities must be tried.

In Chapter 8, we have in addition discussed a number of techniques for finding a preferred set of Pareto-optimal solutions, instead of the complete set. Although some preference information about different objectives is needed, MOEAs require more qualitative rather than quantitative information. For example, if a user is sure of emphasizing cost more than comfort, such preferred MOEAs can be used to find a preferred set of Pareto-optimal solutions with more solutions crowded near the minimum-cost solution. This chapter has demonstrated a number of implementations to find a partial Pareto-optimal set based on the user's qualitative information. It is important to realize the difference between this preferred approach and the preference-based approach commonly used with classical methods. In the proposed preferred approach, no one Pareto-optimal solution is targeted. A biased distribution of Pareto-optimal solutions is the target. As discussed in the text, there are two advantages of such a preferred approach. First, the computational time in finding the distribution may be reduced because of the nature of a focused search. Secondly, more trade-off solutions can be obtained in the desired region of interest. However, more research in this area is needed to develop better techniques where the two advantages mentioned above are better exploited.

Most of the examples used in this book and in other MOEA studies have chosen two-objective optimization problems. Two objectives make the objective space two-dimensional, thereby making it easier to demonstrate the working of an algorithm. However, since most MOEAs lack a rigorous mathematical proof of convergence and a proof of maximum diversity in the obtained solutions, the scalability issue must be kept in mind while developing a new MOEA. An MOEA should not have a serious limitation in handling more than two objectives. On a particular matter, the implementation of the diversity-preserving strategy (niching or clustering) must be paid special attention. Since the dimension of the Pareto-optimal set goes up with the number of objectives, distance computations and identification of neighboring solutions must be made generic so that the computational complexity involved in achieving these tasks is reasonably low. Another matter of interest is the size of the random initial population required in solving problems having more than two objectives. It has been shown in this book that the size of the best non-dominated front in a random population increases with the number of objectives. Thus, an MOEA starting with a random population will have an unavoidable difficulty in maintaining lateral diversity, simply because not many fronts can exist in a finite population with many objectives. Adequate population sizing or the introduction of highly disruptive search operators may be necessary to make progress towards the true Pareto-optimal front in these problems. However, whether such special considerations are needed, or whether the existing diversity in the best non-dominated front can drive the search towards a multi-dimensional Pareto-optimal front, would be interesting topics for future research.

Next, we have discussed a number of MOEAs which have guaranteed convergence to the Pareto-optimal solution. Although the time complexity estimate for a convergence depends on the chosen EA parameter setting, the algorithms have features that guarantee that the MOEA population always proceeds towards the Pareto-optimal front and attempts to maintain the widest possible spread among its members. However, attempts to find MOEAs with the above features and with a time complexity estimate would be worthwhile future research. In any event, it is necessary to realize the importance of diversity preservation among the obtained solutions, while working out a proof of convergence.

Implementing elitism in an MOEA in any form may cause undesirably high selection pressure to the currently non-dominated solutions in a population. Since all solutions in the best non-dominated front are elites, in principle, an elitist strategy must copy all of these to the new population. This means a loss of lateral diversity in the population, a matter which may cause an elitist MOEA to prematurely converge to a sub-optimal front. Thus, the extent of elitism must be controlled in an elitist MOEA. Instead of accepting all individuals from the best non-dominated set, a few can be accepted according to a particular criterion. In this present book, we have suggested an adaptive control strategy where solutions from each front are accepted according to a geometric distribution. On a number of difficult problems the controlled elitist NSGA-II has been able to converge better near the true Pareto-optimal front than an uncontrolled elitist NSGA-II. However, more research is necessary in this direction to better control the selection pressure of elite solutions so that a better balance between exploitation and exploration issues is established.

Although these are some of the issues related to the development and understanding of MOEAs, there exist many other matters which also need immediate attention. Multi-objective optimization should be extended to solve multi-objective scheduling and other multi-objective combinatorial optimization problems. A number of studies in this direction have been discussed in this book. However, more studies are needed. The domination concept will simply carry over to these problems, because it only requires objective function information. If niching is performed in the objective space, the MOEAs suggested in Chapters 5 and 6 can also be applied to these problems in a straightforward manner. The only change required is in the choice of adequate crossover and mutation operators. However, if niching is desired in the decision variable space, an appropriate distance metric to find the similarity between two schedules must be used. With availability of the efficient elitist MOEAs discussed in this book, more applications in scheduling and other combinatorial optimization problems must be attempted in the near future.

All studies related to the development of efficient MOEAs are not complete unless these algorithms are applied to real-world problems. In Chapter 9, we have presented a number of case studies where MOEAs are applied to engineering and space applications. In all of these case studies, one aspect has clearly emerged. Existing MOEAs can be directly used to find multiple trade-off solutions involving at least two or three objective functions. The trade-off solutions span a wide range of values of objective functions. In most cases, the trade-off solutions include optimum solutions of an individual objective function and many other trade-off solutions within these extreme solutions. Finally, a hybrid MOEA, where a local search method is used in coordination with MOEAs, is suggested. Since a local search method requires only one objective function, a composite objective, derived by using the information concerning the location of each solution in the obtained non-dominated set, is suggested. Optimizing such a composite objective not only requires minimal iterations to converge to the nearest optimal solution, but it also reduces the cardinality of the obtained non-dominated set by finding identical optimal solutions from neighboring trade-off solutions, in the case of discrete search space problems. In order to make the approach more practical, a clustering technique is suggested with the obtained trade-off solutions. By specifying the number of desired trade-off solutions, the clustering technique will deliver exactly that many trade-off solutions, and with a maximum spread between them. The efficacy of such a technique has been demonstrated on a number of engineering shape design problems.

Thus far, most MOEAs have been developed on a genetic algorithm or on an evolution strategy platform. Although there exist a few studies using genetic programming (Bleuler et al., 2001; Hinchcliffe et al., 1998; Langdon, 1995) and evolutionary programming (Kim and Kim, 1997), the MOEAs presented in this book may be easily extended with GP, EP, and other EA paradigms. In this regard, despite a few studies, more research on the parallel implementation of MOEAs must be paid attention (Aherne et al., 1997; Mäkinen et al., 1996; Marco et al., 1999; Poloni et al., 2000; Quagliarella and Vicini, 1998; Rowe et al., 1996; Stanley and Mudge, 1995). The concept of using non-dominated sorting and a niche-preservation strategy can also be extended into other search and optimization techniques, such as in tabu search (Balicki and Kitowski, 2001; Hansen, 1997), in fuzzy-logic based evolutionary algorithms (Goméz-Skarmeta et al., 1998; Gueugniaud et al., 1997; Ishibuchi and Murata, 1997, 1998b), in ant colony based search algorithms (Iredi et al., 2001), etc.

This book and the numerous other research studies carried out so far have amply demonstrated the usefulness and power of MOEAs in finding multiple yet well-diverse trade-off solutions with good convergence properties in many test problems and application case studies. For practical problem solving, the number of trade-off solutions usually found by an MOEA are often too many in number. In a real-world scenario, it is desired to have not more than five to ten different candidate solutions from which one could be selected. Since an MOEA needs to use an adequate population size for the underlying EA operators to work correctly, an EA with five or ten population members is not expected to work well in complex problems. Although a simple clustering technique can be used to reduce the large set of trade-off solutions to only a handful, this suggests a wasted effort in simply throwing away the rest of the trade-off solutions. In a more sophisticated approach, the population size can be gradually reduced after no improvement in the obtained non-dominated front is recorded after a few generations. Other ideas are also certainly possible and must be worked out carefully.

Most MOEAs discussed in this book can also be applied in a straightforward manner to solve single-objective optimization problems. In such a case, the domination check between two solutions simply favors the solution with the better objective function value. As a byproduct, if the single-objective optimization problem has multiple optimal solutions (local and global), many of them can be found simultaneously. Thus, the single-objective optimization becomes a degenerate application case of multi-objective optimization using MOEAs. Moreover, the MOEAs of this book can also be applied to solve multi-objective optimization problems with non-conflicting objectives. In such applications, the resulting set would contain only the sole optimal solution which optimizes all objectives.

However, having been able to adequately solve the first step of the ideal approach of multi-objective optimization described in the first chapter, researchers should now look for more comprehensive strategies for solving the tasks in both steps. Such a strategy would be an interactive one and would optimally use an MOEA and a local search technique in such a way that the best features of both techniques are utilized. In this respect, the concepts used in the interactive classical multi-objective optimization may be combined with an MOEA. What these MOEAs can achieve is a part (hopefully a major part) of the complete ideal approach to multi-objective optimization. Laying out a strategy of choosing one desired solution from the MOEA-obtained non-dominated set of solutions, a process which may require involvement of a decision-maker in an iterative manner, is the remaining part, which also needs immediate attention. Hopefully, this book has laid the foundation for developing such a complete strategy, which would make the whole gamut of multi-objective optimization less subjective, more logical and closer to practice.

References

Aguirre, H. E., Tanaka, K., Sugimura, T. and Oshita, S. (2001). Halftone image generation with improved multiobjective genetic algorithm. In *Proceedings of the First International Conference on Evolutionary Multi-Criterion Optimization (EMO-2001)*, pp. 510–515.

Aherne, F. J., Thacker, N. A. and Rockett, P. I. (1997). Automatic parameter selection for object recognition using a parallel multiobjective genetic algorithm. In *Proceedings of the 7th International Conference on Computer Analysis of Images and Patterns*, pp. 559–566.

Antonisse, J. (1989). A new interpretation of schema notation that overturns the binary encoding constraint. In *Proceedings of the Third International Conference on Genetic Algorithms*, pp. 86–91.

Arora, J. S. (1989). *Introduction to Optimum Design*. New York: McGraw-Hill.

Arslan, T., Horrocks, D. H. and Ozdemir, E. (1996). Structural synthesis of cell-based VLSI circuits using a multi-objective genetic algorithm. *IEE Electronic Letters 32*(7), 651–652.

Bäck, T. (1992). The interaction rate of mutation rate, selection, and self-adaptation within a genetic algorithm. In *Proceedings of Parallel Problem Solving from Nature II (PPSN-II)*, pp. 85–94.

Bäck, T. (1996). *Evolutionary Algorithms in Theory and Practice*. New York: Oxford University Press.

Bäck, T. (1997). Self-adaptation. In T. Bäck, D. Fogel and Z. Michalewicz (Eds), *Handbook of Evolutionary Computation*, pp. C7.1:1–15. Bristol: Institute of Physics Publishing and New York: Oxford University Press.

Bäck, T. and Schwefel, H.-P. (1993). An overview of evolutionary algorithms for parameter optimization. *Evolutionary Computation Journal 1*(1), 1–23.

Bäck, T., Fogel, D. and Michalewicz, Z. (Eds) (1997). *Handbook of Evolutionary Computation*. Bristol: Institute of Physics Publishing and New York: Oxford University Press.

Bagchi, T. (1999). *Multiobjective Scheduling by Genetic Algorithms*. Boston: Kluwer Academic Publishers.

Baker, J. E. (1985). Adaptive selection methods for genetic algorithms. In *Proceedings of an International Conference on Genetic Algorithms and Their Applications*, pp. 101–111.

Balicki, J. and Kitowski, Z. (2001). Multicriteria evolutionary algorithm with tabu search for task assignment. In *Proceedings of the First International Conference on Evolutionary Multi-Criterion Optimization (EMO-2001)*, pp. 373–384.

Banzhaf, W., Nordin, P., Keller, R. and Francone, F. D. (1998). *Genetic Programming: An Introduction*. San Fransisco, CA: Morgan Kaufmann Publishers Inc.

Barba, P. D., Farina, M. and Savini, A. (2000). Vector shape optimization of an electrostatic micromotor using a genetic algorithm. *COMPEL 19*(12), 576–581.

Barbosa, H. J. C. (1996). A genetic algorithm for min-max problems. In *Proceedings of the First International Conference on Evolutionary Computation and Its Application (EvCA'96)*, pp. 99–109.

Belegundu, A. D., Murthy, D. V., Salagame, R. R. and Constants, E. W. (1994). Multiobjective optimization of laminated ceramic composites using genetic algorithms. In *Proceedings of the Fifth AIAA/USAF/NASA Symposium on Multidisciplinary Analysis and Optimization*, pp. 1015–1022.

Ben-Tal, A. (1980). Characterization of Pareto and lexicographic optimal solutions. In G. Fandel and T. Gal (Eds), *Multiple Criteria Decision Making: Theory and Applications*, pp. 1–11. Berlin: Springer-Verlag.

Benayoun, R., de Montgolfier, J., Tergny, J. and Laritchev, P. (1971). Linear programming with multiple objective functions: Step method (STEM). *Mathematical Programming 1*(3), 366–375.

Benson, H. P. (1978). Existence of efficient solutions for vector maximization problems. *Journal of Optimization Theory and Applications 26*(4), 569–580.

Bentley, P. J. and Wakefield, J. P. (1997). Finding acceptable solutions in the Pareto-optimal range using multiobjective genetic algorithms. In P. K. Chawdhry, R. Roy and R. K. Pant (Eds), *Soft Computing in Engineering Design and Manufacturing, Part 5*, pp. 231–240. London, UK: Springer-Verlag.

Beyer, H.-G. (1995a). Toward a theory of evolution strategies: On the benefit of sex – $(\mu/\mu, \lambda)$-theory. *Evolutionary Computation Journal 3*(1), 81–111.

Beyer, H.-G. (1995b). Toward a theory of evolution strategies: Self-adaptation. *Evolutionary Computation Journal 3*(3), 311–347.

Beyer, H.-G. and Deb, K. (2000). On the desired behaviour of self-adaptive evolutionary algorithms. In *Parallel Problem Solving from Nature VI (PPSN-VI)*, pp. 59–68.

Bhatia, D. and Aggarwal, S. (1992). Optimality and duality for multiobjective nonsmooth programming. *European Journal of Operational Research 57*(3), 360–367.

Binh, T. T. and Korn, U. (1997). MOBES: A multiobjective evolution strategy for constrained optimization problems. In *The Third International Conference on Genetic Algorithms (Mendel 97)*, pp. 176–182.

Bleuler, S., Brack, M. and Zitzler, E. (2001). Multiobjective genetic programming: Reducing bloat using SPEA2. In *Proceedings of the Congress on Evolutionary Computation (CEC-2001)*, pp. 536–543.

Blumel, A. L., Hughes, E. J. and White, B. A. (2000). Fuzzy autopilot design using a multi-objective evolutionary algorithm. In *Proceedings of the Congress on Evolutionary Computation (CEC-2000)*, pp. 4–5.

Booker, L. B. (1982). *Intelligent Behavior as an Adaptation to the Task Environment*. Ph. D. Thesis, Ann Arbor, MI: University of Michigan. Dissertation Abstracts International 43(2), 469B.

Box, G. E. P. (1957). Evolutionary operation: method for increasing industrial productivity. *Applied Statistics 6*(2), 81–101.

Box, M. J. (1965). A new method of constrained optimization and a comparison with other methods. *Computer Journal 8*(1), 42–52.

Branke, J., Kaußler, T. and Schmeck, H. (2000). Guiding multi-objective evolutionary algorithms towards interesting regions. Technical Report No. 399, Institute AIFB, University of Karlsruhe, Germany.

Buchanan, J. T. (1997). A naive approach for solving MCDM problems: The GUESS method. *Journal of the Operational Research Society* 48(2), 202–206.

Cavicchio, D. J. (1970). *Adaptive Search Using Simulated Evolution*. Ph. D. Thesis, Ann Arbor, MI: University of Michigan.

Chang, C. S., Wang, W., Liew, A. C., Wen, F. S. and Srinivasan, D. (1995). Genetic algorithm based bi-criterion optimization for traction in DC railway system. In *Proceedings of the Second IEEE International Conference on Evolutionary Computation*, pp. 11–16.

Chankong, V. and Haimes, Y. Y. (1983). *Multiobjective Decision Making Theory and Methodology*. New York: North-Holland.

Chankong, V., Haimes, Y. Y., Thadathil, J. and Zionts, S. (1985). Multiple criteria optimization: A state-of-the-art review. In *Proceedings of the Sixth International Conference on Multiple-Criteria Decision Making*, pp. 36–90.

Charnes, A., Cooper, W. and Ferguson, R. (1955). Optimal estimation of executive compensation by linear programming. *Management Science* 1(2), 138–151.

Chipperfield, A. J. and Fleming, P. (1996). Multiobjective gas turbine engine controller design using genetic algorithms. *IEEE Transactions on Industrial Electronics* 43(5), 583–587.

Clayton, E. R., Weber, W. E. and Taylor, B. W. (1982). A goal programming approach to the optimization of multi-response simulation models. *IIE Transactions* 14(4), 282–287.

Cleveland, W. S. (1994). *Elements of Graphing Data*. Murray Hill, NJ: AT & T Bell Laboratories.

Coello, C. A. C. (1999). A comprehensive survey of evolutionary-based multiobjective optimization techniques. *Knowledge and Information Systems* 1(3), 269–308.

Coello, C. A. C. (2000). Treating objectives as constraints for single objective optimization. *Engineering Optimization* 32(3), 275–308.

Coello, C. A. C. and Christiansen, A. D. (1999). MOSES : A multiobjective optimization tool for engineering design. *Engineering Optimization* 31(3), 337–368.

Coello, C. A. C. and Toscano, G. (2000). A micro-genetic algorithm for multi-objective optimization. Technical Report Lania-RI-2000-06, Laboratoria Nacional de Informatica Avanzada, Xalapa, Veracruz, Mexico.

Coello, C. A. C., Christiansen, A. D. and Hernandez, A. (1995). Multiobjective design optimization of counterweight balancing of a robot arm using genetic algorithms. In *Proceedings of the Seventh International Conference on Tools with Artificial Intelligence*, pp. 20–23.

Cohon, J. L. (1985). Multicriteria programming: Brief review and application. In J. S. Gero (Ed.), *Design Optimization*, pp. 163–191. New York: Academic Press.

Cormen, T. H., Leiserson, C. E. and Rivest, R. L. (1990). *Introduction to Algorithms*. New Delhi: Prentice-Hall.

Corne, D., Knowles, J. and Oates, M. (2000). The Pareto envelope-based selection algorithm for multiobjective optimization. In *Proceedings of the Sixth International Conference on Parallel Problem Solving from Nature VI (PPSN-VI)*, pp. 839–848.

Costa, L. and Oliveira, P. (2000). Structural optimization of laminated plates with genetic algorithms. In *Proceedings of the Genetic and Evolutionary Conference (GECCO-2000)*, pp. 621–627.

Coverstone-Carroll, V., Hartmann, J. W. and Mason, W. J. (2000). Optimal multi-objective low-thurst spacecraft trajectories. *Computer Methods in Applied Mechanics and Engineering 186*(2–4), 387–402.

Cunha, A. G., Oliveira, P. and Covas, J. A. (1999). Genetic algorithms in multiobjective optimization problems: An application to polymer extrusion. In *Proceedings of the Workshop on Multi-Criterion Optimization Using Evolutionary Methods held at Genetic and Evolutionary Computation Conference (GECCO-1999)*, pp. 129–130.

Cunha, N. O. D. and Polak, E. (1967). Constrained minimization under vector-evaluated criteria in finite dimensional spaces. *Journal of Mathematical Analysis and Applications 19*(1), 103–124.

Cvetković, D. and Parmee, I. (1998). Evolutionary design and multi-objective optimisation. In *Proceedings of the Sixth European Congress on Intelligent Techniques and Soft Computing (EUFIT)*, pp. 397–401.

Davidor, Y. (1991). A naturally occurring niche and species phenomenon: The model and first results. In *Proceedings of the Fourth International Conference on Genetic Algorithms*, pp. 257–263.

Davis, L. (1991). *Handbook of Genetic Algorithms*. New York: Van Nostrand Reinhold.

Dawkins, R. (1976). *The Selfish Gene*. New York: Oxford University Press.

Dawkins, R. (1986). *The Blind Watchmaker*. New York: Penguin Books.

Deb, K. (1989). Genetic Algorithms in Multi-Modal Function Optimization. Master's Thesis, Tuscaloosa, AL: University of Alabama.

Deb, K. (1991). Optimal design of a welded beam structure via genetic algorithms. *AIAA Journal 29*(11), 2013–2015.

Deb, K. (1995). *Optimization for Engineering Design: Algorithms and Examples*. New Delhi: Prentice-Hall.

Deb, K. (1997a). Mechanical component design using genetic algorithms. In D. Dasgupta and Z. Michalewicz (Eds), *Evolutionary Algorithms in Engineering Applications*, pp. 495–512. New York: Springer-Verlag.

Deb, K. (1997b). Speciation methods. In T. Bäck, D. Fogel and Z. Michalewicz (Eds), *Handbook of Evolutionary Computation*, pp. C6.2:1–4. Bristol: Institute of Physics Publishing and New York: Oxford University Press.

Deb, K. (1999a). Evolutionary algorithms for multi-criterion optimization in engineering design. In K. Miettinen, P. Neittaanmäki, M. M. Mäkelä and J. Périaux (Eds), *Evolutionary Algorithms in Engineering and Computer Science*, pp. 135–161. Chichester, UK: Wiley.

Deb, K. (1999b). An introduction to genetic algorithms. *Sādhanā 24*(4), 293–315.

Deb, K. (1999c). Multi-objective genetic algorithms: Problem difficulties and construction of test problems. *Evolutionary Computation Journal 7*(3), 205–230.

Deb, K. (2000). An efficient constraint handling method for genetic algorithms. *Computer Methods in Applied Mechanics and Engineering 186*(2–4), 311–338.

Deb, K. (2001). Nonlinear goal programming using multi-objective genetic algorithms. *Journal of the Operational Research Society 52*(3), 291–302.

Deb, K. (in press). Multi-objective evolutionary algorithms: Introducing bias among Pareto-optimal solutions. In A. Ghosh and S. Tsutsui (Eds), *Theory and Applications of Evolutionary Computation : Recent Trends*. London: Springer-Verlag.

Deb, K. and Agrawal, R. B. (1995). Simulated binary crossover for continuous search space. *Complex Systems 9*(2), 115–148.

Deb, K. and Agrawal, S. (1999a). A niched-penalty approach for constraint handling in genetic algorithms. In *Proceedings of the International Conference on Artificial Neural Networks and Genetic Algorithms (ICANNGA-99)*, pp. 235–243.

Deb, K. and Agrawal, S. (1999b). Understanding interactions among genetic algorithm parameters. In *Foundations of Genetic Algorithms 5 (FOGA-5)*, pp. 265–286.

Deb, K. and Beyer, H. (2001). Self-adaptive genetic algorithms with simulated binary crossover. *Evolutionary Computation Journal 9*(2), 195–219.

Deb, K. and Goel, T. (2001a). Controlled elitist non-dominated sorting genetic algorithms for better convergence. In *Proceedings of the First International Conference on Evolutionary Multi-Criterion Optimization (EMO-2001)*, pp. 67–81.

Deb, K. and Goel, T. (2001b). A hybrid multi-objective evolutionary approach to engineering shape design. In *Proceedings of the First International Conference on Evolutionary Multi-Criterion Optimization (EMO-2001)*, pp. 385–399.

Deb, K. and Goldberg, D. E. (1989). An investigation of niche and species formation in genetic function optimization. In *Proceedings of the Third International Conference on Genetic Algorithms*, pp. 42–50.

Deb, K. and Goyal, M. (1996). A combined genetic adaptive search (GeneAS) for engineering design. *Computer Science and Informatics 26*(4), 30–45.

Deb, K. and Goyal, M. (1998). A robust optimization procedure for mechanical component design based on genetic adaptive search. *Transactions of the ASME: Journal of Mechanical Design 120*(2), 162–164.

Deb, K. and Horn, J. (2000). Introduction to the special issue: Multicriterion optimization. *Evolutionary Computation Journal 8*(2), iii–iv.

Deb, K. and Kumar, A. (1995). Real-coded genetic algorithms with simulated binary crossover: Studies on multi-modal and multi-objective problems. *Complex Systems 9*(6), 431–454.

Deb, K., Horn, J. and Goldberg, D. E. (1993). Multi-modal deceptive functions. *Complex Systems 7*(2), 131–153.

Deb, K., Agrawal, S., Pratap, A. and Meyarivan, T. (2000a). A fast and elitist multi-objective genetic algorithm: NSGA-II. Technical Report 200001, Indian Institute of Technology, Kanpur: Kanpur Genetic Algorithms Laboratory (KanGAL).

Deb, K., Agrawal, S., Pratap, A. and Meyarivan, T. (2000b). A fast elitist non-dominated sorting genetic algorithm for multi-objective optimization: NSGA-II. In *Proceedings of the Parallel Problem Solving from Nature VI (PPSN-VI)*, pp. 849–858.

Deb, K., Khan, N. and Jindal, S. (2000c). Optimal truss-structure design for multiple objectives. In N. G. R. Iyenger and K. Deb (Eds), *Proceedings of the Tenth National Seminar on Aerospace Structures*, pp. 168–180. New Delhi: Allied Publishers.

Deb, K., Pratap, A. and Moitra, S. (2000d). Mechanical component design for multiple objectives using elitist non-dominated sorting GA. In *Proceedings of the Parallel Problem Solving from Nature VI (PPSN-VI)*, pp. 859–868.

Deb, K., Pratap, A. and Meyarivan, T. (2001). Constrained test problems for multi-objective evolutionary optimization. In *Proceedings of the First International*

Conference on Evolutionary Multi-Criterion Optimization (EMO-2001), pp. 284–298.

DeJong, K. A. (1975). *An Analysis of the Behavior of a Class of Genetic Adaptive Systems*. Ph. D. Thesis, Ann Arbor, MI: University of Michigan. Dissertation Abstracts International 36(10), 5140B (University Microfilms No. 76-9381).

DeJong, K. A. (1999). Personal communication.

Drechsler, R. (1998). *Evolutionary Algorithms for VLSI CAD*. Boston: Kluwer Academic Publishers.

Ehrgott, M. (2000). *Multicriteria Optimization*. Berlin: Springer.

Eklund, N. and Embrechts, M. J. (1999). GA-based multi-objective optimization of visible spectra for lamp design. In C. H. Dagli, A. L. Buczak, J. Ghosh, M. J. Embrechts and O. Ersoy (Eds), *Smart Engineering System Design: Neural Networks, Fuzzy Logic, Evolutionary Programming, Data Mining and Complex Systems*, pp. 451–456. New York: ASME Press.

Eldredge, N. (1989). *Macro-Evolutionary Dynamics: Species, Niches, and Adaptive Peaks*. New York: McGraw-Hill.

Erickson, M., Mayer, A. and Horn, J. (2001). The niched Pareto genetic algorithm 2 applied to the design of groundwater remediation systems. In *Proceedings of the First International Conference on Evolutionary Multi-Criterion Optimization (EMO-2001)*, pp. 681–695.

Eshelman, L. J. (1991). The CHC adaptive search algorithm: How to have safe search when engaging in nontradtional genetic recombination. In *Foundations of Genetic Algorithms 1 (FOGA-1)*, pp. 265–283.

Eshelman, L. J. (February, 2001). Personal communication.

Eshelman, L. J. and Schaffer, J. D. (1993). Real-coded genetic algorithms and interval-schemata. In *Foundations of Genetic Algorithms 2 (FOGA-2)*, pp. 187–202.

Fogel, D. B. (1988). An evolutionary approach to the traveling salesman problem. *Biological Cybernetics 60*(2), 139–144.

Fogel, D. B. (1992). *Evolving Artificial Intelligence*. Ph. D. Thesis, San Diego, CA: University of California.

Fogel, D. B. (1995). *Evolutionary Computation*. Piscataway, NY: IEEE Press.

Fogel, D. B., Fogel, L. J., Atmar, W. and Fogel, G. B. (1992). Hierarchic methods in evolutionary programming. In *Proceedings of the First Annual Conference on Evolutionary Programming*, pp. 175–182.

Fogel, L. J. (1962). Autonomous automata. *Industrial Research 4*(1), 14–19.

Fogel, L. J., Owens, A. J. and Walsh, M. J. (1966). *Artificial Intelligence Through Simulated Evolution*. New York: Wiley.

Fogel, L. J., Angeline, P. J. and Fogel, D. B. (1995). An evolutionary programming approach to self-adaptation on finite state machines. In *Proceedings of the Fourth International Conference on Evolutionary Programming*, pp. 355–365.

Fonseca, C. M. and Fleming, P. J. (1993). Genetic algorithms for multiobjective optimization: Formulation, discussion, and generalization. In *Proceedings of the Fifth International Conference on Genetic Algorithms*, pp. 416–423.

Fonseca, C. M. and Fleming, P. J. (1995). An overview of evolutionary algorithms in multi-objective optimization. *Evolutionary Computation Journal 3*(1), 1–16.

Fonesca, C. M. and Fleming, P. J. (1996). On the performance assessment and comparison of stochastic multiobjective optimizers. In *Proceedings of Parallel Problem Solving from Nature IV (PPSN-IV)*, pp. 584–593.

Fonseca, C. M. and Fleming, P. J. (1998a). Multiobjective optimization and multiple constraint handling with evolutionary algorithms–Part I: A unified formulation. *IEEE Transactions on Systems, Man and Cybernetics, Part A: Systems and Humans 28*(1), 26–37.

Fonseca, C. M. and Fleming, P. J. (1998b). Multiobjective optimization and multiple constraint handling with evolutionary algorithms–Part II: Application example. *IEEE Transactions on Systems, Man, and Cybernetics: Part A: Systems and Humans 28*(1), 38–47.

Fourman, M. P. (1985). Compaction of symbolic layout using genetic algorithms. In *Proceedings of an International Conference on Genetic Algorithms and Their Applications*, pp. 141–153.

Fox, R. L. (1971). *Optimization Methods for Engineering Design*. Reading, MA: Addison-Wesley.

Fujita, K., Hirokawa, N., Akagi, S., Kitamura, S. and Yokohata, H. (1998). Multi-objective optimal design of automotive engine using genetic algorithm. In *Proceedings of 1998 ASME Design Engineering Technical Conferences*. Paper Number DETC98/DAC-5799.

Garg, S. and Gupta, S. K. (1999). Multiobjective optimization of a free radical bulk polymerization reactor using genetic algorithm. *Macromolecular Theory and Simulation 8*(1), 46–53.

Gen, M. and Cheng, R. (1997). *Genetic Algorithms and Engineering Design*. New York: Wiley.

Geoffrion, A. M., Dyer, J. S. and Feinberg, A. (1972). An interactive approach for multi-criterion optimization with an application to the operation of an academic department. *Management Science 19*(4), 357–368.

Goldberg, D. E. (1983). *Computer-Aided Gas Pipeline Operation Using Genetic Algorithms and Rule Learning*. Ph. D. Thesis, Ann Arbor, MI: University of Michigan. Dissertation Abstracts International 44(10), 3174B (University Microfilms No. 8402282).

Goldberg, D. E. (1989). *Genetic Algorithms for Search, Optimization, and Machine Learning*. Reading, MA: Addison-Wesley.

Goldberg, D. E. (1991). Real-coded genetic algorithms, virtual alphabets, and blocking. *Complex Systems 5*(2), 139–168.

Goldberg, D. E. (1992). Personal communication.

Goldberg, D. E. (1999). Optimizating global-local search hybrids. In *Proceedings of the Genetic and Evolutionary Computation Conference (GECCO-1999)*, pp. 220–228.

Goldberg, D. E. and Deb, K. (1991). A comparison of selection schemes used in genetic algorithms. In *Foundations of Genetic Algorithms 1 (FOGA-1)*, pp. 69–93.

Goldberg, D. E. and Richardson, J. (1987). Genetic algorithms with sharing for multimodal function optimization. In *Proceedings of the First International Conference on Genetic Algorithms and Their Applications*, pp. 41–49.

Goldberg, D. E. and Samtani, M. P. (1986). Engineering optimization via genetic algorithms. In *Proceedings of the Ninth Conference on Electronic Computations, ASCE*, pp. 471–482.

Goldberg, D. E. and Wang, L. (1998). Adaptive niching via coevolutionay sharing. In D. Quagliarella, J. Périaux, C. Poloni, and G. Winter (Eds), *Genetic Algorithms and Evolution Strategies in Engineering and Computer Science*, pp. 21–38. Chichester, UK: Wiley.

Goldberg, D. E., Korb, B. and Deb, K. (1989). Messy genetic algorithms: Motivation, analysis and first results. *Complex Systems 3*(5), 493–530.

Goldberg, D. E., Deb, K. and Korb, B. (1990). Messy genetic algorithms revisited: Nonuniform size and scale. *Complex Systems 4*(4), 415–444.

Goldberg, D. E., Deb, K. and Clark, J. H. (1992). Genetic algorithms, noise, and the sizing of populations. *Complex Systems 6*(4), 333–362.

Goldberg, D. E., Deb, K., Kargupta, H. and Harik, G. (1993a). Rapid, accurate optimization of difficult problems using messy genetic algorithms. In *Proceedings of the Fifth International Conference on Genetic Algorithms*, pp. 56–64.

Goldberg, D. E., Deb, K. and Thierens, D. (1993b). Toward a better understanding of mixing in genetic algorithms. *Journal of the Society of Instruments and Control Engineers (SICE) 32*(1), 10–16.

Goméz-Skarmeta, A. F., Jiménez, F. and Ibanez, J. (1998). Pareto-optimality in fuzzy modeling. In *6th European Congress on Intelligent Techniques and Soft Computing*, pp. 694–700.

Grefenstette, J. J. (1993). Deception considered harmful. In *Foundation of Genetic Algorithms 2 (FOGA-2)*, pp. 75–91.

Gueugniaud, P. Y., Bertin-Maghit, M., Hirschauer, C., Bouchard, C., Vilasco, B., Petit, P., Gen, M., Ida, K., Lee, J. and Kim, J. (1997). Fuzzy nonlinear goal programming using genetic algorithm. *Computers and Industrial Engineering 33*(1), 39–42.

Guvenir, H. A. and Erel, E. (1998). Multicriteria inventory classification using a genetic algorithm. *European Journal of Operational Research 105*(1), 29–37.

Hahn, S. Y. (1996). Application of vector optimization employing modified genetic algorithm to permanent magnet motor design. In *Proceedings of the IEEE Conference on Electromagnetic Field Computation*, pp. OD1–7.

Haimes, Y. Y., Lasdon, L. S. and Wismer, D. A. (1971). On a bicriterion formulation of the problems of integrated system identification and system optimization. *IEEE Transactions on Systems, Man, and Cybernetics 1*(3), 296–297.

Hajela, P. and Lin, C.-Y. (1992). Genetic search strategies in multi-criterion optimal design. *Structural Optimization 4*(2), 99–107.

Hajela, P., Lee, E. and Lin, C. Y. (1993). Genetic algorithms in structural topology optimization. In *Proceedings of the NATO Advanced Research Worskhop on Topology Design of Structures*, pp. 117–133.

Hamda, H. and Schoenauer, M. (2000). Adaptive techniques for evolutionary topological optimum design. In I. Parmee (Ed.), *Evolutionary Design and Manufacture*, pp. 123–136. London: Springer.

Hansen, M. P. (1997). Tabu search in multiobjective optimization: MOTS. Paper presented at The Thirteenth International Conference on Multi-Criterion Decision Making (MCDM'97), University of Cape Town.

Hansen, M. P. and Jaskiewicz, A. (1998). Evaluating the quality of approximations to the non-dominated set. Technical Report IMM-REP-1998-7, Lyngby: Institute of Mathematical Modelling, Technical University of Denmark.

Hansen, N. and Ostermeier, A. (1996). Adapting arbitrary normal mutation distributions in evolution strageties: The covariance matrix adaptation. In *Proceedings of the IEEE International Conference on Evolutionary Computation*, pp. 312–317.

Harik, G. and Goldberg, D. E. (1996). Learning linkages. In *Foundations of Genetic Algorithms 4 (FOGA-4)*, pp. 247–262.

Harik, G., Cantú-Paz, E., Goldberg, D. E. and Miller, B. L. (1999). The gambler's ruin problem, genetic algorithms, and the sizing of populations. *Evolutionary Computation Journal 7*(3), 231–254.

Haug, E. J. and Arora, J. S. (1989). *Introduction to Optimal Design*. New York: McGraw Hill.

Herdy, M. (1992). Reproductive isolation as strategy parameter in hierarchically organized evolution strategies. In *Parallel Problem Solving from Nature II (PPSN-II)*, pp. 207–217.

Herrara, F., Lozano, M. and Verdegay, J. L. (1995). Fuzzy connectives based crossover operators to model genetic algorithms population diversity. Technical Report DECSAI-95110, Granada: ETS de Ingenieria Informatica, Universidad de Granada, Spain.

Herrera, F., Lozano, M. and Verdegay, J. L. (1998). Tackling real-coded genetic algorithms: Operators and tools for behavioural analysis. *Artificial Intelligence Review 12*(4), 265–319.

Himmelblau, D. M. (1972). *Applied Nonlinear Programming*. New York: McGraw-Hill.

Hinchcliffe, M., Willis, M. and Tham, M. (1998). Chemical process systems modelling using multi-objective genetic programming. In *Proceedings of the Third Annual Conference on Genetic Programming*, pp. 134–139.

Hiroyasu, T., Miki, M. and Watanabe, S. (1999). Distributed genetic algorithms with a new sharing approach in multiobjective optimization problems. In *Proceedings of the Congress on Evolutionary Computation (CEC-1999)*, pp. 69–76.

Holland, J. H. (1975). *Adaptation in Natural and Artificial Systems*. Ann Arbor, MI: MIT Press.

Hollstein, R. B. (1971). *Artificial Genetic Adaptation in Computer Control Systems*. Ph. D. Thesis, Ann Arbor, MI: University of Michigan. Dissertation Abstracts International 32(3), 1510B.

Homaifar, A., Lai, S. H.-V. and Qi, X. (1994). Constrained optimization via genetic algorithms. *Simulation 62*(4), 242–254.

Horn, J. (1997). Multicriterion decision making. In T. Bäck, D. Fogel and Z. Michalewicz (Eds), *Handbook of Evolutionary Computation*, pp. F1.9:1–15. Bristol: Institute of Physics Publishing and New York: Oxford University Press.

Horn, J., Nafploitis, N. and Goldberg, D. (1994). A niched Pareto genetic algorithm for multi-objective optimization. In *Proceedings of the First IEEE Conference on Evolutionary Computation*, pp. 82–87.

Hwang, C.-L. and Masud, A. S. M. (1979). *Multiple Objective Decision Making – Methods and Applications: A State-of-the-art Survey*. Berlin: Springer-Verlag.

Ignizio, J. P. (1976). *Goal Programming and Extensions*. Lexington, MA: Lexington Books.

Ignizio, J. P. (1978). A review of goal programming: A tool for multiobjective analysis. *Journal of Operations Research Society 29*(11), 1109–1119.

Imai, K. and Schmit, L. A. (1981). Configuration optimization of trusses. *Journal of Structural Division, ASCE 107*(ST5), 745–756.

Iredi, S., Merkle, D. and Middendorf, M. (2001). Bi-criterion optimization with multi colony ant algorithms. In *Proceedings of the First International Conference on Evolutionary Multi-Criterion Optimization (EMO-2001)*, pp. 359–372.

Ishibuchi, H. and Murata, T. (1997). Minimizing the fuzzy rule and maximizing its performance by a multi-objective genetic algorithm. In *Proceedings of the Sixth International Conference on Fuzzy Systems*, pp. 259–264.

Ishibuchi, H. and Murata, T. (1998a). A multi-objective genetic local search algorithm and its application to flowshop scheduling. *IEEE Transactions on Systems, Man and Cybernetics – Part C: Applications and reviews 28*(3), 392–403.

Ishibuchi, H. and Murata, T. (1998b). Multi-objective genetic local search for minimizing the number of fuzzy rules for pattern classification problems. In *Proceedings of the Sixth IEEE International Conference on Fuzzy Systems*, pp. 1100–1105.

Ishibuchi, H., Nakashima, T. and Murata, T. (2001). Multiobjective optimization in linguistic rule extraction from numerical data. In *Proceedings of the First International Conference on Evolutionary Multi-Criterion Optimization (EMO-2001)*, pp. 588–602.

Jakiela, M. J., Chapman, C., Duda, J., Adewuya, A. and Saitou, K. (2000). Continuum structural topology design with genetic algorithms. *Computer Methods in Applied Mechanics and Engineering 186*(2–4), 339–356.

Jaskiewicz, A. (1998). Genetic local search for multiple objective combinatorial optimization. Technical Report RA-GL4/98, Institute of Computing Science, Poznan University of Technology, Poland.

Jaszkiewicz, A. and Slowinsky, R. (1994). The light beam search over a non-dominated surface of a multiple-objective programming problem. In *Proceedings of the Tenth International Conference: Expand and Enrich the Domain of Thinking and Application*, pp. 87–99.

Jelasity, M. (1998). UEGO, an abstract niching technique for global optimization. In *Proceedings of the Parallel Problem Solving from Nature V (PPSN-V)*, pp. 378–387.

Jiménez, F. and Cadenas, J. M. (1995). An evolutionary program for the multiobjective solid transportation problem with fuzzy goals. *Operations Research and Decisions 2*, 5–20.

Jiménez, F., Verdegay, J. L. and Goméz-Skarmeta, A. F. (1999). Evolutionary techniques for constrained multiobjective optimization problems. In *Proceedings of the Workshop on Multi-Criterion Optimization Using Evolutionary Methods held at Genetic and Evolutionary Computation Conference (GECCO-1999)*, pp. 115–116.

Joines, J. A. and Houck, C. R. (1994). On the use of nonstationary penalty functions to solve nonlinear constrained optimization problems with GAs. In *Proceedings of the International Conference on Evolutionary Computation*, pp. 579–584.

Kannan, B. K. and Kramer, S. N. (1994). An augmented lagrange multiplier based method for mixed integer discrete continuous optimization and its applications to mechanical design. *ASME Journal of Mechanical Design 116*(2), 405–411.

Kargupta, H. (1997). SEARCH, computational processes in evolution, and preliminary development of the gene expression messy genetic algorithm. *Complex Systems 11*(4), 233–287.

Keeney, R. L. and Raiffa, H. (1976). *Decisions with Multiple Objectives: Preferences and Value Tradeoffs*. New York: Wiley.

Kim, J.-H. and Kim, K.-C. (1997). Multicriteria fuzzy control using evolutionary programming. *Infomation Sciences* 103(1), 71–86.

Kirsch, U. (1989). Optimal topologies of truss structures. *Computer Methods in Applied Mechanics and Engineering* 72(1), 15–28.

Kita, H., Yabumoto, Y., Mori, N. and Nishikawa, Y. (1996). Multi-objective optimization by means of thermodynamical genetic algorithm. In *Proceedings of Parallel Problem Solving from Nature IV (PPSN-IV)*, pp. 504–512.

Kita, H., Ono, I. and Kobayashi, S. (1998). The multi-parent unimodal normal distribution crossover for real-coded genetic algorithms. Tokyo Institute of Technology, Japan.

Knowles, J. D. and Corne, D. W. (2000). Approximating the non-dominated front using the Pareto archived evolution strategy. *Evolutionary Computation Journal* 8(2), 149–172.

Knowles, J. D., Watson, R. A. and Corne, D. W. (2001). Reducing local optima in single-objective problems by multi-objectivization. In *Proceedings of the First International Conference on Evolutionary Multi-Criterion Optimization (EMO-2001)*, pp. 269–283.

Köppen, M., Teunis, M. and Nicholay, B. (1997). NESSY − an evolutionary learning neural network. In *Proceedings of the Second International ICSC Symposium on Soft Computing, SOCO'97*, pp. 243–248.

Koza, J. R. (1992). *Genetic Programming : On the Programming of Computers by Means of Natural Selection*. Cambridge, MA: MIT Press.

Koza, J. R. (1994). *Genetic Programming II: Automatic Discovery of Reusable Programs*. Cambridge, MA: MIT Press.

Koziel, S. and Michalewicz, Z. (1998). A decoder-based evolutionary algorithm for constrained parameter optimization. In *Parallel Problem Solving from Nature V (PPSN-V)*, pp. 231–240.

Krause, M. and Nissen, V. (1995). On using penalty functions and multicriteria optimization techniques in facility layout. In J. Biethahn and V. Nissen (Eds), *Evolutionary Algorithms in Management Applications*. Berlin, Springer-Verlag.

Krishnakumar, K. (1989). Micro-genetic algorithms for stationary and non-stationary function optimization. In *SPIE Proceedings: Intelligent Control and Adaptive Systems*, pp. 289–296.

Kumar, R. and Rockett, P. (1998). Decomposition of high dimensional pattern spaces for hierarchical classification. *Kybernetika* 34(4), 435–442.

Kumar, R. and Rockett, P. (in press). Improved sampling of the Pareto-front in multi-objective genetic optimizations by neo-stationary evolution: A Pareto converging genetic algorithm. *Evolutionary Computation Journal*.

Kung, H. T., Luccio, F. and Preparata, F. P. (1975). On finding the maxima of a set of vectors. *Journal of the Association for Computing Machinery* 22(4), 469–476.

Kursawe, F. (1990). A variant of evolution strategies for vector optimization. In *Parellel Problem Solving from Nature I (PPSN-I)*, pp. 193–197.

Lahanas, M., Milickovic, N., Baltas, D. and Zamboglou, N. (2001). Application of multiobjective evolutionary algorithms for dose optimization problems in brachytherapy. In *Proceedings of the First International Conference on Evolutionary Multi-Criterion Optimization (EMO-2001)*, pp. 574–587.

Langdon, W. B. (1995). Pareto, population partitioning, price, and genetic programming. Technical Report Research Note RN/95/29, University College London, London, UK.

Laumanns, M., Rudolph, G. and Schwefel, H. P. (1998). A spatial predator-prey approach to multi-objective optimization: A preliminary study. In *Proceedings of the Parallel Problem Solving from Nature V (PPSN-V)*, pp. 241–249.

Laumanns, M., Zitzler, E. and Thiele, L. (2001). On the effects of archiving, elitism, and density based selection in evolutionary multi-objective optimization. In *Proceedings of the First International Conference on Evolutionary Multi-Criterion Optimization (EMO-2001)*, pp. 181–196.

Laumanns, N., Laumanns, M. and Neunzig, D. (2001). Multi-objective design space exploration of road trains with evolutionary algorithms. In *Proceedings of the First International Conference on Evolutionary Multi-Criterion Optimization (EMO-2001)*, pp. 612–623.

Lee, D. (1997). Multiobjective design of a marine vehicle with aid of design knowledge. *International Journal for Numerical Methods in Engineering 40*(14), 2665–2677.

Lee, M. A. and Esbensen, H. (1997). Fuzzy/multiobjective genetic systems for intelligent systems design tools and components. In W. Pedrycz (Ed.), *Fuzzy Evolutionary Computation*, pp. 57–80. Boston: Kluwer Academic Publishers.

Lee, S. M. (1972). *Goal Programming for Decision Analysis*. Philadelphia: Auerbach Publishers.

Leung, K. S., Zhu, Z. Y., Xu, Z. B. and Leung, Y. (1998). Multiobjective optimization using non-dominated sorting in annealing genetic algorithms. Department of Geography and the Centre for Environmental Studies, Chinese University of Hong Kong, Hong Kong.

Li, X., Jiang, T. and Evans, D. J. (2000). Medical image reconstruction using a multiobjective evolutionary algorithm. *International Journal of Computer Mathematics 75*(3), 301–314.

Lichtfuss, H. J. (1965). Evolution eines Rohrkrümmers. Diplomarbeit, Technische Universität Berlin, Deutschland.

Lis, J. and Eiben, A. E. (1996). A multi-sexual genetic algorithm for multiobjective optimization. In *Proceedings of the 1996 International Conference on Evolutionary Computation*, pp. 59–64.

Liu, B., Haftka, R. T., Akgin, M. A. and Todoroki, A. (2000). Permutation genetic algorithm for stacking sequence design of composite laminates. *Computer Methods in Applied Mechanics and Engineering 186*(2–4), 357–372.

Loughlin, D. H. and Ranjithan, S. (1997). The neighborhood constraint method: A multiobjective optimization technique. In *Proceedings of the Seventh International Conference on Genetic Algorithms*, pp. 666–673.

Mañas, M. S. (1982). Graphical methods of multi-criterion optimization. *Zeitschrift für Angewandte Methematik und Mechanik 62*(5), 375–377.

Mahfoud, S. W. (1992). Crowding and preselection revisited. In *Parallel Problem Solving from Nature II (PPSN-II)*, pp. 27–36.

Mahfoud, S. W. (1995). *Niching Methods for Genetic Algorithms*. Ph. D. Thesis, Urbana, IL: University of Illinois at Urbana-Champaign. (Available as IlliGAL Report No. 95001).

Mäkinen, R., Neittaanmäki, P., Periaux, J., Sefrioui, M. and Toivanen, J. (1996). Parellel genetic solution for multiobjective MDO. In *Parellel Computational Fluid Dynamics Conference (CFD'96)*, pp. 352–359.

Marco, N., Lanteri, S., Desideri, J.-A. and Periaux, J. (1999). A parallel genetic algorithm for multi-objective optimization in computational fluid dynamics. In K. Miettinen,

P. Neittaanmäki, M. M. Mäkelä and J. Périaux (Eds), *Evolutionary Algorithms in Engineering and Computer Science*, pp. 445–456. Chichester, UK: Wiley.

Mardle, S., Pascoe, S. and Tamiz, M. (1998). An investigation of genetic algorithms for the optimization of multiobjective fisheries bioeconomic models. Technical Report 136, Portsmouth, UK: Centre for the Economics and Management of Aquatic Resources, University of Portsmouth.

Mariano, C. E. and Morales, E. F. (2000). Distributed reinforcement learning for multiple objective optimization problems. In *Proceedings of the Congress on Evolutionary Computation (CEC-2000)*, pp. 188–195.

Matthews, K. B., Craw, S., Elder, S., Sibbald, A. R. and MacKenzie, I. (2000). Applying genetic algorithms to multi-objective land use planning. In *Proceedings of the Genetic and Evolutionary Computation Conference (GECCO-2000)*, pp. 613–620.

Meisel, W. L. (1973). Tradeoff decision in multiple criteria decision making. In J. L. Cochrane and M. Zeleny (Eds), *Multiple Criteria Decision Making*, pp. 461–476. Columbia, SC: University of South Carolina Press.

Menzcer, F., Degeratu, M. and Street, W. N. (2000). Efficient and scalable Pareto optimization by evolutionary local selection algorithms. *Evolutionary Computation Journal* 8(2), 125–148.

Meunier, H., Talbi, E. G. and Reininger, P. (2000). A multiobjective genetic algorithm for radio network optimization. *2000 Congress on Evolutionary Computation* 1, 317–324.

Michalewicz, Z. (1992). *Genetic Algorithms + Data Structures = Evolution Programs*. Berlin: Springer-Verlag.

Michalewicz, Z. and Attia, N. (1994). Evolutionary optimization of constrained problems. In *Proceedings of the Third Annual Conference on Evolutionary Programming*, pp. 98–108.

Michalewicz, Z. and Janikow, C. Z. (1991). Handling constraints in genetic algorithms. In *Proceedings of the Fourth International Conference on Genetic Algorithms*, pp. 151–157.

Michalewicz, Z. and Schoenauer, M. (1996). Evolutionary algorithms for constrained parameter optimization problems. *Evolutionary Computation Journal* 4(1), 1–32.

Michalewicz, Z., Deb, K., Schmidt, M. and Stidsen, T. (2000). Test-case generator for nonlinear continuous parameter optimization techniques. *IEEE Transactions on Evolutionary Computation* 4(3), 197–215.

Miettinen, K. (1999). *Nonlinear Multiobjective Optimization*. Boston: Kluwer.

Miettinen, K. and Mäkelä, M. M. (1995). Interactive bundle-based method for nondifferentiable multiobjective optimization: NIMBUS. *Optimization* 34(3), 231–246.

Mitchell, M. (1996). *Introduction to Genetic Algorithms*. Ann Arbor, MI: MIT Press.

Mitra, K., Deb, K. and Gupta, S. K. (1998). Multiobjective dynamic optimization of an industrial nylon 6 semibatch reactor using genetic algorithms. *Journal of Applied Polymer Science* 69(1), 69–87.

Moudani, W. E., Cosenza, C., Coligny, M. and Mora-Camino, F. (2001). A bi-criterion approach for the airlines crew rostering problem. In *Proceedings of the First International Conference on Evolutionary Multi-Criterion Optimization (EMO-2001)*, pp. 486–500.

Mühlenbein, H. and Mahnig, T. (1999). FDA: A scalable evolutionary algorithm for the optimization of additively decomposed functions. *Evolutionary Computation Journal* 7(4), 353–376.

Murata, T. and Ishibuchi, H. (1995). MOGA: Multi-objective genetic algorithms. In *Proceedings of the Second IEEE International Conference on Evolutionary Computation*, pp. 289–294.

Neef, M., Thierens, D. and Arciszewski, H. (1999). A case study of a multiobjective recombinative genetic algorithm with coevolutionary sharing. In *Proceedings of the Congress on Evolutionary Computation (CEC-1999)*, pp. 796–803.

Nomura, T. (1997). An analysis for crossovers for real number chromosomes in an infinite population size. In *Proceedings of International Joint Conference on Artificial Intelligence (IJCAI-97)*, pp. 936–941.

Nomura, T. and Miyoshi, T. (1996). Numerical coding and unfair average crossover in GA for fuzzy rule extraction in dynamic environments. In Y. Uchikawa and T. Furuhashi (Eds), *Fuzzy Logic, Neural Networks, and Evolutionary Computation, (Lecture Notes in Computer Science 1152)*, pp. 55–72. Berlin: Springer.

Oates, M., Corne, D. and Loader, R. (1999a). Skewed crossover and the dynamic distributed database problem. In *Proceedings of the International Conference in Artificial Neural Networks and Genetic Algorithms (ICANNGA-99)*, pp. 280–287.

Oates, M., Corne, D. and Loader, R. (1999b). Variation in EA performance characteristics on the adaptive distributed database management problems. In *Proceedings of the Genetic and Evolutionary Computation Conference (GECCO-1999)*, pp. 480–487.

Obayashi, S., Takahashi, S. and Takeguchi, Y. (1998). Niching and elitist models for MOGAs. In *Proceedings of Parallel Problem Solving from Nature V (PPSN-V)*, pp. 260–269.

Oei, C. K., Goldberg, D. E. and Chang, S.-J. (1991). Tournament selection, niching, and the preservation of divrsity. IlliGAL Report No. 91011, Urbana, IL: University of Illinois at Urbana-Champaign.

Ono, I. and Kobayashi, S. (1997). A real-coded genetic algorithm for function optimization using unimodal normal distribution crossover. In *Proceedings of the Seventh International Conference on Genetic Algorithms*, pp. 246–253.

Osyczka, A. and Kundu, S. (1995). A new method to solve generalized multicriteria optimization problems using the simple genetic algorithm. *Structural Optimization 10*(2), 94–99.

Oyman, A. I., Deb, K. and Beyer, H.-G. (1999). An alternative constraint handling method for evolution strategies. In *Proceedings of Congress on Evolutionary Computation (CEC-1999)*, pp. 612–619.

Paechter, B., Rankin, R. C., Cumming, A. and Fogerty, T. (1998). Timetabling the classes of an entire university with an evolutionary algorithm. In *Parallel Problem Solving from Nature V (PPSN-V)*, pp. 865–874.

Palli, N., Azram, S., McCluskey, P. and Sundararajan, R. (1999). An interactive multistage ϵ-inequality constraint method for multiple objectives decision making. *ASME Journal of Mechanical Design 120*(4), 678–686.

Paredis, J. (1994). Coevolutionary constraint satisfaction. In *Parallel Problem Solving from Nature III (PPSN-III)*, pp. 46–55.

Parks, G. T. and Miller, I. (1998). Selective breeding in a multi-objective genetic algorithm.

In *Proceedings of the Parallel Problem Solving from Nature V (PPSN-V)*, pp. 250–259.

Parmee, I. C., Cevtković, D., Watson, A. W. and Bonham, C. R. (2000). Multiobjective satisfaction within an interactive evolutionary design enviornment. *Evolutionary Computation Journal 8*(2), 197–222.

Pelican, M., Goldberg, D. E. and Cantu-Paz, E. (1999). BOA: The bayesian optimization algorithm. In *Proceedings of the Genetic and Evolutionary Conference, (GECCO-1999)*, pp. 525–532.

Perry, A. Z. (1984). *Experimental Study of Speciation in Ecological Niche Theory Using Genetic Algorithms*. Ph. D. Thesis, Ann Arbor, MI: University of Michigan. Dissertation Abstracts International 45(12), 3870B.

Petrovski, A. and McCall, J. (2001). Multi-objective optimization of cancer chemotherapy using evolutionary algorithms. In *Proceedings of the First International Conference on Evolutionary Multi-Criterion Optimization (EMO-2001)*, pp. 531–545.

Poloni, C., Giurgevich, A., Onesti, L. and Pediroda, V. (2000). Hybridization of a multiobjective genetic algorithm, a neural network and a classical optimizer for complex design problem in fluid dynamics. *Computer Methods in Applied Mechanics and Engineering 186*(2–4), 403–420.

Powell, D. and Skolnick, M. M. (1993). Using genetic algorithms in engineering design optimization with nonlinear constraints. In *Proceedings of the Fifth International Conference on Genetic Algorithms*, pp. 424–430.

Press, W. H., Teuklosky, S. A., Vellerling, W. T. and Flannery, B. P. (1988). *Numerical Recipes in C*. Cambridge, UK: Cambridge University Press.

Prügel-Bennett, A. and Shapiro, J. L. (1994). An analysis of genetic algorithms using statistical mechanics. *Physics Review Letters 72*(9), 1305–1309.

Qi, X. and Palmieri, F. (1993). Adaptive mutation in the genetic algorithms. In *Proceedings of the Second Annual Conference on Evolutionary Programming*, pp. 192–196.

Quagliarella, D. and Vicini, A. (1998). Coupling genetic algorithms and gradient based optimization techniques. In D. Quagliarella, J. Périaux, C. Poloni and G. Winter (Eds), *Genetic Algorithms and Evolution Strategies in Engineering and Computer Science: Recent Advances and Industrial Applications*, pp. 289–309. Chichester, UK: Wiley.

Radcliffe, N. J. (1991). Forma analysis and random respectful recombination. In *Proceedings of the Fourth International Conference on Genetic Algorithms*, pp. 222–229.

Rajan, S. D. (1995). Sizing, shape and topology optimization of trusses using genetic algorithm. *Journal of Structural Engineering 121*(10), 1480–1487.

Rao, S. S. (1984). *Optimization: Theory and Applications*. New York: Wiley.

Rao, S. S. (1993). Genetic algorithmic approach for multiobjective optimization of structures. In *Proceedings of the ASME Annual Winter Meeting on Structures and Controls Optimization*, Volume 38, pp. 29–38.

Rauwolf, G. A. and Coverstone-Carroll, V. L. (1996). Near-optimal low thrust orbit transfers generated by a genetic algorithm. *Journal of Spacecraft and Rockets 33*(6), 859–862.

Ray, T., Tai, K. and Seow, K. C. (2000). An evolutionary algorithm for constrained optimization. In *Proceedings of the Genetic and Evolutionary Computation Conference (GECCO-2000)*, pp. 771–777.

Ray, T., Tai, K. and Seow, K. C. (2001). An evolutionary algorithm for multiobjective optimization. *Engineering Optimization 33*(3), 399–424.

Rechenberg, I. (1965). Cybernetic solution path of an experimental problem. Royal Aircraft Establishment, Library Translation Number 1122, Farnborough, UK.

Rechenberg, I. (1973). *Evolutionsstrategie: Optimierung Technischer Systeme nach Prinzipien der Biologischen Evolution.* Stuttgart: Frommann-Holzboog Verlag.

Reeves, C. (1993a). *Modern Heuristic Techniques for Combinatorial Problems.* London: Blackwell Scientific Publications.

Reeves, C. R. (1993b). Using genetic algorithms with small populations. In *Proceedings of the Fifth International Conference on Genetic Algorithms,* pp. 92–99.

Reklaitis, G. V., Ravindran, A. and Ragsdell, K. M. (1983). *Engineering Optimization Methods and Applications.* New York : Wiley.

Ringertz, U. T. (1985). On topology optimization of trusses. *Engineering Optimization 9*(3), 209–218.

Rogers, A. and Prügel-Bennett, A. (1998). Modelling the dynamics of steady-state genetic algorithms. In *Foundations of Genetic Algorithms 5 (FOGA-5),* pp. 57–68.

Romero, C. (1991). *Handbook of Critical Issues in Goal Programming.* Oxford, UK: Pergamon Press.

Rosenberg, R. S. (1967). *Simulation of Genetic Populations with Biochemical Properties.* Ph. D. Thesis, Ann Arbor, MI, University of Michigan.

Rosenthal, R. E. (1985). Principles of multiobjective optimization. *Decision Sciences 16*(2), 133–152.

Rowe, J., Vinsen, K. and Marvin, N. (1996). Parallel GAs for multiobjective functions. In *Proceedings of the Second Nordic Workshop on Genetic Algorithms and their Applications,* pp. 61–70.

Rudolph, G. (1994). Convergence analysis of canonical genetic algorithms. *IEEE Transactions on Neural Network 5*(1), 96–101.

Rudolph, G. (1996). Convergence of evolutionary algorithms in general search spaces. In *Proceedings of the Third IEEE Conference on Evolutionary Computation,* pp. 50–54.

Rudolph, G. (1998a). Evolutionary search for minimal elements in partially ordered finite sets. In *Proceedings of the 7th Annual Conference on Evolutionary Programming,* pp. 345–353. Berlin: Springer.

Rudolph, G. (1998b). On a multi-objective evolutionary algorithm and its convergence to the pareto set. In *Proceedings of the 5th IEEE Conference on Evolutionary Computation,* pp. 511–516.

Rudolph, G. (2001). Evolutionary search under partially ordered fitness sets. In *Proceedings of the International Symposium on Information Science Innovations in Engineering of Natural and Artificial Intelligent Systems (ISI 2001),* pp. 818–822.

Rudolph, G. and Agapie, A. (2000). Convergence properties of some multi-objective evolutionary algorithms. In *Proceedings of the Congress on Evolutionary Computation (CEC-2000),* pp. 1010–1016.

Ruy, W. S., Yang, Y. S., Kim, G. H. and Yeun, Y. S. (2000). Topology design of truss structures in a multicriteria environment. Department of Mechanical Engineering, Daejin University, Republic of Korea.

Sandgren, E., Jensen, E. and Welton, J. (1990). Topological design of structural components using genetic optimization methods. In *Proceedings of the Winter Annual Meeting of the American Society of Mechanical Engineers*, pp. 31–43.

Sannier, A. V. and Goodman, E. D. (1987). Genetic learning procedures in distributed environments. In *Proceedings of the Second International Conference on Genetic Algorithms*, pp. 162–169.

Sarvanan, N. N., Fogel, D. B. and Nelson, K. M. (1995). A comparison of methods for self-adaptation in evolutionary algorithms. *BioSystems 36*(2), 157–166.

Sasaki, D., Morikawa, M., Obayashi, S. and Nakahashi, K. (2001). Aerodynamic shape optimization of supersonic wings by adaptive range multiobjective genetic algorithms. In *Proceedings of the First International Conference on Evolutionary Multi-Criterion Optimization (EMO-2001)*, pp. 639–652.

Sayyouth, M. H. (1981). Goal programming: A new tool for optimization in petroleum reservoir history matching. *Applied Mathematics Modelling 5*(4), 223–226.

Sbalzarini, I. F., Müller, S. and Koumoutsakos, P. (2001). Microchannel optimization using multiobjective evolution strategies. In *Proceedings of the First International Conference on Evolutionary Multi-Criterion Optimization (EMO-2001)*, pp. 516–530.

Schaffer, J. D. (1984). *Some Experiments in Machine Learning Using Vector Evaluated Genetic Algorithms*. Ph. D. Thesis, Nashville, TN: Vanderbilt University.

Schaffer, J. D. (1985). Multiple objective optimization with vector evaluated genetic algorithms. In *Proceedings of the First International Conference on Genetic Algorithms*, pp. 93–100.

Schlemmer, E., Harb, W., Lichtenecker, G. and Müller, F. (2000). Multicriterion optimization of electrical machines using evolutionary algorithms and regular expressions. VATech Hydro GmbH and Company, Weiz, Austria.

Schoenauer, M. and Xanthakis, S. (1993). Constrained GA optimization. In *Proceedings of the Fifth International Conference on Genetic Algorithms*, pp. 573–580.

Schott, J. R. (1995). Fault Tolerant Design Using Single and Multi-Criteria Genetic Algorithms. Master's Thesis, Boston, MA: Department of Aeronautics and Astronautics, Massachusetts Institute of Technology.

Schwefel, H.-P. (1968). Projekt MHD-Staustrahlrohr: Experimentelle optimierung einer zweiphasendüse, Teil I. Technical Report 11.034/68, 35, AEG Forschungsinstitut, Berlin.

Schwefel, H.-P. (1981). *Numerical Optimization of Computer Models*. Chichester, UK: Wiley.

Schwefel, H.-P. (1987a). Collective intelligence in evolving systems. In W. Wolff, C. J. Soeder and F. Drepper (Eds), *Ecodynamics – Contributions to Theoretical Ecology*, pp. 95–100. Berlin: Springer.

Schwefel, H.-P. (1987b). Collective phenomena in evolutionary systems. In P. Checkland and I. Kiss (Eds), *Problems of Constancy and Change – the Complementarity of Systems Approaches to Complexity*, pp. 1025–1033. Budapest: International Society for General System Research.

Schwefel, H.-P. (1995). *Evolution and Optimum Seeking*. New York: Wiley.

Schwefel, H.-P. and Bäck, T. (1998). Artificial evolution: How and why? In D. Quagliarella, J. Périaux, C. Poloni and G. Winter (Eds), *Genetic Algorithms and Evolution*

Strategies in Engineering and Computer Science: Recent Advances and Industrial Applications, pp. 1–19. Chichester, UK: Wiley.

Sefrioui, M. and Periaux, J. (2000). Nash genetic algorithms: Examples and applications. In *Proceedings of the Congress on Evolutionary Computation (CEC-2000)*, pp. 509–516.

Shaw, K. J., Nortcliffe, A. L., Thompson, M., Love, J., Fonseca, C. M. and Fleming, P. J. (1999). Assessing the performance of multiobjective genetic algorithms for optimization of a batch process scheduling problem. In *1999 Congress on Evolutionary Computation*, pp. 37–45.

Smith, R. E., Forrest, S. and Perelson, A. S. (1993). Searching for diverse, cooperative populations with genetic algorithms. *Evolutionary Computation Journal 1*(2), 127–149.

Spears, W. M. (1998). *The Role of Mutation and Recombination in Evolutionary Algorithms*. Ph. D. Thesis, Fairfax, VA: George Mason University.

Srinivas, N. and Deb, K. (1994). Multi-objective function optimization using non-dominated sorting genetic algorithms. *Evolutionary Computation Journal 2*(3), 221–248.

Stanley, T. J. and Mudge, T. (1995). A parallel genetic algorithm for multiobjective microprocessor design. In *Proceedings of the Sixth International Conference on Genetic algorithms*, pp. 597–604.

Starkweather, T., McDaniel, S., Mathias, K., Whitley, D. and Whitley, C. (1991). A comparison of genetic sequencing operators. In *Proceedings of the Fourth International Conference on Genetic Algorithms*, pp. 69–76.

Steuer, R. E. (1986). *Multiple Criteria Optimization: Theory, Computation and Application*. New York: Wiley.

Surry, P. D., Radcliffe, N. J. and Boyd, I. D. (1995). A multi-objective approach to constrained optimisation of gas supply networks: The COMOGA method. In *Evolutionary Computing. AISB Workshop*, pp. 166–180.

Tamaki, H., Mori, M., Araki, M., Mishima, Y. and Ogai, H. (1995). Multi-criteria optimization by genetic algorithms. In *Proceedings of the Third Conference of the Association of Asian-Pacific Operational Research Societies within IFORS (APORS'94)*, pp. 374–381.

Tamaki, H., Nishino, E. and Abe, S. (1999). A genetic algorithm approach to multi-objective scheduling problems with earliness and tardiness penalties. In *1999 Congress on Evolutionary Computation*, pp. 46–52.

Tan, K. C., Lee, T. H. and Khor, E. F. (2001). Incrementing multi-objective evolutionary algorithms: Performance studies and comparisons. In *Proceedings of the First International Conference on Evolutionary Multi-Criterion Optimization (EMO-2001)*, pp. 111–125.

Tanaka, M. (1995). GA-based decision support system for multi-criteria optimization. In *Proceedings of the International Conference on Systems, Man and Cybernetics*, Volume 2, pp. 1556–1561.

Thompson, M. (2001). Application of multi-objective evolutionary algorithms to analogue filter tuning. In *Proceedings of the First International Conference on Evolutionary Multi-Criterion Optimization (EMO-2001)*, pp. 546–559.

Todd, D. S. and Sen, P. (1997). A multiple criteria genetic algorithm for containership loading. In *Proceedings of the Seventh International Conference on Genetic*

Algorithms, pp. 674–681.

Topping, B. H. V. (1983). Shape optimization of skeletal structures: A review. *Journal of Structural Engineering 11*(8), 1933–1951.

Tsutsui, S., Yamamura, M. and Higuchi, T. (1999). Multi-parent recombination with simplex crossover in real-coded genetic algorithms. In *Proceedings of the Genetic and Evolutionary Computation Conference (GECCO-1999)*, pp. 657–664.

Vanderplaat, G. N. and Moses, F. (1972). Automated design of trusses for optimal geometry. *Journal of Structural Division 98*(ST3), 671–690.

Veldhuizen, D. V. (1999). *Multiobjective Evolutionary Algorithms: Classifications, Analyses, and New Innovations*. Ph. D. Thesis, Dayton, OH: Air Force Institute of Technology. Technical Report No. AFIT/DS/ENG/99-01.

Veldhuizen, D. V. and Lamont, G. B. (1998). Multiobjective evolutionary algorithm research: A history and analysis. Technical Report TR-98-03, Dayton, OH: Department of Electrical and Computer Engineering, Air Force Institute of Technology.

Veldhuizen, D. V. and Lamont, G. B. (2000). Multiobjective evolutionary algorithms: Analyzing the state-of-the-art. *Evolutionary Computation Journal 8*(2), 125–148.

Vicini, A. and Quagliarella, D. (1999). Airfoil and wing design using hybrid optimization strategies. *AIAA journal 37*(5), AIAA Paper 98-2729.

Viennet, R. (1996). Multicriteria optimization using a genetic algorithm for determining the Pareto set. *International Journal of Systems Science 27*(2), 255–260.

Voigt, H.-M., Mühlenbein, H. and Cvetković, D. (1995). Fuzzy recombination for the Breeder Genetic Algorithm. In *Proceedings of the Sixth International Conference on Genetic Algorithms*, pp. 104–111.

Vose, M. D. (1999). *Simple Genetic Algorithm: Foundation and Theory*. Ann Arbor, MI: MIT Press.

Vose, M. D. and Rowe, J. E. (2000). Random heuristic search: Applications to GAs and functions of unitation. *Computer Methods and Applied Mechanics and Engineering 186*(2–4), 195–220.

Vose, M. D. and Wright, A. H. (1996). The Walsh transform and the theory of the simple genetic algorithm. In S. K. Pal and P. P. Wang (Eds), *Genetic Algorithms and Pattern Recognition*, pp. 25–44. Boca Raton, FL: CRC Press.

Waagen, D., Diercks, P. and McDonnell, J. (1992). The stochastic direction set algorithm: A hybrid technique for finding function extrema. In *Proceedings of the First Annual Conference on Evolutionary Programming*, pp. 35–42.

Weile, D. S., Michielssen, E. and Goldberg, D. E. (1996). Genetic algorithm design of Pareto-optimal broad band microwave absorbers. *IEEE Transactions on Electromagnetic Compatibility 38*(4), 518–525.

Whitley, D. (1989). The GENITOR algorithm and selection pressure: Why rank-based allocation of reproductive trials is best. In *Proceedings of the Third International Conference on Genetic Algorithms*, pp. 116–121.

Whitley, D. (1992). An executable model of a simple genetic algorithm. In *Foundations of Genetic Algorithms 2 (FOGA-2)*, pp. 45–62.

Wierzbicki, A. P. (1980). The use of reference objectives in multiobjective optimization. In G. Fandel and T. Gal (Eds), *Multiple Criteria Decision Making Theory and Applications*, pp. 468–486. Berlin: Springer-Verlag.

Wright, A. (1991). Genetic algorithms for real parameter optimization. In *Foundations of Genetic Algorithms 1 (FOGA-1)*, pp. 205–218.

Yu, P. L. (1973). A class of solutions for group decision problems. *Management Science 19*(8), 936–946.

Zeleny, M. (1973). Compromise programming. In J. L. Cochrane and M. Zeleny (Eds), *Multiple Criteria Decision Making*, pp. 262–301. Columbia: SC: University of South Carolina Press.

Zeleny, M. (1982). *Multiple Criteria Decision Making*. New York: McGraw-Hill.

Zhou, F. B., Gupta, S. K. and Ray, A. K. (2000). Multiobjective optimization of the continuous casting process for polymethyl methacrylate using an adapted genetic algorithm. *Journal of Applied Polymer Science 78*(7), 1439–1458.

Zhou, G. and Gen, M. (1997). Evolutionary computation on multicriterion production process planning problem. In *Proceedings of the 1997 IEEE International Conference on Evolutionary Computation*, pp. 419–424.

Zitzler, E. (1999). *Evolutionary Algorithms for Multiobjective Optimization: Methods and Applications*. Ph. D. Thesis, Zürich, Switzerland: Swiss Federal Institute of Technology (ETH) (Dissertation ETH No. 13398).

Zitzler, E. and Thiele, L. (1998a). An evolutionary algorithm for multiobjective optimization: The strength Pareto approach. Technical Report 43, Zürich, Switzerland: Computer Engineering and Networks Laboratory (TIK), Swiss Federal Institute of Technology (ETH).

Zitzler, E. and Thiele, L. (1998b). Multiobjective optimization using evolutionary algorithms – A comparative case study. In *Parallel Problem Solving from Nature V (PPSN-V)*, pp. 292–301.

Zitzler, E., Deb, K. and Thiele, L. (2000). Comparison of multiobjective evolutionary algorithms : Empirical results. *Evolutionary Computation Journal 8*(2), 125–148.

Zitzler, E., Deb, K., Thiele, L., Coello, C. A. C. and Corne, D. (Eds) (2001). *Evolutionary Multi-Criterion Optimization (Lecture Notes in Computer Science 1993)*. Heidelberg: Springer.

Zydallis, J. B., Veldhuizen, D. V. and Lamont, G. B. (2001). A statistical comparison of multiobjective evolutionary algorithms including the MOMGA-II. In *Proceedings of the First International Conference on Evolutionary Multi-Criterion Optimization (EMO-2001)*, pp. 226–240.

Index

Printed and bound by CPI Group (UK) Ltd, Croydon, CR0 4YY

27/10/2024

14580297-0002